BRAIN PLASTICITY

Development and Aging

ADVANCES IN EXPERIMENTAL MEDICINE AND BIOLOGY

Recent Volumes in this Series

A Continuation Order Plan is available for this series. A continuation order will bring delivery of each new volume immediately upon publication. Volumes are billed only upon actual shipment. For further information please contact the publisher.

BRAIN PLASTICITY

Development and Aging

Edited by

Guido Filogamo

University of Torino
Torino, Italy

Antonia Vernadakis

University of Colorado Health Sciences Center
Denver, Colorado

Fulvia Gremo

University of Cagliari
Cagliari, Italy

Alain M. Privat

Université Montpellier II
Montpellier, France

and

Paola S. Timiras

University of California
Berkeley, California

SPRINGER SCIENCE+BUSINESS MEDIA, LLC

Library of Congress Cataloging-in-Publication Data

Conference on Recent Advances in Neurobiology : Plasticity and Regeneration (1995 : Aosta, Italy)
Brain plasticity : development and aging / edited by Guido Filogamo, ... [et al.].
p. cm. -- (Advances in experimental medicine and biology ; v. 429)
"Proceedings of the Conference on Recent Advances in Neurobiology : Plasticity and Regeneration, held October 9-11, in St. Vincent, Aosta, Italy"--T.p. verso.
Includes bibliographical references and index.
ISBN 978-1-4757-9553-0 ISBN 978-1-4757-9551-6 (eBook)
DOI 10.1007/978-1-4757-9551-6
1. Neuroplasticity--Congresses. 2. Brain--Growth--Congresses. 3. Brain--Aging--Congresses. 4. Neurotoxicology--Congresses. I. Filogamo, Guido. II. Title. III. Series.
QP363.3.C675 1995
612.8'2--dc21 97-35346
CIP

Proceedings of the Conference on Recent Advances in Neurobiology: Plasticity and Regeneration, held October 9 – 11, 1995, in St. Vincent, Aosta, Italy

ISBN 978-1-4757-9553-0

Originally published by Plenum Press, New York in 1997
Softcover reprint of the hardcover 1st edition 1997

http://www.plenum.com

10 9 8 7 6 5 4 3 2 1

PREFACE

This book is dedicated to the memory of two colleagues and friends, Amico Bignami and Hendrick Van der Los. They were both pioneers in their fields: Bignami on ontogenesis and function of neuroglia, and Van der Los on brain plasticity and neuronal circuitry. Their ideas are further pursued by the authors in this book. Some of the chapters are products of a conference dedicated to these two scientists entitled "Recent Advances in Neurobiology: Plasticity and Regeneration." The conference was organized by the Institute of Developmental Neuroscience and Aging and sponsored by the Region della Valle d'Aosta. Also, several chapters are written by colleagues who knew well either Amico or Hendrick and were invited to contribute to this dedication.

The book is divided in four sections. The first part covers neurons, neuroglia including microglia, their plasticity and phenotypic expression, and specific functions and interactions. It is now established that neuroglia are an intimate component of the neuronal environment and thought to regulate several neuronal functions. More recently microglia have become prominent as the immune cells in the CNS. This part contributes new information for these cellular interactions. The second part deals with neuronal and glial cell plasticity as it relates to regeneration and neurodegeneration, more or less an extension of Part I. In recent years the role of transplantation in regeneration has become promising. This is also covered in Part II. Neuroprotection has been a subject of great interest and importance for understanding not only exogenous excitotoxins but more importantly endogenous excitotocity and more specifically glutamate neurotoxicity. The subject of neuronal cell death and neuroprotection is well covered in Part III. The fourth part is in a way an extension of Part III and covers the role of neurohormones including glucocorticoids and sex hormones, neural cell adhesion molecule (NCAM) during development and aging. In this Part the subject is approached both in using *in vitro* models and at the organismic level.

The overall theme of *Brain Plasticity: Development and Aging* is covered from different directions and points of view. It illustrates our considerable knowledge in this expanding field where both Amico Bignami and Hendrick Van der Los were giants. This book will be of interest not only to basic neurobiologists but to those interested in possible mechanisms and factors involved in neurodegenerative diseases.

The Editors

CONTENTS

Part II. Neuronal Circuitry, Synaptic Plasticity, Regeneration

Part III. Neuronal Cell Death, Neuroprotection

Part IV. Neurotrophins and Neuromodulators, Neurohormones, Cell Adhesion Molecules

Part I

Regulatory Mechanisms of Expression of Neuronal and Glial Phenotypes

1

GENES CONTROLLING NEURAL FATE AND DIFFERENTIATION

Rebecca Matsas

Department of Biochemistry
Hellenic Pasteur Institute
127 Vasssilis Sofias Avenue
11521 Athens, Greece

1. INTRODUCTION

All neural functions—from simple sensory responses and motor commands to elaborate cognitive behaviours—depend on the assembly of neural circuits, a process initiated during embryonic development. An early and fundamental step in this process is the generation of distinct classes of neurons at precise locations within a primitive neural epithelium (reviewed in Tanabe and Jessel, 1996). Over the past decade, many of the mechanisms that control the identity of specific neural cell types have been defined, in large part through the application of molecular genetics in invertebrate organisms such as *Drosophila* and *Caenorhabditis elegans* but also through cellular and biochemical approaches in vertebrates. Collectively, the study of these diverse systems has provided considerable insight into the relative contributions of environmental signaling and lineage restrictions in neural development and has revealed the identity of many of the extracellular signalling factors and intracellular proteins that direct cell fate. In this chapter we a) present a brief overview of recent progress made in defining positive and negative regulators of neurogenesis in vertebrates and their similarities with invertebrate organisms, b) discuss how genes regulating cell-cycle are also essential participants in neural differentiation and c) review recent data implicating neuron-specific markers in neuronal differentiation.

2. EARLY DEVELOPMENT, PRONEURAL GENES AND NEURAL INDUCTION

Much of our understanding of the genetic control of early development of the nervous system comes from studies in *Drosophila* and in amphibia. In *Drosophila*, mutagenesis experiments have provided evidence that activation of certain genes at appropriate

Brain Plasticity, edited by Filogamo *et al.*
Plenum Press, New York, 1997

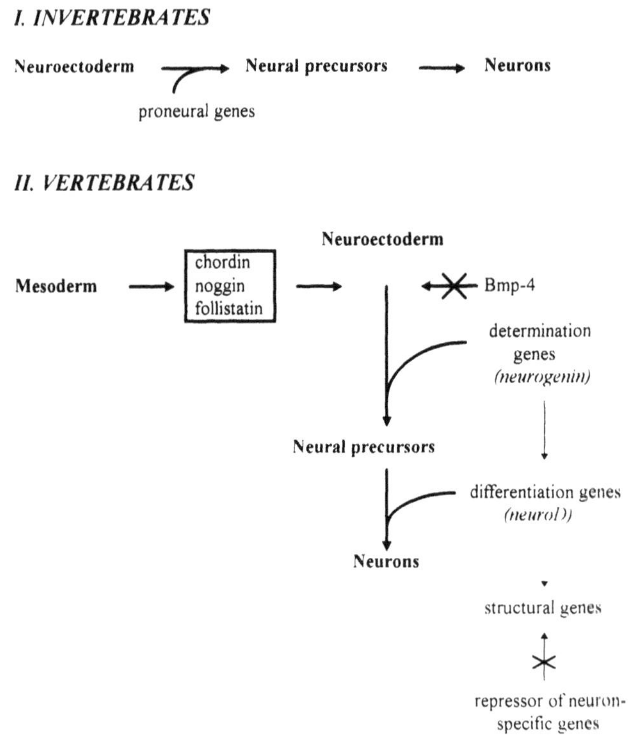

Figure 1. Shematic representation of neurogenesis in invertebrates and vertebrates.

places in the ectoderm is the principal force controlling neurogenesis (Fig. 1). Thus expression of the proneural genes of the *achaete-scute* complex and *atonal* which encode basic helix-loop-helix transcriptional regulators, is thought to provide ectodermal cells with neural potential (reviewed in Jan and Jan, 1993). In the absence of expression of proneural genes, cells differentiate as epidermis (Heitzler and Simpson, 1991). Similarly, ectopic expression of *achaete-scute* within the ectoderm leads to the development of ectopic neural precursors. Functional homologues of these genes have also been found in *Caenorhabditis elegans* (Zhao and Emmons, 1995).

Whereas *Drosophila* has been a crucial system for identifying genes controlling neurogenesis, the amphibian embryo has also been instrumental for the study of neural fate determination. The main difference between *Drosophila* and vertebrates is that, in the latter, the nervous system forms as the result of an induction from a different tissue type, the dorsal mesoderm (Fig. 1). During Xenopous development, ectoderm can differentiate into either epidermal or neural tissue. The classical grafting experiments of Spemann and Mangold established that the formation of neural tissue depends on signals provided by prospective axial mesodermal cells (reviewed Tanabe and Jessel, 1996). Thus neural induction was considered to be an instructive process for the activation of neural-specific genes, and that in the absence of this induction the ectoderm would differentiate into its default state, the epidermis. This concept emerged from experiments with isolated animal cap explants derived from the dorsal ectoderm of blastula-stage Xenopous embryos, which take on an epidermal fate (reviewed in Simpson, 1995). More recent experiments, however, have suggested that the default state may be the neural fate and that neural induction may result from the removal of endogenous inhibitors of neural differentiation that are usually present in the explants (Godsave and Slack, 1991; Hemmati-Brinvalou and Melton, 1994). Indeed, dissociation of blastula-stage ectoderm into single cells, presumably preventing intercellular signaling, was sufficient to elicit the formation of neuronal phenotype (reviewed in Green, 1994). This observation suggests that cell-cell communication is required to suppress neural-specific genes and maintain epidermal fate, whilst neural induction antagonizes an inhibitor of neural fate.

In the past few years spectacular progress has been made towards the delineation of the molecular machinery underlying the phenomenon of neural induction in vertebrates. Studies in Xenopous embryos now suggest that in one major pathway of neural induction, factors antagonize the action of a member of the transforming growth factor-β family, bone morphogenetic protein-4 (Bmp-4), which represses neural and promotes epidermal cell fate (Wilson and Hemmati-Brivanlou, 1995). Three candidate neural inducers have been identified: follistatin, noggin and chordin (Sasai et al., 1995; Yamashita et al., 1995; Zimmerman et al., 1996). All three proteins, although structurally unrelated, bind to and antagonize the action of Bmp-4. It has been argued that, in particular, chordin and Bmp-4 constitute an evolutionary conserved mechanism that controls ectodermal patterning (Sasai et al., 1995; Piccolo et al., 1996). Indeed, previous studies in *Drosophila* have established that two secreted molecules, decapentaplegic (dpp) and short-gastrulation (sog), homologues of Bmp-4 and chordin respectively, act antagonistically to define the dorsal limit of the neuroectoderm (Ferguson and Anderson, 1992; Holley et al., 1995). It is therefore possible to assume that neural induction in vertebrates is the result of a dorsalization of the ectoderm induced by antagonism of Bmp-4.

3. DETERMINATION AND DIFFERENTIATION GENES

As discussed above, proneural genes of the basic-helix-loop-helix (bHLH) family play a central role in cell fate determination in *Drosophila*. Moreover, such bHLH transcription factors have been shown to be instrumental for cell type determination in several other tissues and organisms (reviewed in Weintraub, 1993; Jan and Jan, 1994). Within a given lineage multiple, functionally interchangeable bHLH proteins often act in cascades (Jan and Jan, 1993). For example, at least four different bHLH proteins are sequentially expressed during murine muscle development: MyoD/myf5, myogenin and MRF4 (Olson and Klein, 1994). It has been suggested that early-acting bHLH proteins control cell fate determination, whilst later-acting ones control differentiation (Weintraub, 1993).

An alternative approach to understand the development of the nervous system in vertebrates has been to search for bHLH proteins homologous to the proneural gene products described in *Drosophila*. Thus, the murine *MASH1* and Xenopous *XASH1* and *XASH3* genes, homologous to the *achaete-scute* genes of *Drosophila*, were initially identified inferring a functional analogy from insects to vertebrates (Ferreiro et al., 1992; Guilemot and Joyner, 1993; Zimmerman et al., 1993). As in *Drosophila*, their expression in the vertebrate nervous system is transient; it precedes that of neuron-specific markers and ceases as obvious differentiation proceeds. However, it has been argued that none of these vertebrate genes has real proneural activity as their *Drosophila* counterparts. For example, although elimination of *MASH1* by gene targeting in mice resulted in a severe loss of olfactory sensory neurons and sympathetic autonomic neurons in the PNS (Guillemot et al., 1993), the precursors of these peripheral neurons did form and, additionally, there was no effect in the CNS of mutant mice. Similarly, ectopic expression of the Xenopous Achaete-Scute homologue-3 (XASH-3) was able to induce neural plate expansion or ectopic neurogenesis within the neural plate but was incapable of converting epidermal cells into neurons (Turner and Weintraub, 1994). It is possible that, in order to accommodate the increased complexity of their nervous system, additional tactics may have evolved in vertebrates. Truthful as it may be, subsequent description of two other proteins of the bHLH family, Neurogenin and NeuroD, have lent further support to the remarkable conservation of the mechanisms of neurogenesis between flies and vertebrates (Lee et al., 1995; Ma et al., 1996). Endogenous Neurogenin is expressed in regions of the Xenopous neural plate destined to generate primary neurons. Overexpression of Neurogenin induces ectopic neurogenesis and expression of a second and, normally, later appearing bHLH protein, NeuroD, which, in turn, can also induce ectopic neuronal differentiation of non-neural ectoderm. Most important, the conversion of ectoderm elicited by NeuroD can occur in the absence of neural induction. Furthermore overexpression of NeuroD in cells of the CNS which normally express the gene causes premature differentiation of dividing neural precursors into neurons. Thus NeuroD seems to control neuronal differentiation directly. These and other results (Lee et al., 1995; Ma et al., 1996) suggest that Neurogenin and NeuroD are important regulators of neurogenesis and more generally that neurogenesis in vertebrates, as in *Drosophila*, involves the sequential activation of distinct bHLH factors that either determine neuronal fate (determination genes, like *neurogenin*) or promote later aspects of neronal differentiation (differentiation genes, like *neuroD*).

Another striking example of evolutionary conserved systems during neurogenesis is the operation of the *Delta/Notch*-mediated signalling pathway by which a limited number of equipotential cells within a proneural domain are selected to become neural precursors. It has been established that in the embryonic ectoderm of *Drosophila*, proneural genes are activated by anterior-posterior and dorsal-ventral patterning genes in precise domains. Cells within these proneural domains are predisposed to a neural fate. However only a subset of these cells will eventually become neural precursors. This restriction is due to the *Delta/Notch*-mediated cell communication, a process known as lateral inhibition. This is a type of inhibitory cell signaling by which a cell committed to a neural cell fate forces its neighbours either to remain uncommitted or to enter a non-neural pathway (Artavanis-Tsakonas et al., 1995). Recently, homologues of the *Drosophila Delta* gene were identified both in Xenopous and the chick (Chitnis et al., 1995; Henrique et al., 1995) and evidence was provided that the process of the *Delta/Notch*-mediated lateral inhibition operates in vertebrates too. That is, overexpression of either an activated form of Delta or of an activated form of Notch was shown to inhibit the generation of primary neurons. Conversely, expression of a dominant-negative form of Delta resulted in the generation of additional primary neurons. It is noteworthy that the *neurogenin* pathway of vertebrate neural fate determination described above, is under genetic control of the *Delta-Notch*-mediated signalling (Ma et al., 1996).

4. THE MAMMALIAN MODEL OF GENERAL REPRESSION OF NEURAL FATE

Most interestingly, an alternative strategy of regulation of neural fate has emerged from studies in mammals. The default model of vertebrate neural development implies the existence of repressors of neuron-specific genes and that all embryonic tissues would be subject to such a repression before neurogenesis (Hemmati-Brivanlou and Melton, 1994).

A protein that has properties appropriate for such a repressor has been identified by two independent groups (Schoenherr and Anderson, 1995; Chong et al., 1995) and is known as REST (RE1-silencing transcription factor) or NRSF (neuron-restrictive silencer factor). The isolation of this factor resulted from the prior description of a 23bp silencer element, called RE1 or NRSE, which was initially found upstream of several neuron-specific genes, including SCG 10, synapsin I and the type II sodium channel (Kraner et al., 1992; Mori et al., 1992; Li et al., 1993). This silencing activity was shown to control the expression of all three genes in CNS neurons by restricting their expression everywhere else. Furthermore, deletion or mutation of this silencing element caused aberrant expression in non-neuronal tissues of genes normally restricted to CNS neurons. Data base searches revealed altogether 18 neuron-specific genes that contained consensus sequences corresponding to the silencer element, thus pointing out to a general mechanism of repression of neuronal gene transcription outside the CNS. The identified REST/NRSF protein is present in non-neuronal but absent from neuronal cell extracts and forms complexes with the silencer element. Its pattern of expression during embryogenesis and in the adult is consistent with a regulatory role in restricting neuronal gene expression. Moreover, expression of the recombinant REST/NRSF protein confers the ability to silence neuron-specific genes in PC12 cells lacking the native protein. Thus, REST/NRSF appears to be a master negative regulator of the neuronal phenotype.

5. CELL-CYCLE REGULATION AND NEURAL DIFFERENTIATION

It has long been recognized that the balance between cellular proliferation and cell death during embryogenesis is a key factor in formation of the CNS. Cell division during embryogenesis is known to play an integral part in formation of the CNS, but the molecular interactions that co-ordinate cell-cycle regulation with CNS-pattern formation and neural differentiation remain largely unknown. It is proposed that not only is the cell-division cycle acted upon by developmentally controlled molecular programs but gene products that drive the cell-cycle are also essential participants in neural differentiation (reviewed in Ross, 1996).

The importance of mitotic control to development is particularly evident in the CNS, which is among the earliest organotypic specializations in the embryo. In mammals, regulation of mitotic rate coincides with major events in CNS development. For example, there is a progressive lengthening of cell-cycle times within the neuroepithelium over the period of neural tube closure. Later in development, variation in mitotic rates within the ventricular germinal epithelium appears to contribute to the regional differences in neuronal number in the cerebral cortex (Dehay et al., 1993).

Current evidence suggests that cell proliferation is controlled by both intrinsic characteristics of progenitor cells and their response to signals provided by surrounding cells. There are several examples of interactions between neurons and glia that can influence their relative proliferation. In Drosophila, the *anachronism* locus encodes a glia-secreted glycoprotein that inhibits proliferation of neighboring neuroblasts (Ebens et al., 1993). Similarly in mouse, the addition of glia or glial cell membranes to cultures of cerebellar granule precursor cells arrests proliferation of the precursor cells (Gao et al., 1991). It is possible that a protein analogous to the *anachronism* gene product might be responsible for the arrest of neuronal proliferation imposed by glia.

With few notable exceptions, it is generally held that the bulk of morphological differentiation of neural progenitors occurs after cells have ceased proliferating. However, several observations suggest that cell-cycle regulation directly influences differentiation. Heterochronic transplantation studies in mammalian cerebral cortex indicate that the laminar fate of neuroblasts is determined in the final round of mitosis. This conclusion is based on the observation that cells that have completed mitosis before transplantation migrate to the lamina appropriate to the "birthdate" of the neuroblast, while those that undergo an addtional round of division after transplant adopt a laminar fate commensurate with the age of the host neuroepithelium into which they are placed (McConnel and Kaznowski, 1991). Consistent with this view, double-label studies in chick retina that combine incorporation of BrdU with expression of a mature neuronal antigen, called RA4, have shown that retinal ganglion cells differentiate within minutes of exit from the cell-cycle (Waid and McLoon, 1995). This suggests that crucial events in differentiation of retinal neurons occur during late mitoses. More particular, there appears to be an intimate relationship between the final division of the neural progenitor and the acquisition of at least some of the characteristics of its ultimate fate. Further support for this idea has come from recent studies in the cerebral cortex where it was shown that neuronal-fate determination might be influenced in the final division of the neuroblast through an assymetric partitioning of the neurogenic gene product, Notch-1 (Chenn and McConnel, 1995). In this assymetric division, the Notch-1-negative daughter cell remains attached to the ventricular surface and continues to proliferate, while the vertically displaced, Notch-1-expressing daughter cell exits the cycle and leaves the ventricular epithelium to differentiate into a neuron. Studies in *Drosophila* have also shown that in certain neural cells, the Notch-mediated control of neurogenesis is itself subject to regulation by proteins that are assymetrically inherited during the division of the progenitor cell. Notable amongst these is the *Drosophila* protein Numb, which confers neuronal identity to cells that inherit the protein by inhibiting the intracellular transduction of Notch-mediated signals (Campos-Ortega, 1996). Numb-related proteins have now been isolated in vertebrates and, especially, in the ventricular zone of the mammalian cerebral cortex Numb-, like Notch-proteins, are localized assymetrically during certain progenitor cell divisions (Chenn and McConnel, 1995; Zhong et al., 1996). Further analysis of Numb and Notch function should help to elucidate the extent by which proteins segregated during cell division control neuronal identity in the vertebrate CNS.

Another major force in brain histogenesis that ensures the proper match of neuronal interconnections and neuron-glia interactions is programmed cell death. It has been estimated that approximately half of the sympathetic and sensory neurons and oligodendrocytes of optic nerve that are initially produced die during a well-defined period of programmed cell death that starts with synaptogenesis (reviewed in Raff et al., 1993). After this period, neuronal cell death is considered to be pathological. In vertebrates, the extent of cell death depends in large part on epigenetic factors amongst which retrograde trophic factors play an important role (Oppenheim, 1991). It is believed that neurons are in competition for specific trophic factors that are supplied in limiting amounts by the target cells they innervate. Neurons that receive enough trophic factors survive whilst the others activate a cell death program.

The best defined genetic pathway of cell death has come from studies in *Caenorhabditis elegans*. At least 14 cell-death-defective (*ced*) genes have been found in *C. elegans* and there is now evidence for similar genes in mammals (Raff et al., 1993; Steller, 1995). Positive effectors of apoptosis include *ced-3* (which is homologous to the mammalian interleukin-1β-converting enzyme family) and *ced-4* proteases, which are required for cell-death to occur (Yuan and Horvitz, 1992; Yuan et al., 1993). *ced-9* is a repressor gene that protects cells from apoptosis by interfering with the function of *ced-3* and *ced-4* (Hengartner and Horvitz, 1994). The *bcl2* proto-oncogene is the mammalian functional homologue of *ced-9;* overexpression of the Bcl-2 protein in transgenic mice protected neurons from naturally occuring cell death (Martinou et al., 1994). It has recently been shown that Bcl-2 directly affects neuronal differentiation, thereby providing additional evidence for the close relationship between cell-cycle, apoptosis and differentiation. The challenge ahead—to understand the interplay among cell-cycle regulation, neural differentiation and apoptosis during brain development—requires a more complete description of the interactions of growth, differentiation and death signals with cell-cycle proteins and their subsequent molecular targets.

6. LATE DIFFERENTIATION GENES

Upon exit from the cell-cycle, post-mitotic neurons begin to migrate to their appropriate places and this process is accompanied by overt neuronal differentiation and the generation of distinct neuronal phenotypes. At this point, neuronal circuits are assembled and involve processes such as axonal elongation and pathfinding. Studies in the past two decades have provided a detailed understanding of the cellular interactions between growth cones and their surroundings and the emerging picture is that axonal growth and pathfinding are directed by the coordinate action of multiple guidance forces that are mediated by evolutionary conserved ligand-receptor systems (reviewed in Tessier-Lavigne and Goodman, 1996). In addition, recent data on activity-dependent modifications of the nervous system suggest a possible convergence on the same molecular mechanisms to control development of neural connections and synaptic plasticity in the adult (reviewed in Fields and Itoh, 1996). The ability to recognize multiple environmental cues and to undergo specific adhesion is critical to each of these complex cellular functions. Such recognition and adhesion processes are mediated by cell adhesion molecules (CAMs) expressed on the surface of neural cells. CAMs function through homophilic or heterophilic interactions with molecules on the surface of neighbouring cells or in the extracellular matrix (ECM).

CAMs are classified into two distinct functional groups: the Ca^{2+}-dependent and Ca^{2+}-independent molecules (Fields and Ito, 1996). Cadherins are Ca^{2+}-dependent molecules that interact through their cytoplasmic domain with cytoskeletal proteins and signal transduction pathways to regulate cell adhesion (Tacheichi, 1990). Integrins constitute a large family of Ca^{2+}-independent, heterodimeric proteins that mediate cell-cell interactions by binding to ECM molecules, such as fibronectin, laminin and tenascin (Jones, 1996). They are very important in modulating cell migration. The second and largest class of Ca^{2+}-independent CAMs are members of the immunoglobulin (Ig) superfamily and possess a number of Ig motifs and fibrinectin type III repeats in their exracellular domain. These molecules have been extensively studied (Brummendorf and Rathjen, 1995) and it is clear that beyond their adhesive properties they also play a role in delivering intracellular signals (reviewed in Doherty and Walsh, 1994, 1996). For many of the Ig superfamily molecules, multiple isoforms with distinct functional properties and patterns of expression are known. Their method of attachment to the cell membrane and the length of their cytoplasmic tail are important structural features with significant functional consequences. Oligosaccharides, which consist of polysialic acids, regulate adhesive interactions and their proportion varies with molecular species and with post-translational modification. The oldest known representative of this family is the neural cell adhesion molecule

NCAM (Cunningham et al., 1987). Several other glycoproteins from this family, which are localized predominantly on axons of vertebrate neurons, have been characterized and shown to play pivotal role in growth cone guidance and axonal elongation (Brummendorf and Rathjen, 1993). Among them are L1, (Moos et al., 1988), rat TAG1 (Furley et al., 1990) and its chick homolog axonin-1 (Zuellig et al., 1992), mouse F3 (Gennarini et al., 1989) and its chick homolog F11 (Brummendorf et al., 1989)/ contactin (Ranscht, 1988). Their restricted expression, both temporally and spatially, suggests that they are involved in the guidance of axons and/or the maintenance of synapses.

Another family of axon-guidance molecules that belongs to the Ig superfamily comprises the mammalian DCC (depleted in colorectal cancer) genes which have been characterized as netrin-receptors (Keino-Masu et al., 1996). The netrins are a small family of diffusible proteins with bifunctional guidance capacity: they attract some axons and repel others (Culotti and Kolodkin, 1996). Analogous molecules exist in *C. elegans*. Genetic analysis in this organism has implicated UNC-5, a transmembrane protein that defines a distinct branch of the Ig superfamily, in mediating repulsive actions of the netrin UNC-6 (Chan et al.,1996). It remains to be determined whether the receptors for semaphorins, another large family of cell-surface or secreted proteins that mediate chemorepulsion (reviewed in Tessier-Lavigne and Goodman, 1996), belong to the Ig superfamily as well.

Another group of molecules which appears to influence neuronal differentiation has emerged through molecular and cellular approaches in mammals. More particular, recent studies using neuronal or fibroblast cell lines have implicated a number of intracellular neural molecules in the differentiation of these cells. Such molecules include the growth associated protein GAP-43 (Zuber et al., 1989; Morton and Buss, 1992; Kumagai-Tohda et al., 1993), synaptotagmin (Feany and Buckley, 1993), the microtubule associated protein MAP 2C (LeClerk et al., 1993), a novel neuron-specific marker referred to as the BM88 antigen (Mamalaki et al., 1995) and the PTP20 protein-tyrosine phosphatase (Aoki et al., 1996). These observations suggest that a large number of late differentiation markers acting independently or in concert in the same or alternative pathways, should contribute to the mechanisms leading to terminal neuronal differentiation. As one such example we will review here work performed in our laboratory on the BM88 antigen. This molecule, first identified by means of a monoclonal antibody, is a neuron-specific protein widely distributed in the mammalian nervous system including that of mouse, rat, rabbit, pig and human (Patsavoudi et al.,1989; 1992; 1995). The BM88 antigen appears in the rat brain at the time during which neurons are generated whilst its expression increases with age and remains at high levels in the mature animal. It is an integral membrane protein composed of two ~22kD, apparently not glycosylated, polypeptide chains linked together by disulfide bridges. Electron microscopic observations in the adult brain have shown that the BM88 antigen is mainly associated with the limiting membrane of a number of intracellular organelles within the neuron, such as the endoplasmic reticulum, small electron-lucent vesicles and the outer membrane of mitochondria, but is also present at the plasma membrane. Recently, cDNA cloning and stable overexpression of the molecule in mouse neuroblastoma cells (Neuro 2a) demonstrated that the BM88 antigen slows down the proliferation and enhances the differentiation of these cells into a neuronal phenotype (Mamalaki et al., 1995 and Fig. 2). This biological activity of the BM88 antigen suggests that it may play a role in the differentiation of neuronal cells. In this context it is interesting to note that another neural protein associated with the outer membrane of mitochondria (Janiak et al., 1994), the proto-oncogene Bcl-2, is a well recognized regulator of cell-cycle, apoptosis and neuronal differentiation (for refs. see the section on cell-cycle regulation and neural differentiation in this chapter).

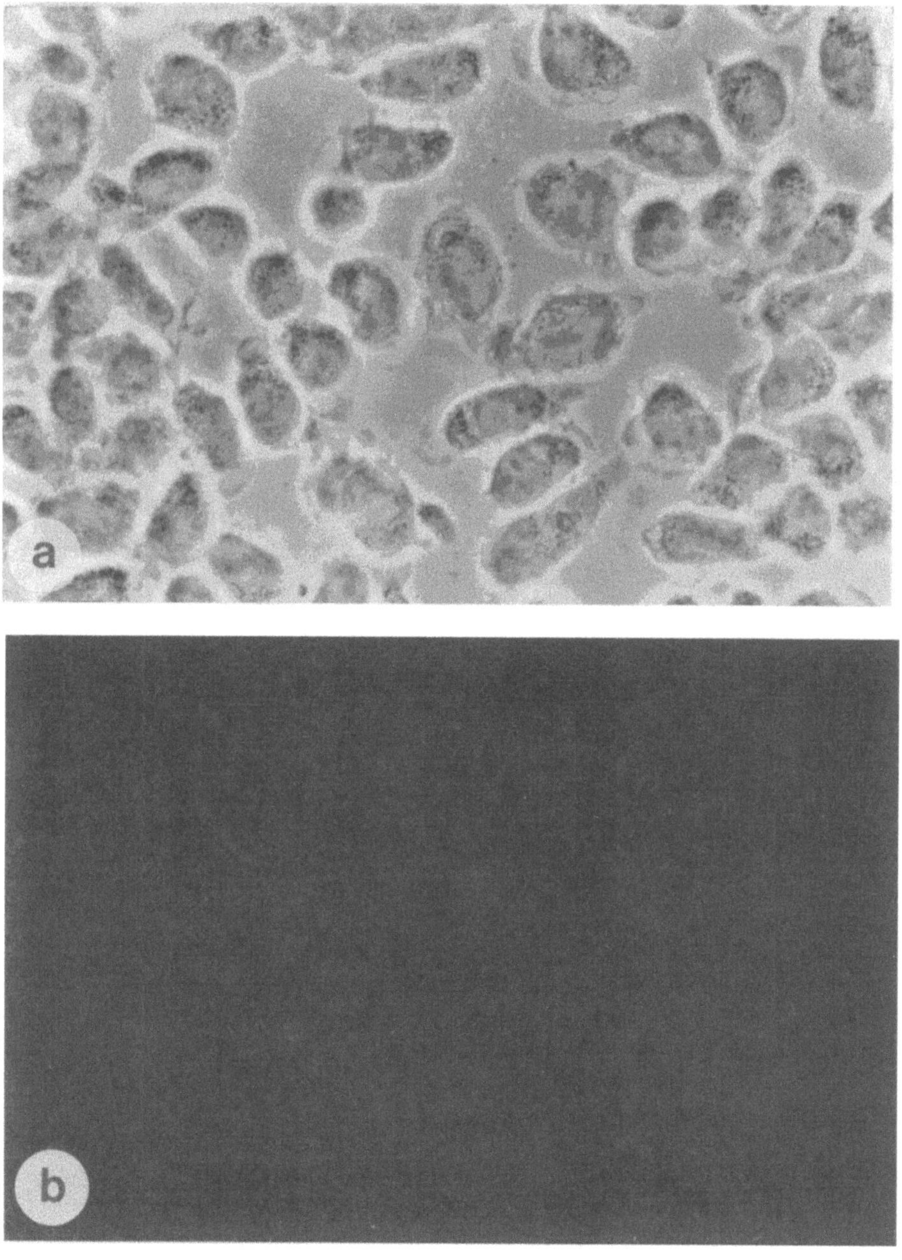

Figure 2a and b. Expression of the BM88 antigen in transfected Neuro 2a cells alters their morphology and proliferation rate. (a,b) parental Neuro 2a and (c,d) Neuro 2a-BM88 transfected cells cultured for 7 days in the absence of differentiation agents. Note the absence of endogenous BM88 antigen in the parental cells as shown by immuno-fluorescence labelling (b) and the high expression of the molecule in the transfected cells (d). Neuro 2a-BM88 transfectants develop elaborate processes, whilst the parental cells are essentially round.

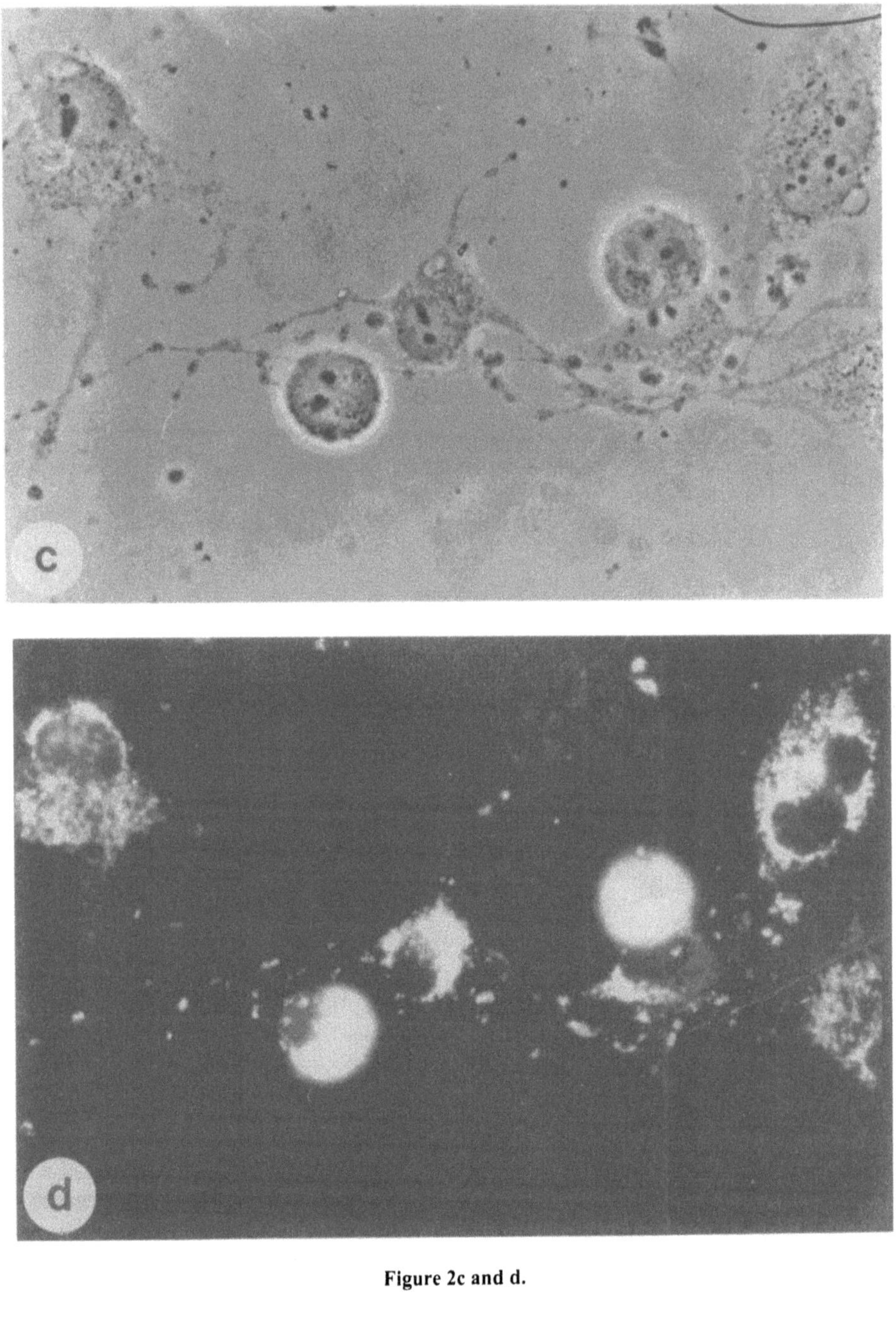

Figure 2c and d.

To further explore the functional role of the BM88 antigen in a system which closely resembles early murine development *in vivo*, we have used P19 embryonal carcinoma cells. These cells resemble embryonic stem cells in that they are multipotential and, depending upon the treatment used, they can differentiate in culture into derivatives of all three germ layers—endoderm, mesoderm and ectoderm (Bain et al., 1994). In particular, aggregated P19 cells exposed to retinoic acid differentiate into cells that resemble neurons, glia and fibroblast-like cells, whilst aggregated P19 cells treated with DMSO differentiate into cells with many of the characteristics of cardiac and skeletal muscle. Therefore, they are eminently suitable for developmental studies and the identification of regulatory molecules essential for neuronal differentiation. The BM88 antigen is not expressed in undifferentiated P19 cells. However, its expression is induced in aggregated cells driven to differentiate towards a neuronal phenotype by retinoic acid (Boutou E., Hurel C, Mamalaki A. and Matsas R., unpublished observations). Of interest, the appearence of the BM88 antigen in these cells coincides with that of polysialylated NCAM, an early marker conferring neuronal identity in neural precursor cells, and precedes that of neurofilament protein (Fig. 3). These data further support our view that the BM88 antigen may participate in mechanisms underlying neuronal differentiation. An *in vivo* model of ectopic expression of the BM88 antigen in transgenic mice, which is currently being developed in our laboratory, should certainly contribute in our understanding of the function of this molecule in the nervous system.

7. CONCLUDING REMARKS

Progress in elucidating the mechanisms that regulate neural fate and differentiation has accelerated appreciably over the past few years and has enriched our concept of how neurogenesis occurs. The major conclusion that has emerged from studies in vertebrate

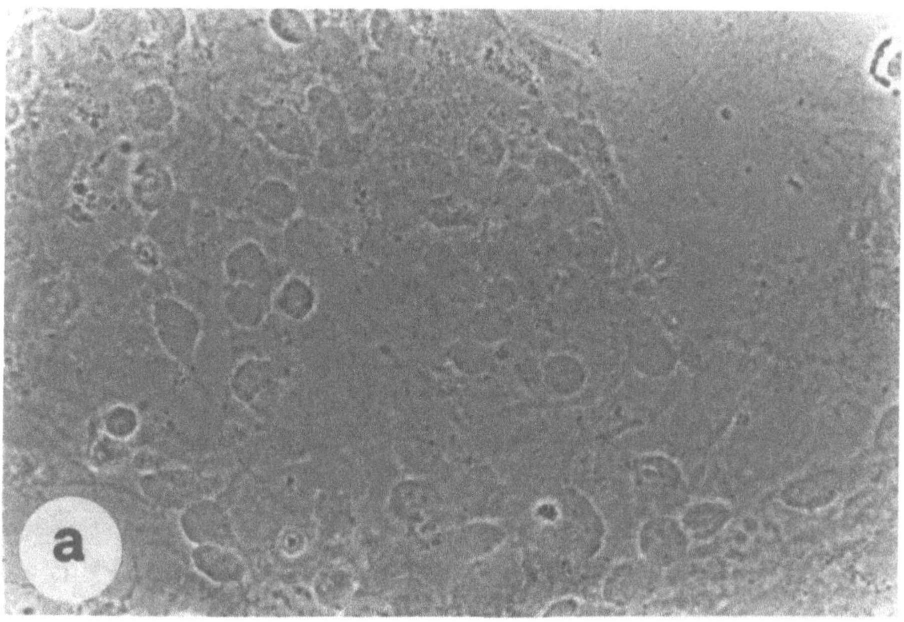

Figure 3a. Immunofluorecence double-labelling of P19 embryonal carcinoma cells driven to differentiate into a neuronal phenotype with retinoic acid. Cells were left to aggregate in the presence of RA for 4 days and were then harvested and cultured for a further 5 days on glass coverslips. (a), phase-contrast; (b), fluorescein optics and (c), rhodamine optics to visualize PSA and BM88 antigen, respectively. The dotted line in (c) demarkates an area of BM88 antigen-immunofluorescence corresponding to the intense PSA-labelling in (b).

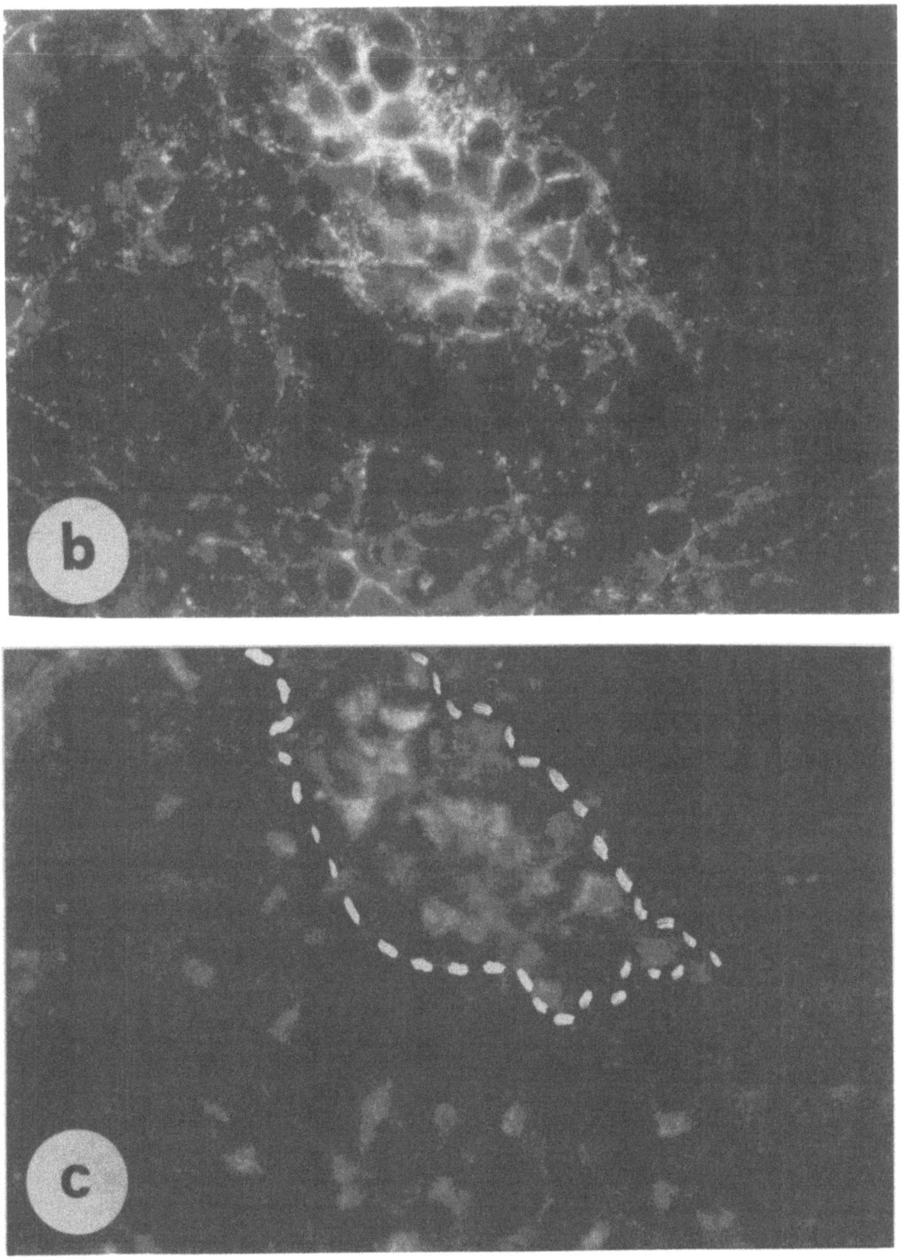

Figure 3b and c.

and invertebrate organisms is that despite the increased complexity of the former, there is an amazing evolutionary conservation of the mechanisms and strategies employed at the cellular level in controlling neural fate determination and differentiation. An increasing number of genes has been identified and characterized to unravel the complex mosaic of cellular and molecular interactions that take place during neural development. However, it is evident that enough issues still remain unresolved to leave room for the identification and functional characterization of new molecules that would contribute to our understandimg of the fine tuning mechanisms underlying each of the processes during nervous system development.

ACKNOWLEDGMENTS

We thank E. Patsavoudi for comments on the manuscript.

REFERENCES

Aoki N., Yamaguchi-Aoki Y. and Ulrich A. (1996) The novel protein-tyrosine phosphatase PTP20 is a positive regulator of PC12 cell neuronal differentiation. J. Biol. Chem. 271: 29422–29426.

Artavanis-Tsaconas S., Matsuno K. and Fortini M. E. (1995) Notch signalling. Science 268: 225–232.

Bain G., Ray W. J., Yao M. and Gottlieb D. I. (1994) Bioessays 16: 343–348.

Brummendorf T., Wolff J.M., Frank R., and F.G. Rathjen (1989) Neural cell recognition molecule F11: homology with fibronectin type III and immunoglobulin C domains. Neuron, 2: 1351–1361.

Brummendorf, T., and Rathjen, F. (1995) Cell adhesion molecules 1: immunoglobulin superfamily. Protein Profile 2: 963–1108.

Campos-Ortega J. A. (1996) Numb diverts notch pathway off the tramtrack. Neuron 17: 1–4.

Chan S. S., Zheng H., Su M. W., Wilk R., Killeen M. T., Hedgecock E. M. and Culotti J. G. (1996) UNC-40, a C. elegans homolog of DCC (Deleted in Colorectal Cancer), is required in motile cells responding to UNC-6 netrin cues. Cell 87: 187–195.

Chenn A. and McConnel S. K. (1995) Cleavage orientation and the asymmetric inheritance of Notch-1 immunoreactivity in mammalian neurogenesis. Cell 82: 631–641.

Chitnis A., Henrique D., Lewis J., Ish-Horowicz D. and Kintner C. (1995) Primary neurogenesis in *Xenopous* embryos regulated by a homologue of the *Drosophila* neurogenic gene *Delta*. Nature 375: 761–766.

Chong J. A., Tapia-Ramirez J., Kim S., Toledo-Aral J., Zheng Y., Boutros M. C., Altshiller Y., Frohman M. A. Kraner S. D. and Mandel G. (1995) REST: a mammalian silencer protein that restricts sodium channel gene expression to neurons. Cell 80: 949–957.

Culotti J. G. and Kolodkin A. L. (1996) Functions of netrins and semaphorins in axon guidance. Curr. Opin. Neurobiol. 6: 81–88.

Cunningham B. A., Hemperly J. J., Murray G. A., Prediger E. A., Brackenbury B. and Edelman G. M. (1987) Neural cell adhesion molecule: structure, immunoglobulin-like domains, cell surface modulation, and alternative RNA splicing. Science 236: 799–806.

Dehay C., Giroud P., Berland M. Smart I. and Kennedy H. (1993) Modulation of the cell cycle contributes to the parcellation of the primate visual cortex. Nature 366: 464–466.

Doherty P. and Walsh, F. (1994) Signal transduction events underlaying neurite outgrowth stimulated by cell adhesion molecules. Curr. Opin. Neurobiol. 4: 49–55.

Doherty P. and Walsh, F. (1996) CAM-FGF interaction: amodel for axonal growth. Mol. Cell. Neurosci., 8: 99–111.

Ebens A. J., Garren H., Cheyette B. N. and Zipursky S. L. (1193) The Drosophila anachronism locus: a glycoprotein secreted by glia inhibits neuroblast proliferation. Cell 74: 15–27.

Feany, M. B. and Buckley, K. M. (1993) The synaptic vesicle protein synaptotagmin promotes formation of filopodia in fibroblasts. Nature 364: 537–540.

Ferguson E. L. and Anderson K. V. (1992) *decapentaplegic* acts as a morphogen to organize dorsal-ventral pattern in the Drosophila embryo. Cell 71: 451–461.

Ferreiro B., Skoglund P., Bailey A., Dorsky R. and Harris W. (1992) *XASH-1*, a Xenopous homolog of *achaete-scute:* a proneural gene in anterior regions of the vertebrate CNS. Mech. Dev. 40: 25–36.

Fields, D., and Itoh, K. (1996) Neural cell adhesion molecules in activity-dependent development and synaptic plasticity. Trends Neurosci. 19: 473–480.

Furley A.J., Morton S.B., Manalo D., Karagogeos D., Dodd J. and T.M. Jessell (1990) The axonal glycoprotein TAG-1 is an immunoglobulin superfamily member with neurite outgrowth promoting activity. Cell, 61: 157–170.

Gao W. O., Heintz N. and Hatten M. E. (1991) Cerebellar granule cell neurogenesis is regulated by cell-cell interactions in vitro. Neuron 6: 705–715.

Gennarini, G., G. Cibelli, G. Rougon, M Mattei, G. and Goridis, C. (1989) The mouse neuronal cell surface protein F3: a phosphatidylinositol-anchored member of the immunoglobulin superfamily related to the chicken contactin. J. Cell Biol. 109: 775–788.

Godsave S. F. and Slack J. M. Single cell analysis of mesoderm formation in the Xenopous embryo. Development 111, 523–530.

Green J. B. A. (1994) Roads to neuralness: embryonic neural induction as depression of a default state. Cell 77: 317–320.

Guillemot F. and Joyner A. L. (1993) Dynamic expression of the murine *achaete-scute* homologue *Mash-1* in the developing nervous system. Mech. Dev. 42: 171–185.

Guillemot F., Lo L. C., Johnson J. E., Auerbach A. Anderson D. J. and Joyner A. L. (1993) Mammalian achaete-scute homolog-1 is required for early development of olfactory and autonomic neurons. Cell 75: 463–476.

Heitzler P. and Simpson P. (1991) The choice of cell fate in the epidermis of Drosophila. Cell 64: 1083–1092.

Hemmati-Brinvalou A. and Melton D. A. (1994) Inhibition of activin receptor signaling promotes neuralization in Xenopous. Cell 77: 273–281.

Hengartner M. O. and Horvitz H. R. (1994) C. elegans cell survival gene *ced-9* encodes a functional homolog of the mammalian proto-oncogene *bcl-2*. Cell 76: 655–676.

Henrique D., Adam J., Myat A., Chitnis A., Lewis J. and Ish-Horowicz D. (1995) Expression of a *Delta* homologue in prospective neurons in the chick. Nature 375: 787–790.

Holley S. A., Jackson P. D., Sasai Y., Lu B., De Robertis E. M., Hoffmann F. M. and Ferguson E. L. (1995) A conserved system for dorsal-ventral patterning in insects and vertebrates involving *sog* and *chordin*. Nature 376: 249–253.

Jan Y. N. and Jan L. Y. (1993) HLH proteins, fly neurogenesis, and vertebrate myogenesis. Cell 75: 827–830.

Jan Y. N. and Jan L. Y. (1994) Genetic control of cell fate specification in the Drosophila peripheral nervous system. Ann. Rev. Genet. 28: 373–393.

Janiak F., Leber B. and Andrews D. W. (1994) Assembly of Bcl-2 into microsomal and outer mitochondrial membranes. J. Biol. Chem. 269: 9842–9849.

Jones L. S. (1996) Integrins: possible functions in the adult CNS. Trends Neurosci. 19: 68–72.

Keino-Masu K., Masu M., Hinck L., Leonardo E. D., Chan S. S., Culotti J. G. and Tessier-Lavigne M. (1996) Deleted in Colorectal Cancer (DCC) encodes a netrin receptor. Cell 87: 175–185

Kraner S. D., Chong J. A., Tsay H. J. and Mandel G. (1992) Silencing the type II sodium channel gene: a model for neural specific gene regulation. Neuron 9: 37–44.

Kumagai-Tohda, C., Tohda, M. and Nomura, Y. (1993) Increas in formation and acetylcholine release by transfection of growth-associated protein GAP-43 cDNA into NG108–15 cells. J. Neurochem. 61: 526–532.

LeClerc N., Kosik K. S., Cowan, N. Pienkowski, T. P. and Baas, P. W. (1993) Process formation in Sf9 cells induced by the expression of a microtubule-associated protein 2C-like construct. Proc. Natl. Acad. Sci. USA 90: 6223–6227.

Lee J. E., Hollenberg S. M., Snider L., Turner D. L., Lipnick N and Weintraub H. (1995) Conversion of *Xenopous* ectoderm into neurons by NeuroD, a basic Helix-Loop-Helix protein. Science 268: 836–844.

Li L., Suzuki T., Mori N. and Greengard P. (1993) Identification of a functional silencer element involved in neuron-specific expression of the synapsin I gene. Proc. Natl. Acad. Sci. USA 90, 1460–1464.

Ma Q., Kintner C. and Anderson D. J. (1996) Identification of *neurogenin*, a vertebrate neuronal determination gene. Cell 87: 43–52.

Mamalaki A. Boutou E. Hurel C. Patsavoudi E. Tzartos S. and Matsas R. (1995) The BM88 antigen, a novel neuron-specific molecule, enhances the differentiation of mouse neuroblastoma cells. J. Biol. Chem. 270: 14201–14208.

Martinou J. C., Dubois-Dauphin M., Staple J. K., Rodriguez I., Frankowski H., Missotten M., Albertini P., Talabot D., Katsikas S., Piera C. and Huarte J. (1994) Overexpression of BCL-2 in transgenic mice protects neurons from naturally occuring cell death and experimental ischemia. Neuron 13: 1017–1030.

McConnel S. K. and Kaznowski C. E. (1991) Cell cycle dependence of laminar determination in developing cerebral cortex. Science 254: 282–285.

Moos M., Tacke R., Schere H., Teploe D., Fruh K. and Schachner M. (1988) Neural adhesion molecule L1 as a member of the immunoglobulin superfamily with binding domains similar to fibronectin Nature 334: 701–703.

Mori N., Schoenherr C., Vandenbergh D. J. and Anderson D. J. (1992) A common silencer element in the *SCG10* and type II Na^+ chanel genes binds a factor present in non-neuronal cells but not in neuronal cells. Neuron 9: 45–54.

Morton, A. J. and Buss, T. N. (1992) Accelerated differentiation in response to retinoic acid after retrovirally mediated gene transfer of GAP-43 into mouse neuroblastoma cells. Eur. J. Neurosci. 4: 910–916.
Olson E. N. and Klein W. H. (1994) bHLH factors in muscle development: dead lines and commitments, what to leave in and what to leave out. Genes Dev. 8: 1–8.
Oppenheim R. W. (1991) Cell death during development of the nervous system. Ann. Rev. Neurosci. 14: 453–501.
Patsavoudi E., Hurel, C. and Matsas, R. (1989) Neuron and myelin specific monoclonal antibodies recognizing cell surface antigens of the central and peripheral nervous system. Neuroscience 30: 463–478.
Patsavoudi E., Hurel C. and Matsas R. (1991) Purification and characterization of neuron-specific surface antigen defined by monoclonal antibody BM88. J. Neurochem. 56: 782–788.
Patsavoudi E., Merkouri E., Thomaidou D., Sandillon F., Alonso G. and Matsas, R. (1995) Characterization and localization of the BM88 antigen in the developing and adult rat brain. J. Neurosci. Res., 40: 506–518.
Piccolo S., Sasai Y., Lu B. and De Robertis E.M. (1996) Dorsoventral patterning in Xenopus: inhibition of ventral signals by direct binding of chordin to BMP-4. Cell 86: 589–598.
Raff M. C., Barres B. A., Burne J. F., Coles H. S., Ishizaki Y. and Jacobson M. D. (1993) Programmed cell death and the control of cell survival: lessons from the nervous system. Science 262: 695–700.
Ranscht, B. (1988) Sequence of contactin, a 130 kD glycoprotein concentrated in areas of interneural contact, defines a new member of the Ig superfamily in the nervous system. J. Cell Biol. 104: 343–353.
Ross E. M. (1996) Cell division and the nervous system: regulating the cell-cycle from neural differentiation to death. Trends Neurosci. 19: 62–68.
Sasai Y., Lu B., Steinbeisser H. and De Robertis E. M. (1995) Regulation of neural induction by the chordin and Bmp-4 antagonistic patterning signals in Xenopous. Nature 376: 333–336.
Schoenherr C. J. and Anderson D. J. (1995) The neuron-restrictive silencer factor (NRSF): a coordinate repressor of multiple neuron-specific genes. Science 267: 1360–1363.
Simpson P. (1995) Positive and negative regulators of neural fate. Neuron 15: 739–742.
Steller H. (1995) Mechanisms and genes of cellular suicide. Science 267: 1445–1449.
Tacheichi M. (1990) Cadherins: a molecular family important in selective cell-cell adhesion. Ann. Rev. Bioch. 59: 237–252.
Tanabe Y. and Jessel T. M. (1996) Diversity and pattern in the developing spinal cord. Science 274: 115–1123.
Tessier-Lavigne M and Goodman C. (1996) The molecular biology of axon guidance. Science 274: 1123–1132.
Turner D. L. and Weintraub H. (1994) Expression of*achaete-scute* homolog 3 in Xenopous embryos converts ectodermal cells to a neural fate. Genes Dev. 8: 1434–1447.
Waid D. K. and McLoon C. (1995) Immediate differentiation of ganglion cells following mitosis in the developing retina. Neuron 14: 117–124.
Weintraub H. (1993) The MyoD family and myogenesis: redundancy, networks and thresholds. Cell 75: 1241–1244.
Wilson P. A. and Hemmati-Brivanlou A. (1995) Induction of epidermis and inhibition of neural fate by Bmp-4. Nature 376: 331–333.
Yamashita H., ten-Dijke P., Huylebroeck D., Sampath T. K., Andries M., Smith J. C., Heldin C. H. and Miyazono K. (1995) Osteogenic protein-1 binds to activin type II receptors and induces certain activin-like effects. J. Cell Biol. 130, 217–226.
Yuan J. and Horvitz H. R. (1992) The ceanorhabditis elegans cell death gene ced-4 encodes a novel protein and is expressed during the period of extensive programmed cell death. Development 116: 309–320.
Yuan J., Shaham S., Ledoux S., Ellis H. M. and Horvitz H. R. (1993) The C. elegans cell death gene *ced-3* encodes a protein similar to mammalian interleukin-1β-converting enzyme. Cell 75: 641–652.
Zhao C. and Emmons S. W. (1995) A transcription factor controlling development of peripheral sense organs in C. elegans. Nature 373: 74–78.
Zhong W., Feder J. N., Jiang M. M., Jan L. Y. and Jan Y. N. (1996) Asymmetric localization of a mammalian numb homolog during mouse cortical neurogenesis. Neuron 17: 43–53.
Zimmerman K., Shih J., Bars J., Collazo A. and Anderson D. J. (1993) *XASH-3*, a novel *Xenopous achaete-scute* homolog, provides an early marker of planar neural induction and position along the medio-lateral axis of the neural plate. Development 119: 221–232.
Zimmerman L. B., Jesus-Escobar J. M. and Harland R. M. (1996) The Spemann organizer signal noggin binds and inactivates bone morphogenetic protein 4. Cell 86, 599–606.
Zuber M. X., Goodman D. W., Karns L. R. and Fishman M. C. (1989) The neuronal growth-associated protein GAP-43 induces filopodia in non-neuronal cells. Science 244: 1193–1195.
Zuellig R., Rader C., Schroeder A., Kalousek F., vonBohlen und Halbach F., Osterwalder T., Inan C., Stoeckli E., Affolter U., Fritz A., Hafen, and P. Sonderegger (1992) The axonally secreted cell adhesion molecule, axonin-1: primary structure, Ig- and fibronectin type III-like domains, and glycosyl phosphatidylinositol anchorage. Eur. J. Biochem., 204: 453–463.

2

NEURONAL PLASTICITY IN DEVELOPMENT: LESSONS FROM ETHANOL NEUROTOXICITY DURING EMBRYOGENESIS

Susan Kentroti

Departments of Pharmacology and Psychiatry
University of Colorado Health Sciences Center
4200 East Ninth Avenue
Denver, Colorado 80262

1. INTRODUCTION

Conceptually, it can be said that development of the embryonic central nervous system begins with a population of undifferentiated, multipotent cells and ends up with a highly organized arrangement of mature neurons each of which expresses a functional neurotransmitter phenotype. This ultimate, functional phenotype is a product of the interaction of numerous influences impinging on multipotential neuroblasts. Physical factors such as cell-cell contacts, migratory pathways and a plethora of soluble factors all profoundly influence establishment of neuronal phenotype during neuroembryogenesis. For example, definitive studies by LeDouarin and coworkers beginning in the early 1970's (Le Douarin, 1973; 1980; Le Douarin and Teillet, 1974; Le Douarin et al, 1975; Le Douarin and Smith, 1988; Teillet and Le Douarin, 1983) demonstrated *in vivo* that neuronal phenotypes are neither rigidly predetermined nor static. They found that the migratory destination of autonomic neuroblasts of neural crest origin determines their ultimate neurotransmitter phenotype. Furthermore, they demonstrated that exposure of neural crest-derived presumptive adrenergic neuroblasts to cells from another germ cell layer, splanchnic mesoderm, results in a phenotypic shift to fully differentiatied cholinergic neurons (Le Douarin et al, 1975). Collectively, these studies have introduced the idea of neuronal plasticity in the developing brain.

The notion of plasticity can encompass several areas of neuronal function. For example, neuronal plasticity as it applies to cellular function may relate to the "adaptability" of a single cell under a variety of microenvironmental situations, all involving normal functioning. It is, thus, important to define neuronal plasticity in the context of this chapter. Neuronal plasticity refers to the "ability of the neuron to alter its functional phenotype". In general, this capacity of the cell decreases with advancing age and/or stage of differentiation. Although the CNS is composed of many neuronal phenotypes, this discus-

Brain Plasticity, edited by Filogamo *et al.*
Plenum Press, New York, 1997

sion is confined to three primary phenotypic classifications for which much work on the subject of neuronal plasticity has been done: 1) cholinergic neurons which utilize acetylcholine as their neurotransmitter; 2) catecholaminergic neurons, of which epinephrine, norepinephrine and dopamine are the representative neurotransmitters; and 3) GABAergic neurons, which utilize GABA as neurotransmitter. All are widely represented throughout the CNS from early stages of embryonic development.

The ontogenesis of CNS structures has been studied in great detail, particularly in the avian and murine embryos, and several facts have emerged from early reports (Hamburger and Hamilton, 1951). Studies have shown that the CNS is derived from an early embryonic structure known as "neural tube" while the peripheral nervous system arises from a transient structure of migrating cells known as "neural crest". From the earliest stages of CNS development, the neuroepithelium of the neural tube consists of multipotent cells with the potential for adopting numerous phenotypes, both neuronal and non-neuronal. (for review, see Sanes, 1989; McKay, 1989; McConnell, 1991). Through a series of restrictive decisions proliferating cells commit, first to a neural lineage (neuroblast), then to an immature phenotype (e.g. cholinoblast) and, finally to a mature phenotype (e.g. cholinergic neuron). In general, cholinergic traits appear prior to adrenergic in developing neural crest (Fauquet et al, 1981). The early process of neurogenesis is characterized by intense cell proliferation such that far more neuroblasts are produced than can be supported at maturation (Oppenheim, 1991). Subsequently, only cells which can form target connections survive, establishing the cytoarchitecture of the mature CNS (see reviews, Oppenheim, 1991; Raff et al, 1993). During the transition from neuroblast to mature neuron, cells retain a significant degree of phenotypic plasticity and are highly sensitive to microenvironmental manipulations. This sensitivity is reflected in the immense number of published studies demonstrating the effects of factors, both neurotrophic and neurotoxic, to affect phenotypic expression.

2. TROPHIC FACTORS AND NEURONAL PLASTICITY

Subsequent studies by many investigators have emphasized the importance of the cellular microenvironment in the determination of phenotype within the neural crest, embryonic cells which give rise to a number of neural derivatives in the adult. Studies using cultured neural crest cells have been helpful in elucidating some of the factors which influence their phenotypic expression. For example, Howard and Bronner-Fraser (1985) and Ziller (1987) find that altering specific components of the culture media results in differential expression of cholinergic and catecholaminergic phenotypes. Low concentrations of chick embryo extract combined with 15% fetal bovine serum predisposes these cultures to exhibit cholinergic properties while high concentrations of chick embryo extract (15%) results in catecholaminergic expression (Ziller and Smith, 1982; Ziller, 1987). Howard and Bronner-Fraser (1985) also demonstrated that extracts from the neural tube (presumptive CNS) influences cultured neural crest cells to exhibit catecholaminergic properties. Sieber-Blum and Kahn (1982) reported that soluble factors contained in heart cell-conditioned medium promote cholinergic differentiation in neural crest cell cultures.

More recently, investigators are identifying specific factors with evident roles in determination of neuronal phenotype. Brain-derived neurotrophic factor (BDNF) promotes peptidergic expression (Carnahan and Nawa, 1995); combined exposure of striatal neurons to acidic FGF (aFGF) and catecholamine promotes catecholaminergic expression (Du and Iacovitti, 1995) while FGF-5 promotes cholinergic expression in septal cholinergic and raphe serotonergic neurons (Lindholm et al, 1994); and interferon-γ is reported to promote cholinergic differentiation in septal nuclei and adjacent basal forebrain (Jonakait et al, 1994). Evidence indicates that endogenous opioids may have a differential effect on neuronal differentiation in that κ-receptor *agonists* have been shown to promote morphological differentiation of chick embryonic forebrain neurons at 7 days of development (Maderspach and Nemeth, 1993) whereas μ- and δ-receptor *antagonists* promote both cholinergic and catecholaminergic expression at earlier stages (3 days of development) (Vernadakis and Kentroti, 1990).

3. NEURON-GLIA INTERRELATIONSHIPS AND NEURONAL PLASTICITY

To add further complexity to the issue, the interrelationship between neurons and neuroglia, in terms of both cell-cell interactions and elaboration of soluble factors, is emerging as an integral component of phenotypic determination in developing neurons (See reviews Vernadakis, 1988; 1996; Vernadakis and Kentroti, 1996). There is abundant evidence that glial cells play a decisive role in neuronal plasticity. This subject has recently been reviewed by Abbott, 1991; Fedoroff and Vernadakis, (1986, eds of 3 Volumes on "Astrocytes"; Muller, 1992; and Vernadakis, 1988; 1996). Glial cells influence such parameters as **neuronal phenotypic expression** (Landis 1978, 1980; Patterson and Chun 1977a,b; Le Douarin 1980), **neuronal migration and guidance** (classic study by Rakic, 1971; Hatten, 1984); and **neuritic growth** (Varon, 1975); regulation of the **ionic microenvironment** (Walz and Hertz, 1984; Ransom and Carlini, 1986; White and Woodbury, 1987); and **neurotransmitter uptake** (Vernadakis, 1974; Kimelberg, 1986; Massarelli et al., 1986; Hertz and Schousboe, 1987). More recently, research has been focusing on both the responsiveness of glial cells *to* various growth factors as well as production of growth factors *by* glial cells. It is now clear that various neurotrophins including nerve growth factor (NGF), epidermal growth factor (EGF), basic- and acidic- fibroblast growth factors (bFGF, a-FGF), and transforming growth factor beta (TGF*β*) influence astroblast differentiation and proliferation (Chao et al., 1992; Han et al., 1992; Huff and Schreier, 1989; Keningsbery et al., 1992; Loret et al., 1989; Perraud et al., 1988, 1990; Selmaj et al., 1991). Moreover, several neurotrophins including NGF, IGF-I, CNTF, BDNF, NT-3, are known to be secreted by astrocytes (Furukawa et al., 1986; Houlgatte et al., 1989; Olson et al., 1991; Rudge et al., 1992; Yamakumi et al., 1987; Yoshida and Gage, 1991).

Investigators have reported the presence of both neuronal precursors and glial precursors within the brain through late developmental stages (Gray and Sanes, 1992; Landis and Keefe, 1983; Landis et al 1988) and even into adulthood (Barami et al, 1994; Kirschenbaum et al, 1994; Morshead et al, 1994; Schwab 1996). A recent study has shown that glial cells derived from *aged* mouse brain exhibit enhanced activity of glutamine synthetase (GS; the astrocytic marker) in response to both NGF and EGF, suggesting the presence of neurotrophin-responsive glial precursors (Lee et al, 1992). This implies a substantial plasticity in both cell types within the CNS. However, there are clearly differences in the properties of both neurons and glia derived from different stages within the lifespan of the organism. These properties tend to limit cellular phenotypic plasticity with time.

The stage-specific synergy between developing neurons and glia can be seen in the following series of studies. Characterization of neuroglia from an early (E3) stage of neurogenesis in the avian embryo reveals a population of largely undifferentiated radial glia (Kentroti and Vernadakis, 1996a). When maintained in culture, these cells exhibit interesting properties. 1) With advancing cell passage, these glioblast cultures follow the normal cell lineage previously established for developing glia *in vivo*. Thus, glioblasts do not cease their process of phenotypic maturation and remain static following dissociation from the developing brain. And, 2) following each successive cell passage, early on, cells "revert back" to a less differentiated state before proceeding through the cell lineage with advancing days *in vitro* (Kentroti and Vernadakis, 1996a). Contemporary wisdom has established that immature radial glia arise from pluripotential progenitors and transform into mature astrocytes (Cameron and Rakic, 1991; Levitt and Rakic, 1980; Misson et al, 1990; Hirano and Goldman, 1988; Voigt 1989; Voigt and deLima, 1985). In comparison to glioblasts from early neuroembryogenesis, glia derived from a later stage of development (E15) exhibit distinctly different properties. At this stage of development expression of *neuronal* phenotypes is quite advanced (Figure 1). Likewise, mature glial phenotypes are also expressed (Kentroti and Vernadakis, 1997). Primary cultures from E15CH contain a greater diversity of cell types including neurons and glia at several stages of maturation, as evidenced by immunocytochemical staining (Kentroti and Vernadakis, 1997). The striking observation is the apparent necessity for neurons to be present in order for mature astrocytes to develop, and vice versa. The presence of neurofilament-positive cells is not observed after cell passage and mature astrocytes disappear soon after, illustrating the co-dependent relationship between glia and neurons.

Of major importance to the study of CNS development is the differential action of glia derived from these two distinct embryonic stages on *neuronal* development. In preliminary studies, early- (E3) and late-stage (E15) glia when co-cultured with neuroblasts from early development (E3) exhibit differential properties. In neuroblast co-cultures with early-stage glia, astrocytic differentiation was promoted with little or no effect on cholinergic expression. In neuroblast co-cultures with late-stage glia cholinergic expression was

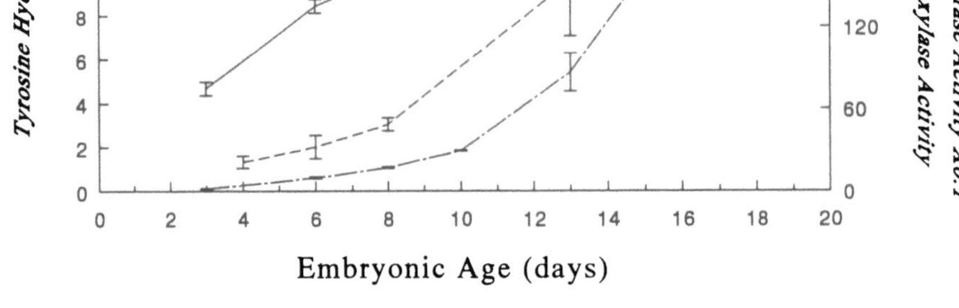

Figure 1. Developmental profiles of three neuronal enzyme markers expressed in chick embryonic brain. Tyrosine hydroxylase (TH) activity is the representative catecholaminergic marker, choline acetyltransferase (ChAT), and glutamate decarboxylase (GAD) activities are the respective markers for cholinergic and GABAergic phenotypes. Data taken from Kentroti and Vernadakis, 1989; 1990a; 1991a.

promoted with little or no effect on astrocytic expression (Kentroti and Vernadakis, in preparation). These data further emphasize the reciprocal relationship between developing neurons and glia, a relationship which appears to be highly stage-specific.

4. ETHANOL NEUROTOXICITY AND PHENOTYPIC EXPRESSION

Just as neurotrophins influence expression in emerging neuronal phenotypes during development, neurotoxic substances can also affect phenotypic determination. This laboratory has focused intense effort on the study of one such neurotoxin, **ethanol**, and its effects on neuronal plasticity. This project has uncovered much information regarding the cellular effects of prenatal ethanol exposure and revealed important details regarding the early process of neurogenesis itself.

The developmental profile of three major neuronal phenotypes within the developing chick brain was reported in early studies and those findings are summarized in Figure 1. Several significant observations are revealed in this Figure. First, choline acetyltransferase (ChAT) activity is in excess of 10-fold greater than that observed for tyrosine hydroxylase (TH) or glutamate decarboxylase (GAD). This observation reflects the earlier ontogenesis of the cholinergic phenotype as well as the greater number of cholinergic neurons in the developing chick brain as compared to catecholaminergic or GABAergic neurons (Kentroti and Vernadakis, 1989; 1990a; 1991a).

Initial studies reveal that exposure of chick embryos to ethanol *in ovo* significantly alters expression of emerging neuronal phenotypes in a dose-dependent manner (Brodie and Vernadakis, 1990). Treatment of embryos with ethanol (10mg/50μl/day) from embryonic day 1 to 3 (E1-E3) results in a dose-dependent decrease in cholinergic expression at 8 days (E8) of embryogenesis (Figure 2A) while catecholaminergic and GABAergic phenotypes are enhanced (Kentroti and Vernadakis, 1990b; 1991b; Brodie and Vernadakis, 1990). *In contrast*, when embryos are exposed to the same dose of ethanol from E4 to E6, expression in the three phenotypes is unaffected at E8 (Figure 2B). This study establishes the *critical window of sensitivity* to ethanol with respect to phenotypic expression as embryonic days 1–3.

Subsequent experiments show that these effects of ethanol to diminish cholinergic expression are reversible. NGF, EGF and aqueous extracts from chick embryonic skeletal muscle (SME) restore cholinergic activity *in ovo* in both brain and spinal cord tissue using the same experimental paradigm. In fact, the ethanol effects on ChAT activity are reversed with either concomitant, or subsequent administration of trophic substances (E4-E6) with ethanol exposure. These significant findings demonstrate the plasticity of the cholinergic population and provide a basis for studies into the mechanism whereby these and other factors can ameliorate the effects of prenatal ethanol exposure (Brodie and Vernadakis, 1992; Brodie et al, 1991). Similar to the *in ovo* effects, neuroblast-enriched cultures derived from early neurogenesis (E3) in the chick embryo, exhibit a dose-dependent cholinotoxicity to ethanol concentrations ranging from 12.5 to 50mM. In addition, the morphology of these cultures is also strikingly different from control cultures, a difference which can be expressed quantitatively (Kentroti and Vernadakis, 1991c). The most profound morphological difference is the appearance of a broad field of randomly-growing neurites as opposed the the normal pattern of compactly bundled neuritic processes interconnecting neuronal aggregates (Figure 3A-D), an observation which implicates the various cell adhesion molecules (including NCAM) as a target for ethanol neurotoxicity.

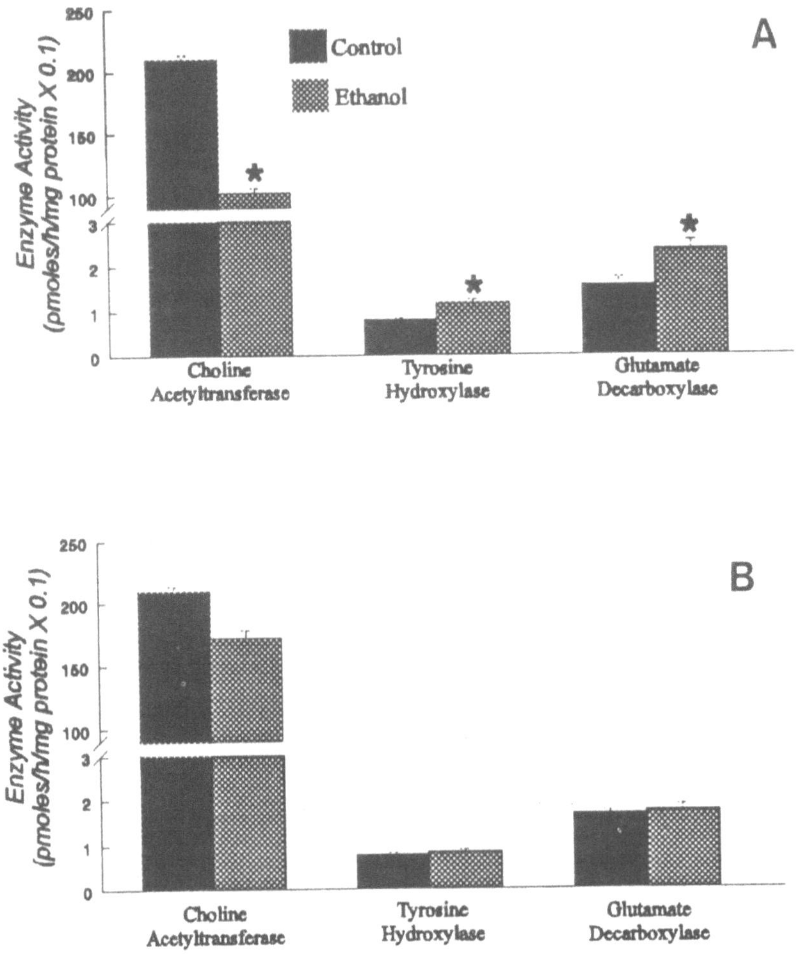

Figure 2. This graph shows the relative activity of three neuronal enzyme markers in whole brain homogenates from 8-day-old (E8) chick embryos. Embryos received 0.9% saline (control) or ethanol (10mg/50μl/day) at either 1–3 days (**A**), or 4–6 days (**B**) of embryogenesis before being sacrificed at E8. Whole brains were removed and tissue homogenates assayed for activity of choline acetyltransferase as the marker for cholinergic expression; tyrosine hydroxylase, for catecholaminergic expression; and glutamate decarboxylase for GABAergic expression. Enzyme activity is expressed as picomoles of product formed/h/mg protein and barograms represent the mean ± s.e.m. of at least 8 samples. *P<0.005 versus respective controls.

These findings lead to the formulation of several hypotheses concerning the mechanisms of action of ethanol on neurotransmitter phenotypic expression during neurogenesis: 1) ethanol selectively enhances naturally-occurring death within the cholinergic phenotype, while preserving catecholaminergic and GABAergic populations; 2) ethanol exposure elicits a shift in the expression of multipotential neuroblasts away from cholinergic, in favor of catecholaminergic and GABAergic phenotypes; 3) ethanol selectively retards maturation of cholinergic neuroblasts arresting them at an immature stage of development. In all likelihood, these hypotheses are not mutually exclusive. Some potential mechanisms for the actions of ethanol include: a) interference with local microenvironmental signals

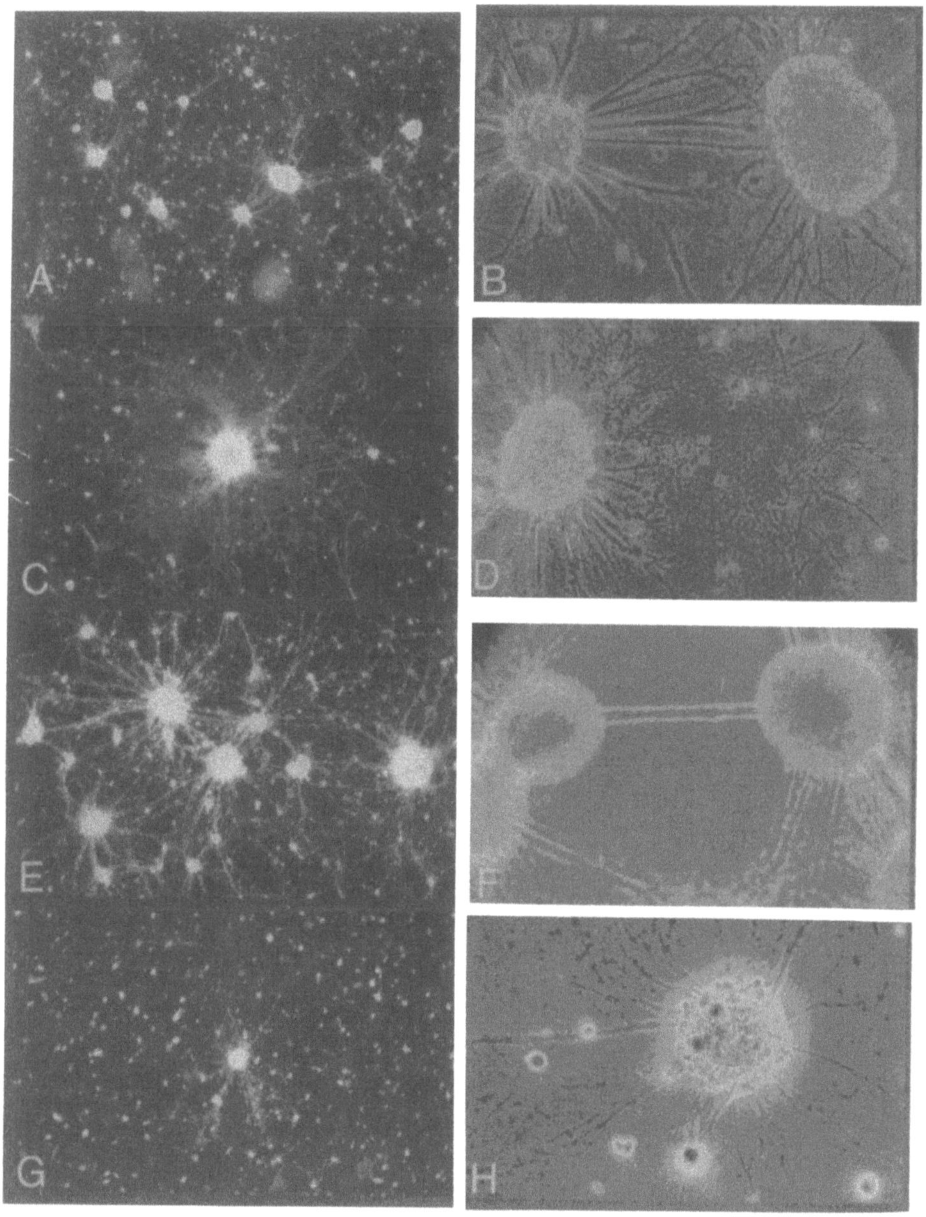

Figure 3. Photomicroraphs of neuroblast-enriched cultures derived from 3-day-old chick embryos (E3WE) at 10DIV. Control cultures (A,B) were grown in DMEM+5%FBS. Note the characteristic growth pattern of E3WE, exhibiting extensive neuritic outgrowth and fasciculation with complex aggregate interconnection. Cultures exposed to 50mM ethanol (C,D) from 0 to 3DIV exhibit extensive neuritic outgrowth but lack the neuritic bundles and fasciculation. Cultures growth in medium supplemented with 50ng/ml NGF from 3 to 10DIV (E,F) exhibit marked neuritic faciculation exceeding that observed with controls. Also observed are more and larger aggregates with thick neuritic bundles interconnecting. Cultures grown in the presence of 50mM ethanol in the medium from 0 to 3DIV followed by removal of ethanol and replacement from 3–10DIV with medium supplemented with 50ng/ml NGF (G,H) exhibit a restoration of neuritic fasciculation similar to that observed with controls. Thus, even in culture, NGF rescues ethanol-insulted neurons. Magnification=60x (A,C,E,G); 300X (B,D,F,H). Data taken Rahman et al, 1994a.

provided by cell-cell contact accounting for the alterations in neuronal growth patterns observed in culture; b) interference with signals provided by soluble growth and trophic factors, resulting in a shift in phenotypic "balance"; c) interference with signal transduction cascades involved in determination of neurotransmitter phenotypes.

4.1. Ethanol Neurotoxicity: Selective Death of Cholinergic Neurons

Evidence in support of the hypothesis that the shift from cholinergic expression described earlier is a result of selective death of cholinergic neuronal populations comes from the following study. Neuroblast-enriched cultures were established from 3-day-old chick embryos which had been treated *in ovo* with either ethanol or saline (as control). Staining for bromodeoxyuridine (BrdU) incorporation into DNA established the birthdate of neuroblasts at 0–3 days *in vitro* (DIV). A second co-staining for one of four primary

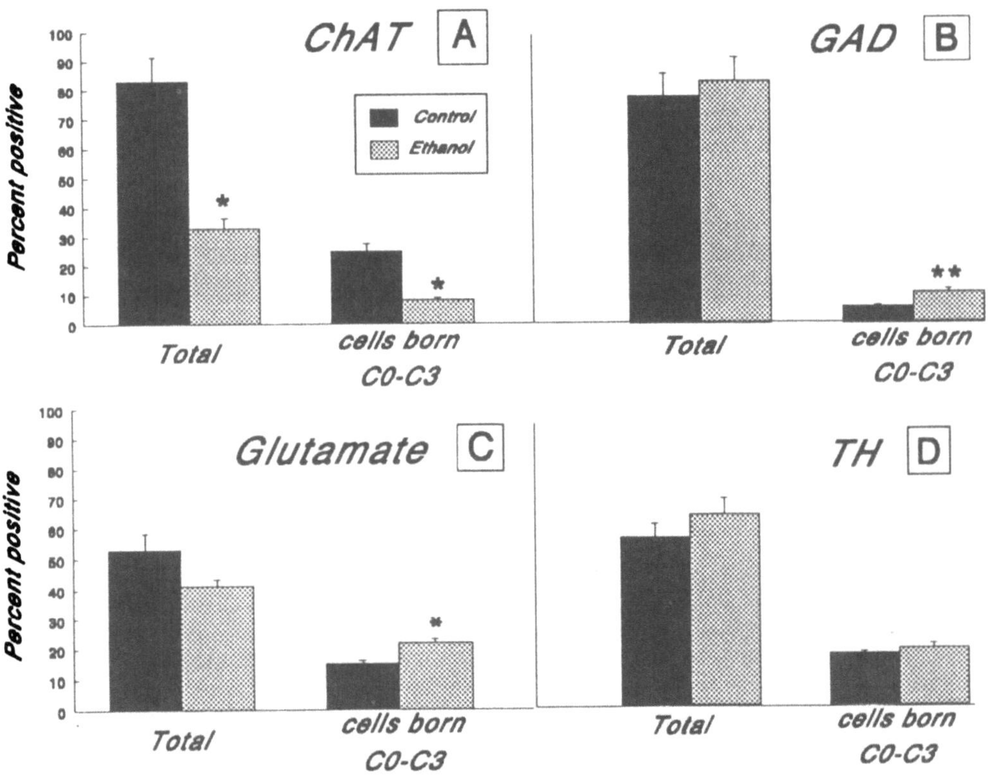

Figure 4. Selective survival of four neuronal phenotypes following ethanol exposure *in vitro*. Experimental paradigm is described above. The left-hand bars in each frame (A-D) labelled "Total" represent the survival within the total population of neurons which were positive for their respective neuronal marker regardless of date of birth. The right-hand bars labelled "cells born C0-C3" represent that portion of the "total" population born between 0 and 3 DIV and surviving through 6 DIV. From this graph it is clear that *in ovo* ethanol exposure during the predetermined critical period of sensitivity (E1-E2) reduces selectively the overall survival within the cholinergic population. Other phenotypes are only slightly affected. Within the population of new neurons generated between 0 and 3 days *in vitro*, again, there is a selective reduction in the cholinergic population (A) whereas within GABAergic (B) and glutamatergic (C) neurons, survival is enhanced by pre-treatment with ethanol. *P<0.001 vs respective control; **P<0.005 vs respective control. Data taken from Kentroti and Vernadakis, 1996b.

neuronal markers (ChAT, TH, GAD, glutamate) established the specific neuronal phenotype of individual cells. Cells staining positive for both BrdU and the neuronal marker (i.e."born" between 0 and 3DIV) or cells negative for BrdU but positive for the neuronal marker were quantitated. The results (shown in Figure 4) clearly demonstrate the selective sensitivity of cholinergic neurons to the effects of prenatal ethanol exposure. Longevity of all cholinergic neuroblasts in culture, regardless of birthdate, is significantly diminshed by *in ovo* exposure to ethanol. Other phenotypes (i.e. GABAergic and glutamatergic) exhibit an enhancement in survival following ethanol exposure (Kentroti and Vernadakis, 1996b).

4.2. Ethanol-Neurotrophin Interactions during Neurogenesis: *In Ovo* and in Culture

Among the possible mechanisms for the action of ethanol is the hypothesis that it interferes with signals provided by soluble growth and trophic factors, resulting in a shift in phenotypic "balance". Early evidence from this laboratory has shown that NGF and EGF restore cholinergic expression *in ovo* in ethanol-exposed embryos whether given concomitantly, or subsequent to ethanol treatment.

Neuroblast-enriched cultures derived from 3-day-old chick embryos under control conditions exhibit growth patterns consisting of highly organized neuronal aggregates interconnected by tightly bundled neuritic processes (Figure 3 A,B). Growth of cultures in medium supplemented with 50mM ethanol results in a decrease in the overall number of aggregates accompanied by a defasciculated pattern of neuritic outgrowth (Figure 3 C,D). Supplementation of the medium with NGF under very specific conditions not only reverses the morphological effects of ethanol (Figure 3 G,H), but restores the cholinergic expression impaired by ethanol. In culture, as *in ovo* the cholinotoxic effects of ethanol as well as the cholinotropic effects of NGF are confined to critical periods of sensitivity which occurs between 0 and 4 DIV, with 3DIV being the most significant. In order to reverse the cholinotoxic effects of ethanol(Figure 3 G,H), NGF may be given concomitantly from 0–3DIV or, the ethanol-supplemented medium *replaced* at 3DIV with medium containing (50ng/ml) NGF (Rahman et al, 1993a; 1994a,b).

Growth factors have also been assessed for their effectiveness to ameliorate the effects of ethanol on neuronal survival. Supplementation of the culture medium with NGF (50ng/ml) restores to control levels the lifespan of neurons exposed to ethanol *in ovo* at embryonic days 1–2. In culture, the marked decline in survival of neuroblasts in ethanol-supplemented medium (50mM) is again reversed by NGF as well as other trophic factors (i.e. EGF and bFGF) with a relative efficacy NGF>EGF>bFGF (Figure 5) (Rahman et al, 1994a;b).

Other studies have examined the effects of lesser-studied trophic factors using similar experimental paradigms. GABA is one of the primary inhibitory neurotransmitters of the vertebrate CNS. It has also received recent attention as an important modulator of neuronal development, affecting parameters such as neuronal proliferation, differentiation and synaptogenesis (Eins, et al, 1983; Spoerri and Wolff, 1981; Spoerri 1988). When GABA is examined for its neurotrophic effects, as seen in Figure 6, cholinergic expression is significantly enhanced. In addition, the neurotoxic effects of ethanol on cholinergic expression are counteracted in culture by addition of GABA to the growth medium (Figure 6) (Spoerri et al, 1995a,b).

The trophic effects of GABA on cholinergic expression can be further localized to activation of the $GABA_A$ receptor. Figure 6 summarizes the effects of several GABA receptor agonists/antagonists on ethanol-induced depression of cholinergic expression (as

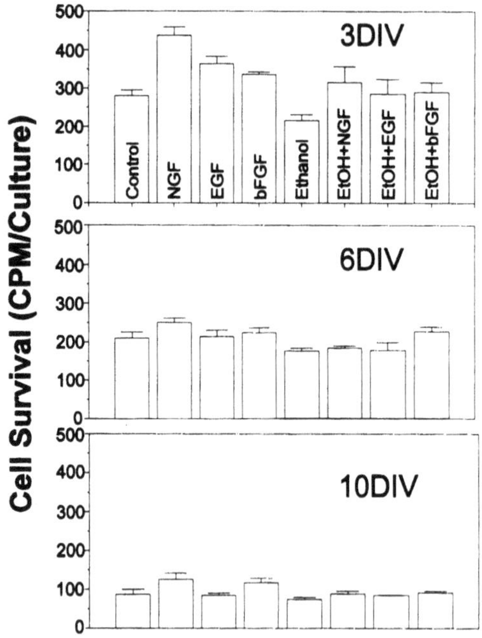

Figure 5. Relative effectiveness of three different growth factors to ameliorate the effects of ethanol on cell survival at three ages in culture. E3WE cultures were prepared from untreated embryos and grown as indicated. The doses were as follows: ethanol (50mM); NGF and bFGF (50ng/ml); EGF (25ng/m)l. Survival is measured by uptake of ^{3}H-thymidine and expressed as DPM/culture. At 3 DIV all three growth factors significantly restore cell survival to control levels as compared with ethanol alone. Data taken from Rahman et al, 1994b.

assessed by ChAT activity) in E3WE cultures. The $GABA_A$ receptor agonist muscimol potentiates cholinergic expression and reverses the effects of ethanol on ChAT, as does GABA. The $GABA_B$ receptor agonist baclofen has no significant effect on ChAT activity (data not shown) either alone or on the ethanol-induced depression of ChAT activity.

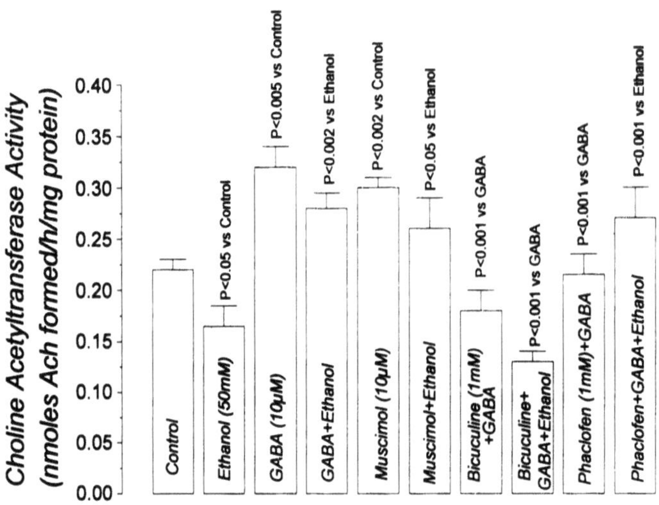

Figure 6. Effects of several GABA receptor agonists/antagonists on ethanol-induced depression of cholinergic expression. Neuroblast-enriched cultures derived from 3-day-old chick embryos (E3WE) were grown in the presence of ethanol or GABA agonists/antagonists as indicated, harvested at 7DIV and the cell pellets assayed for biochemical activity of choline acetyltransferase. Both GABA and the GABAA agonist GABA=$GABA_{A\,B}$ receptor agonist; muscimol=$GABA_A$ agonist; bicuculine=$GABA_A$ antagonist; phaclofen=$GABA_B$ antagonist. Data taken from Spoerri et al, 1996.

Similarly, the The $GABA_B$ receptor antagonist phaclofen cannot prevent the reversal of ethanol effects by GABA (Figure 6). Thus, both the potentiation of cholinergic expression by GABA as well as the GABA-mediated attenuation of ethanol insult are mediated through the $GABA_A$ receptor (Spoerri et al, 1996).

4.3. Effects of Ethanol on Cell-Cell Contacts

The profound morphological changes elicited by exposure of neuroblasts in culture to ethanol suggests a shift away from normal cholinergic development by interfering with cell-cell contact signals involved in phenotypic expression. Within the developing nervous system, intercellular contacts provide fundamental signals for specific developmental events. For example, homophilic binding between neuronal perikarya is a function of the neural cell adhesion molecule (NCAM), an integral membrane whose presence on neuritic processes also supports fasciculation. One or more isoforms of this molecule are expressed by virtually all neuronal phenotypes as well as non-migrating neural crest cells and embryonic myotubes. Functionally, NCAM is implicated in many processes integral to CNS development including neuritic fasciculation, guidance of optic axons to the tectum, organization of cell and plexiform layers of the retina, innervation of skeletal myotubes, preservation of neural tube integrity and (permissively) emigration of neural crest, aggregation of neural crest cells to form ganglia and even reorganization within nervous structures in the adult.

The normal profile of NCAM ontogenesis during development in chick embryonic whole brain as well as in cerebral hemispheres, and cerebellum has been studied in detail (Rutishauser, 1983; Kentroti et al, 1996) and it has been shown that the NCAM profile shifts from a high to low molecular weight isoform with advancing days of embryonic development (Figure 7). This shift represents the transition from highly sialated (and thus lower binding affinity) to less sialated forms of NCAM (Kentroti et al, 1995). Ethanol administration (10mg/50μl/day) *in ovo,* during early embryogenesis (E1-E3) evokes a shift in NCAM expression (in both whole brain and cerebral hemisphere homogenates) towards higher molecular weight isoforms when examined at E8 and E10 (Figure 7). Interpretation of the results suggests that ethanol exposure during the period of abundant neuroblast migration (E3-E4) does not immediately effect NCAM expression in either whole brain or cerebral hemispheres. However, during the period corresponding to axonal and dendritic generation, synaptogenesis, and cortical stratification (E8-E10), ethanol-treatment results in an increase in high molecular weight (polysialated) forms of NCAM.

Examination of the effects of ethanol on the distribution of NCAM *in vitro* in neuroblast-enriched cultures demonstrates widespread NCAM immunoreactivity in these cultures (at 3, 6 and 9DIV) both in the absence and presence of ethanol in the growth medium. When interpreted together with the *in ovo* results showing an increase in expression of high molecular weight isoforms of NCAM with prenatal ethanol exposure, evidence suggests that ethanol delays the embryonic-adult shift or interferes with NCAM homophilic binding to uncouple neuritic fasciculation and/or neuronal aggregation (Kentroti et al, 1995).

4.4. Existence of Multipotential Neuroblasts

One of the more striking revelations about CNS development provided by the study of ethanol neurotoxicity regards the heterogeneity of developing neuronal populations within specific phenotypes. Within discrete regions of the developing CNS ethanol has a differential effect on emerging phenotypes depending on their site of origin. Figure 8 il-

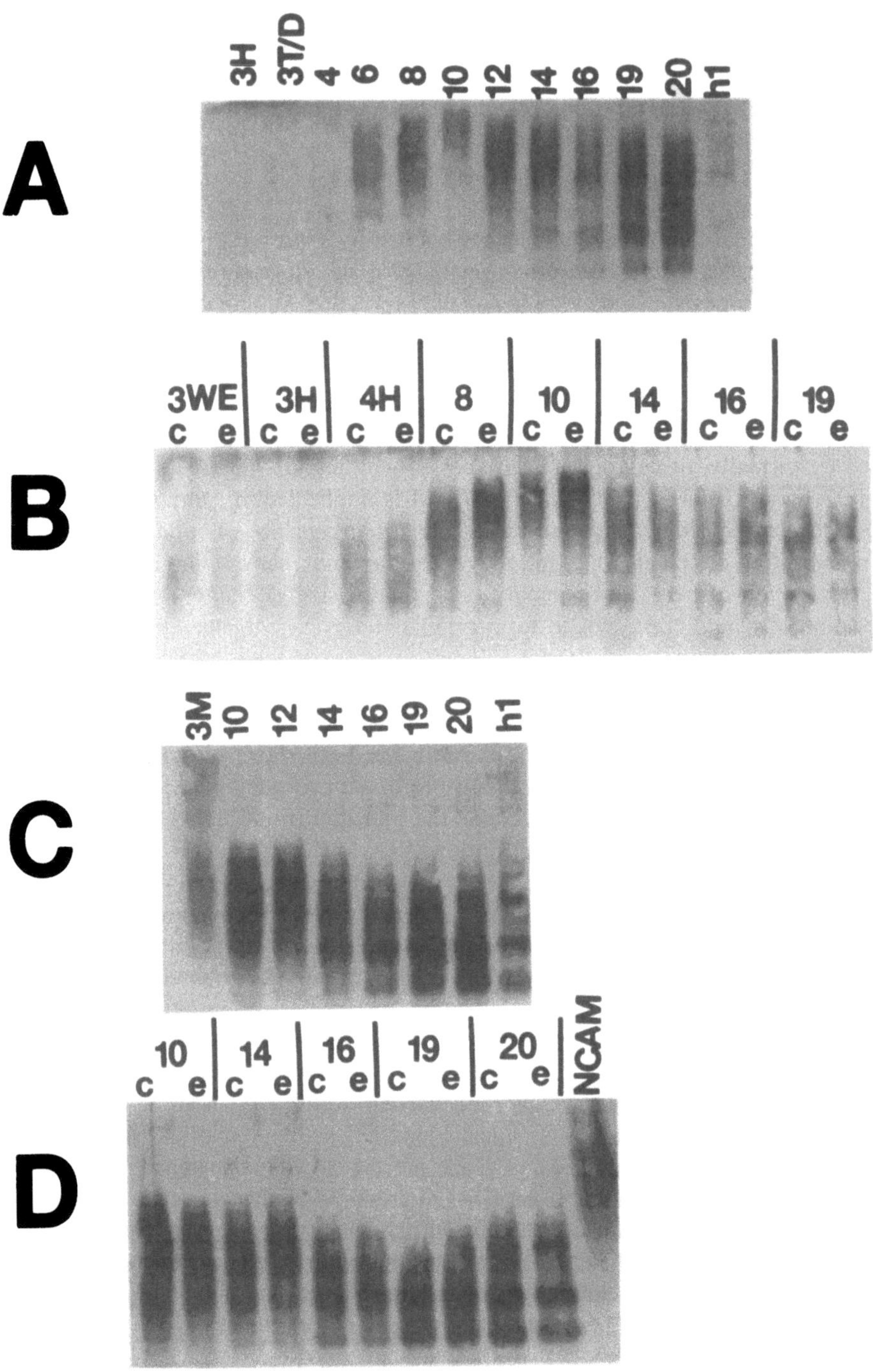

Figure 7. Western blots showing the developmental profile (from embryonic day 3 through post-hatching day 1; h1) of NCAM in cerebral hemispheres (A,B) and cerebellar (C,D) homogenates derived from embryonic chick brain. Panels B and D show the shift in NCAM expression in the same brain regions in control (c) versus ethanol-treated (e) embryos at corresponding developmental ages. Ethanol exposure at E1-E2 shifts the NCAM profile to a higher molecular weight isoform in cerebral hemispheres at 8 and 10 days of development (B). WE=whole embryo; M=mesencephalon; H=head; T/D=telencephalon/ diencephalon. Data from Kentroti et al. 1995.

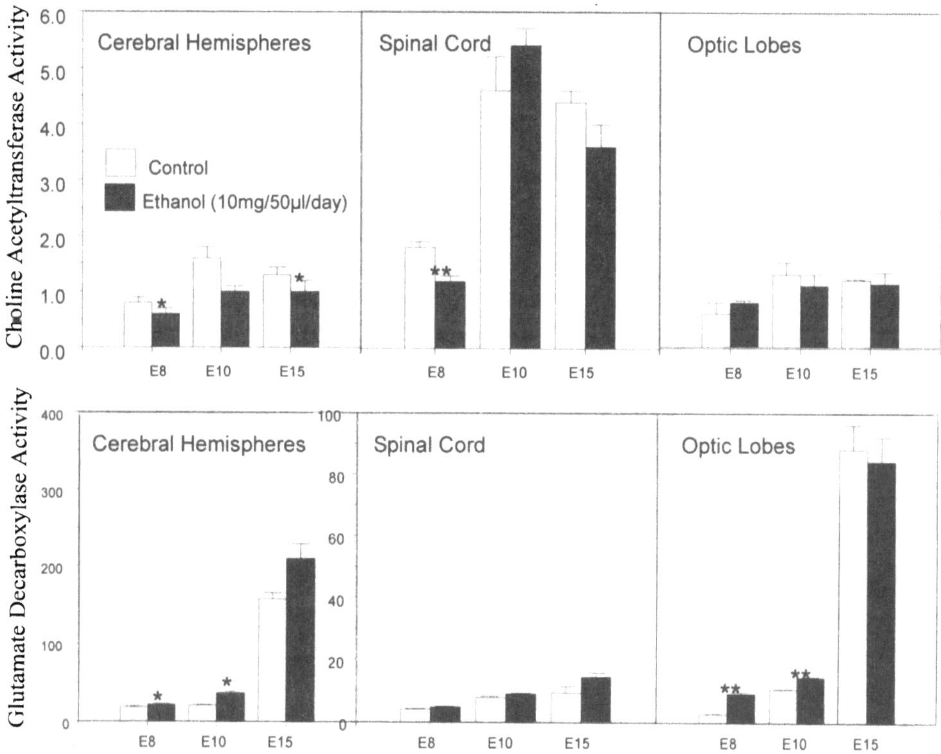

Figure 8. Effect of *in ovo* ethanol on cholinergic (top row) and GABAergic (bottom row) expression in 3 regions of the developing chick embryo CNS. Ethanol (10mg/50µl/day) was administered at E1-E3. Embryos were sacrificed at E8, E10 or E15, cerebral hemispheres (CH), spinal cord (SC) and optic lobes (OL) were assayed for enzyme activity. Choline acetyltransferase activity (ChAT), expressed as nmoles acetylcholine formed/h/mg protein, was used as the cholinergic biochemical marker and glutamate decarboxylase activity (GAD), expressed as pmoles CO_2 formed/h/mg protein, was used as the GABAergic biochemical marker. There is a differential effect of ethanol on cholinergic expression when comparing different regions of the CNS. ChAT activity in CH was markedly reduced at all ages, whereas in SC, the decrease in ChAT activity at E8 is reversed by E10. In OL, cholinergic neurons were unaffected by ethanol at all ages examined. GAD activity was markedly enhanced in CH and OL at all ages. In contrast, activity in SC was unaffected by ethanol. *$P<0.02$; **$P<0.002$ vs controls. Data from Kentroti and Vernadakis, 1992.

lustrates the differential effects of ethanol on cholinergic expression in 3 distinct regions of the CNS (Kentroti and Vernadakis, 1992). Cholinergic expression within the cerebral hemispheres is significantly impaired by prenatal exposure to ethanol from E1 to E3 and this effect persists through E15. Whereas cholinergic expression in optic lobes, a structure more advanced in its development, is unaffected by ethanol exposure during the same period. Finally, intermediate to the effects in cerebral hemispheres versus optic lobes, cholinergic expression in spinal cord neurons is affected by ethanol exposure (E1-E3) only through E8, after which time cholinergic expression in ethanol-exposed neurons is indistinguishable from controls. GABAergic neurons, which exhibit an increase in expression, also respond differentially to ethanol exposure depending upon their site of origin in the CNS (Figure 8). Both cerebral hemisphere and optic lobe GABAergic expression are enhanced through 10 days of development by ethanol exposure from E1 to E3 while expression in spinal cord is unaffected. The results from these studies provide further evidence in support of the theory that ethanol shifts neuronal fates in multipotential neuroblasts during early neurogenesis. Furthermore, it underscores the importance of the local microenvironment in directing expression of emerging neuronal phenotypes.

The most compelling evidence in support of the presence of multipotential neuroblasts comes from colocalization studies on neuroblasts in cultures derived from 3-day-old embryos. Specific antibodies to ChAT (cholinergic), TH (catecholaminergic), GABA (GABAergic), and somatostatin (SRIF; representitive peptidergic) are found colocalized with one another in all possible combinations (Kentroti and Vernadakis, 1995). Furthermore, each phenotype is exhibited by <50% of cells within each culture/field. Thus, the summation of all positive cells is in excess of 100%. The only explanation for this number is that cells are positive for more than two phenotypes. This finding implies that neuroblasts possess the machinery necessary to adopt multiple phenotypes and supports the idea that microenvironmental factors (i.e. ethanol, neurotrophins) provide signals which affect emerging phenotoypes (Kentroti and Vernadakis 1992; 1995).

4.5. Ethanol Arrests Cholinergic Differentiation at an Immature Stage

High-affinity choline reuptake by cholinergic neurons is the primary source of choline in the synthesis of acetylcholine and is one of the most reliable indices of neuronal maturation within this phenotype. Ethanol exposure of neuron-enriched cultures derived from 8-day-old chick embryonic cerebral hemispheres (E8CH) has a significant effect on high-affinity choline uptake. Determination of K_m and V_{max} values using the Lineweaver-Burke analysis reveals that ethanol exposure results in two distinct values for V_{max}, one similar to the single control value and one which is significantly lower. No differences are observed in K_m values between control and ethanol-treated cultures. The low secondary V_{max} value is interpreted to represent that population of cholinergic neurons which have been arrested at an immature stage in response to ethanol exposure (Rahman et al, 1993b).

4.6. Neuroprotective Role of Glia in Ethanol Neurotoxicity

The importance of the neuron-glia interrelationship in the process of neurogenesis has already been discussed. In addition, the presence and maturational stage of glia is showing some importance in protecting sensitive neuroblasts from the deleterious effects of prenatal ethanol exposure.

The neuron-glia coculture paradigm has proven to be an extremely useful tool to examine their interrelationship. This paradigm allows for the dissociation of the microenvironmental effects of the neuron-glia interrelationship from those provided by direct cell-cell contact by physically separating the two cell populations. Soluble factors elaborated by either cell type are readily permeable across a separating partition but cells do not cross. With a neuroblast source of 3-day-old chick embryo (E3WE) preliminary data has been obtained defining stage-specific differences in the neuroprotective properties of glia. Neuroblasts pretreated at E1 and E2 *in ovo* with saline (control) or ethanol (10mg/50μl/day) prior to neuroblast dissociation and cocultured with early- (E3) or late-stage (E15) glia yielded the following observations: 1) coculture with neuroblasts promotes astrocytic expression in early-stage glia (E3H), and 2) coculture of ethanol-insulted neurons with late-stage glia but not with early-stage glioblasts ameliorates the effects of ethanol exposure on cholinergic expression in neuroblasts following *in ovo* ethanol exposure. These observations suggest the presence of stage-specific soluble factors produced by mature glia which can be trophic to neurons at earlier stages of development (Kentroti and Vernadakis, in preparation).

5. CONCLUSIONS

From studies on the neurotoxic effects of prenatal ethanol exposure, several pertinent facts have come to light. Ethanol has a significant effect to influence the dynamic process of phenotypic determination in the embryonic CNS. Its effects are felt in at least three neuronal phenotypes (cholinergic, catecholaminergic and GABAergic) and are confined to a discrete critical window of sensitivity occurring between 1 and 3 days of embryogenesis in the chick embryo. The effects of ethanol on phenotypic expression are differential with respect to neuronal phenotype (cholinergic is depressed while catecholaminergic and GABAergic are enhanced). The fact that neuronal phenotypic balance can be restored, following ethanol insult, by various trophic substances, identified (i.e. NGF, EGF, bFGF, GABA) and unidentified (i.e. skeletal muscle extract, glial-conditioned medium) further emphasizes the plasticity of neurons at early stages of development. Some of the potential cellular mechanisms for the effects of ethanol to shift phenotypic expression are summarized in Figure 9. Positive evidence exists for the selective death of cholinergic neuroblasts (Figure 9A). On the other hand, multipotent neuroblasts are prevalent during early stages of neurogenesis and preliminary data supports the notion that ethanol promotes a shift from cholinergic to catecholaminergic and GABAergic expression in this

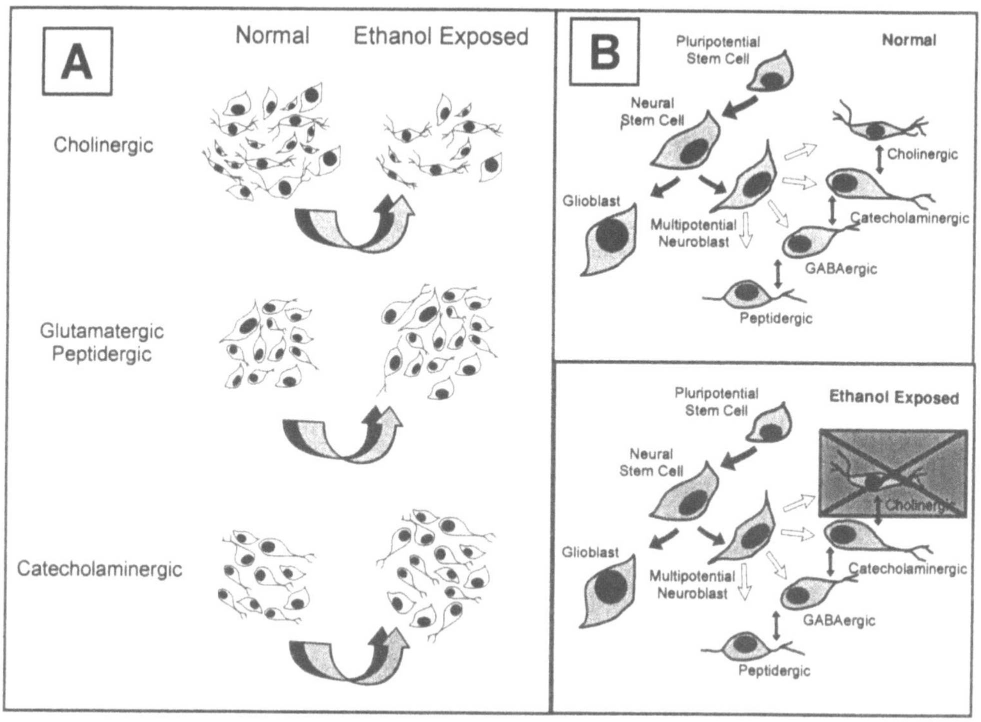

Figure 9. Frame **A** represents the theory of "selective survival" of developing neuroblasts following ethanol exposure. Normal neuroblast populations are shown above the line. Following ethanol exposure cholinergic neurons, being more sensitive to the neurotoxic effects of ethanol, exhibit an enhancement of naturally-occurring cell death relative to the other phenotypes as shown by distribution of cell populations below the line. Frame **B** represents the hypothetical "shift in phenotype" following ethanol exposure. Ethanol diminishes the cholinergic population by inducing multipotential neuroblasts to express other phenotypes.

population or at least maintains cholinergic neuroblasts at an immature stage of differentiation (Figure 9B). Finally, the subcellular signalling mechanisms which mediate phenotypic expression and are subject to ethanol insult are beginning to be revealed. In view of the fact that the presence of precursor cells in adult CNS has now been accepted (Barami et al, 1994; Kirschenbaum et al, 1994; Morshead et al, 1994), a complete understanding of the factors which influence phenotypic expression could have profound implications in treatment of many neuropathological conditions ranging from Alzheimer's disease to spinal cord injury.

ACKNOWLEDGMENTS

Studies from this laboratory and reported in this review were supported by grant AA08026 from the National Institutes of Health.

REFERENCES

Abbot NJ (ed) (1991) *Glia-Neuronal Interactions*. Vol 633, Ann N.Y. Acad Sci, New York.

Barami K, Kirschenbaum B, Lemmon V, Goldman SA (1994): N-cadherin and Ng-CMA/8D9 are involved serially in the migration of newly generated neurons into the adult songbird brain. *Neuron* 13:567–582.

Brodie C, Vernadakis A (1992) Ethanol increases cholinergic and decreases GABAergic neuronal expression in cultures derived from 8-day-old chick embryo cerebral hemispheres: interaction of ethanol and growth factors. *Devl Brain Res* 65:253–257.

Brodie C, Kentroti S, Vernadakis A (1991) Growth factors attenuate the cholinotoxic effects of during early neuroembryogenesis in the chick embryo. *Int J Devl Neurosci* 9:203–213.

Brodie C, Vernadakis A (1990) Critical periods to ethanol exposure during early neuroembryogenesis in the chick embryo: cholinergic neurons. *Devl Brain Res* 56:223–228.

Cameron RS, Rakic P (1991) Glial cell lineage in the cerebral cortex: A review and synthesis. *GLIA* 4:124–137.

Carnahan J, Nawa H (1995) Regulation of neuropeptide expression in the brain by neurotrophins. *Mol Biol* 10:135–149.

Chao CC, Hu S, Tsang M, Weatherbee J, Molitor TW, Anderson WR, Peterson K (1992) Effects of transforming growth factor-ß on murine astrocyte glutamine synthetase activity: Implications in neuronal injury. *J Clin Invest* 90:1786–1793.

Du X, Iacovitti L (1995) Synergy between growth factors and transmitters required for catecholamine differentiation in brain neurons. *J Neurosci* 15:5420–5427.

Eins E, Spoerri PE and Heyder E, (1983) GABA or sodium bromide induced plasticity of neurites of mouse neuroblastoma cells: A quantitative study. *Cell Tiss Res* 229:457–460.

Fauquet M, Smith J, Ziller C, LeDouarin NM (1981) Differentiation of autonimic neuron precursors *in vitro*: Cholinergic and adrenergic traits in culture neural crest cells. *J Neurosci* 1:478–492.

Fedoroff S, Vernadakis A (1986) "*Astrocytes*" Vol. 1–3. Academic Press, Orlando, Florida.

Furukawa S, Furukawa Y, Satogoshi E, Hayashi K (1986) Synthesis and secretion of nerve growth factor by mouse astroglial cells in culture. *Biochem Res Comm* 136:57–63.

Gray EG, Sanes JR (1992) Lineage of radial glia in the chicken optic tectum. *Development* 114:271–285.

Hamburger V, Hamilton HL (1951) A series of normal stages in the development of the chick embryo. *J Morph* 88:49–92.

Han VKM, Smith A, Myint W, Mygard K, Bradshaw S (1992) Mitogenic activity of epidermal growth factor on newborn rat astroglia: Interaction with insulin-like factors. *Endocrinol* 131:1134–1142.

Hatten ME (1984) Embryonic cerebellar astroglia *in vitro*. *Dev Brain Res* 13:309–313.

Hertz L, Schousboe A (1987) Primary cultures of GABAergic and glutamatergic neurons as model systems to study neurotransmitter functions. I. Differentiated cells. . *In: Model Systems of Development and Aging of the Nervous System* (Vernadakis A, Privat A, Lauder JM, Timiras PS, Giacobini E, Eds) Martinus Nijhoff, Boston, pp 19–31.

Hirano M, Goldman JE (1988) Gliogenesis in rat spinal cord: Evidence for origin of astrocytes and oligodendrocytes from radial precursors. *J Neurosci Res* 21:155–167.

Houlgatte R, Mallat M, Brachet P, Prochiantz A (1989) Secretion of NGF in culture of glial cells and neurons derived from different regions of the mouse brain. *J Neurosci Res* 24:143–152.

Howard MJ, Bronner-Fraser M (1985) The influence of neural tube-derived factors on differentiation of neural crest cells *in vitro*. *J Neurosci* 5:3302–3309.

Huff RK, Schreier WC (1989) Fibroblast growth factor pretreatment reduces epidermal growth factor-induced proliferation in rat astrocytes. *Life Sci* 45:1515–1520.

Jonakait GM, Wei R, Sheng ZL, Hart RP, Ni L (1994) Interferon-γ promotes cholinergic differentiation of embryonic septal nuclei and adjacent basal forebrain. *Neuron* 12:1149–1159.

Keningsbery RL, Mazzoni JE, Collier B, Cuello AC (1992) Epidermal growth factor affects both glia and cholinergic neurons in septal cell culture. *Neurosci* 50:85–97.

Kentroti S, Vernadakis A (1997) Differential expression in glial cells derived from chick embryo cerebral hemispheres at an advanced stage of development. *J Neurosci Res* 47:822–331.

Kentroti S, Vernadakis A (1996a) Immunocytochemical and biochemical characterization of glial phenotypes in normal and immortalized cultures derived from 3-day-old chick embryo encephalon. *GLIA* 18:79–91.

Kentroti S, Vernadakis A (1996b) Ethanol neurotoxicity in culture: Selective loss of cholinergic neurons. *J Neurosci Res* 44:577–585.

Kentroti S, Vernadakis A (1995) Early neuroblasts are pluripotential: Colocalization of neurotransmitters and neuropeptides. *J Neurosci Res* 41:696–707.

Kentroti S, Grove J, Rahman H, Vernadakis A (1995) Ethanol neuronotoxicity in the embryonic chick brain *in ovo* and in culture: Interaction of the neural cell adhesion molecule (NCAM). *Int J Devl Neurosci* 13:859–870.

Kentroti S, Vernadakis A (1992) Ethanol administration during early embryogenesis affects neuronal phenotypes at a time when neuroblasts are pluripotential. *J Neurosci Res* 33:617–625.

Kentroti S, Vernadakis A (1991a) Growth hormone-releasing hormone and somatostatin influence neuronal expression in the developing chick brain: III. GABAergic neurons. *Brain Res.* 562:34–38.

Kentroti S, Vernadakis A (1991b) Effects of early *in ovo* administration of ethanol on expression of the GABAergic neuronal phenotype in the chick embryo. *Dev. Brain Res.* 61:290–292.

Kentroti S, Vernadakis A (1991c) Correlation between morphological and biochemical effects of ethanol on neuroblast-enriched cultures derived from 3-day-old chick embryos. *J. Neurosci. Res.* 30:484–492.

Kentroti S, Vernadakis A (1990a) Growth hormone-releasing hormone and somatostatin influence neuronal expression in the developing chick brain: II. Cholinergic neurons. *Brain Res.* 512:297–303.

Kentroti S, Vernadakis A (1990b) Neuronal plasticity in the developing chick brain: Interaction of ethanol and neuropeptides. *Develop. Brain Res.* 56:205–210.

Kentroti S, Vernadakis A (1989) Growth hormone-releasing hormone influences neuronal expression in the developing chick brain: I. Catecholaminergic neurons. *Develop. Brain Res.* 49:275–280.

Kimelberg HK (1986) Catecholamine and serotonin uptake in astrocytes. *In: Astrocytes* Vol 2, (Fedoroff S, Vernadakis A, Eds) Academic Press, Orlando, Florida, pp 107–131.

Kirschenbaum B, Nedergaard M, Preuss A, Barami K, Fraser RAR, Goldman SA (1994) *In vitro* neuronal production and differentiation by precursor cells derived from the adult human forebrain. *Cerebral Cortex* 6:576–589.

Landis SC (1980) Developmental changes in the neurotransmitter properties of dissociated sympathetic neurons: a cytochemical study of the effect of medium. *Devl Biol* 77:349–361.

Landis SC (1978) Growth cones of cultured sympathetic neurons contain adrenergic vesicles. *J Cell Biol* 78:8–14.

Landis SC, Keefe D (1983) Evidence for neurotransmitter plasticity *in vivo:* Developmental changes in properties of cholinergic sympathetic neurons. *Dev Biol* 98:349–372.

Landis SC, Siegel RE, Schwab M (1988) Evidence for neurotransmitter plasticity *in vivo* II. Immunocytochemical studies of rat sweat gland innervation during development. *Dev Biol* 126: 129–140.

Le Douarin N (1980) Migration and differentiation of neural crest cells. In: *Current Topics in Developmental Biology, Vol. 16* Academic Press, Oxford. pp31–86.

Le Douarin N (1973) A biological cell labeling technique and its use in experimental embryology. *Dev Biol* 30:217–222.

Le Douarin N, Smith J (1988) Development of the peripheral nervous system from the neural crest. *Ann Rev Cell Biol* 4:375–404.

Le Douarin N, Teillet M-AM (1974) Experimental analysis of the migration and differentiation of neuroblasts of the autonomic nervous system and of neurectodermal mesenchymal derivatives, using a biological cell marking technique. *Dev Biol* 41:162–184.

Le Douarin N, Renaud D, Teillet M-AM and Le Douarin GH (1975) Cholinergic differentiation of presumptive adrenergic neuroblasts in interspecific chimeras after heterotopic transplantation. *Proc Natl Acad Sci, USA* 72:728–732.

Lee K, Kentroti S, Vernadakis A (1992) Comparative biochemical, morphological and immunocytochemical studies between C6 glial cells of early and late passage and advanced passages of glial cells derived from aged mouse cerebral hemispheres. *GLIA* 6:245–257.

Levitt P, Rakic P (1980) Immunoperoxidase localization of glial fibrillary acidic protein in radial glial cells and astrocytes of the developing rhesus monkey brain. *J. Comp. Neurol.* 193–417:448.

Lindholm D, Harikka J, da Penha Berzaghi M, Castren E, Tzimagiorgis G, Highes RA, Thoenen H (1994) Fibroblast growth factor-5 (FGF-5) promotes differentiation of cultured rat septal cholinergic and raphe serotonergic neurons: comparison with the effects of neurotrophins. *Eur J Neurosci* 6:244–252.

Loret C, Sensenbrenner M, Labourdette G (1989) Differential phenotypic expression induced in cultured rat astroblasts by acidic fibroblast growth factor, epidermal growth factor and thrombin. *J Biol. Chem* 264:8319–8327.

Maderspach K, Nemeth K (1993) Immunocytochemical visualization of κ-opioid receptors on chick embryonic neurons differentiating in vitro. *Neurosci* 57:459–465.

Massarelli R, Mykita S, Sorrentino G (1986) The supply of choline to glial cells. *In: Astrocytes* Vol 2, (Fedoroff S, Vernadakis A, Eds) Academic Press, Orlando, Florida, pp 155–178.

McConnell SK (1991) The generation of neuronal diversity in the central nervous system. *Annu Rev Neurosci 1991* 14:269–300.

McKay RDG (1989) The origins of cellular diversity in the mammalian central nervous system. *Cell* 58:815–821.

Misson J-P, Takahashi T, Caviness VS Jr (1991) Ontogeny of radial and other astroglial cells in murine cerebral cortex. *Glia,* 4:138–148.

Morshead CM, Reynolds BA, Craig CG, McBurney MW, Staines WA, Morassutti D, Weiss S, van der Kooy D (1994) Neural stem cells in the adult mammalilan forebrain: A relatively quiescent subpopulation of subependymal cells. *Neuron* 13:1071–1082.

Muller CM (1992) A role for glial cells in activity-dependent central nervous plasticity? Review and hypothesis. *Int Res Neurobiol* 34:215–221.

Olson JA, Shiverick KT, Ogilvie S, Buhi WC, Raizada MK (1991) Developmental expression of rat IGF binding protein-2 by astrocytic glial cells in culture. *Endocrinol* 129:1066–1074.

Oppenheim RW (1991) Cell death during development of the nervous system. *Annu Rev Neurosci, 1991* 14:453–501.

Patterson PH, Chun LLY (1977a) The induction of acetylcholine synthesis in primary cultures of dissociated rat sympathetic neurons. *Devl Biol* 56:263–280.

Patterson PH, Chun LLY (1977b) The induction of acetylcholine synthesis in primary cultures of dissociated rat sympathetic neurons. *Devl Biol* 60:473–481.

Perraud F, Labourdette G, Eclancher F, Sensenbrenner M (1990) Primary cultures of astrocytes from different brain areas of newborn rats and effects of basic fibroblast growth factor. *Devl Neurosci* 12:11–21.

Perraud F, Besnard F, Pettman B, Sensenbrenner M, Labourdette G (1988) Effects of aFGF and bFGF on the proliferation and glutamine synthetase expression of rat astroblasts in culture. *GLIA* 1:124–131.

Raff MC, Barres BA, Burne JF, Coles HS, Ishizaki Y, Jacobson MD (1993) Programmed cell death and the control of cell survival: lessons from the nervous system. *Science* 262:695–700.

Rahman H, Kentroti S, Vernadakis A (1994a) The critical period for ethanol effects on cholinergic neuronal expression in neuroblast-enriched cultures derived from 3-day-old chick embryo: NGF ameliorates the cholinotoxic effects of ethanol. *Int J Devl Neurosci* 12:397–404.

Rahman H, Kentroti S, Vernadakis A (1994b) Neuroblast cell death *in ovo* and in culture: interaction of ethanol and neurotrophic factors. *Neurochem Res* 19:1495–1502.

Rahman H, Kentroti S, Vernadakis A (1993a) Early *in ovo* exposure of chick embryos to ethanol prevents the neuronotrophic effects of intracerebral NGF administration on cholinergic phenotypic expression. *Dev Brain Res* 76:256–259.

Rahman H, Lee K, Kentroti S, Vernadakis A (1993b) Ethanol exerts differential effects on high affinity choline uptake in neuron-enriched cultures from 8-day-old chick embryo cerebral hemispheres. *Neurochem Res* 18:551–557.

Rakic PC (1971) Neuron-glia relationship during granule cell migration in developing cerebellar cortex; a Golgi and electron microscopic study in Macacus rhesus. *J Comp Neurol* 141:283–312.

Ransom BR, Carlini WG (1986) Electrophysiological properties of astrocytes. *In: Astrocytes,* Vol 2, Academic Press, Orlando, Florida. pp1–47.

Reynolds BA, Weiss S (1992) Generation of neurons and astrocytes from isolated cells of the adult mammalian central nervous system. *Science* 255:1707–1710.

Reynolds BA, Weiss S (1993) EGF-responsive stem cells in the mammalian central nervous system. In: (A.C. Cuello, Ed) *Restorative Neurology and Neuroscience, Vol. 6, Neuronal Cell Death and Repair*, Elsevier, Amsterdam. pp. 247–255.

Rudge JS, Alderson RF, Pasnikowski E, McClain J, TP NY, Lindsay RM (1992) Expressionof CNTF and neurotrophins-NGF, BDNF and NT3 in cultured rat hippocampal astrocytes. *Eur J Neurosci* 4:459–471.
Rutishauser U (1983) Molecular and biological properties of a neural cell adhesion molecule. *Cold Spring Harbor Symp Quant Biol* 48:501–514.
Sanes JR (1989) Analysing cell lineage with a recombinant retrovirus. *Trends Neurosci* 12:21–28.
Schwab ME (1996) Structural plasticity of the adult CNS. Negative control by neurite growth inhibitory signals. *Int J Dev Neurosci* 14:379–385.
Selmaj K, Shafit-Zagardo B, Aquino DA, Farooq M, Raine CS, Norton WT, Brossan CF (1991) Tumor necrosis factor-induced proliferation of astrocytes from mature brain is associated with down-regulation of GFAP mRNA. *J Neurochem* 57:823–830.
Sieber-Blum M, Kahn CR (1982) Suppression of catecholamine and melanin synthesis and promotion of cholinergic differentiation of quail neural crest cells by heart cell conditioned medium. *Stem Cells* 2:344–353.
Spoerri PE, Srivastava N, Vernadakis A (1996) Role of $GABA_A$ receptors in the GABA attenuation of ethanol neurotoxicity. *J Neurosci Res* 44:499–506.
Spoerri PE, Srivastava N, Vernadakis A (1995a) GABA attenuates the neurotoxic effects of ethanol in neuron-enriched cultures derived from 8-day-old chick embryo cerebral hemispheres. *Dev Brain Res* 86:94–100.
Spoerri PE, Srivastava N, Vernadakis A (1995b) Ethanol neurotoxicity on neuroblast-enriched cultures from 3-day-old chick embryo is attenuated by the neuronotrophic action of GABA. *Int J Devl Neurosci* 13:539–544.
Spoerri PE (1988) Neurotrophic effects of GABA in cultures of embryonic chick brain and retina. *Synapse* 2:11–22.
Spoerri PE and Wolff JR (1981) Effect of GABA-administration on murine neuroblastoma cells in culture. I. Increased membrane dynamics and formation of specialized contacts. *Cell Tiss Res* 218:567–579.
Teillet M-A, Le Douarin NM (1983) Consequences of neural tube and notochord excision on the development of the peripheral nervous system in the chick embryo. *Dev Biol* 98:192–211.
Varon S (1975) Neurons and glia in neural cultures. *Exp Neurol* 48:93–134.
Vernadakis A (1996) Glia-neuron intercommunications and synaptic plasticity. *Prog Neurobiology* 49:185–214.
Vernadakis A (1988) Neuron-glia interrelations. *Int Rev Neurobiol* 30:149–224.
Vernadakis A (1987) Neuron-glia interrelations. In: (J.R. Smythies and R.J. Bradley, Eds) *International Review of Neurobiology, Vol 30.* Academic Press, New York, NY. pp 149–224.
Vernadakis A (1974) Uptake and storage of 3H-norepinephrine in the cerebral hemispheres and cerebellum of chicks during embryonic development and early posthatching. *In: Drugs and the Developing Brain.* (Vernadakis A, Weiner N, Eds) Plenum Publishing, New York, pp 138–148.
Vernadakis A, Kentroti S (1996) Glia models to study glial cell cytotoxicity. In: (J.R. Perez-Polo, Ed.) *Methods in Neurosciences, Vol 30. "Paradigms of Neural Injury"* Academic Press, New York, NY. pp 55–80.
Vernadakis A, Kentroti S(1990) Opioids influence neurotransmitter phenotypic expression in chick embryonic neuronal cultures. *J Neurosci Res* 26:342–348.
Voigt T (1989) Development of glial cells in the cerebral wall of ferrets: direct tracing of their transformation from radial glia into astrocytes. *J Comp Neurol* 289:74–88.
Voigt T, de Lima AD (1985) The role and fate of radial glial cells during development of the mammalian cortex. In: *Neuron-Glia Interrelations During Phylogeny.* A. Vernadakis and B.I. Roots, eds. Humana Press, New Jersey, pp. 59–78.
Walz W, Hertz L (1984) Sodium transport in astrocytes. *J Neurosci Res* 11:231–239.
White S, Woodbury DM (1987) Electrophysiological and ionic transport properties of glial cells in culture. *In: Model Systems of Development and Aging of the Nervous System* (Vernadakis A, Privat A, Lauder JM, Timiras PS, Giacobini E, Eds) Martinus Nijhoff, Boston, pp 171–191.
Yamakumi T, Ozawa F, Hashimoto F, Kuwano R, Takahashi Y, Amano T (1987) Expression of ßNGF mRNA in rat glioma cells and astrocytes from rat brain. *FEBS Lett* 223:117–121.
Yoshida K, Gage FH (1991) FGF stimulates NGF synthesis and secretion by astrocytes. *Brain Res* 538:118–126.
Ziller C (1987) Emergence of neuronal phenotypes in neural crest derived cells: and *in vitro* study. In: *Model Systems in Neurotoxicology: Alternative Approaches to Animal Testing.* Alan R. Liss, Inc. New York, NY. pp59–68.
Ziller C, Smith J (1982) Migration and differentiation of neural crest cells and their derivatives: *in vivo* and *in vitro* studies on the early development of the avian peripheral nervous system. *Reprod Nutr Develop* 22(1B):153–162.

3

PLASTICITY IN ASTROCYTIC PHENOTYPES

A Role For Protein Kinase C, Tyrosine Kinases, and Cytoskeleton Signaling

D. Mangoura,* C. Pelletiere, D. Wang, N. Sakellaridis, and V. Sogos

Department of Pediatrics and Committee on Neurobiology
The University of Chicago
Chicago, Illinois 60637

1. SECRETED FACTOR SIGNAL TRANSDUCTION PATHWAYS REGULATE SHAPE AND FUNCTION

Neurons and astrocytes derive from common progenitor ectodermal cells. The neuronal progenitors actively proliferate early in development, as the development of the neural tube and the CNS vesicles progresses. Neuroblast proliferation ceases quite early, and in most species it precedes the burst of astroblastic proliferation. During the massive proliferation of neurons, astrocytes exist in small numbers and with one identifiable phenotype, namely radial glia (Cameron and Rakic, 1994). After their final mitotic division in the subventricular zone, neurons migrate, populate specific laminae in the developing brain, elaborate processes, and form functional synapses. While neurons are migrating for more precise formation of CNS layers, astroblasts proliferate, so that in the adult brain the ratio is 9 astrocytes to 1 neuron. During the course of differentiation from a glioblast to a mature astrocyte, astrocytes undergo dynamic shape-function changes. Most intermediate and differentiated phenotypes of astrocytes are characterized by expression of specific cytoskeletal proteins and the acquisition of specific shape (reviewed in Cameron and Rakic, 1991). After final positioning and cell programmed death of neurons and astrocytes, the patterning of the brain remains a very dynamic process. It now includes constant remodeling of synapses, and continuous differentiation and proliferation of astrocytes, or differentiation of neurons and oligodendrocytes.

Fully differentiated astrocytes may re-enter the cell cycle in response to various stimuli, mediated by membrane receptors for hormones, neurotransmitters, and growth

* Correspondence to: Dimitra A. Mangoura, MD, Ph.D., Kennedy Mental Retardation Research Center MC 5058, Division of Biological Sciences, The University of Chicago, 5841 South Maryland Avenue, Chicago, Illinois 60637. Telephone: (773) 702–1136; FAX: (773) 702–9234; e-mail: dm36@midway.uchicago.edu.

factors. In a series of previous studies, we have shown that neuron secreted factors regulate both astrocytic proliferation and differentiation. This was examined in chick, rodent and neural cell lines (Mangoura et al., 1989 Mangoura, 1995; Sakellaridis et al., 1986; Sakellaridis et al., 1984). From our studies and studies from several groups, it has become clear that as the microenvironment regulates the physiological state of neurons and astrocytes the cytoskeleton is constantly regulated as well, and reflects both dynamic cell shape changes and the dynamic process of homeostasis of a cell shape phenotype. A specifically important role for the cytoskeleton is in secretion, that is regulated (synaptic) and constitutive (non-synaptic vesicle release) (Chavez et al., 1996; Janz and Sudhof, 1995. Constitutive secretion, present in all cells, may change the microenvironment and extracellular stimuli supplied to the cells. Neurotransmitter secretion from neurons may occur at synapses or from varicosities, distant from synapses. These varicosities are important for actions of neurotransmitter receptors known to be expressed on adjacent astrocytes. Furthermore, activation at synapses is acutely reversed, while extra-synaptically located receptor activation may be more sustained. Thus, non-synaptic diffusion neurotransmission may be an important mechanism of information transmission among non-synaptically coupled cells (Nedergaard, 1994) in the nervous system, including information on cell lineage.

What drives each cell lineage and gene regulation throughout generation of intermediate phenotypes is not known. A common pathway of gene regulation involves the sequential activation of a network of protein kinases and the phosphorylation of transcription factors, which act in concert to induce or repress specific genes. Among the regulated genes, there are several cytoskeleton proteins, as reflected in the acquisition of different cytoskeleton proteins as cells move on in each neural lineage. Each cell lineage exhibits several intermediate phenotypes and each subtype secretes factors in a constitutive or in regulated fashion in the milieu of the growing cells (Vernadakis and Mangoura, 1988). These extrinsic stimuli, currently classified as growth factors, hormones, neurotransmitters, and extracellular matrix (ECM) molecules, bind to integral membrane proteins (receptors) to activate a signal transduction cascade that leads ultimately to the expression of both common and cell type-specific genes (Fig. 1). The exact sequence of events that follow membrane receptor activation in astrocytes and the role of their signaling cascades in regulating both proliferation and differentiation is not yet known. We therefore study specific signal transduction pathways of membrane receptor agonists to define the precise biochemical and molecular mechanism of plasticity during brain mor-

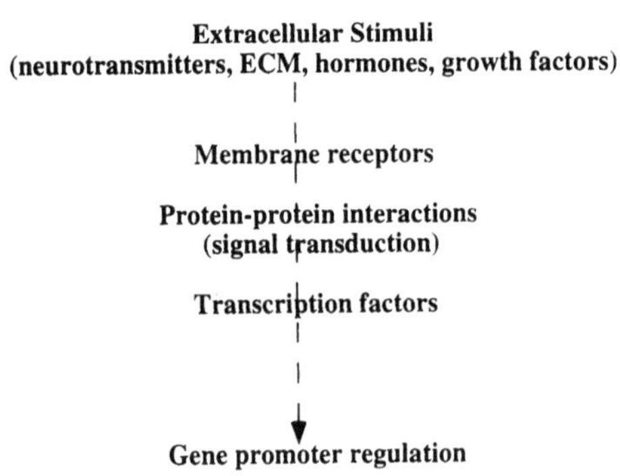

Figure 1. Extracellular stimuli generate signal transduction cascades to regulate the intrinsic cell program.

phogenesis, a period of intense proliferation and differentiation of astrocytes (Mangoura et al., 1995, Mangoura et al., 1995; Pelletiere et al., 1997).

2. THE CHICK EMBRYO AS A MODEL OF DEVELOPMENT

The development of the chick embryo CNS has been carefully and extensively studied the last 120 years. As a result, the temporal and spatial patterns of the generation of the chick embryo CNS are well established (Tsai HM et al., 1981 and refs). The precise timing of the proliferation of neurons and astrocytes and the final lamination of the chick brain are known for each region along the neural tube. The extent of the mapping of the telecephalon has allowed us to customize tissue culture systems according to our experimental needs. Over the years we have developed, optimized and extensively characterized a series of culture systems (Mangoura et al., 1986; Mangoura, 1995, Mangoura et al., 1988, Mangoura et al., 1990, Mangoura and Vernadakis, 1988; Sakellaridis et al., 1986). We have succeeded to develop both homogeneous cultures consisting >98% of the same cell type and identify the time periods in culture, where the in vivo cell type specific-properties are intact in the corresponding culture system: *E3WE*, a culture system from 3-day-old chick embryo consisting of dividing **neuroblasts**; *E6CH*, primary cultures derived from 6 or 7-day-old chick embryo cerebral hemispheres consisting of early post mitotic, **differentiating**, and presynaptic neurons. *E8CH*, which offers a system of **differentiated** neurons after 3 days in culture. *E15CH* astroblasts, which develop in fully differentiated astrocytes after 6 days in culture. These systems have been invaluable models in understanding cell-type specific signaling pathways and their developmental regulation.

3. PKC ISOFORMS, THEIR CENTRAL ROLE IN CNS DEVELOPMENT

As described above neurons and astrocytes depend on many extracellular mediators for survival, growth and differentiation. These agonists (growth factors, hormones, neurotransmitters) activate or inhibit transmembrane signaling systems that control the production of second messenger molecules. Second messengers, in turn, modulate the activity of protein kinases. The protein kinases phosphorylate serine, threonine or tyrosine residues of other proteins-substrates to change the substrate conformation and function. Among the principal signaling systems, protein kinase C (PKC) is a key effector molecule in the regulation of growth and differentiation. PKC is actually a family of 13 isoforms of phospholipid-dependent kinases, classified as calcium-dependent (α, β_I, β_{II}, γ), calcium independent (δ, ε, η, θ, μ), and atypical (ζ, ι, λ)(Nishizuka, 1995). PKC acts via ser/thr phosphorylation to stimulate a number of kinases, exert feedback control on other signaling pathways, and change the function of structural proteins. Each isoform has different substrate specificities, a possible mechanism for cell type or developmental state specificity for a single agonist. Therefore we investigated the roles of specific PKC isoforms in the embryonic chick CNS.

For this, we used isoform-specific antibodies, Western blotting, immunocytochemistry, a combination of CNS region-specific dissections throughout embryogenesis (from day 3 of incubation to day one post-hatching), and cell type-specific neural culture systems derived from the chick CNS, i.e. neuroblasts, neurons and astrocytes. During embryogenesis of the chick, neuronal and glial lineage progenitor cells, which may enter the

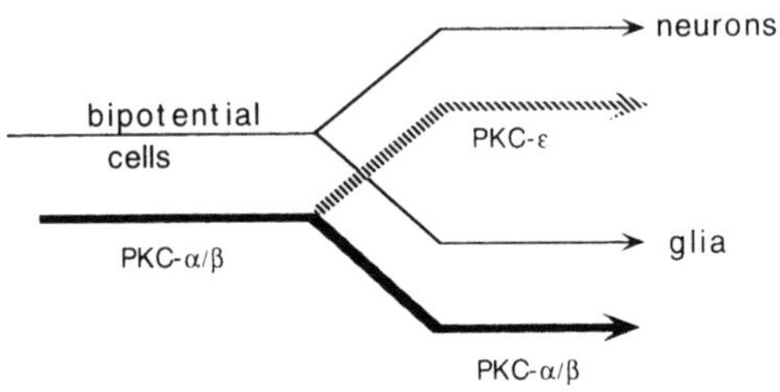

Figure 2. PKC isoform expression correlates with specific cell lineage in the chick embryo CNS.

cell cycle express, PKC-α/β isoforms. However, neurons that have entered their final G phase and have started to differentiate express the Ca^{2+}-independent PKC isoform PKC–ε. The atypical isoforms ζ and ι are ubiquitous. Therefore, we have suggested that the heavy distribution of PKC-ε in neurons, right after completion of their final division, is essential to accommodate receptor mediated information, pertinent to neuronal differentiation (Mangoura et al., 1993). We thus showed for the first time that (1) PKC-ε is a developmentally regulated, neuron specific and differentiation-related isoform; and (2) the expression of specific PKC isoforms correlates with cell fate in neural chicken embryo cells (Fig. 2). Further work by other groups has verified these predictions in other species. For example, overexpression of PKC isoforms in PC12 cells demonstrated that only PKC-ε promoted neuritic outgrowth and differentiation related responses to NGF (Hundle et al., 1995), while rat astrocytoma cells switch PKC isoforms as they undergo through the phases of the cell cycle (Baltuch et al., 1995).

4. PKC SIGNALING IN ASTROCYTES

We have shown that the expression of specific PKC isoforms correlates with cell fate in neural chicken embryo cells. Cell fate is the acquisition of a specific morphological and functional phenotype. Both parameters had been shown regulated by PKC in astrocytes (Mobley et al., 1986; Neary et al., 1986; Bender et al., 1991). To understand the signaling pathways that include different PKC isoforms and regulate proliferation, differentiation, and fate of neural cells, we employed chick embryo cortical astrocytes derived from 15-day-old chick embryo cerebral hemispheres (E15CH). As previously reported, astrocytes in the E15CH culture system develop from a few flat $A2B5^{+}/GFAP^{+-}$ cells into fully differentiated astrocytes (90%) by culture day 10 (C10) (Kentroti and Vernadakis, 1997, Mangoura et al., 1993, Sakellaridis et al., 1983, Sakellaridis, et al., 1986). Oligodendrocytes, neurons and microglial constitute about 1% of the total cell number (Fig. 3). The system faithfully represents the corresponding events in ovo as shown by earlier studies on the developmental profiles of the astrocytic marker glutamine synthetase (Martinez-Hernandez et al., 1977; Parker et al., 1980; Sakellaridis et al., 1983) activity and more recent studies on the expression of Glia Fibrillary Acidic Protein (GFAP). Western blot analysis of samples and of purified GFAP was used to quantitate and compare the levels of GFA protein expression between cerebral hemispheres (CH) and cerebellum (CB) and astrocytic cultures derived from these two regions (Fig. 4). Total cell protein from E15CH and E15CB in astrocytic cultures, and from E6–18CH and E12–18CB in tissue were analyzed. In each case GFA protein expression increased with development in ovo and in culture (Fig. 4). The increase in E15CH reached a 6-fold maximum expression

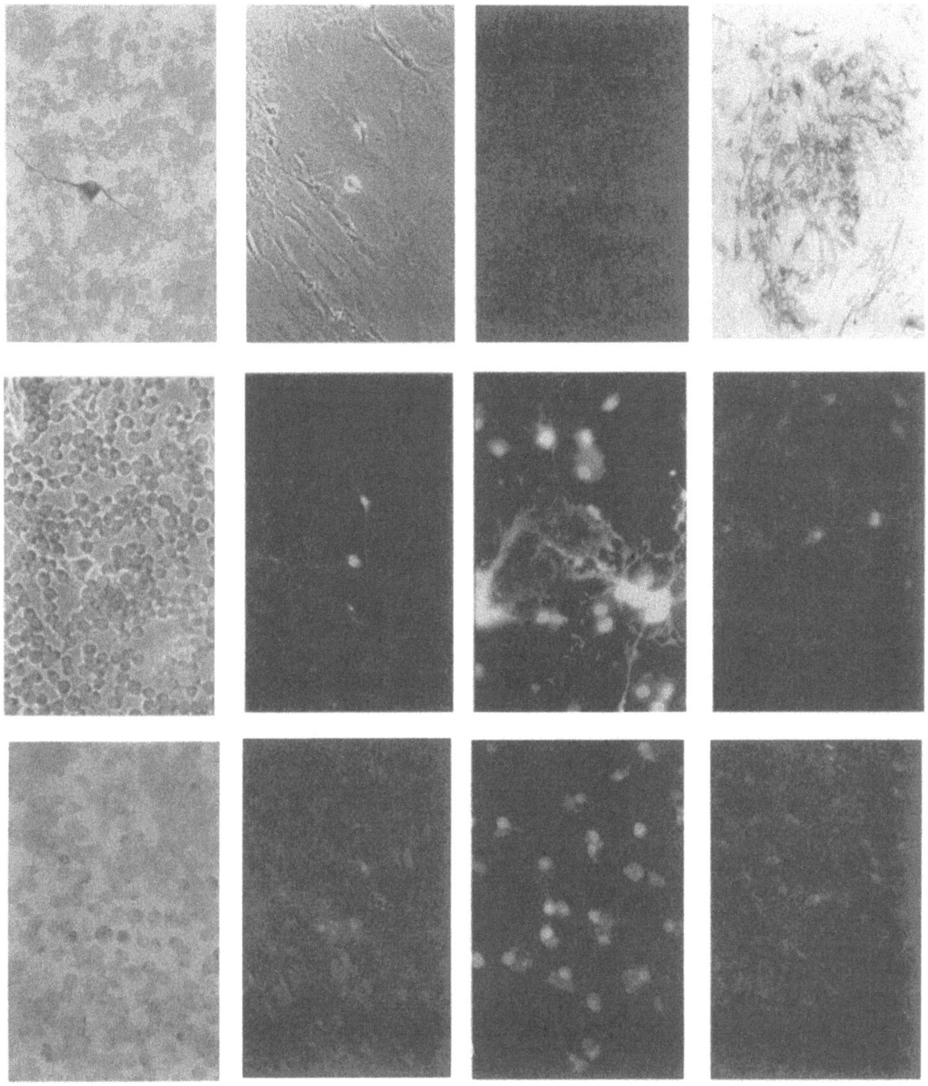

Figure 3. E15CH astrocytic culture system. Culture day 8 to 12. (A) E15CH stained for GFAP (antibodies from Dr. D. Dahl, detection with peroxidase reaction). (B) E15CH stained for myelin basic protein (antibodies from Dr. T. Campagnoni, detection with fluorescein (FITC) immunofluorescence), 0–2 cells per four optic fields are positive. (C) Same field as in b, viewed under rhodamine filter to reveal counter stain with propidium iodide. (D) E15CH stained for GalCer (antibodies Dr. C. Dyer, detection with FITC), 0–1 cells per optic field are positive. (E) Positive control for b: rat oligodendrocytes stained for myelin basic protein (FITC) and counter stained with propidium iodide (brighter staining on nuclei). (F) Positive control for d: rat oligodendrocytes stained for GalCer (FITC) and counter stained with propidium iodide (brighter staining on nuclei). (G) Phase photomicrograph of E15CH cultures depicting two phase bright, small round cells with processes (microglia). (H) The same field viewed under fluorescein filter reveals that the two small cells are positive for α_1 trypsin (antibodies from Dr. F. Gremo), a microglia marker. (I) Same field as in h, viewed under rhodamine filter to reveal counter stain with propidium iodide. (J) Human microglia cultures (from Dr. F. Gremo) stained for GFAP, reveals positive staining of an astrocyte, while all microglia are negative. (K) The same field as in j under phase for better viewing of the microglia cell processes. (L) Positive control for H: human microglia cells are positive for α_1 trypsin (detection with peroxidase). Notes: 1) Similar results were obtained with E15CB (cerebellum) astrocytic cultures. 2) a-d, g-i were photographed under a 25x Neofluar lens; j-l were under a 40x Neofluar. 3) g-l: similar results were obtained with staining for lysozyme or RCA, two other microglia markers.

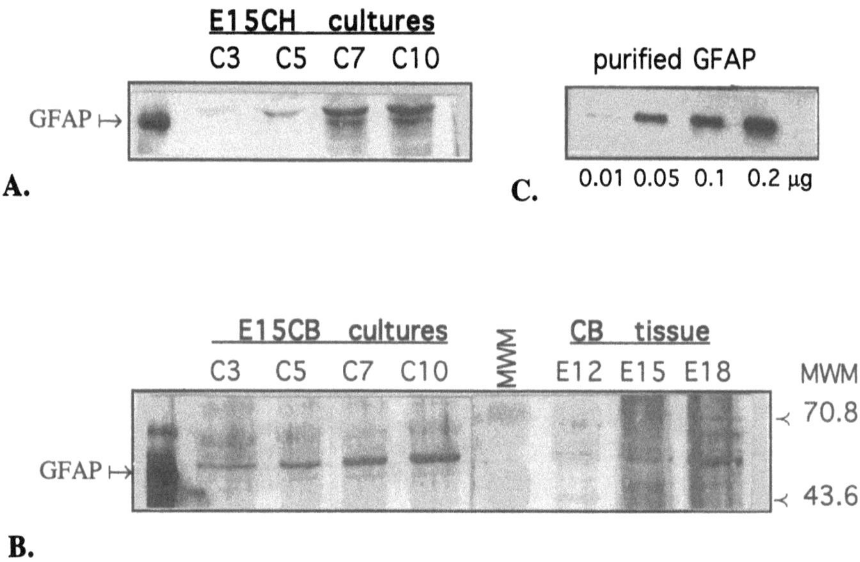

Figure 4. GFAP expression is developmentally regulated in E15CH astrocytic cultures from cerebral hemispheres (A, E15CH), increasing from culture day 3 (C3) to culture day 10 (C10). Similarly, GFAP expression increased in cerebellar astrocytes with days in culture (B: E15CB cultures C3 to C10) and the cerebellum (CB) with days of embryonic incubation (B: CB from embryonic day 12 (E12) to embryonic day 18 (E18)). Samples, normalized per protein content, were analyzed by Western blot with GFAP antibodies and by densitometry of purified GFAP Western blots, to quantitate levels of expression (C).

by day 10 in culture (C10), as compared to C3 (Fig. 4, A). The slope of the increase was higher, however, as compared to the one observed in tissue (Kentroti and Vernadakis, 1997). Analysis of GFAP expression in homogenized cerebellar tissue from E12 to 18 revealed that cerebellum is enriched by a 4-fold over the cerebral hemispheres (Fig. 4B). Similarly, GFAP levels in E15CB cultures were higher at day 3 than in the E15CH cultures and reached a 4-fold increase by C10. Therefore, developmentally regulated increases of GFAP in E15 cerebral hemispheres and cerebellum are precisely reflected in astrocytic cultures cortical (CH) or cerebellar (CB). GFAP positive glial phenotypes appear after several intermediate phenotypes from GFAP negative cells in vivo and therefore the E15CH and E15CB systems inherently contain the intermediate astrocytic phenotypes. Indeed, these culture systems have allowed us to investigate in detail the signal transduction molecules that are involved in the acquisition of astrocytic phenotypes.

Using the E15CH cultures as a model and the phorbol ester 12-tetradecanoylphorbol-13-acetate (TPA) as a prototype activator of PKC, we have shown that activation of the PKC signaling pathway in astrocytes caused the mitosis and, according to several criteria, the acquisition of an earlier phenotype (Mangoura, 1994, Mangoura et al., 1995, Pelletiere et al., 1995). In the short term, TPA causes a rapid association of PKC with the cytoskeleton and phosphorylation of many cytoskeletal proteins in E15CH astrocytes, in particular GFAP and vimentin. Both proteins are substrates for Ca^{2+}/phospholipid-dependent protein kinases (PKC) (Harrison and Mobley, 1992). We used pulse-chase labeling with [^{35}S] methionine and immunoprecipitation with an anti-vimentin mAb from extractable and cytoskeletal fractions to show that TPA regulates (increases) both the post-translational assembly (rate of assembly of newly synthesized full-length) and the cotranslational assembly (rate of assembly of nascent vimentin) of vimentin (Mangoura et al., 1995). From studies in myoblasts and fibroblasts it is known that assembly of vimentin into the cytoskeleton depends on cell type and state of differentiation (Isaaks et al., 1989). Several other cytoskeletal proteins appeared to be PKC substrates, as seen in vitro phosphorylation of plasma membrane-free and organelle-free cytoskeleton preparations after short term exposure to TPA (Mangoura et al., 1995). SDS-PAGE analysis of cytoskeletons revealed that the protein composition changed rapidly upon exposure to TPA and PKC activation (Fig. 5). When these assays were performed in cytoskeletons that had been treated for 24 hours with TPA and PKC had already been downregulated, several other proteins were phosphorylated by other than PKC kinases. While in short term TPA, we observed primarily increases in the level of baseline phosphorylation, with 24 h of TPA the profile of phosphoproteins included proteins that did not appear except after hours of continuous exposure. The profile continued to change and after 48 hr of treatment it reverted to control levels, although the baseline phosphorylation of several proteins was higher (Fig. 5).

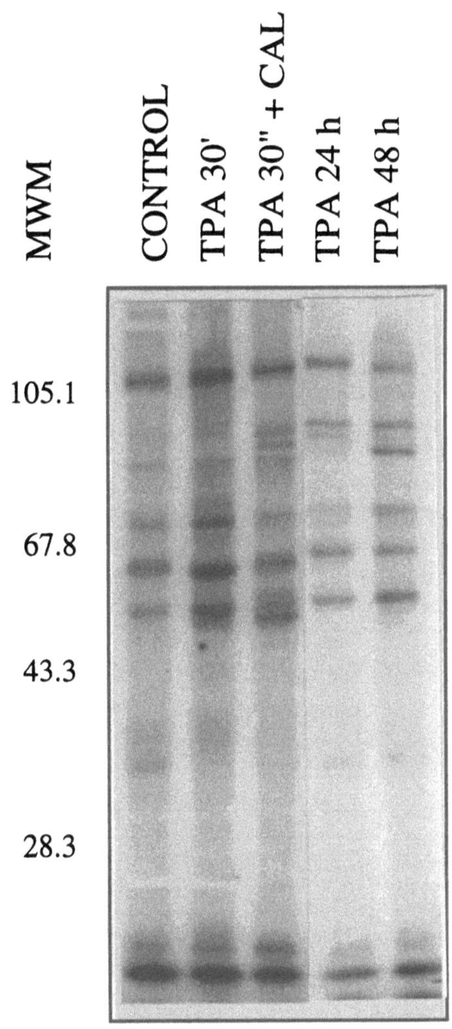

Figure 5. Cytoskeleton protein phosphorylation with time of TPA treatment in E15CH astrocytic cultures (autoradiogram of a 5–15% gradient gel). Cells were pretreated for times indicated with vehicle, 100nMM TPA or TPA. Cytoskeleton preparations were incubated with $[^{32}P]$-γ-ATP, phosphatidylserine, diolein, $CaCl_2$ in the presence or absence of 1 nM calyculin (Cal). Several different proteins were phosphorylated after PKC activation (see text and Mangoura et al., 1995).

In addition, these studies further strengthen the evidence that ser/thr kinases, other than PKC which are preserved in this cytoskeleton preparation, phosphorylate cytoskeleton proteins. It is reasonable to assume that activation of these kinases depends on prior activation of PKC.

The effect of PKC activation (TPA) on E15CH cytoskeleton protein regulation was reflected in the astrocytic morphology, initially observed by 3h of treatment, when astrocytes begin to lose their flat lamellipodia and first exhibit thin, long processes (Fig. 6, a). By 24h of treatment, elongation of astrocytic processes is rather dramatic as compared to controls (Fig. 6, b). Double staining with anti-vimentin and propidium iodide as a counter stain showed both the increase of nuclei/field, and the changes in cell and nuclei polarization (Mangoura et al., 1995). By 48h, all astrocytes still possessed the TPA-induced fibrous appearance. By 72 h almost all cells had recovered their original morphology with both flat and fibrous shapes (Fig. 7: a-control, c-72 hr of TPA). As we have observed several times, GFAP immunoreactivity develops initially in the core of cell colonies, and extends radially with development in culture (Fig. 7, b), possibly indicating that the cells mature also in a radial way. After 24 h of TPA treatment only few cells were GFAP positive. After 72 h, when cells had regained their original morphology, the number of GFAP

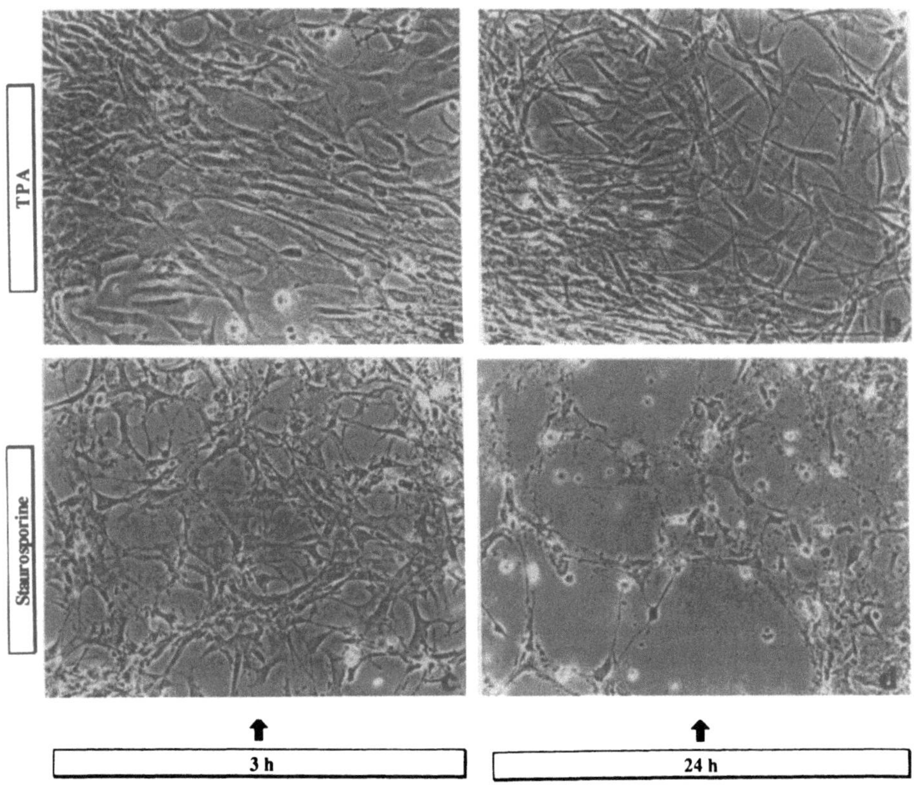

Figure 6. Astrocytic morphology is regulated by PKC activators and inhibitors in chick embryo cultures. Phase photomicrographs of E15CH cultures. (A) After a 3h treatment with 0.1 mm TPA, cells appear polarized and their processes narrow and elongated; (B) by 24h of TPA, the majority of astrocytes have assumed a fibrous morphology; (C) staurosporine alone resulted in a smaller, polygonal astrocytic shape with short processes bearing varicosities; (D) after 24h of staurosporine, the "differentiated" morphology results in extensive cell death of E15CH astrocytes.

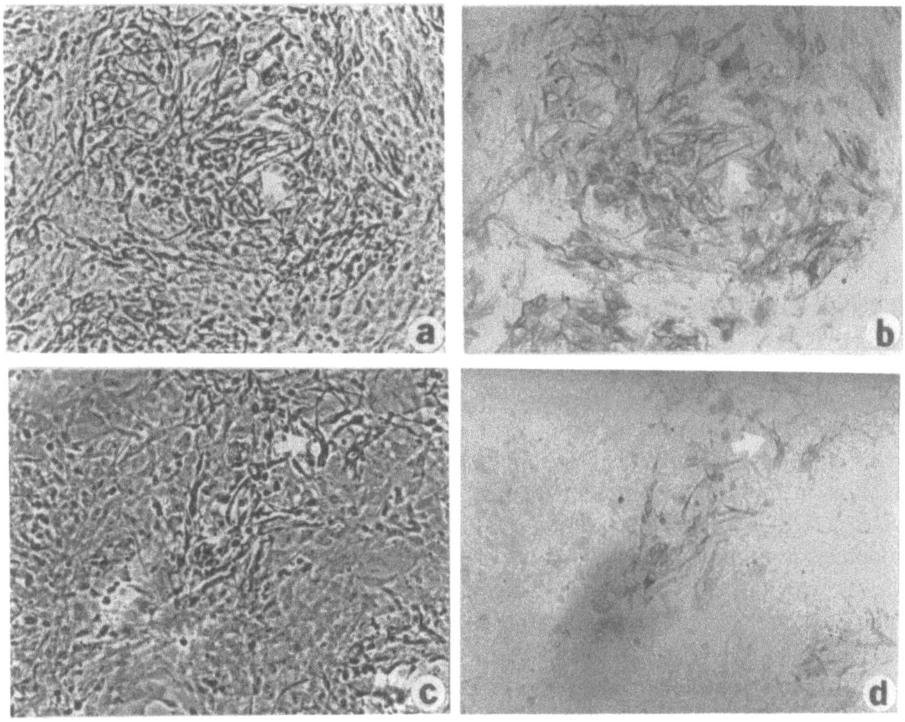

Figure 7. GFAP expression is downregulated by TPA in E15CH astrocytic cultures. Phase (left column) and light (right column) photomicrographs of E15CH cultures fixed and stained for GFAP (white arrows point to identical cells of the field) a and b: untreated (control vehicle); c and d: after 72h of TPA, while cells morphology has recovered, GFAP positive cells were still scarce.

positive cells was still low (Fig. 7, d) as was the activity of glutamine synthetase, another established marker of astrocytic differentiation (Bignami and Dahl, 1976). The differences depicted in the immunocytochemical studies were reflected in concomitant biochemical studies of the abundance of GFAP protein in control and treated astrocytes (Bignami and Dahl, 1976). Taken altogether, activation of the PKC signaling pathway in astrocytes results in the acquisition of a cytoskeleton, which resembles an immature or earlier phenotype (Cameron and Rakic, 1991). Therefore, PKC activation is pivotal for the acquisition and maintenance of phenotypes in embryonic astrocytes.

This finding was further supported by studies on phenotypic changes after PKC inhibition (Mangoura et al., 1993; Mangoura et al., 1995). Two PKC inhibitors staurosporine and H8 were used to block the TPA effects. Both inhibitors exert effects of their own, and would result in a different morphologic phenotype expression. Exposure to 100 nM staurosporine caused distinct cell polarity changes and apoptosis. The TPA effects were modulated considerably in the presence of this PKC inhibitor, but as shown by other groups, staurosporine elicited unique morphological and pharmacological effects quite different from that of other PKC inhibitors (Rasouly, et al., 1994 Mangoura, et al., 1993). One of the most striking effects following chronic staurosporine treatment was the reduction in cell numbers and the induction of a morphologically differentiated cell (Mangoura and Dawson, 1997). Specifically within the first 3h of treatment with 100 nM staurosporine, astrocytes had acquired a stellate shape, with a polygonal body, and numerous, short, thin, and varicosed processes. By 24 h, the cell loss was over 30% and by 48 h over 60%. The cause of reduced cell numbers was apoptosis, as evident by demonstration of nuclear DNA fragmentation. The "apoptotic phenotype" was characterized by prior induction of differentiation (increased GS activity)[56]. Apoptosis or programmed cell death is part of normal brain development, for example >40% of neurons in the CNS undergo apoptosis over a period of a few days in developing vertebrate embryos as part of the natural pattern formation process (Oppenheim, 1991). Apoptosis in astrocytes had been assumed (Morrison-Bogorad et al., 1994), but not experimentally verified and it is generally believed these cells are resistant to apoptosis. Neurons become differentiated before they become apoptotic (Oppenheim, 1991). Similarly, staurosporine first initiates differentiation of embryonic chick brain astrocytes (increases in GFAP, glutamine synthetase activity) and then apoptosis. Furthermore, apoptosis in the embryonic astrocyte was associated with the stimulation of a specific protein kinase (PK60) and increased formation of ceramide. The activation of PK60 appears to be upstream of ceramide formation since ceramide did not activate PK60 (Mangoura and Dawson, 1997). In addition we showed a role for the cytoskeleton in the cellular program of apoptosis. PK60 activity associates with the cytoskeleton with a sustained activation up to 24h, the time it takes for an appreciable level of apoptosis to occur (Mangoura and Dawson, 1997). The importance of protein kinase association with the cytoskeleton has become apparent, following the identification and cloning of RACKS (the PKC-binding proteins), and the raf-binding 14–3–3 family of proteins, fashioned after RACKS (Fantl et al., 1994; Ron et al., 1994. Staurosporine causes changes in the synthesis and assembly of many cytoskeletal proteins in astrocytes (Mangoura, 1994, Mangoura and Dawson, 1997, Mangoura et al., 1993, Mangoura et al., 1995). The prolonged activation of PK60 may regulate the phosphorylation state of these proteins, as they in turn regulate the cellular and nuclear shape and subsequent the function of astrocytes. Staurosporine is not a physiological regulator of apoptosis, but since the pharmacology of staurosporine is consistent with a specific uptake mechanism (receptor) for staurosporine (Rasouly et al., 1994), the possibility of an endogenous ligand with similar signaling is likely. With these studies we have created models of distinct physiological states for astrocytes (phenotypes summarized on Fig. 8), and we can readily ask questions concerning proliferation, differentiation, or apoptosis.

5. TYROSINE KINASES SIGNALING IN ASTROCYTES

We used the E15CH tissue cell culture model to investigate signaling events upstream to PKC activation in astrocytes, starting with the source of diacylglycerol for PKC activation. Protein kinase C (PKC) is activated by both G-protein-coupled and tyrosine kinase membrane receptors. Endogenous diacylglycerol (DG) was originally thought to be primarily formed by hydrolysis of membrane polyphosphoinositides by phospholipase C (PLC). However, the release of DG is biphasic, and DG can also be generated by activation of phospholipase D (Billah et al., 1989; Billah et al., 1989b). Phospholipase D catalyzes the hydrolysis of phosphatidylcholine and phosphatidylethanolamine to form phosphatidic acid (Exton, 1990; Liscovitch and V., 1996). A coupled action of phosphatidate phosphohydrolase (PPH) then generates diacylglycerol which can in turn activate PKC (Billah et al., 1989b). PLD activation has been proposed to be coupled to either G-protein membrane receptors, direct activation by PKC (mimicked by phorbol esters) Gustavsson and Hansson, 1990, calcium ionophores (Billah et al., 1989b, or receptors

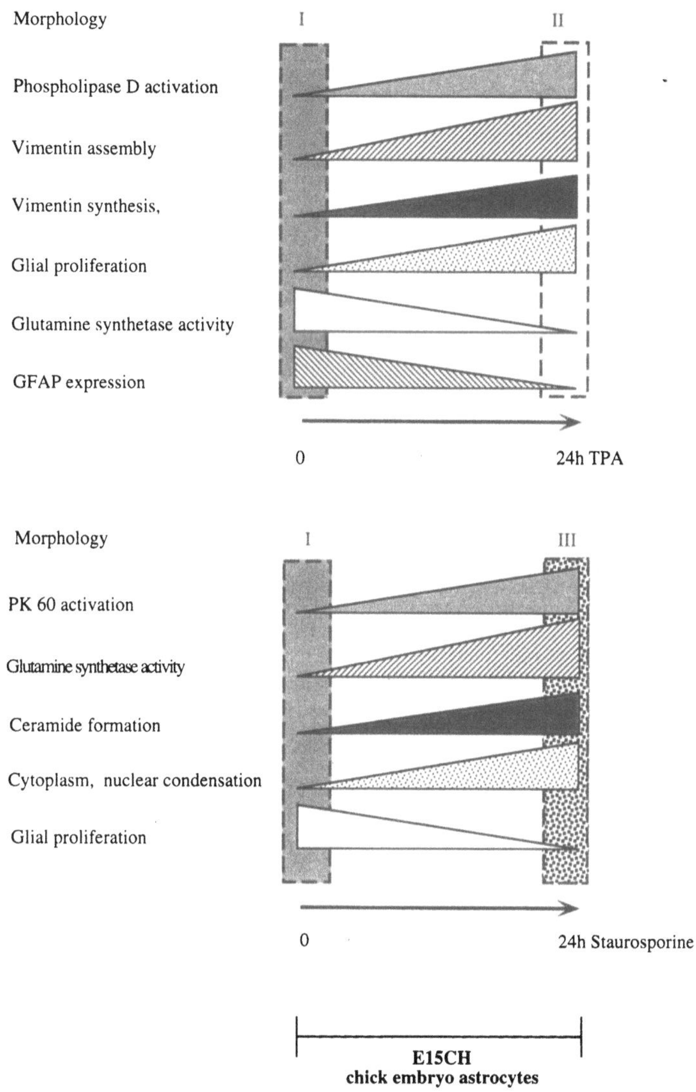

Figure 8. PKC regulation results in distinct phenotypes in chick embryo astrocytes.

possessing tyrosine kinase activity (i.e., EGF) (Cook and Wakelam, 1992). We observed a dramatic 10-fold activation of PLD by TPA in astrocytes (Mangoura et al., 1995). TPA activation of PLD was totally blocked by staurosporine and by genistein, a tyrosine kinase inhibitor. We then screened a number of physiological activators in astrocytes, i.e., glutamate, norepinephrine and bradykinin, and found that in each case PLD activation was blocked by the tyrosine kinase inhibitor, independent of the duration or magnitude of the response (Mangoura et al., 1995). The central role of PKC and PLD activation in proliferation and differentiation of chick embryo astrocytes is emphasized, and the hypothesis that kinases may be key points in cell lineage processes is further strengthened. These data also suggested that activation of phospholipase D, an important effector in the signaling of mitosis and differentiation in astrocytes, is dependent on tyrosine kinase activation.

To test this prediction, we investigated the second messenger coupling of prolactin (PRL). Prolactin receptor activation may stimulate mitosis in various cell types via signal transduction pathways that depend on tyrosine kinase activation. This neuroendocrine hormone stimulated cell proliferation in cultured astrocytes, as assessed by a 3-fold higher BrdU uptake in astrocytes incubated with 1nM PRL for 18h as compared to controls. PLD activation was a constituent of the PRL signaling cascade and its activity was significantly increased acutely (Pelletiere et al., 1995). To further elucidate the pathway, kinase inhibitors were used to block PLD activation. 10^{-7}M staurosporine only partially inhibited the PRL effect, suggesting some PKC involvement. However, genistein, a specific tyrosine kinase inhibitor, blocked the activation, indicating that the PRL receptor in astrocytes is coupled to non-receptor tyrosine kinases. We then identified the tyrosine kinases in the PRL-stimulated activation of PLD and mitogenesis by studying the *src* family of tyrosine kinases (Jiang et al., 1995). Using autokinase assays with avian specific anti-*src*($pp60^{src}$) and anti -*fyn*($pp59^{fyn}$) immunoprecipitates from astrocytes treated with 100 nM PRL for varying times, we observed that PRL stimulates $pp60^{src}$ and $pp59^{fyn}$ in a time dependent manner, but with differential duration of stimulation (Pelletiere et al, submitted). These protein-protein interactions are most probably upstream to PKC activation. Activation of src and fyn resulted in phosphorylation of several cytoskeleton proteins on tyrosine residues, a common mechanism of acute regulation. In the long term prolactin had no effects on the cytoskeleton, in contrast to the effects of TPA. Furthermore we observed no significant differences in the GFAP levels or glutamine synthetase activity after completion of prolactin-stimulated mitosis in astrocytes. These data provide evidence for a novel, PTK-dependent coupling of prolactin to PLD for developing astrocytes, which signals both mitosis and maintenance of differentiation (Fig. 9).

6. CONCLUSIONS AND FUTURE DIRECTIONS

Our studies have indicated that the view that proliferation and differentiation are two mutually exclusive processes and that the cell may be undergoing one or the other may be

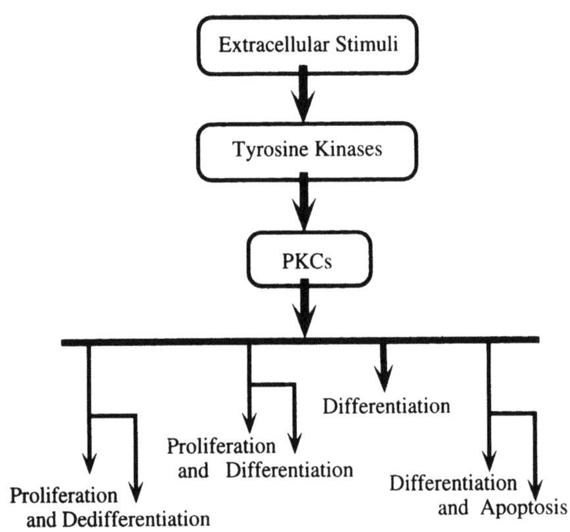

Figure 9. Extracellular stimuli initiate signal transduction cascades to regulate cell fate in astrocytes.

overly simplistic, especially for highly regulated tissues. It is becoming increasingly obvious from studies on knockout models of several receptor or effector molecules in mitogenic pathways that differentiation is an ongoing process, and that mitosis may in fact take place in highly differentiated cells that may retain or return rapidly to their specialized functions. Furthermore, signal transduction pathways are not really as linear as commonly presented, because several feedback mechanisms exist and several of the pathways are coupled and interactive. The cytoskeleton is always part of the signaling, and very often it is the morphology which is used as a marker for differentiation. Our studies on cytoskeleton proteins have outlined this fact. The same cytoskeleton protein may be differentially regulated (posttranslational modifications, synthesis, subcellular localization) during the course of action of different membrane receptor agonists, mitogenic or not. Another example is the small GTPases of the Rho family, which are involved primarily in cell shape regulation events, such as the formation of focal adhesions (Barry and Critchley, 1994, Kozma et al., 1995, Luo et al., 1996). Small GTPases and focal adhesion proteins are discontinuously expressed throughout the cell membrane. In any case, they regulate actin polymerization in several cell types and may be activated either by G-protein coupled receptors or tyrosine kinase-receptors associated with mitosis or differentiation. Also, the ras protein of the low molecular weight GTP-binding protein superfamily can cause cell transformation, which is characterized equally by repeated mitosis and by dramatic cytoskeleton changes. In E15CH cultures, tyrosine kinase (PTK) activation occurs upon homologous mitogenic ligand binding with tyrosine kinase receptors (EGF, FGF), PTK-associate receptors (prolactin), or G-protein coupled receptors (bradykinin) (Mangoura D, unpublished). Phosphorylation/activation of effector enzymes transduce the mitogenic signal to the nucleus. In addition, each of these agonists stimulated tyrosine phosphorylation of several cytoskeleton proteins and associated kinases. The duration and magnitude of the phosphorylation for each protein differed greatly among agonists, resulting in differential shape regulation. Our data suggest that PKC regulation is sufficient to alter the astrocytic phenotype, and that the PKC-dependent signal transduction pathways include tyrosine kinase activation (Fig. 9). Furthermore, the mitogenic pathways in astrocytes may mediate concurrently important information on regulation of astrocytic morphological/functional phenotypes (supported by HD09402 and a Brain Research Foundation award to DM).

REFERENCES

Baltuch, G. H., N. P. Dooley, K. M. Rostworowski, J. G. Villemure, and V. W. Yong. 1995. Protein kinase C isoforms alpha overexpression in C6 glioma cells and its role in cell proliferation. *J. Neuro Oncol.* 24:241–250.

Barry, S. T., and C. D.R. 1994. The RhoA-dependent assembly of focal adhesions in Swiss 3T3 cells is associated with increased tyrosine phosphorylation and the recruitment of both pp125FAK and protein kinase C-δ to focal adhesions. *J. Cell Sci.* 107:2033–2045.

Bender, A. S., J. T. Neary, J. Blicharska, L. O. Norenberg, and M. D. Norenberg. 1991. Role of calmodulin and protein kinase C in astrocytic volume regulation. *J. Neurochem.* 58:1874–1882.

Bignami, A., and D. Dahl. 1976. Immunofluorescence studies with antibodies to astrocyte-specific protein (GFA) in mammalian and submammalian vertebrates. *Neuropathol Appl. Neurobiol.* 2:99–111.

Billah, M. M., S. Eckel, T. J. Mullmann, R. W. Egan, and M. I. Siegel. 1989b. Phosphatidylcholine hydrolysis by phospholipase D determines phosphatidate and diglyceride levels in chemotactic peptide-stimulated human neutrophils. Involvement of phosphatidate phosphohydrolase in signal transduction. *J. Biol. Chem.* 264: 17069–77.

Billah, M. M., J. K. Pai, T. J. Mullmann, R. W. Egan, and M. I. Siegel. 1989. Regulation of phospholipase D in HL-60 granulocytes. Activation by phorbol esters, diglyceride, and calcium ionophore via protein kinase-independent mechanisms. *J. Biol. Chem.* 264:9069–76.

Cameron, R. S., and P. Rakic. 1991. Glial cell lineage in the cerebral cortex: a review and synthesis. *Glia.* 4:124–137.

Cameron, R. S., and P. Rakic. 1994. Identification of membrane proteins that comprise the plasmalemmal junction between migrating neurons and radial glial cells. *J Neuroscience.* 14:3139–3155.

Chavez, R. A., r. S. Mille, and H. Moore. 1996. A biosynthetic regulated secretory pathway in constitutive secretory cells. *J. Cell Biol.*:1177–91.

Cook, S. J., and M. J. O. Wakelam. 1992. Epidermal growth factor increases sn-1,2-diacylglycerol levels and activates phospholipase D-catalysed phosphatidylcholine breakdown in Swiss 3T3 cells in the absence of inositol-lipid hydrolysis. *Biochem. J.* 285:247–253.

Exton, J. H. 1990. Signalling through phosphatidylcholine breakdown. *J. Biol. Chem.* 265:1–4.

Fantl, W., A. Muslin, A. Kikuchi, J. Martin, A. MacNicol, R. Gross, and L. Williams. 1994. Activation of Raf-1 by 14–3–3 proteins. *Nature.* 371:612–614.

Gustavsson, L., and E. Hansson. 1990. Stimulation of phospholipase D activity by phorbol esters in cultured astrocytes. *J. Neurochem.* 54737–42:737–42.

Harrison, B. C., and P. L. Mobley. 1992. Phosphorylation of glial fibrillary acidic protein and vimentin by cytoskeletal-associated intermediate filament protein kinase activity in astrocytes. *J. Neurochem.* 58:320–327.

Hundle, B., T. McMahon, J. Dadgar, and R. Messing. 1995. Overexpression of PKC-ε enhances nerve growth factor-induced phosphorylation of mitogen-activated protein kinases and neurite outgrowth. *J. Biol. Chem.* 260:30134–30140.

Isaaks, W. B., R. K. Cook, J. C. Van Atta, C. M. Redmond, and A. B. Fulton, J. Biol. Chem., 264 17953–17960. 1989. Assembly of vimentin in culture varies with cell types. *J. Biol. Chem.* 264:17953–17960.

Janz, R., and T. Sudhof. 1995. A systematic approach to studying synaptic function in vertebrates. *Cold Spring Harbor Symposia on Quantitative Biology.* 60:309–14.

Jiang, H., J. Luo, T. Urano, P. Frankel, Z. Lu, D. Foster, and L. Feig. 1995. Involvement of Ral GTPase in v-Src-induced phospholipase D activation. *Nature.* 378:409–12.

Kentroti, S., and A. Vernadakis. 1997. Differential expression in glial cells derived from chick embryo cerebral hemispheres at an advance stage of development. *J. Neurosci. Res.* 47:322–331.

Kozma, R., S. Ahmed, A. Best, and L. Lim. 1995. The Ras-related Cdc42Hs and Bradykinin promote formation of peripheral actin microspikes and filopodia in Swiss 3T3 fibroblasts. *Mol. Cell. Biol.* 15: 1942–1952.

Liscovitch, M., and C.-C. V. 1996. Enzymology of mammalian phospholipases D: in vitro studies. *Chemistry & Physics of Lipids 80.* 80:37–44.

Luo, L., Hensch, T.K.,, L. Ackerman, S. Barbels, L. Y. Jan, and Y. N. Jan. 1996. Differential effects of the Rac GTPase on Purkinje cell axons and dendritic trunks and spines. *Nature.* 379:837–840.

Mangoura, D. 1994. PKC-dependent regulation of glia phenotypes: effects on vimentin assembly. *Int. J. Devl. Neurosci.* 12:1:80.

Mangoura, D. 1995. *Brain Res.* In press.

Mangoura, D., and G. Dawson. 1997. Programmed cell death in cortical astrocytes from chick embryo cerebral hemisphere cultures is associated with activation of protein kinase pk60 and ceramide formation. *J. Neurochem.*

Mangoura, D., C. Pelletiere, and G. Dawson. 1993. Astrocytic phenotype regulation by PLD and kinases in chick embryo culture. *Soc. Neurosci Abstract.* 28:3:43.

Mangoura, D., N. Sakellaridis, J. Jones, and A. Vernadakis. 1989. Early and late passage C-6 glial cell growth: similarities with primary glial cell in culture. *J. Neurosci.* 14:941–947.

Mangoura, D., N. Sakellaridis, and A. Vernadakis. 1986. Factors influencing neuronal growth in primary cultures derived from 3- day-old chick embryos. *Int. J. Devl. Neurosci.* 6:89–102.

Mangoura, D., N. Sakellaridis, and A. Vernadakis. 1988. Cholinergic neurons in cultures derived from three, six or eight-day-old chick embryos: a biochemical and immunocytochemical study. *Brain Res.* 40:25–46.

Mangoura, D., N. Sakellaridis, and A. Vernadakis. 1990. Evidence for plasticity in phenotypic neurotransmitter expression in culture. *Dev. Brain Res.* 51:93–101.

Mangoura, D., V. Sogos, and G. Dawson. 1993. PKC-ε is a developmentally regulated, neuronal isoform inthe chick embryo CNS chick embryo. *J. Neurosci. Res.* 35:488–498.

Mangoura, D., V. Sogos, and G. Dawson. 1995. Phorbol ester and PKC signalling regulate proliferation, vimentin cytoskeleton assembly and glutamine synthetase activity of chick embryo cerebrum astrocytes in culture. *Brain Res.* 87:1–11.

Mangoura, D., V. Sogos, C. Pelletiere, and G. Dawson. 1995. Differential regulation of phospholipases C and D by phorbol esters and the physiological activators carbachol and glutamate in astrocytes from chick embryo cerebrum and cerebellum. *Brain Res.* 87:12–21.

Mangoura, D., and A. Vernadakis. 1988. Gabaergic neurons in cultures derived from three, six or eight-day-old chick embryos: a biochemical and immunocytochemical study. *Dev. Brain Res.* 40:37–46.

Martinez-Hernandez, A., K. P. Bell, and M. D. Norenberg. 1977. Glutamine-synthetase-glial localization in the brain. *Science*. 195:1356–1358.

Mobley, P. L., S. L. Scott, and E. G. Cruz. 1986. Protein kinase C in astrocytes: a determinant of cell morphology. *Brain Res.* 398:366–369.

Morrison-Bogorad, M., S. Pardue, D. McIntire, and E. Miller. 1994. Cell size and the heat-shock response in rat brain. *Journal of Neurochemistry*. 63:857–867.

Neary, J. T., L.-O. -. B. Norenberg, and M. D. Norenberg. 1986. Calcium-activated, phospholipid-dependent protein kinase and protein substrates in primary cultures of astrocytes. *Brain Res.* 385:420–424.

Nedergaard, M. 1994. Direct signalling from astrocytes to neurons in cultures of mammalian brain cell. *Science*. 263:1768–1771.

Nishizuka, Y. 1995. Protein kinase C and lipid signaling for sustained cellular responses. *FASEB Journal*. 9:484–96.

Oppenheim, R. 1991. Cell death during development of the nervous system. *Annual Reviews of Neuroscience*. 14:453–501.

Parker, K. K., M. D. Norenberg, and A. Vernadakis. 1980. "Transdifferentiation" of C-6 glial cells in culture. *Science*. 208:179–181.

Peletiere, Wang, and Mangiura. 1997.

Pelletiere, C., S. Leung, N. Sakellaridis, and D. Mangoura. 1995. Coupling of the prolactin receptor to tyrosine kinases regulate s activation of PLD, STAT91 and mitosis. *Soc. Neuroscience*,. 21:562.

Rasouly, D., E. Rahamim, I. Ringel, I. Ginzburg, C. Muarakata, Y. Matsuda, and P. Lazarovici. 1994. Neurites induced by staurosporine in PC12 cells are resistant to colchicine and express high levels of tau proteins. *Molecular Pharmacology*. 45:29–35.

Ron, D., C. Chen, J. Caldwell, L. Jamieson, E. Orr, and D. Mochly-Rosen. 1994. Cloning of an intracellular receptor for protein kinase C: a homolog of the beta subunit of G proteins. *Proceedi.Nati.Acad.Sci. USA*. 91:839–843.

Sakellaridis, N., D. Bau, D. Mangoura, and A. Vernadakis. 1983. Developmental profiles of glial enzyme sin the chick embryo; in vivo and in culture. *Neurochem. Intern.* 5:685–689.

Sakellaridis, N., Mangoura D, and V. A. 1984. Glial cell growth in culture: influence of living cell substrata. *Neurochem. Res.* 9:1469–1483.

Sakellaridis, N., D. Mangoura, and A. Vernadakis. 1986. Effects of neuron conditioned medium and fetal calf serum content on glial growth in dissociated cultures. *Develop. Brain Res.* 27:31–41.

Sakellaridis, N., D. Mangoura, and A. Vernadakis. 1986. Effects of opiates on the growth of neuron-enriched cultures from chick embryonic brain. *J. Develop. Neurosci.* 4:293–303.

Tsai HM, Garber BB, and L. LMH. 1981. ^{3}H-thymidine autoradiographic analysis of telencephalic histogenesis in the embryo. I Neuronal birthdays of telencephalic compartments in situ. *J. Comp. Neurol.* 198: 275–292.

Vernadakis, A., and D. Mangoura. 1988. Factors influencing glial growth in culture: Nutrients and cell-secreted factors. *In* Nutrition, Growth and Cancer. G. P. Tryfiates and K. N. Prasad, editors. Alan R. Liss, Inc., New York. 57–79.

4

NEURON-GLIA CROSS TALK IN RAT STRIATUM AFTER TRANSIENT FOREBRAIN ISCHEMIA

Michele Zoli,* Giuseppe Biagini, Rosaria Ferrari, Patrizia Pedrazzi, and Luigi F. Agnati

Section of Physiology
Department of Biomedical Sciences
University of Modena, Italy
Interuniversity Center for the Study of Aging
Milan, Italy

1. INTRODUCTION

Striatum is highly vulnerable to transient forebrain ischemia induced by the 4 vessel occlusion (4VO) method (Brierley 1976, Pulsinelli et al. 1982, Zini et al. 1990a). Massive degeneration and loss of Nissl-stained neurons occur within 24 hr from an ischemia of long duration (30 min) (Pulsinelli et al. 1982). Neuronal loss is mainly restricted to the lateral part of caudate-putamen (Pulsinelli et al. 1982, Zini et al. 1990a). Cellular alterations include loss of medium-size spiny projection neurons (Pulsinelli et al. 1982, Francis and Pulsinelli 1982), largely corresponding to dopaminoceptive neurons (Benfenati et al. 1989, Zoli et al. 1989), and increase in reactive astrocytes (Pulsinelli et al. 1982, Grimaldi et al. 1990) and microglia (Gehrmann et al. 1982). On the other hand, large cholinergic (Francis and Pulsinelli 1982) and medium-size aspiny somatostatin (SS)/neuropeptide Y (NPY)-containing interneurons are resistant to the ischemic insult (Pulsinelli et al. 1982, Grimaldi et al. 1990). In a few instances, such as in the case of SS and NPY immunoreactivity (IR), the initial loss is followed by full recovery within 7 (SS) or 40 (NPY) days post-ischemia (Grimaldi et al. 1990). However, it is not known whether some kind of recovery is present for the bulk of medium-size spiny projections neurons after the first days post-ischemia.

This paper is divided into two parts: in the first, we report a study of the spatiotemporal relationships between neuron loss and recovery and astrocyte activation in the is-

* Address correspondence to: Dr. Michele Zoli, Sezione di Fisiologia, Dipartimento di Scienze Biomediche, Università di Modena, via Campi 287, 41100, Modena, Italy. Tel: +39 59 374179; FAX: +39 59 367372; E-mail agnati@c220.unimo.it.

Brain Plasticity, edited by Filogamo *et al.*
Plenum Press, New York, 1997

chemic striatum. In the second, we show some data on the characterization a possible molecular mechanisms linking neuronal lesion with astrocyte activation (i.e., the activation of the polyamine system) after striatal lesion.

2. SPATIOTEMPORAL RELATIONSHIPS BETWEEN NEURON LOSS AND RECOVERY AND ASTROCYTE ACTIVATION AFTER TRANSIENT FOREBRAIN ISCHEMIA

Three striatal cellular populations have been examined by means of immunocytochemistry coupled to computer-assisted image analysis:

a. vulnerable spiny neurons (labelled for their content of DA and cyclic AMP regulated phosphoprotein mr32, DARPP-32, a phosphoprotein related to the transduction of D1 receptor (Hemmings et al. 1987)) which constitute more than 90% of the entire neuronal population.
b. resistant aspiny neurons (labelled for their content of SS and NPY) which constitute around 1% of the entire neuronal population. They also contain high levels of nitric oxide synthase (NOS) (Vincent and Kimura 1992, Kawaguchi et al. 1995) and are known to be spared in a number of models of brain injury (Beal et al. 1986) and human brain pathology (Ferrante et al. 1986).
c. reactive astrocytes, labelled for their increased content of glial fibrillary acidic protein (GFAP), an intermediate filament protein specific for differentiated astrocytes (Eng et al. 1985, Eng and De Armond 1982). Its increase is a marker of astroglial reaction to various types of brain injury (Grimaldi et al. 1990, Eng and De Armond 1982, Ludwin 1985, Bignami et al. 1980, Mathewson and Berry 1985, Brock and O'Callaghan 1987).

The 4VO method of Pulsinelli and Brierley (1982) was used to induce transient forebrain ischemia, as previously described (Zini et al. 1990a). Only those animals immediately losing their righting reflex, being unresponsive for 20 to 30 min after bilateral carotid occlusion, with an isoelectric electroencephalographic activity within 2–3 min after carotid artery occlusion and without recovery throughout the ischemic period, were studied. All the procedures used were in accordance with institutional (italian Ministero della Sanità) guidelines for animal care. Several times (4 hr, 1, 7 and 40 days, 8 months) after 30 min of forebrain ischemia, induced by 4VO were investigated.

Immunocytochemistry was performed as previously described (Agnati et al. 1988), using the avidin-biotin technique. Six sections/animal for each antiserum were taken at various coronal levels of the pre-commissural striatum (regularly spaced from bregma 1.7 to 0.2 mm, according to Paxinos and Watson 1982). The following primary antisera were used: mouse monoclonal antibody against DARPP-32 (16), rabbit polyclonal antiserum against GFAP (Dako, Glostrup, Denmark, lot no. 015), rabbit policlonal antiserum against SS (Johansson et al. 1984), rabbit polyclonal antiserum against NPY (Peninsula, Merseyside, U.K., lot no. 006802–3) which have been previously characterized. The antisera were used in a dilution of 1:2000 for DARPP-32, 1:300 for GFAP, 1:1500 for NPY and SS.

Morphometric and microdensitometric analyses of the histological preparations were performed by means of an automatic image analyzer (IBAS I-II, Zeiss Kontron, Munich, FRG) (Zoli et al. 1990). Both a global microdensitometric analysis of DARPP-32 IR and manual count of DARPP-32 ir cells in 40x microscope fields (153x117 μm) were per-

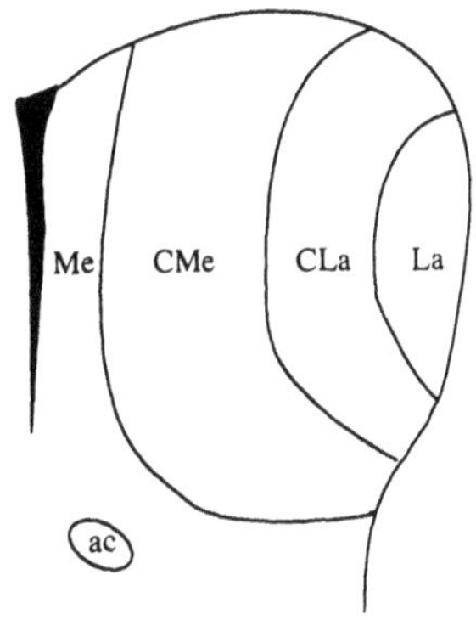

Figure 1. Subdivision of the caudate-putamen in 4 mediolateral subregions. Abbreviations: ac = anterior commissure, CLa = centrolateral region, CMe = centromedial region, La = lateral region, Me = medial region.

formed in the 4 striatal subregions. Both microdensitometry and manual count of GFAP ir cells were carried out in 40x microscope fields in the 4 striatal subregions (for further details, see Zoli et al. 1997).

2.1. Subdivision of Caudate-Putamen into 4 Regions with Different Post-Ischemic Fate

The analysis of the patterns of change in DARPP-32 and GFAP IR showed that the caudate-putamen can be subdivided into 4 subregions with different fate after 30 min 4VO (figs.1 and 2, panel A-C; for a full description see Zoli et al. 1997).

Region 1. A medial subregion (around 20% of control caudate-putamen area) with no significant change in both number and staining intensity of DARPP-32 immunoreactive (ir) neurons and GFAP ir astrocytes at any post-ischemic time studied. Note that GFAP IR in this region is already much higher than in the rest of striatum in control animals.

Region 2. A centromedial and dorsal subregion (around 40% of control caudate-putamen area) in which DARPP-32 ir neurons start to disappear at 4 hr and are totally lost at 1 day but recover their ir within 40 days (probably the recovery is already complete within 10 days). DARPP-32 ir disappearance at 1 day is likely to be a consequence of the depression of protein synthesis, which is known to occur in the early phase after ischemia (Hossmann 1993, Kogure and Kato 1993, Krause and Tiffany 1993). In this subregion reactive astrocytes are already significantly increased at 4–24hr, remain elevated at 7 days and return to control levels at later time intervals studied. The maximal increase in the number of GFAP ir astrocytes is already observed 4–24 hr post-ischemia. As previous studies have shown that astrocyte hyperplasia is not present at early post-ischemic times (du Bois et al. 1985, Petito et al. 1990), reactive astrocytes in region 2 likely correspond to hypertrophic resident glia.

Region 3. A centrolateral subregion (around 20% of control caudate-putamen area) in which DARPP-32 ir neurons are not detected at 1 day and do not recover within 40 days. GFAP ir astrocytes are low or absent at 4–24hr, but are moderately to highly concentrated from 7 days to 8 months post-ischemia. It is likely that astroglia remain activated until the subregion is completely reabsorbed. Accordingly, GFAP IR is elevated even 21 months after 4VO in the small DARPP-32 IR-negative areas still present in some of these

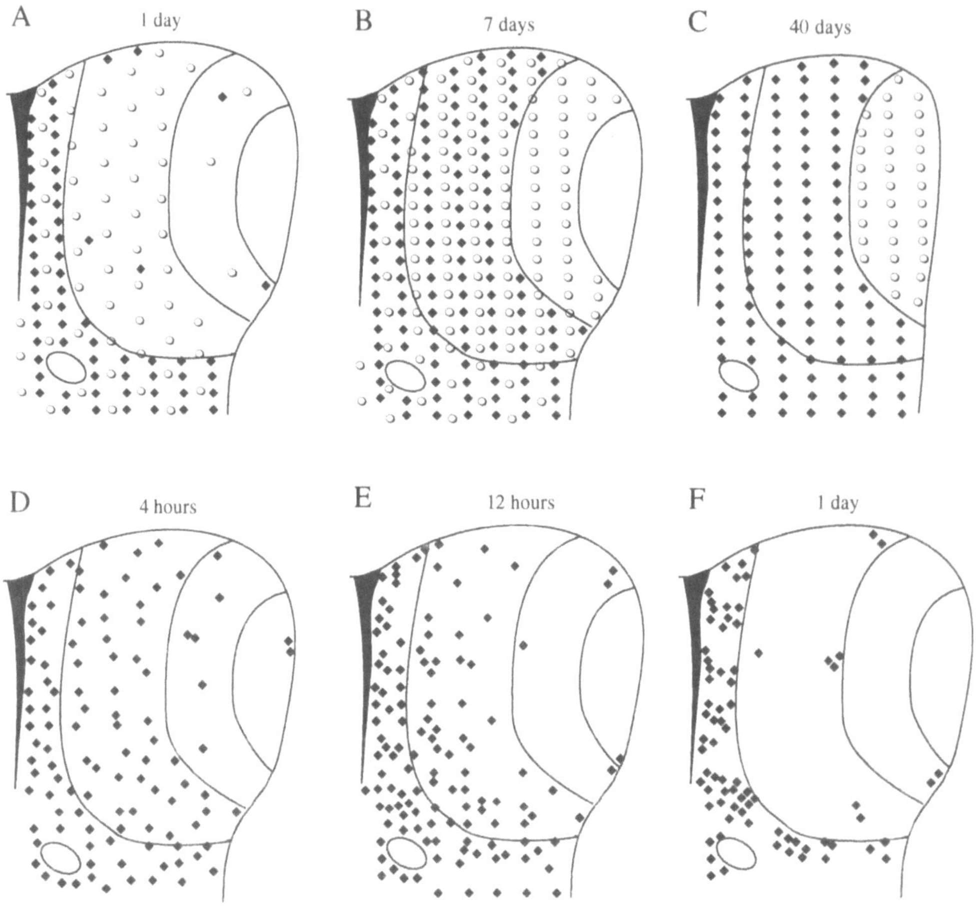

Figure 2. Panels A-C: Schematic drawings representing the loss and recovery of DARPP-32 ir neurons (◆) and appearance of reactive astrocytes (O) in the caudate-putamen after 30 min 4VO. Note the reduction of caudate-putamen total area at 40 days with the disappearance of the lateral subregion. Panels D-F: Schematic drawings representing the distribution of neurons with increased SSAT mRNA levels in the caudate-putamen after 30 min 4VO. Before 4 hr and after 1 day post-reperfusion, neurons have control levels of SSAT mRNA (◆).

animals (Zoli and Grimaldi, unpublished observations). Thus, during the "maturation" of the ischemic lesion (Pulsinelli et al. 1982) in neurons of region 3, astrocytes are degenerated and/or biosynthetically impaired. Long-term persistence of reactive astrocytes in areas of DARPP-32 ir neuronal loss tallies well with previous evidence obtained in the hippocampal formation, where GFAP IR remains elevated for more than 2 months in areas where neuronal loss has occurred (i.e., CA1 field), but returns towards basal levels within 40 days in areas where no neuronal loss has occurred (i.e., CA3 field) (Grimaldi et al. 1990, Kondo et al. 1995).

Region 4. A lateral and caudal subregion (around 20% of control caudate-putamen area), corresponding to the core of large ischemic lesions, in which DARPP-32 ir neurons and GFAP ir astrocytes are not detected from 4 hr to 7 days post-ischemia (area of pan-necrosis). This subregion is no longer detected at 40 or more days post-ischemia and may be therefore reabsorbed within this time.

The fate of the resistant NPY/SS ir interneurons was different from that of DARPP-32 ir neurons. NPY and SS ir cell bodies disappeared at early time intervals (4 hr and 1 day after reperfusion) and fully recovered between 7 and 40 days post-reperfusion (Grimaldi et al. 1990). Eight months after ischemia NPY and SS ir neurons were still detected in areas of DARPP-32 ir neuron loss and contained markedly higher levels of the two neuropeptides (for further details, see Zoli et al. 1997). Increased levels of these peptides may derive from altered trans-synaptic regulation of their biosynthesis and release after ischemia (see, e.g., Kerkerian et al. 1989). The long-term survival of SS/NPY interneurons in areas devoid of the vast majority of other neuronal types indicate that they are capable of maintaining trophic links with surviving neuronal afferents to striatum and local glia.

The results described above indicate that early activation of astroglia and neuronal survival are correlated and point to the hypothesis that the high density of reactive astroglia present during the maturation of ischemic lesion (first post-ischemic day) in regions 1 and 2 contribute to the protection or rescue of neurons of these areas. Astroglia is known to exert a neuroprotective action, e.g., thanks to neurotrophic factor release as well as maintenance of extracellular glutamate, K^+ concentration and pH (Giulian et al. 1993, Manthorpe et al. 1986, Needels et al. 1986, Nieto-Sampedro et al. 1982).

Regions 3 and 4 are devoid of DARPP-32 ir neurons at all time intervals studied starting from 1 day post-ischemia and are progressively reabsorbed, so that 8 months after ischemia caudate-putamen is almost exclusively constituted by regions 1 and 2. GFAP ir astroglia are either detected at low density or totally absent in regions 3 and 4 at early times post-ischemia when they might exert some neuroprotective action (see above). They are instead highly concentrated at late times (7 days or more) in the parts of regions 3 and 4 which have not yet been reabsorbed. It has been shown that reactive astrocytes appearing in lesioned tissue can have a phagocytic function complementary to that of microglia (Gehrmann et al. 1982, Raisman and Field 1973, Matthews et al. 1976, Cheng et al. 1994). Therefore, the high density of reactive astrocytes in areas devoid of the bulk of neurons may contribute to the progressive reabsorption of the lesioned tissue. In agreement with this hypothesis, it has been shown that high density of reactive astrocytes and microglia was detected in the hippocampus 2 months after ischemia, when neuronal debris are still present and the reabsorption processes ongoing. Instead, reactive astrocytes and microglia were no longer detected 6 months after ischemia when neuronal debris are absent and tissue reabsorption process completed (Kondo et al. 1995).

Region 2, and likely also 3, corresponds to the ischemic penumbra (Astrup et al. 1981, Hossmann 1994, Heiss and Graf 1994). This term refers to the brain tissue surrounding the ischemic core which is transiently inactivated, but remains viable and can undergo spontaneous and/or pharmacologically-induced recovery. In region 2 there is spontaneous recovery in present conditions. It can be hypothesized that therapeutical interventions administered even a few hours after ischemia can rescue neurons in region 3, whereas those administered before or immediately after the ischemic insult may also rescue neurons in region 4.

3. CHARACTERIZATION OF POSSIBLE NEURONAL SIGNALS RESPONSIBLE FOR ASTROCYTE ACTIVATION AFTER ISCHEMIA: THE ROLE OF ENDOGENOUS POLYAMINES

Natural polyamines (putrescine, spermidine and spermine) are ubiquitous organic polycations known to play a fundamental role in the regulation of cellular growth processes, including pre- and postnatal growth, and cell regeneration (Jänne et al. 1978, Pegg 1986, Pegg and McCann 1982,1988). Polyamines are also important intra- and extra-cellular signals regulating the function of voltage- and ligand-gated ion channels (Scott et al. 1993). Accordingly, release and uptake of polyamines have been shown in several tissues (Seiler and Deckardt 1978).

In mammals (Pegg 1986, Pegg and McCann 1982,1988), putrescine biosynthesis occurs through the decarboxylation of ornithine via the enzyme ornithine decarboxylase (ODC, EC 4.1.1.17). The subsequent addition of aminopropyl groups forms spermidine and spermine (catalysed by the enzyme spermidine synthase and spermine synthase, respectively). The aminopropyl groups derive from the decarboxylation of S-adenosyl-methionine catalysed by the enzyme S-adenosyl-methionine decarboxylase (AdoMetDC, EC 4.1.1.50). Spermine and spermidine can be converted back to spermidine and putrescine, respectively, through the combined action of spermidine/spermine N^1-acetyltransferase (SSAT) and polyamine oxidase. The intermediate of these reactions is the N^1acetylated form of the polyamine.

ODC, AdoMetDC and SSAT are regulated by a great number of stimuli (for review see Casero and Pegg 1993, Pegg and McCann 1988) while the other enzymes of the pathway are only regulated by their substrates. Regulation of ODC, SSAT and AdoMet DC activity appears to consist in changes in enzyme concentration rather than catalytic efficacy. The three enzymes have a fast turnover rate (10–60 min), resulting in a continuous and precise regulation of activity.

Considerable evidence points to an involvement of polyamines in trophic regulation of brain tissue. Increases in polyamine tissue levels have been consistently observed after mechanical, toxic or ischemic lesion (see e.g., Agnati et al. 1985b,c, Desiderio et al. 1988, Paschen et al. 1987,1991, Porcella et al. 1991). Putrescine levels begin to increase a few hours after the lesion (4–8 hr, see e.g., Desiderio et al. 1988, Paschen et al. 1987), whereas spermidine and spermine respond more slowly and are often initially decreased. For instance, after both mesodiencephalic hemitransection (Desiderio et al. 1988) and transient forebrain ischemia (Paschen et al. 1987), striatal spermidine is decreased 24 hr after the insult. Accordingly, rapid (within 4 hr) post-lesion increases in ODC activity were observed in all lesion models studied, including transient ischemia (Agnati et al. 1985a, Desiderio et al. 1988, Dienel and Cruz 1984, Kleihues et al. 1975, Müller et al. 1991, Paschen et al. 1988,1991, Porcella et al. 1991). The evidence that increased ODC activity is due to increased ODC protein concentration (Müller et al. 1991), is particularly remarkable in view of the fact that the biosynthesis of the vast majority of proteins is depressed in the early post-ischemic phase (see above and Hossmann 1993, Kogure and Kato 1993, Krause and Tiffany 1993). These data indicate that polyamines are possible mediators of some early neuroprotective or neurotoxic events occurring in the nervous tissue after lesion.

3.1. Astrocyte Activation and Polyamines: Role of Putrescine

As stated above ODC is the enzyme necessary to enter new molecules in the polyamine cycle. α-difluoromethylornithine (DFMO) is a selective irreversible inhibitor of ODC. Accordingly, DFMO can decrease by around 50% putrescine basal levels and almost completely putrescine lesion-induced increase (Agnati et al. 1985c). Tissue levels of higher polyamines are instead less affected by DFMO. In a first series of experiments, we studied the possible changes in the early lesion-induced activation of astrocytes after blockade of ODC activity. With this aim, DFMO (400 mg/kg, sc) was injected 3 hr before and 1 hr after 30 min 4VO or partial mesencephalic hemitransection (a mechanical lesion of the pathways linking the brainstem and the striatum).

Partial hemitransection was performed under deep ketamine hydrochloride (Ketalar, Parke-Davis, 100 mg/kg, i.p.) anaesthesia according to a previously published procedure (Agnati et al. 1985c, Desiderio et al. 1988). Briefly, a thin metal blade was inserted in the right cerebral hemisphere at the level of bregma with an inclination of 20° in order to reach the nigrostriatal pathway at the mesodiencephalic junction (approximately bregma level -4.3 mm according to Paxinos and Watson, 1982). GFAP immunocytochemistry was performed as described above. The amount of GFAP IR was evaluated by means of a microdensitometric procedure in the entire caudate-putamen.

In both animal models, striatal GFAP IR was markedly reduced in DFMO-treated animals (see fig.3). Based on the known effects of ODC on the different polyamine species, a decrease in putrescine appeared to be the most likely molecular event underlying this phenomenon.

In order to test this hypothesis, we performed the following experiment. A small mechanical lesion was performed in striatum by means of the insertion of a microdialysis probe and the change in GFAP IR was measured 24 hr later. In addition, extracellular polyamine levels were monitored via the dialysis probe during 5 hr after probe implantation.

The microdialysis probe (500 μm outer diameter, with a 2 mm length membrane, Carnegie Medicine) was implanted in the striatum (coordinates: 1.0 mm anterior, 2.5 mm lateral and 5.5 mm vertical from dura mater) and continuously perfused at 2 μl/min with Ringer's solution. After 5 hr of perfusion the probe was removed from the brain and the animals allowed to recover (for further details see Zini et al. 1990b). Polyamines were assayed by HPLC with fluorescence detection after precolumn derivatization with dansyl chloride (Desiderio et al. 1987). They were injected either after extraction from the tissues with perchloric acid or without modifications when sampled in brain perfusate. Several groups were compared: control, probe insertion, probe insertion + DFMO (400mg/kg sc, 2 hr before and 3 hr after probe implantation), probe insertion + DFMO + putrescine (administered through the probe at a concentration of 20μM during 5 hr post-implantation).

Figure 3. Effects of DFMO on GFAP IR increase evaluated 5 hr post-lesion. DFMO (400 mg/kg sc) was administered 3 hr before and 1 hr after the lesion. Statistical analysis according to Mann-Whitney U-test, ** = $p<0.01$; saline (clear bars), a-DFMO (shaded).

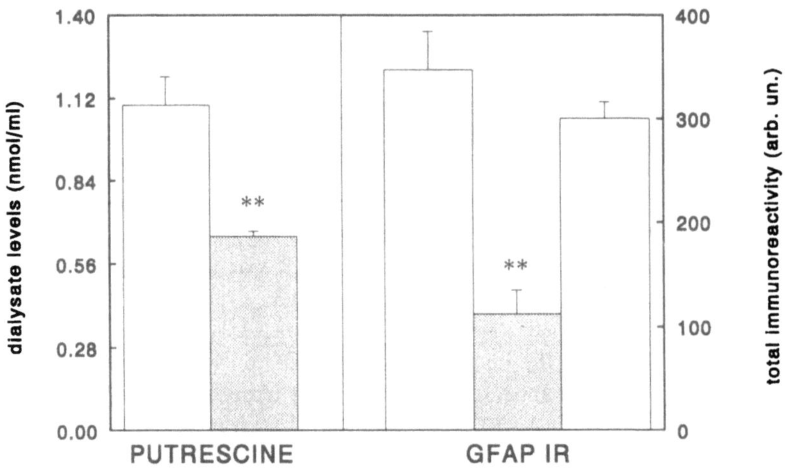

Figure 4. Effects of DFMO (400 mg/kg sc, 2 hr before and 3 hr after probe insertion) on GFAP IR (left panel) and extracellular putrescine (right panel) increase induced by the insertion of a microdialysis probe in the rat caudate-putamen. In the experiment of the right panel, putrescine (20 μM) was administered through the microdialysis probe. Statistical analysis according to Mann-Whitney U-test (left panel) or non-parametric test for multiple comparisons (right panel), ** = $p<0.01$; saline (clear bars), a-DFMO (shaded), a-DFMO + putrescine (cross hatched).

DFMO treatment significantly decreased both putrescine perfusate levels (fig.4, panel A) and GFAP post-lesion increase (fig.4, panel B). Interestingly, local administration of putrescine through the microdialysis probe was able to restore lesion-induced GFAP IR increase (fig.4, panel B).

Overall, these experiments establish a causal link between activation of the polyamine system (in particular, increased putrescine extracellular levels) and early astroglia activation after lesion.

3.2. Changes in Striatal SSAT mRNA Levels after 4VO Ischemia

The experiments described above point to the fact that increased putrescine level is an important early signal for astroglia activation. The early post-lesion changes in polyamine levels are usually attributed to a rapid (within 4 hr post-lesion) increase in ODC activity (Agnati et al. 1985a, Desiderio et al. 1988, Dienel and Cruz 1984, Kleihues et al. 1975, Müller et al. 1991, Paschen et al. 1988,1991, Porcella et al. 1991). Also changes in AdoMetDC activity can modify polyamine concentration. After transient ischemia, AdoMetDC activity is decreased at both early and late time intervals (Röhn et al. 1992), and may contribute to increases in putrescine concentration. Much less is known about SSAT, the third regulated enzyme in the polyamine system. However, the early decrease in spermidine after transient forebrain ischemia and hemitransection (Desiderio et al. 1988, Paschen et al. 1987) as well as the partial maintenance of putrescine levels after ODC irreversible blockade after lesion (Agnati et al. 1985c, Paschen et al. 1988) suggest that (similarly to some peripheral organs, Casero and Pegg 1993) SSAT may be up-regulated in the early post-lesion phase. Accordingly, an increase in SSAT activity has been recently shown after kainic acid-induced seizures (Baudry and Najm 1994) and mesodiencephalic hemitransection (Corti, Desiderio, Zoli, in preparation).

We have, therefore, investigated the levels of SSAT mRNA by means of *in situ* hybridization histochemistry at several times (1, 4, 10, 24 hr and 4 days) after transient forebrain ischemia induced by 4VO (Pulsinelli and Brierley 1982, Zini et al. 1990a). This technique allowed us to detect the changes in the 4 striatal subregions described in section 2.1. Moreover, it was possible to distinguish between SSAT mRNA changes in neuronal and non-neuronal cells.

As the sequence of rat SSAT mRNA is unknown, the probes for *in situ* hybridization histochemistry were taken from regions in which hamster (Pegg et al. 1992) and mouse (Fogel-Petrovic et al. 1993) cDNAs show complete identity. Two oligodeoxynucleotides were selected. For probe specificity see Zoli et al. 1996. The oligonucleotide probes were labelled at the 3' end using ^{35}S-dATP (Amersham) as described in Zoli et al. 1995. *In situ* hybridization histochemistry was carried out according to Zoli et al. 1995. The slides were exposed for 10–15 days to Hyperfilm-3H (Amersham) and then to a photographic emulsion (K5, Ilford, or NTB2, Kodak) for 1–2 months.

The semiquantitative evaluation of film autoradiograms was performed according to a previously published microdensitometric method (Zoli et al. 1991) using an automatic image analyzer (VIDAS, Zeiss Kontron, Munich, FRG). Non-specific labelling was assessed in adjacent sections incubated with an oligonucleotide of same length and GC content but unrelated sequence. In the analysis at the cellular level, grain counting was performed in 63x microscope fields by means of the automatic image analyzer VIDAS (for further details, see Zoli et al. 1996).

A limited increase of SSAT mRNA levels was observed in 20–50% of non-neuronal cells in all striatal subregions starting from 1 hr post-lesion. Instead, increase in SSAT mRNA levels in neurons was marked and heterogeneous.

In regions 1 and 2, both at 4 and 10 hr post-reperfusion, large patches or stripes of neurons showed significantly increased labelling (4.2±0.1 in control animals, 22.7±0.9 grains/100 μm^2 cytoplasm/cell at 4 hr, 30.4±1.8 at 10 hr) (fig.2, panels D-F). At 24 hr, neurons with increased labelling (34.8±2.4 grains/100 μm^2 cytoplasm/cell) were limited to region 1. At 4 days, the pattern of labelling was similar to that observed in control animals.

In regions 3 and 4, scattered neurons showed increased labelling (16.8±1.2 grains/100 μm^2 cytoplasm/cell) at 4 hr. Starting from 10 hr, only small (2–10 cells) dispersed clusters of neurons especially located in the periphery of the area of neuronal degeneration (e.g., below the corpus callosum) were still recognized. Most of these neurons showed very high SSAT mRNA labelling at both 10 and 24 hr time intervals (42.4±2.7 and 44.3±0.9 grains/100 μm^2 cytoplasm/cell, respectively) (fig.2, panels D-F). At 4 days the few scattered surviving neurons showed control levels of SSAT mRNA.

These experiments show that marked increases in SSAT mRNA levels at early times post-ischemia selectively occur in regions 1 and 2, the areas where astroglia is also precociously activated. Therefore, in these regions polyamine metabolism is oriented towards putrescine production. In addition, the cellular analysis show that SSAT mRNA is especially increased in neurons. This evidence suggests that putrescine is mainly produced (and possibly subsequently released, see chapter 3.1) by neurons in a fashion spatially and temporally related to the appearance of reactive astrocytes.

4. GENERAL DISCUSSION

Overall, the data presented in this paper favour the following hypothesis of post-lesion events in the ischemic striatum. During the first few hours following an ischemic insult, polyamine metabolism in lesioned neurons is actively oriented towards hyperproduction of putrescine. Among other effects, putrescine released by these neurons and diffusing in the local extracellular fluid represents a signal to activate quiescent resident astrocytes. In turn, reactive astrocytes, if activated during the maturation period of ischemic lesion, can exert a positive effect on neuron survival. Instead, in areas where neuronal signalling is insufficient and/or astroglia is unable to react properly, neither rapid astroglia activation nor neuronal survival is present.

It must be noticed that putrescine is not the only astroglial activating factor, although its action seems of particular relevance in the crucial first post-ischemic hours. In fact, DFMO subchronic treatment partially blocks early astroglial activation but cannot prevent the long term astroglial reaction (see above and Zoli et al. 1993). Other factors contributing to the triggering of astroglial reaction in the early post-lesion phase may be neurotransmitters (e.g., glutamate and serotonin), gliotransmitters (e.g., taurine), trophic factors (e.g., Transforming Growth Factor and basic Fibroblast Growth Factor), and the local raise in extracellular ions (Junier 1994. Reilly 1996). Other glial activating factors may intervene at later post-lesion times. For instance, many cytokines released by inflammatory mononuclear cells have been shown to increase astrogliosis (Balasingam 1994).

Astroglia likely play different roles at different post-lesion phases. Our data indicate that a correlation between early astroglial activation and neuronal survival is present in the post-ischemic striatum. The overall relevance of astrocytic neuroprotective function has been underlined by a recent study. It has been shown that spreading depolarization waves, which are thought to be responsible for damage of neurons lying close to infarcted areas in the post-ischemic period (Hossmann 1994), do not injury neurons unless glial cell function is impaired (Largo et al. 1996). Astrocytes are thought to exert a neurotrophic action through a number of mechanisms which include maintenance of extracellular ion and pH homeostasis, uptake of excitotoxic transmitters, release of neurotrophic factors, and supplementation of energetic precursors (Giulian et al. 1993, Manthorpe et al. 1986, Nieto-Sampedro et al. 1982, Largo et al. 1996, Tsacopoulos and Magistretti 1996). Besides these classical actions, astroglia may also intervene in determining the fate of lesioned brain tissue through an indirect action on the extracellular compartment. In fact, astroglia-induced modifications of the brain extracellular matrix may importantly contribute to the reactive changes observed in post-ischemic brain. For instance, it has recently been observed that increased proteoglycan mRNA levels are expressed in astroglia after nervous tissue injury (Snyder 1996). Moreover, modifications of the cell volume are known to occur in activated astrocytes (astrocyte swelling and hypertrophy) (Hansson and Röhnback 1995, Cooper 1995).

Both glial volumetric changes and alteration of the extracellular matrix lead to changes in the characteristics (volume fraction and/or tortuosity) of the brain extracellular fluid (ECF), thus altering the dynamics of molecule diffusion (so called volume transmission, Agnati et al. 1995). This phenomenon has in fact been observed in a number of pathological states of the central nervous system, including experimental ischemia (Sikova et al. 1994, Perez-Pinon et al. 1995). Besides global changes in molecule diffusion in brain ECF, morphological changes in astroglial processes and extracellular matrix may also specifically control the access of neurotrophic or neurotoxic molecules to lesioned neurons, e.g., by sealing/opening portions of the tissue or parts of the neurons. This kind of phenomenon has been shown in development, when astrocytes can form barriers in brain tissue to guide migration of cells and growth of axons (Steindler 1993).

In a subsequent phase, reactive astrocytes are highly concentrated in areas devoid of neurons and likely orient their function towards the reorganization of the lesioned tissue.

Whereas astroglia and microglia may have neuroprotective and neurotoxic actions, respectively, in the early post-lesion phase (Giulian et al. 1993, Manthorpe et al. 1986, Needels et al. 1986, Nieto-Sampedro et al. 1982), these cell types exert a complementary action on reorganization of lesioned tissue. For instance, both microglia and reactive astrocytes have been shown to have a phagocytic function in lesioned tissue (Gehrmann et al. 1982, Raisman and Field 1973, Matthews et al. 1976, Cheng et al. 1994). However, at least in mechanical lesions where the different neuronal compartments can be distinguished, astroglia and microglia are responsible for removal of degenerating axon terminals and myelinated axons, respectively (Cheng et al. 1994). Moreover, astrocytes can guide the sprouting of regenerating axons (Raisman and Field 1973, Beck et al. 1993, Cheng et al. 1994). These observations suggest that while the function of microglia is mainly to remove the bulk of degenerated tissue, the activity of astroglia is mainly directed at reconstituting functional networks.

This brief discussion emphasizes the relevance of a therapeutical approach aimed at activating astroglia in the early post-lesion phase. This may be obtained by stimulating production of astrocyte activating factors from neurons and other cells, and/or directly supplementing astrocyte activating factors. Interestingly, drugs (called astrocyte-kinetic) with the property of activating astrocytes selectively in lesioned brain tissue have been recently proposed (Biagini et al. 1994). The relatively long maturation time of neuronal lesion after brain ischemia makes feasible such a kind of pharmachological intervention and stimulates further research on this subject.

ACKNOWLEDGMENTS

This paper has been supported by grants from MURST and CNR.

REFERENCES

Agnati L.F., Fuxe K., Davalli P., Zini I., Corti A., Zoli M. Striatal ornithine decarboxylase activity following neurotoxic and mechanical lesions of the mesostriatal dopamine system of the male rat. Acta Physiol. Scand. 125 (1985a) 173–175

Agnati L.F., Fuxe K., Zoli M., Davalli P., Corti A., Zini I., Toffano G. Effects of neurotoxic and mechanical lesions of the mesostriatal dopamine pathway on striatal polyamine levels in the rat: modulation by chronic ganglioside GM1 treatment. Neurosci. Lett. 61 (1985b) 339–344

Agnati L.F., Fuxe K., Zini I., Davalli P., Corti A., Calzà L., Toffano G., Zoli M., Piccinini G., Goldstein M. Effects of lesions and ganglioside GM1 treatment on polyamine levels and nigral DA neurons. A role of putrescine in the neurotrophic activity of gangliosides. Acta Physiol. Scand. 124 (1985c) 499–506

Agnati L.F., Fuxe K., Zoli M., Zini I., Härfstrand A., Toffano G., Goldstein M. Morphometrical and microdensitometrical studies on phenylethanolamine-N-methyltransferase and neuropeptide Y immunoreactive neurons in the rostral medulla oblongata of the adult and old male rats. Neuroscience 26 (1988) 461–478

Agnati L.F., Zoli M., Strömberg I., Fuxe K. Intercellular communication in the brain: Wiring versus volume transmission. Neuroscience 69 (1995) 711–726.

Astrup J., Siesjo B.K., Symon L. Threshold in cerebral ischemia - the ischemic penumbra. Stroke 12 (1981) 723–725

Balasingam V., Tejada-Berges T., Wright E., Bouckova R., Yong V.W. Reactive astrogliosis in the neonatal mouse brain and its modulation by cytokines. J. Neurosci. 14 (1994) 846–856

Baudry M., Najm I. Kainate-induced seizure activity stimulates the polyamine interconersion pathway in rat brain. Neurosci. Lett. 171 (1994) 151–154

Beal M.F., Kowall N.W., Ellison D.W., Mazureck M.F., Swartz K.J., Martin J.B. Replication of the neurochemical characteristics of Huntington's disease by quinolinic acid. Nature 321 (1986) 168–171.

Beck K.D., Lamballe F., Rudiger K., Barbacid M., Schauwecker PE., McNeill T.H., Finch C.E., Hefti F., Day J.R. Induction of non-catalytic trkB neurotrophin receptors during axonal sprouting in the adult hippocampus. J Neurosci. 13 (1993) 4001–4014

Benfenati F., Merlo Pich E., Grimaldi R., Zoli M., Fuxe K., Toffano G., Agnati L.F. Transient forebrain ischemia produces multiple deficits in dopamine D1 transmission in the lateral neostriatum of the rat. Brain Res. 498 (1989) 376–380

Biagini G., Frasoldati A., Fuxe K., Agnati L.F. The concept of astrocyte-kinetic drug in the treatment of neurodegenerative diseases: Evidence for L-deprenyl-induced activation of reactive astrocytes. Neurochem. Int. 25 (1994) 17–22

Bignami A., Dahl D. Differentiation of astrocytes in the cerebellar cortex and the pyramidal tracts of the newborn rat. An immunofluorescence study with antibodies to a protein specific to astrocytes. Brain Res. 49 (1973) 393–402

Bignami A., Dahl D., Rueger DG. Glial fibrillary acidic (GFA) protein in normal neuronal cells and in pathological conditions, in Federoff S, Hertz L (eds): Advances in cellular neurobiology, vol 1, New York, Academic Press, 1980, pp 285–310

Brierley J. Cerebral hypoxia, in Blackwood W, Corsellis J (eds): Grenfield's Neuropathology, London, Edward Arnold, 1976, pp 43–85

Brock T.O., O'Callaghan J.P. Quantitative changes in the synaptic vesicle proteins, synapsin I and p38, and the astrocyte specific protein, glial fibrillary acidic protein, are associated with chemical-induced injury to the rat central nervous system. J Neurosci. 7 (1987) 931–942

Casero R.A., Pegg A.E., Spermidine/spermine N^1-acetyltransferase - the turning point in polyamine metabolism. FASEB J. 7 (1993) 653–661

Cheng H-W., Jiang T., Brown S.A., Pasinetti G.M., Finch C.E., McNeill H. Response of striatal astrocytes to neuronal deafferentiation: an immunocytochemical an ultrastructural study. Neuroscience 62 (1994) 425–439

Cooper S.M. Intercellular signaling in neuronal-glial networks. Biosystems 34 (1995) 65–85

Desiderio M.A., Davalli P., Perin A. Simultaneous determination of γ-aminobutyric acid and polyamines by high performance liquid chromatography. J. Chromatograph. 419 (1987) 285–290

Desiderio M.A., Zini I., Davalli P., Zoli M., Corti A., Fuxe K., Agnati L.F. Polyamines, ornithine decarboxylase, and diamine oxidase in the substantia nigra and striatum of the male rat after hemitransection. J. Neurochem. 51 (1988) 25–31

Dienel G.A., Cruz N.F., Induction of brain ornithine decarboxylase during recovery from metabolic, mechanical, thermal or chemical injury. J. Neurochem. 42 (1984) 1053–1061.

du Bois M., Bowman P.D., Goldsten G.W. Cell proliferation after ischemic injury in gerbil brain. An immunocytochemical and autoradiographic study. Cell Tissue Res. 242 (1985) 17–23

Eng L.F. Glial fibrillary acidic protein (GFAP): the major protein of glial intermediate filaments in differentiated astrocytes. J. Neuroimmunol. 8 (1985) 203–214

Eng L.F., De Armond S.J. Immunocytochemical study of astrocytes, in normal development and diseas, in Federoff S, Hertz L (eds): Advances in cellular neurobiology, vol 3, New York, Academic Press, 1982, 145–171

Ferraguti F., Zoli M., Aronsson M., Agnati L.F., Goldstein M., Filer D. Fuxe K. Distribution of glutamic acid decarboxylase messenger RNA containing nerve cell populations of the male rat brain. J. Chem. Neuroanat. 3 (1990) 377–396

Ferrante R.J., Kowall L.W., Beal M.F., Richardson E.P., Bir E.D., Martin J.B. Selective sparing of a class of striatal neurons in Huntington's disease. Science 230 (1986) 561–563

Francis A., Pulsinelli W.A. Response of GABAergic and cholinergic neurons to transient cerebral ischemia. Brain Res. 243 (1982) 271–278

Gehrmann J., Bonnekoh P., Miyazawa T., Hossmann K-A., Kreutzberg G.W. Immunocytochemical study of an early microglial activation in ischemia. J. Cereb. Blood Flow Metab. 12 (1982) 257–269

Giulian D., Vaca K., Corpuz M. Brain glia release factors with opposing actions upon neuronal survival. J. Neurosci. 13 (1993) 29–37

Grimaldi R., Zoli M., Agnati L.F., Ferraguti F., Fuxe K., Toffano G., Zini I. Effects of transient forebrain ischemia on peptidergic neurons and astroglial cells. Evidence for recovery of peptide immunoreactivities in neocortex and striatum but not hippocampal formation. Exp. Brain Res. 82 (1990) 123–136

Hansson E., Rönnback L. Astrocytes in glutamate neurotransmission. FASEB J. 9 (1995) 343–350

Heiss W-D., Graf R. The ischemic penumbra. Curr. Op. Neurol. 7 (1994) 11–19

Hemmings H. Jr., Nestler E.J., Walaas I.S., Ouimet C.C., Greengard P. Protein phosphorylation and neuronal function: DARPP-32, an illustrative example, in Edelman GM, Gall WE, Cowan WM (eds): Synaptic function, New York, John Wiley, 1987, pp 213–240

Hossmann K.A. Disturbances of cerebral protein synthesis and ischemic cell death. Prog. Brain Res. 96 (1993) 161–177

Hossmann K.A. Viability thresholds and the penumbra of focal ischemia. Ann. Neurol. 36 (1994) 557–565

Jänne J., Williams-Ashman H.G. On the purification of L-ornithine decarboxylase from rat prostate and effects of thiol compounds on the enzyme. J. Biol. Chem 246 (1971) 1725–1732

Jänne J., Pösö H., Raina A. Polyamines in rapid growth and cancer. Biochem. Biophys. Acta 473 (1978) 241–293

Johansson O., Hökfelt T., Elde R.P. Immunohistochemical distribution of somatostatin-like immunoreactivity in the central nervous system of the adult rat. Neuroscience 13 (1984) 265–339

Junier M.-P., Coulpier M., Le Forestier N., Cadusseau J., Suzuki F., Peschanski M., Dreyfus P.A. Transforming Growth Factor α (TGFα) expression in degenerating motoneurons of the murine mutanet *wobbler*: a neuronal signal for astrogliosis? J. Neurosci. 14 (1994) 4206–4216

Kawaguchi Y., Wilson C.J., Augood S.J. Emson PC: Striatal interneurones: chemical, physiological and morphological characterization. Trends Neurosci. 18 (1995) 527–535

Kerkerian L., Salin P., Nieoullon A. Cortical regulation of striatal neuropeptide Y (NPY)-containing neurons in the rat. Eur. J. Neurosci. 2 (1989) 181–189

Kleihues P., Hossmannn K.A., Pegg A.E., Kobayashi K., Zimmermann V. Resuscitation of the monkey brain after one hour complete ischemia. III. Indications of metabolic recovery. Brain Res. 95 (1975) 61–73

Kogure K., Kato H. Altered gene expression in cerebral ischemia. Stroke 24 (1993) 2121–2127

Kondo Y., Ogawa N., Asanuma M., Ota Z., Mori A. Regional differences in late-onset iron deposition, ferritin, transferrin, astrocytes proliferation and microglial activation after transient forebrain ischemia. J. Cereb. Blood Flow Metab. 15 (1995) 216–226

Krause G.S., Tiffany B.R. Suppression of protein synthesis in the reperfused brain. Stroke 24 (1993) 747–755.

Largo C., Cuevas P., Somjen G.G., Martín del Río R., Herreras O. The effect of depressing glial function in rat brain *in situ* on ion homeostasis, synaptic transmission, and neuron survival. J. Neurosci. 16 (1996) 1219–1229

Ludwin S.K. Reaction of oligodendrocytes to trauma and implantation. A combined autoradiographic and immunohistochemical study. Lab. Invest. 52 (1985) 20–30.

Manthorpe M., Rudge J.S., Varon S. Astroglial cell contributions to neuronal survival and neuritic growth, in Federoff S, Vernadakis A (eds): Astrocytes. New York, Academic Press, 1986, pp 315–376

Mathewson A.J., Berry M. Observations on astrocyte response to a cerebral stab wound in adult rats. Brain Res. 327 (1985) 61–69

Matthews D.A., Cotman C.W., Lynch G. An electron microscopic study of lesion-induced synaptogenesis in the dentate gyrus of the adult rat. I. Magnitude and time course of degeneration. Brain Res. 115 (1976) 1–21

Müller M., Cleef M., Röhn G., Bonnekoh P., Pajunen A.E.I., Bernstein H.G., Paschen W. Ornithine decarboxylase in reversible cerebral ischemia. An immunohistochemcial study. Acta Neuropathol. 83 (1991) 39–45

Needels D.L., Nieto-Sampedro M., Cotman C.W. Induction of a neurite-promoting factor in rat brain following injury or deafferentation. Neuroscience 18 (1986) 517–526.

Nieto-Sampedro M., Lewis E.R., Cotman C.W., Manthorpe M., Skaper S.D., Barbin G., Longo F.M., Varon S. Brain injury causes a time-dependent increase in neuronotrophic activity at the lesion site. Science 217 (1982) 860–861

Paschen W., Schmidt-Kastner R., Djuricic B., Meese C., Linn F., Hossmann K.A. Polyamine changes in reversible cerebral ischemia. J. Neurochem. 49 (1987) 35–37

Paschen W., Röhn G., Meese C.O., Djuricic B., Schmidt-Kastner R. Polyamine metabolism in reversible cerebral ischemia: effect of α-difluoromethylornithine. Brain Res. 453 (1988) 9–16

Paschen W., Csiba L., Röhn G., Bereczki D. Polyamine metabolism in transient focal ischemia of rat brain. Brain Res. 566 (1991) 354–357

Paxinos G., Watson C. The rat brain in stereotaxic coordinates. London, Academic Press, 1982

Pegg A.E. Recent advances in the biochemistry of polyamines in eukaryotes. Biochem. J. 234 (1986) 249–262

Pegg A.E., McCann P.P. Polyamine metabolism and function. Am. J. Physiol. 243 (1982) C212-C221

Pegg A.E., McCann P.P. Polyamine metabolism and function in mammalian cells and protozoans. ISI Atlas Sci. Biochem. 1 (1988) 11–18

Perez-Pinon M.A., Tao L., Nicholson C. Extracellular potassium, volume fraction, and tortuosity in rat hippocampal CA1, CA3, and cortical slices during ischemia. J. Neurophysiol. 74 (1995) 565–573

Petito C.K., Morgello S., Felix J.C., Lesser M.L. The two patterns of reactive astrocytosis in postischemic rat brain. J Cereb. Blood Flow Metab. 10 (1990) 850–859

Porcella A., Carter C., Fage D., Voltz C., Lloyd K.G., Serrano A., Scatton, B. The effects of N-methyl-D-aspartate and kainate lesions of the rat striatum on striatal ornithine decarboxylase activity and polyamine levels. Brain Res. 549 (1991) 205–212

Pulsinelli W.A., Brierley J.B. A new model of bilateral hemispheric ischemia on the unanesthetized rat. Stroke 10 (1979) 267–272

Pulsinelli W.A., Brierley J.B., Plum F. Temporal profile of neuronal damage in a model of transient forebrain ischemia. Ann. Neurol. 11 (1982) 491–498

Raisman G., Field P.M. A quantitative investigation of the development of collateral reinnervation after partial deafferentation of the septal nuclei. Brain Res. 50 (1973) 241–264

Reilly J.F., Kumari V.G. Alterations in Fibroblast Growth Factor receptor expression following brain injury. Exp. Neurol. 140 (1996) 139–150

Röhn G., Schlenker M., Paschen W. Activity of ornithine decarboxylase and S-adenosylmethionine decarboxylase in transient cerebral ischemia: relationship to the duration of vascular occlusion. Exp. Neurol. 117 (1992) 210–215

Scott R.H., Sutton K.G., Dolphin A.C. Interactions of polyamines with neuronal ion channels. Trends Neurosci. 16 (1993) 153–160

Seiler N., Bolkenius F.N., Knödgen B. The influence of catabolic reactions on polyamine excretion. Biochem. J. 225 (1985) 219–226

Seiler N., Deckhardt K. Uptake of polyamines and related compounds into nerve endings. Adv. Pol. Res. 2 (1978) 101–107

Snyder S.E., Li J., Schauwecker P.E., McNeill T.H., Salton S.R.J. mRNA expression in the developing and adult rat nervous system and induction of Rptp-Zeta/Beta and phosphacan mRNA following brain injury. Mol. Brain Res. 40 (1996) 79–96

Steindler D.A. Glial boundaries in the developing nervous system. Annu. Rev. Neurosci. 16 (1993) 445–470

Sykova E., Svoboda J., Polak J., Chvatal A. Extracellular volume fraction and diffusion characteristics during progressive ischemia and terminal anoxia in the spinal-cord of the rat. J. Cereb. Blood Flow Metab. 14 (1994) 301–311

Tsacopoulos M., Magistretti P.J. Metabolic coupling between glia and neurons. J. Neurosci. 16 (1996) 877–885

Vincent S.R., Kimura H. Histochemical mapping of nitric oxide synthase in the rat brain. Neuroscience 46 (1992) 755–784

Zini I., Grimaldi R., Merlo Pich E., Zoli M., Fuxe K., Agnati L.F. Aspects of neural plasticity in the central nervous system. V. Studies on a model of transient forebrain ischemia in male sprague dawley rats. Neurochem. Int. 16 (1990a) 451–468

Zini I., Zoli M., Grimaldi R., Merlo Pich E., Biagini G., Fuxe K., Agnati L.F. Evidence for a role of neosynthetized putrescine in the increase of glial fibrillary acidic protein immunoreactivity induced by a mechanical lesion in the rat brain. Neurosci. Lett. 120 (1990b) 13–16

Zoli M., Grimaldi R., Agnati L.F., Zini I., Merlo Pich E., Toffano G., Fuxe K. Neurohistochemical studies on striatal lesions induced by transient forebrain ischemia. Evidence for protective effects of the ganglioside analogue AGF2. Neurosci. Res. Comm. 4 (1989) 153–158

Zoli M., Zini I., Agnati L.F., Guidolin D., Ferraguti F., Fuxe K. Aspects of neural plasticity in the central nervous system. I. Computer-assisted image analysis methods. Neurochem. Int. 16 (1990) 383–418

Zoli M., Bettuzzi S., Ferraguti F., Ingletti M.C., Zini I., Fuxe K., Agnati L.F., Corti A. Regional increase in ornithine decarboxylase mRNA levels in the rat brain after partial mesodiencephalic hemitransection as revealed by in situ hybridization histochemistry. Neurochem. Int. 18 (1991) 347–52

Zoli M., Zini I., Grimaldi R., Biagini G., Agnati L.F. Effects of putrescine synthesis blockade on neuronal loss and astroglial reaction after transient forebrain ischemia. Int. J. Dev. Neurosci. 11 (1993) 175–187

Zoli, M., Le Novère, N., Hill, J.A. jr. and Changeux, J.P., Developmental regulation of nicotinic receptor subunit mRNAs in the rat central and peripheral nervous systems. J. Neurosci. 15 (1995) 1912–1939

Zoli M., Pedrazzi P., Zini I., Agnati L.F. Spermidine/spermine N^1-acetyltransferase mRNA levels show marked and region-specific changes in the early phase after transient forebrain ischemia. Mol. Brain Res. 38 (1996) 122–134

Zoli M., Grimaldi R., Ferrari R., Zini I., Agnati L.F. Short- and long-term changes in striatal neurons and astroglia after transient forebrain ischemia. Stroke (1997), in press

5

REGULATION OF OLIGODENDROCYTE DIFFERENTIATION BY FIBROBLAST GROWTH FACTORS

Rashmi Bansal and S. E. Pfeiffer

Departments of Pharmacology and Microbiology, and Program in Neurological Sciences
University of Connecticut Medical School
Farmington, Connecticut 06030-3205

1. FIBROBLAST GROWTH FACTORS (FGFs)

The interface between proliferation and differentiation is a key point of regulation in development. Among elements of the environment that influence these alternative states is a structurally related family of fourteen polypeptides (see references in Coulier et al., 1997), the fibroblast growth factors, half of which are expressed in the brain. These growth factors play crucial roles in both normal physiological (such as embryonic and postnatal development, nervous system differentiation, angiogenesis and wound repair) and pathological (such as tumorogenesis and metastasis) processes (reviewed in Basilico and Moscatelli, 1992; Eckenstein, 1994). The most extensively studied member of this family, FGF-2 (commonly known as basis fibroblast growth factor), targets a broad spectrum of cell types derived from mesoderm and neuroectoderm. It induces a variety of biological activities, such as the inhibition or stimulation of differentiation, proliferation, chemotaxis, enhanced survival and cytoskeletal alterations.

2. FGF HIGH AFFINITY RECEPTORS

The high affinity FGF receptors (FGFR) comprise an equally diverse family of tyrosine kinases, for which four separate genes (FGFR 1–4) have been identified (reviewed in Johnson and Williams, 1993). FGF receptors consist of an extracellular ligand-binding region with three immunoglobulin (Ig)-like domains, a transmembrane region and a cytoplasmic tyrosine kinase domain. Some FGFRs are activated by multiple FGFs; other members of this family respond more specifically to certain of the FGF (Ornitz and Leder, 1992; Partenen et al., 1991; Johnson and Williams, 1993). In addition, there are multiple splice variants of FGFR-1, -2 and -3 that can exhibit altered ligand specificities and bind-

Brain Plasticity, edited by Filogamo *et al.*
Plenum Press, New York, 1997

ing affinities (Miki et al., 1992; Werner et al., 1992; Johnson et al., 1991). The spatial distribution of FGF receptors changes during embryogenesis (Peters et al., 1992; Orr-Urtreger et al., 1991) and in adult tissues (Asai et al., 1993; Yazaki et al., 1994).

3. HEPARAN SULFATE PROTEOGLYCANS (FGF CO-RECEPTORS)

Differentiating cells undergo alterations in their interactions with other cells and with the matrix as they migrate through a changing series of distinct microenvironments. These interactions have substantial opportunity to alter cell phenotype and physiology in a sequential manner. Heparan sulfate proteoglycans (HSPGs) are thought to be intimately involved in these processes. This is a view consistent with their diversity of structure, developmentally regulated patterns of expression, and their co-receptor function with components of the extracellular matrix, cell adhesion molecules, and growth factors, including FGF.

Heparan sulfate proteoglycans are a diverse group of macromolecules composed of a core protein to which are attached heparan sulfate and related heparin-like glycosaminoglycan (GAG) chains (Kjellén and Lindahl, 1991). They are abundant at the surface of most, if not all, adherent cells. Two major families of heparan sulfate proteoglycans are present on the cell surface. The syndecan family is composed of integral membrane proteins with a strict evolutionary conservation of their cytoplasmic and transmembrane domains, but with substantial diversity in their extracellular domains (reviewed in Bernfield et al., 1992). Four members of the syndecan family have been cloned from a number of tissues and species, including syndecan-1, -2, -3 and -4 (see Bansal et al., 1996b for further review and references). A second family of integral membrane heparan sulfate proteoglycans, characterized by glycosylphosphotidy-inositol (GPI) linkage and conserved structural motifs, includes glypican and cerebroglycan. While glypican is expressed in a variety of tissues, including abundant representation in the nervous system, cerebroglycan is restricted to the nervous system.

Pertinent to the present review, these heparan sulfate proteoglycans appear to be essential for FGF signaling. They are thought to act as high capacity, low affinity co-receptors to the high affinity receptors discussed above (Yayon et al., 1991; Rapraeger et al., 1991).

Therefore, there are fourteen (at least) similar but distinct ligands, four types of high affinity receptors and their multiple variant isoforms, and an extended family of co-receptors. Together these provide the cell with an remarkable opportunity to modulate its response to this family of growth factors by regulating the combinatorial complexity of a diverse set of cell surface receptor complexes.

4. THE OLIGODENDROCYTE LINEAGE

Oligodendrocytes provide an excellent model in which to examine key issues of cell biology in a developmental context (Pfeiffer et al., 1993). Oligodendrocytes form myelin in the central nervous system through an elaboration of their plasma membranes which wrap around neuronal axons to form compact, multilamellar sheaths. The progression of oligodendrocyte progenitors along a carefully regulated lineage, characterized by changes in their morphology, proliferative and migratory capacity, and the ordered expression of oligodendrocyte and myelin-specific proteins and lipids, can be summarized in its final stages as:

Early Progenitors ("O2As") ---->
Late Progenitors ("Pro-OLs") ---->
Maturing Oligodendrocytes ("OLs")

The Early Progenitor stage of the oligodendrocyte lineage is characterized by the proliferative and migratory capacity of the cells. The Late Progenitors are proliferative but have lost their migratory ability (Warrington et al., 1993). Maturing Oligodendrocytes, cells entering terminal differentiation, quickly become post-mitotic, and soon begin to express myelin proteins and lipids. Purified oligodendrocyte progenitors grown in culture follow a developmental pathway consistent with lineage progression *in situ* (Warrington and Pfeiffer, 1992).

5. REGULATION OF THE OLIGODENDROCYTE LINEAGE BY FGFs

The oligodendrocyte developmental lineage is regulated by a number of growth factors, among them FGF-2 (basic FGF). Of particular interest to this review, the physiological response to FGF-2 as a function of the stage of the oligodendrocyte lineage varies dramatically. Specifically, presentation of FGF-2 to Early Progenitors leads to the up-regulation of PDGF-a receptor and, in combination with PDGF, elicits their long-term proliferative activity (McKinnon et al., 1990; Bögler et al., 1990). FGF-2 alone is mitogenic for Late Progenitors, resulting in a reversible block to these cells entry into terminal differentiation (Gard and Pfeiffer, 1993; McKinnon et al., 1990; Mayer et al., 1993; Bansal and Pfeiffer, 1994). Finally, Maturing Oligodendrocytes treated with FGF-2 undergo substantial phenotypic changes which are discussed further below (Grinspan et al., 1993, 1996; Fressinaud et al., 1995; Bansal and Pfeiffer, 1997).

6. HIGH AFFINITY FGF RECEPTOR EXPRESSION DURING OLIGODENDROCYTE DIFFERENTIATION

In light of the variety of responses of oligodendrocytes to FGF-2 as they progress along their developmental lineage, we hypothesized that multiple FGF receptors are expressed in a stage specific manner.

To test this hypothesis, we have studied the developmental expression of the FGF receptors (Bansal et al., 1996a) and some of their known co-receptors as a function of the stages of the oligodendrocyte lineage (Bansal and Pfeiffer, 1994; Bansal et al., 1996b). In addition, we examined the regulation of the FGF receptor expression by ligand- (FGF-2) mediated regulation. We found, in support of the hypothesis, that during lineage progression multiple FGF high affinity receptor transcripts in specific alternatively spliced isoforms are expressed in a developmentally regulated manner, and that FGF-2 itself regulates the mRNA levels of these receptors at specific stages during development (Figure 1).

7. FGF CO-RECEPTOR EXPRESSION (HSPGs) DURING OLIGODENDROCYTE DIFFERENTIATION

In view of the potential for heparan sulfate proteoglycans to act as co-receptors for the FGF high affinity receptors, we hypothesized that in addition, multiple FGF co-receptors are expressed in a stage specific manner during the oligodendrocyte lineage.

Once again, consistent with the hypothesis, we have observed that multiple putative FGF co-receptors (syndecans 1–4, glypican; Bansal et al., 1996b) are also developmentally regulated in these cells (Figure 2), and are also regulated by FGF.

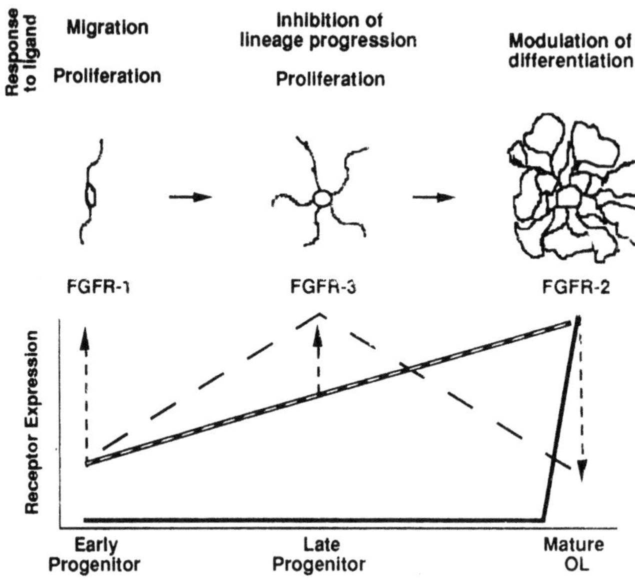

Figure 1. Model for the regulation of oligodendrocyte lineage cells by FGF receptors. Hatched line, FGF receptor-1 mRNA expression. Dashed line, FGF receptor-3 mRNA expression. Solid line, FGF receptor-2 mRNA expression Vertical dashed lines with arrowheads indicate the up- or down-regulation of FGF receptor mRNA expression by FGF-2. The "Receptor Expression" (y-axis) only compares relative mRNA levels among the three stages for a given receptor type, not the abundance of the differ receptors relative to each other.

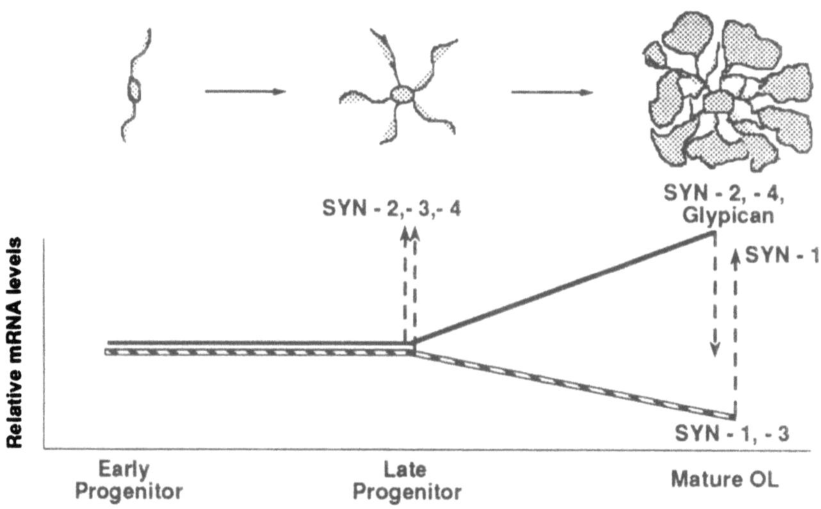

Figure 2. Summary for the developmental and FGF-2 mediated regulation of syndecans and glypican mRNA expression in the oligodendrocyte lineage. Hatched line, syndecan-1 and -3 mRNA expression. Solid line, syndecan-2 and -4 and glypican mRNA expression. Vertical dashed lines with arrowheads indicate the up- or down-regulation of syndecan and glypican mRNA expression by FGF-2.

On the basis of the data briefly summarized above, we conclude that there is a changing repertoire of these signaling molecules on the cell surface as they progress through the developmental lineage. Estimates of the various permutations and combinations of low and high affinity receptors that could be presented by oligodendrocytes as they progress through their lineage suggests that as many as 500 different combinations are theoretically possible! Obviously, even if on a fraction of these combinations are actually used, a vast variety of signaling stimuli are potentially available.

8. FGF-2 CONVERTS MATURE OLIGODENDROCYTES TO A NOVEL PHENOTYPE

Exposure of oligodendrocytes undergoing terminal differentiation to FGF-2 results in the down-regulation of myelin-specific gene expression (*e.g.*, ceramide galactosyltransferase, 2',3'-cyclic nucleotide 3'-phosphohydrolase, myelin basic protein, proteolipid protein). In concert with these changes, there is a dramatic increase in the length of the cellular processes in a time- and dose-dependent manner. In addition, the cells are induced to re-enter the cell cycle, but apparently do not undergo mitosis; they may be induced to activate a apoptotic pathway (Muir and Compston, 1996, Bansal and Pfeiffer, 1997). Finally, there is an alteration of the expression of both low- and high-affinity FGF receptors.

Compared to oligodendrocyte progenitors, the differentiated oligodendrocytes treated with FGF-2 incorporate BrdU at a slower rates, exhibit different patterns of both FGF high and low-affinity (syndecans) receptors, and are morphologically very different. In addition, they do not re-express the progenitor markers A2B5, NG2 or PDGFα receptor. Therefore, although the FGF-treated cells lose their differentiated OL/myelin markers, they do not revert to progenitors and clearly represent a different, apparently novel, phenotype both morphologically and biochemically. We have termed these cells novel or non-myelinating oligodendrocytes, or NOLs.

These data indicate that terminally differentiated oligodendrocytes retain the plasticity to re-program their differentiation fate under the influence of environmental factors. The relevance of these data to normal and pathological physiology is under consideration. In particular, we have predicted the appearance of cells in and around multiple sclerosis plaques with the phenotype O4$^+$, NG2$^-$, A2B5$^-$, O1$^-$, MBP$^-$. All of the necessary components appear to be available in situ for a similar phenomenon to occur. For example, FGF receptor-2 mRNA expression is present in the white matter tracks and mature PLP-expressing OLs (in situ hybridization) in rat brain (Asai et al., 1993; Yazaki et al., 1994; Miyake et al., 1996) and purified myelin (Western blot; our unpublished observations). Further, a number of members of the FGF family, particularly FGF-1 and -2, are expressed at several stages of brain development including the postnatal period when OLs are generated in vivo (Thomas et al., 1991; Giordano et al., 1992). Both astrocytes and neurons are a source for these FGFs (Matsuyama et al., 1992; Kuzis et al., 1995; Gonzalez et al., 1995).

FGF is up-regulated following CNS injury (Gomez-Pinilla et al., 1992; Koshinaga et al., 1993; Logan, 1988; Nieko-Sampedro et al., 1988) and in diseases such as multiple sclerosis, demyelination is accompanied by reactive gliosis (Dahl and Bignami, 1974), resulting in high levels of FGF-2 produced by the reactive astrocytes (Frank and Ragal, 1995; Gomez-Pinella et al., 1992; Liu and Chen, 1994; Lotan and Schwarz, 1992; Takami et al., 1993). In these pathological conditions OLs could respond to increases in FGF-2 at sites of injury and lesions in the brain, and acquire the NOL phenotype, leading to either

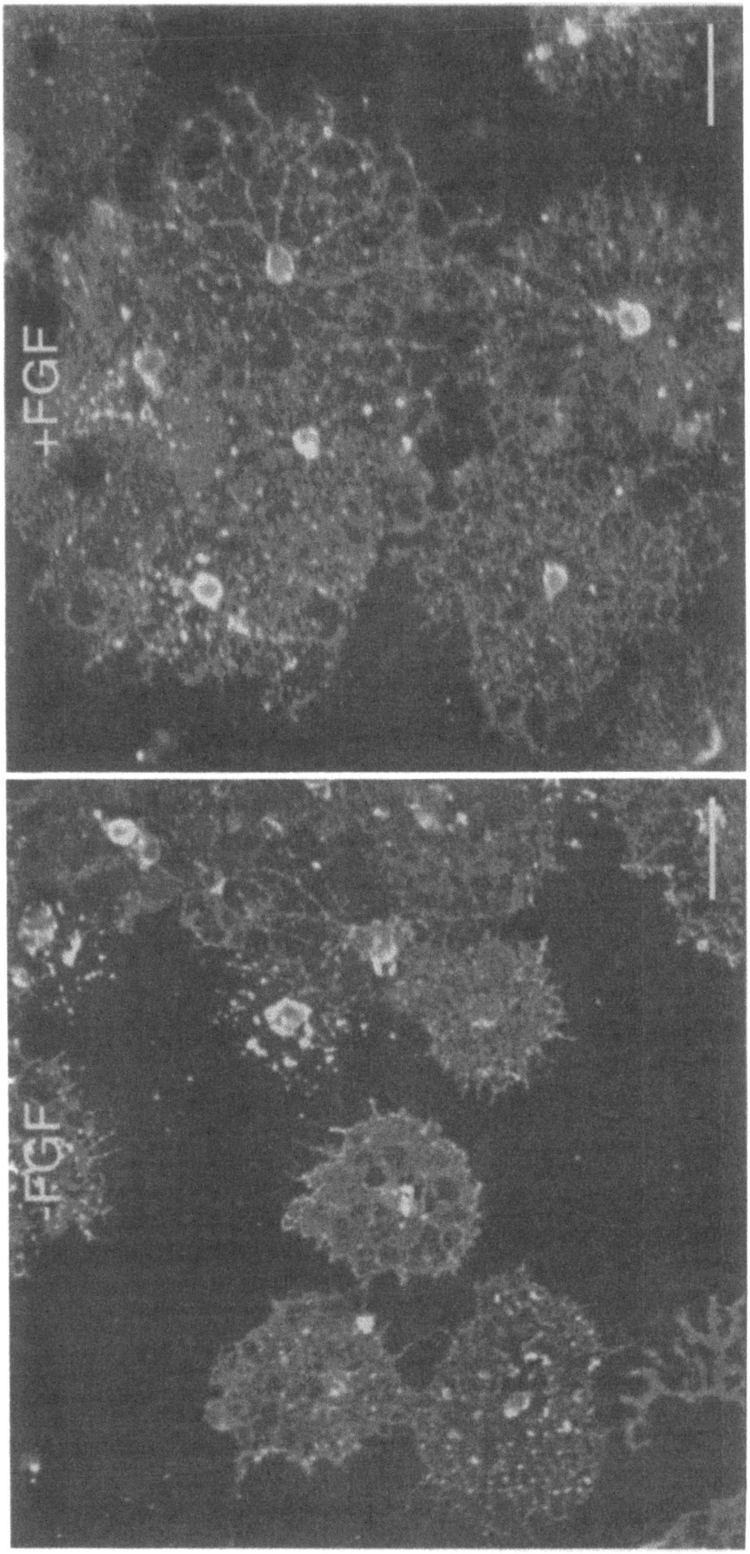

Figure 3. Immunofluorescent photomicrograph of a mature oligodendrocyte treated with FGF-2. The stimulation of extensive process outgrowth is revealed.

positive or negative consequences. That is, NOLs may be either the consequence of the damage leading to demyelination, or in contrast a potential source of cells for myelinative rehabilitation. It is important to decide which is the case so as to know if one should encourage or discourage FGF-2 production in demyelinating lesions. The clarification of these opposing hypotheses may contribute to an informed treatment designed to encourage the remyelination process.

9. PROSPECTUS

Clearly, the study of fibroblast growth factors and their receptors has reached the level of a forefront of molecular cell biology related to a plethora of areas of both normal and pathologic physiology. While this review has concentrated on the effects of FGF-2 on oligodendrocytes, it is becoming clear that many other members of the FGF family are equally important, not only for oligodendrocyte biology, but other types of cells in the nervous system as well. The rapid development of a number of genetically modified mice, including receptor knock-outs and dominant negative overexpressors, offer exciting possibilities of the further investigation for these issues.

REFERENCES

Asai T, Wanaka A, Kato H, Masana Y, Seo M, Tohyama M (1993): Differential expression of two members of FGF receptor gene family, FGFR-1 and FGFR-2 mRNA, in the adult rat central nervous system. Mol Brain Res 17:174–178.

Bansal R, Kumar M, Murray K, Morrison R, Pfeiffer S (1996a): Regulation of FGF receptors in the oligodendrocyte lineage. Mol Cell Neurosci 7:263–275.

Bansal R, Kumar M, Murray K, Pfeiffer S (1996b): Developmental and FGF-2-mediated regulation of syndecans (1–4) and glypican in oligodendrocytes. Mol Cell Neurosci 7:276–288.

Bansal R, Pfeiffer SE (1994): Inhibition of protein and lipid sulfation in oligodendrocytes blocks biological responses tot FGF-2 and retards cytoarchitectural maturation, but not developmental lineage progression. Dev Biol 162:511–524.

Bernfield M, Kokenysesi R, Kato M, Hinkes MT, Spring J, Galo RL, Lose EF (1992): Biology of the syndecans: A family of transmembrane hepanan sulfate proteoglycans. Ann Rev Cell Biol 8:365–398., 1992

Bögler O, Wren D, Barnett SC, Land H, Noble M (1990): Cooperation between two growth factors promotes extended self-renewal and inhibits differentiation of oligodendrocyte-type-2 astrocyte (O-2A) progenitor cells. Proc.Natl Acad Sci USA 87:6368–6372.

Dahl D, Bignami A (1974): Heterogeneity of the glial fibrillary protein in gliosed human brains. J Neurol Sci 23:551–563.

Eckenstein FP (1994): Fibroblast growth factors in the nervous system. J Neurobiol 11:1467–1480.

Frank E, Ragel B (1995): Cortical basic fibroblast factor expression after head injury: preliminary results. Neurol Res 17:129–131.

Fressinaud C, Laeng P, Labourdette G, Durand J, Vallat JM (1993): The proliferation of mature oligodendrocytes in vitro is stimulated by basic fibroblast growth factor and inhibited by oligodendrocyte-type 2 astrocyte precursors. Dev Biol 158:317–329.

Fressinaud C, Vallet JM, Labourdette G (1995): Basic fibroblast growth factor down-regulates myelin basic protein gene expression and alters myelin compaction of mature oligodendrocytes in vitro. J Neurosci Res 40:285–293.

Gard AL, Pfeiffer SE (1993): Glial cell mitogens bFGF and PDGF differentially regulate development of $O4^{+}GalC^{-}$ oligodendrocyte progenitors. Dev Biol 159:618–630.

Giordano S, Sherman I, Lyman W, Morrison R (1992): Multiple molecular weight forms of basic fibroblast growth factor are developmentally regulated in the central nervous system. Dev Biol 152:293–303.

Gogate N, Verma L, Zhou JM, Milward E, Rusten R, O'Connor M, Kufta C, Kim J, Hudson L, Dubois-Dalc M (1994): Plasticity in the adult human oligodendrocyte lineage. J Neurosci 14:4571–4587.

Gomez-Pinilla F, Lee JW, Cotman CW (1992): Basic FGF in adult rat brain: Cellular distribution and response to entorhinal lesion and fimbria-bornix transection. J Neurosci 12:345–355.

Gonzalez AM, Berry M, Maher PA, Logan A, Baird A (1995): A comprehensive analysis of the distribution of fgf-2 and fgf-1 in the rat brain. Brain Res 701:201–226.

Grinspan JB, Reeves MF, Coulaloglou MJ, Nathanson D, Pleasure D (1996): Re-entry into the cell cycle is required for bFGF-induced oligodendroglial dedifferentiation and survival. J Neurosci Res 46:456–464.

Grinspan JB, Stern JL, Franceschini B, Pleasure D (1993): Trophic effect of basic fibroblast growth factor (bFGF) on differentiatied oligodendroglia: A mechanism for regeneration of the oligodendroglial lineage. J Neurosci Res 36:672–690.

Johnson DE, Williams LT (1993): Structural and functional diversity in the FGF receptor multigene family. Adv Cancer Res 60:1–41.

Johnson DE, Lu J, Chen H, Werner S, Williams LT (1991): The human fibroblast growth factor receptor genes: A common structural arrangement underlies the mechanisms for generating receptor forms that differ in their third immunoglobulin domain. Mol Cell Biol 11:4627–4634.

Kjellén L, Lindahl U (1991): Proteoglycans: Structures and interactions. Ann Rev Biochem 60:443–475.

Koshinaga M, Sanon HR, Whittemore SR (1993): Altered acidic and basic fibroblast growth factor expression following spinal cord injury. Exp Neurol 120:32–48.

Kuzis K, Read S, Herry NJ, Woodward WR, Eckenstein FP (1995): Develoopmental time course of acidic and basic fibroblast growth factors expression in distinct cellular populations of teh rat central nervous system. J Comp Neurol 358:142–153.

Liu HM, Chen HH (1994): Correlation between fibroblast growth factor expression and cell proliferation in experimental brain infarct: studied with proliferating cell nuclear antigen immunohistochemistry. J Neuropathol Exper Neurol 53:118–126.

Logan A (1988): Elevation of acidic fibroblast growth factor in lesioned rat brain. Mol Cell Endocrinol 58:275–278.

Lotan M, Schwarz M (1992): Post injury changes in platelet derived growth factor - like activity in fish and rat optic nerves: possible implications in regeneration. J Neurochem 58:1637–1642.

Matsuyama A, Iwata H, Okumura N, Yoshida S, Imaizumi K, Lee Y, Shiraishi S, Shiosaka S (1992): Localization of basic fibroblast growth factor-like immunoreactivity in the brain. Brain Res 587:49–65.

Mayer M, Bögler O, Noble M (1993): The inhibition of oligodendrocytic differentiation of O-2A progenitors caused by basic fibroblast growth factor is overriden by astrocytes. Glia 8:12–19.

McKinnon RD, Matsui T, Dubois-Dalcq M, Aaronson SA (1990): FGF modulates the PDGF-driven pathway of oligodendrocyte development. Neuron 5:603–614.

Miki T, Bottaro DP, Fleming TP, Smith DL, Burgess WH, Chan AM, Aaronson SA (1992): Determionation of ligand-binding activity by alternate splicing: Two distinct growth factor receptors encoded by a single gene. Proc Natl Acad Sci USA 89:246–250.

Miyake A, Hattori Y, Ohta M, Itoh N (1996): Rate oligodendrocytes and astrocytes preferentially express fibroblast growth factor receptor-2 and -3 mRNAs. J Neurosci Res 45:534–541.

Nieko-Sampedro M, Lim R, Hicklin DJ, Cotman CW (1988): Early release of glia maturation factor and acidic fibroblast growth factor after rat brain injury. Neurosci Lett 86:361–365.

Ornitz DM, Leder P (1992): Ligand specificity and heparin dependence of fibroblast growth factor receptors 1 and 3. J. Biol Chem 23:16305–16311.

Orr-Urtreger A, Givol D, Yayon A, Yarden Y, Lonai P (1991): Developmental expression of two fibroblast growth factor receptors, flg and bek. Development 113:1419–1434.

Partanen J, Mäkelä TP, Eerola E, Korhonen J, Hirvonen H, Claesson-Welsh L, Alitalo K (1991): FGFR-4, a novel acidic fibroblast growth factor receptor with a distinct expression pattern. EMBO J 6:1347–1354.

Peters K, Werner S, Chen G, Williams LT (1992): Unique expression pattern of the FGF receptor 3 gene during mouse organogenesis. Dev Biol 155:432–430.

Pfeiffer SE, Warrington AE, Bansal R (1993): The oligodendrocyte and its many cellular processes. Trends Cell Biol 3:191–197.

Rapraeger AC, Krufka A, Olwin B (1991): Requirement of heparan sulfate for bFGF-medicated fibroblast growth and myogenic differentiation. Science 252:1705–1708.

Takami K, Kiyota Y, Iwane M, Miyamoto M, Tsukuda R, Igarashi K, Shino A, Wanaka A, Shiosaka S, Tohyama M (1993): Up-regulation of fibroblast growth factor-receptor messenger RNA expression in rat brain following transient forebrain ischemia. Exp Brain Res 97:185–194.

Thomas D, Caruelle J-P, Lachapelle F, Barritault D, Boully B (1991): Acidic fibroblast growth factor in normal, injured and jimpy mutant developing mouse brain. Ann NY Acad Sci 648:161–166.

Warrington AE, Pfeiffer SE (1992): Proliferation and differentiation of O4+ oligodendrocytes in postnatal rat cerebellum: analysis in unfixed tissue slices using anti-glycolipid antibodies. J Neurosci Res 33:338–353.

Warrington AE, Barbaresc E,Pfeifer SE (1993): Differential myelingoenic capacity of specfic developmental stages of the oligodendrocyte lineage upon transplantation into hypomyelinating hosts. J Neurosci Res 34:1–13.

Werner S, Duan D-S, DeVries C, Peters KG, Johnson DE, Williams LT (1992): Differential splicing in the extracellular region of fibroblast growth factor receptor 1 generates receptor variants with differnt ligand-binding specificities Mol Cell Biol 12:82–88.

Yayon A, Klagsbrun M, Esko JD, Leder P, Ornitz DM (1991): Cell surface, heparin-like molecules are required for biding of basic fibroblast growth factor to its high affinity receptor. Cell 64:841–848.

Yazaki N, Hosoi Y, Kawabata K, Miyake A, Minami M, Satoh M, Ohta M, Kawasaki T, Itoh N (1994): Differential expression patterns of mRNAs for members of the fibroblast growth factor receptor family, FGFR-1-FGFR-4, in rat brain. J Neurosci Res 37:445–452.

FEATURES AND FUNCTIONS OF HUMAN MICROGLIA CELLS

Fulvia Gremo,* Valeria Sogos, Maria Grazia Ennas, Alessandra Meloni, Tiziana Persichini, Marco Colasanti, and Giuliana Maria Lauro

Department of Cytomorphology
School of Medicine, Cagliari, Italy
Department of Cell Biology
III University, Roma, Italy

1. INTRODUCTION

Microglia represent a population of resident cells of the central nervous system (CNS) which have attracted much interest because of their multiple functions. Two principal forms of microglia have been described. "Ameboid" cells, also referred as gitter cells or reactive microglia, which appear during the embryonal period and after brain lesions, are morphologically similar to macrophages (Giulian and Baker, 1986; Giulian, 1987). The second form of microglia, the "ramified" cells (Ling, 1981), appear during the late postnatal period and persist through adult life (Murabe and Sano, 1983). Although the nature of the relationship between ameboid and ramified microglia is controversial, ramified microglia are generally viewed as functional quiescent microglia that lack monocytic properties, but acquire ameboid features and reactivity in the presence of tissue damage (del Rio Hortega, 1919; 1932; Oemichen, 1983).

From a functional point of view, microglia can be seen as "brain macrophages": they interact with cells of the immune system, play a key role in immune and inflammatory reactions, (Fujita et al., 1981; Dolman, 1985; Frei et al., 1987) and may also be the target of several pathogenetic agents, such as, for example, human immunodeficiency virus type 1 (HIV-1) (Watkins et al., 1990). On the other hand, a role of microglia in neural tissue development and remodelling has been suggested (for a review see Thomas, 1992).

In view of these observations, it is of great importance to acquire extensive knowledge of microglia features and functions in the developing human brain. Not many data are available. Hutchins et al. (1990) described the localisation of different forms of microglia in the whole brain of 13–25 week old fetuses and showed that microglia cells were al-

* Send correspondence to: Fulvia Gremo, M.D., Professor, Department of Cytomorphology, Via Porcell 2, 09124 Cagliari, Italy. Tel. +39–70–6758914/8914; Fax +39–70–657972.

ready detectable at the 13th week of gestation. However, in tissue sections, due to the strict relationship between microglia and other glial cells, it is sometimes difficult to distinguish their fine morphological and histochemical features. Thus, more extensive information can be acquired using human fetal microglia tissue cultures as a tool. In 1991 Hassan et al. described microglia-enriched, mixed cultures from human brain of 14–18 week old fetuses. Then, we first demonstrated that pure microglia cells could be obtained from long-term human fetal astrocytic cultures (Lauro et al., 1992). Later on, several authors, including us, used in vitro microglia cultures to study their physiological and pathological functions.

In this chapter, a brief review of this work will be presented.

2. CELL MORPHOLOGY AND CHARACTERISATION

We obtained pure microglia cells from long-term human astrocytic cultures. We initially developed human fetal brain cultures which contained only neurons, astrocytes and their precursors (Presta et al., 1990; Sogos et al., 1990; Torelli et al., 1991). With time in vitro (about 3 months) and cell passages, neurons were lost (Torelli et al., 1991). From approximately the 15th week on, a few round-shaped cells growing on the top of the thick astrocyte layer could randomly be observed (fig. 1a). These cells stained weakly for Fc receptor, but their percentage remained under 2%. None of the other typical microglial markers were present (Table I), except M1 and M5 on glial-fibrillar-acidic protein (GFA-P) positive astrocytes, as already shown (Ennas et al., 1992). After approximately 5 months, pleomorphic, phase- dark cells started to appear in increasing number on the surface of astroglia (fig. 1b). In some areas where the phase-dark cells were found, the astroglia tended to retract (fig. 1c). The above mentioned processes were accelerated by cell passages, so that in two weeks the whole culture was populated by macrophage-like cells. At the same time, the astroglia formed clumps detaching from the substratum and was washed out from the cultures. Immunohistochemical characterisation performed after the completion of the above mentioned phenomena demonstrated the presence of cells positive for all the typical microglial markers, except a few GFA-P^+ cells, lost in the following passages (Table I). After the 4th passage, cells showed the typical "ameboid" morphology, i.e. most of the cells looked round-shaped with two or more short spinous processes arising from the cell body. However, some large, flat cells bearing short processes were observed tightly adhering to the substratum. Flat cells were weakly stained for the metabolic marker and slow in phagocyting. (fig. 2). Through cell passages, the percentage of round-shaped cells decreased so that at the 30th most of the cells were ramified with long and thin processes; finally, at the 50th they represented over 95% of the cell population.

Phagocytosis was shown at both light and electron microscope. At E.M. numerous latex beads could be observed in the cell cytoplasm. Particles were often surrounded by a membrane (fig. 3) and sometimes fused with lysosomes in order to produce large digesting bodies (Lauro et al., 1995).

Figure 1. (a) Scanning electron microscope micrograph of a microglia cell which might randomly be observed on the top of human astrocyte layer after a few months *in vitro*. X 7,300. (b) Light microscope picture of a long-term astrocytic culture where groups of pleomorphic, phase-dark cells (i.e. microglia) could be observed. X 100. (c) After a few cell passages, almost all astroglia formed clumps detaching from the substratum and only microglia could be found (arrows). X 100.

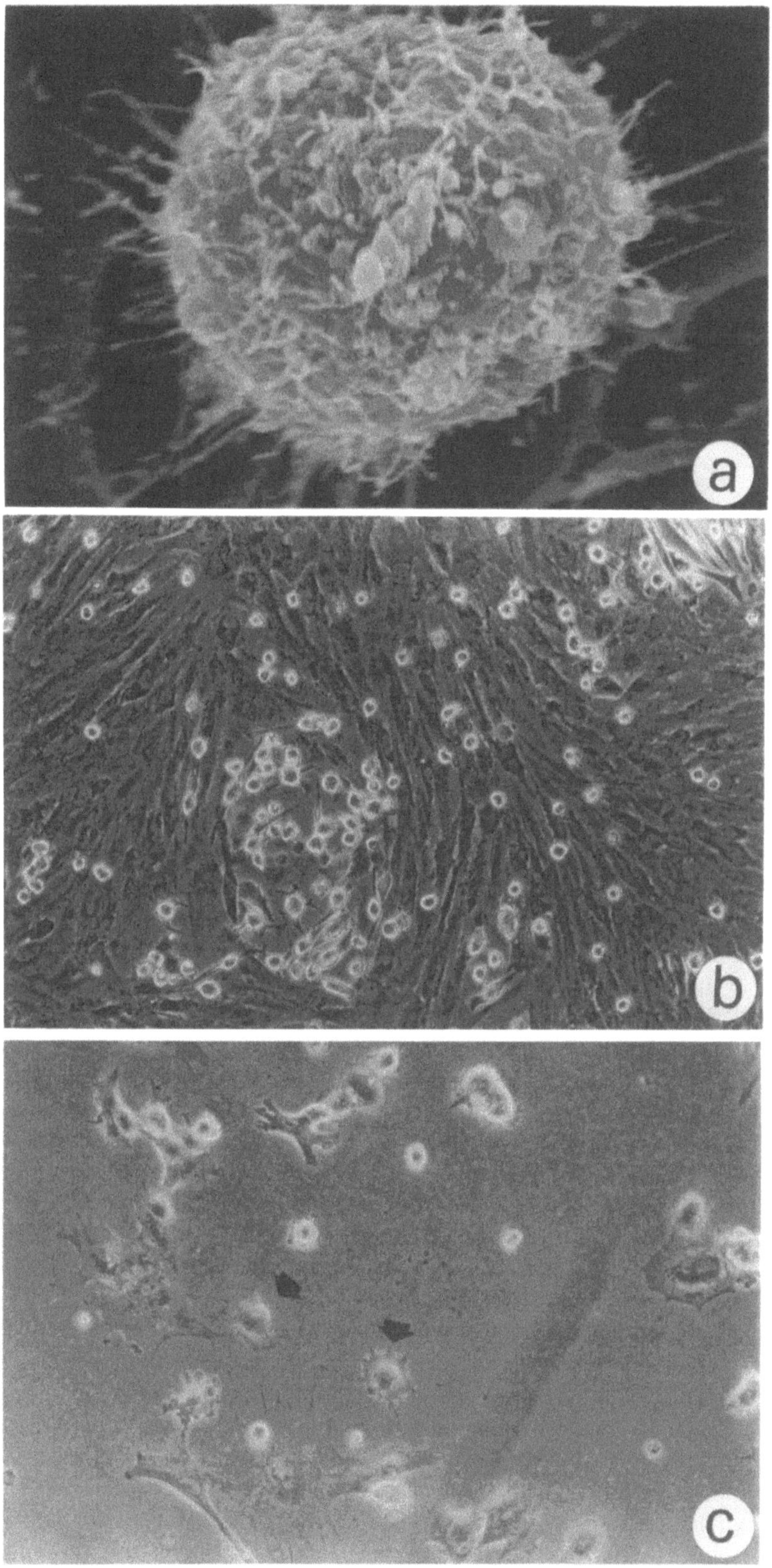
a
b
c

Table I. Immunocytohemical characterisation of human fetal brain cultures at day 150 *in vitro* (GF) and of human microglia (WILL) at the 4th passage *in vitro* (day 28)

Markers	GF	Will
$\beta_2\mu$	+	+
GFAP	+	-(<2%)
S100	+	–
FcR	-(<2%)	+
LYS	–	+
M1	+	+
M5	+	+
GSA	–	+
RCA	–	+
α_1	–	+
NF	–	–
GalC	–	–
VIII	–	–

$\beta_2\mu$: beta2 microglobulin; GFAP: glial-fibrillar-acidic protein; FcR: Fc receptor; LYS: lysozime; GSA: Griffonia semplicifolia agglutinin; RCA: Ricinus communis agglutinin; α1: α1-trypsin and α1-chymotrypsin; NF: neurofilament; GalC: galactocerebroside; VIII: factor VIII.

These microglia cells actively and spontaneously proliferated. Regardless of cell passage and type of assay used, both *early* (up to passage 30) and *late* (from 30th to 80th passage) microglia had the same growth rate: duplication time was 24 hours in the first two days, then plateau was reached between 72 and 96 hours (Lauro et al., 1995). In order to identify cell type/s involved in proliferation, cells were assayed for 5-bromo-2- deoxyuridine incorporation. Results demonstrated the presence of nuclear staining for 5BrdU in several RCA-positive cells (fig. 4) Our results are in line with the findings of other authors. Mennerick et al. (1994) have shown the presence of RCA binding sites on human microglia cells and Ulvestad et al. (1994a) demonstrated that they were negative for monocyte markers such as CD14 and RFD7, but positive for CD11C and CD68. Hassan et al. (1991) reported that after 12 weeks in vitro human microglial cells seemed to arise from underneath the astrocyte cell layer, stained for non-specific esterase and could be maintained in culture for many weeks. Similar observations on newborn mouse brain astroglial cultures have been reported by Fedoroff and Hao (1991) who suggested that astrocytes could regulate microglia precursor differentiation and survival through the production of colony stimulating factor-1. Recently, Sievers et al. (1994) demonstrated the in vitro induction of ramified microglial shape by astrocytes.

Interestingly, we found that our cells spontaneously proliferate in vitro, without any specific stimulation. However, these cells are sensitive to both macrophage colony stimulating factor (M-CSF) and granulocyte-macrophage colony stimulating factor (GM-CSF) (Lee et al., 1994). Moreover, M-CSF can mediate astrocyte-induced microglia ramification in human fetal brain cultures (Liu et al., 1994).

All these data, taken together, demonstrate that human microglial cells can be obtained at high degree of purity in culture, that they have strict similarities, but also differences with respect to macrophages, as suggested by Giulian et al. (1995). Moreover, they

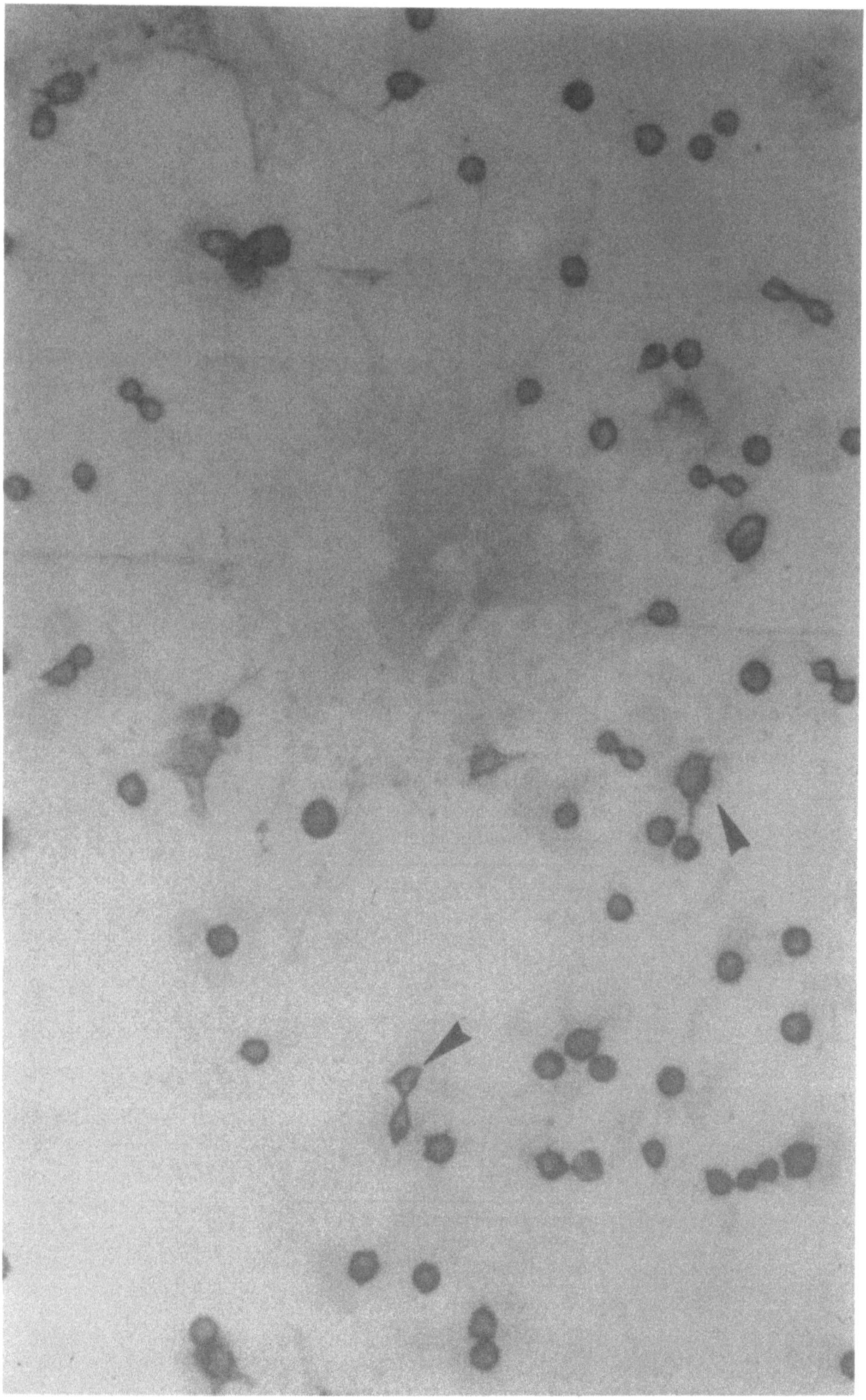

Figure 2. Representative picture of a pure microglia culture containing both ameboid (arrowheads) and large, flat, substratum adhering cells stained with RCA lectin. X 400.

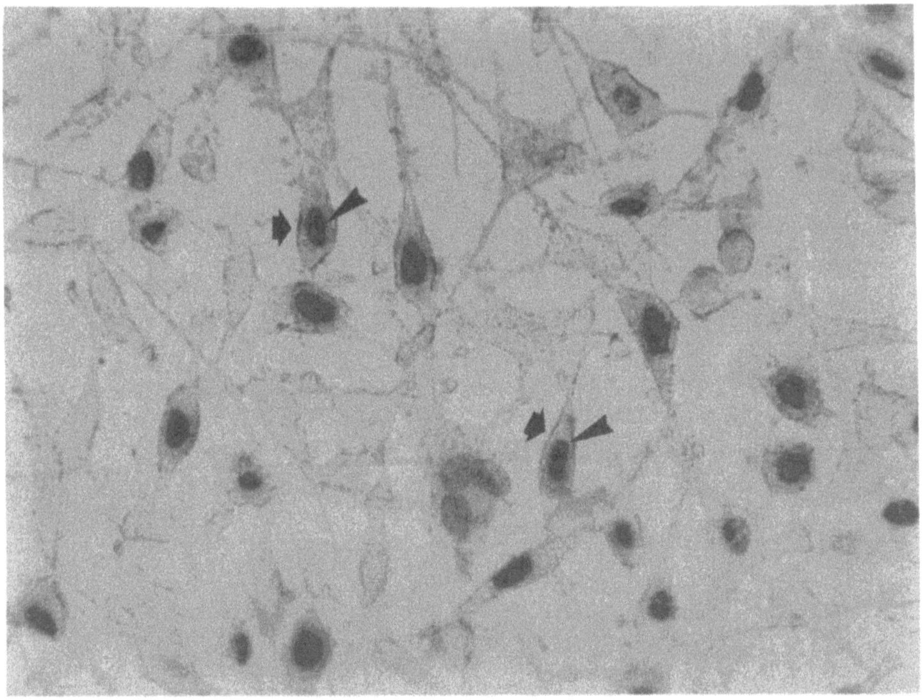

Figure 3. Immunocytochemistry of 5-BrdU in microglia cultures. Most of the nuclei of RCA-positive cells (arrows) are heavily stained with anti 5-BrdU antibody (arrowheads). X 400.

Table II. Immunocytochemical characterisation of human fetal microglia cultures

Marker	Percentage
FcR	98 ± 5
LYS	98 ± 1
VIM	98 ± 3
FN	95 ± 2
RCA-1	94 ± 5
GSA-B4	97 ± 2
TRYP	93 ± 7
CHYM	98 ± 2
M1	93 ± 5
M5	99 ± 5
NSE	98 ± 1
CD_4	4% or less

Percentages of positive cells were calculated on 5,000 cells counted in 20 randomly chosen microscope fields. Positivity ranged from weak to intense.

FcR: Fc receptor; LYS: lysozime: VIM: vimentin; FN: fibronectin; RCA-1: Ricinus communis agglutinin; GSAB4: Griffonia semplicifolia agglutinin; TRYP: α1-trypsin; CHIM: α1-chymotrypsin; NSE: non-specific esterase.

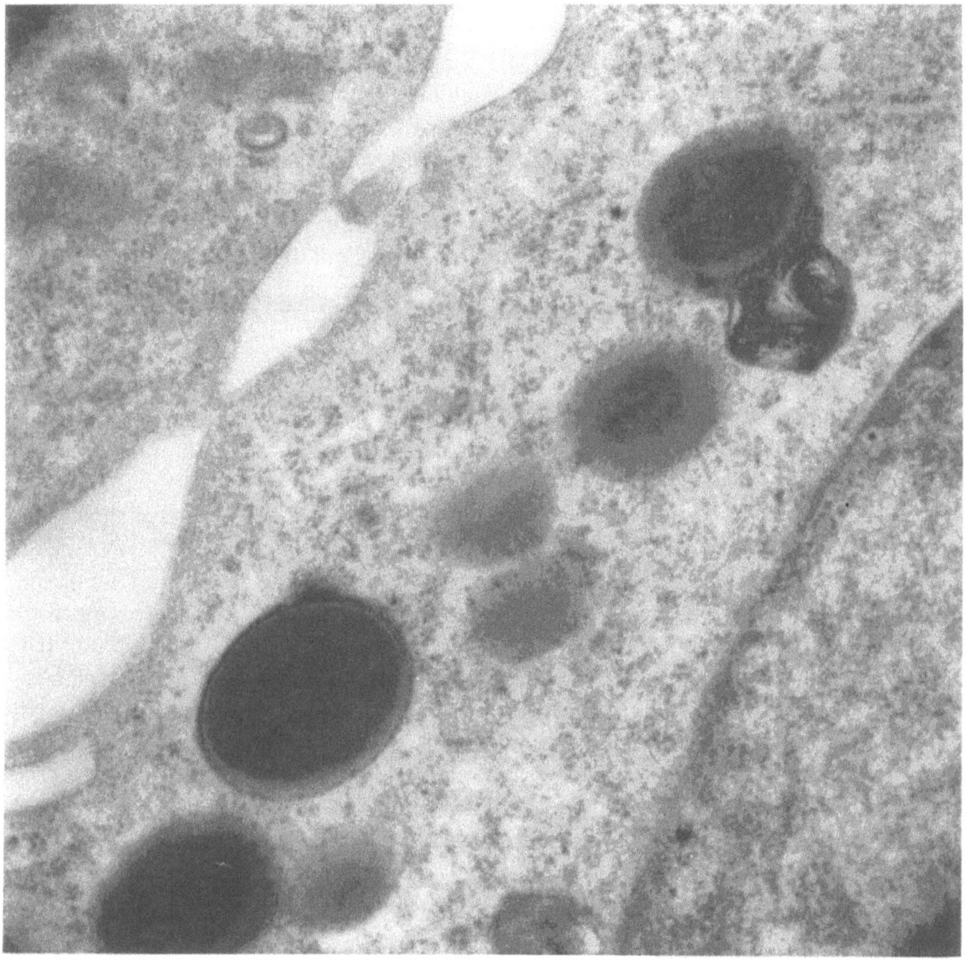

Figure 4. SEM micrograph of a phagocyting microglia cell. The cytoplasm is filled with membrane-bound latex particles. X 42,000.

strongly suggest that a direct influence of astrocytes on microglia occurs in brain so that microglia function(s) might be influenced by macroglia.

3. MICROGLIA AND IMMUNE FUNCTIONS

Microglia are believed to be the most important cell type to be involved in brain pathological dysfunctions. As an example, microglia can be not only the target, but also activated by HIV-I in infected brain (Weis et al., 1994). Microglia are certainly involved in the pathogenesis of Alzheimer's disease (Carpenter et al., 1993; Akiyama, 1994; McGeer et al., 1994), as well as in epilepsy (Sheng et al., 1994). However, the mechanism(s) underlying the role they play in brain immune functions still needs to be elucidated. Thus, we first studied the induction of immunohistocompatibility-complex (MHC) antigens in fetal microglia cultures.

3.1. MHC Antigen Expression and Induction

Major histocompatibility complex (MHC) class I and II antigens were detected analysing by FACS the cell surface expression of stabilised molecules and by ELISA both the surface expression and the cytoplasmic molecules. Analysis was performed in basal conditions and after treatment with the immunomodulant cytokine gamma-interferon (IFN), in *early* and *late* microglia. Under basal conditions (zero U/ml of gamma-IFN), cells never expressed HLA A-B-C antigens, but were weakly positive for HLA DP-DQ-DR. Induction of HLA A-B-C was observed only in *early* at 100–1000 U/ml of gamma-IFN while no induction could be detected in *late*. HLA DP-DQ-DR antigen production significantly arose in *early* after 10, 100 and 1000 U/ml of gamma-IFN administration. Similarly to HLA A-B-C, even the highest concentration of interferon failed to increase HLA DP-DQ-DR in *late*, which were totally unresponsive.

Beta2 microglobulin was measured after 4 days of incubation in serum free medium. Results showed no correlation with HLA A-B-C molecule expression both in *early* and in *late* (Lauro et al., 1995).

These data are in accord with several authors. Hassan et al. (1991) have shown that human microglia, as murine microglia (Giulian, 1987; Suzumura et al., 1987; Woodroofe et al., 1989) expressed HLA-DR under basal conditions and after cytokine stimulation, further supporting the concept that microglia are involved in CNS immune response for which MHC expression is necessary. More recently, Lee et al. (1992) also observed MHC class II expression in human microglia enriched cultures which was enhanced by gamma-interferon. These data were confirmed by Gehrmann et al. (1993) who examined HLA-DR expression immunocytochemically in different areas of adult human brain. They showed a strong constitutive HLA-DR expression on ramified microglia. Also Ulvestad et al. (1994a, b) showed HLA-DR-DP-DQ expression in human microglia which was upregulated after stimulation with interferon-gamma.

3.2. Production of Inflammatory Molecules by Human Microglia

Microglia has been shown to produce several inflammatory molecules such as interleukin-1 (Hetier et al., 1988) and -6 (Frei et al., 1989b) and α-tumor necrosis factor (α-TNF) (Sawada et al., 1989; Hetier et al., 1990). Chao et al. (1995) have shown that human fetal microglia release abundant amount of α-TNF upon stimulation with lipopolysaccharide (LPS). The release was regulated by other cytokines such as IL-6, IL-10 or transforming growth-factor. On the other hand, Sheng et al. (1995) have shown that α-TNF upregulated human microglial cell production of IL-10 in vitro. Cytokine production by adult human microglial cells is induced by human T cell lymphotrophic virus type I Tax protein (Dhiv-Jalbut et al., 1994).

Among the numerous inflammatory molecules, we addressed the question of the in vitro production of platelet activating factor (PAF), a lipid molecule involved in inflammatory processes and in cell-cell communication. PAF is released from activated inflammatory cells such as neutrophils, monocytes, platelets and endothelial cells (Braquet et al., 1987), as well as by tissues such as spleen, liver, kidney and heart (Snyder, 1985).

In the central nervous system (CNS) PAF has been shown to produce a wide variety of effects (for a review see Yue and Feuerstein, 1994; Bussolino et al., 1994). There is growing evidence that PAF is a mediator of neuro-injury caused by stroke and trauma (for a review see Frerichs and Feuerstein, 1990). Moreover, PAF exerts cytotoxic effects on neuronal cells (Kornechi and Ehrlich, 1988), causes cerebral vasoconstriction (Kochanek

Table III. PAF production in microglial cultures

Control	0.487 ± 0.35
IFN	0.2 ± 0.22
PDGF	0.325 ± 0.175
TNF	2.4 ±1.3*
LPS	1.91 ± 1.1**
bFGF	0.90 ± 0.40
TPA	0.62 ± 0.30

Data are expressed as ng PAF/100 μg phosphor (Pi) and represent the mean of three different experiments ± SD from different cultures. Data on supernatants are omitted, since they are close to 0.
**p = 0.004 vs control; *p = 0.02 vs control.
IFN: gamma interferon; PDGF: platelet derived growth factor; TNF: tumour necrosis factor-α; LPS: lipopolysaccharide; bFGF: basic fibroblast growth factor; TPA: phorbol 12-myristate 13-acetate.

et al., 1988), is a mediator of bacterial meningitis (Cabellos et al., 1992) and might play a role in chronic CNS diseases (Brochet et al., 1992). Though CNS cells are certainly targets of PAF (Junier et al. 1988, Marcheselli et al., 1990; Bito et al., 1992), the cellular origin of PAF in brain is still unclear. Neurons have been shown to produce PAF in vitro either spontaneously (Yue et al., 1990) or after specific stimulation (Sogos et al., 1990), whereas purified astrocytic cultures were incapable of PAF synthesis (Sogos et al., unpublished observations). Thus, we attempted to study the production of PAF by pure microglial cultures from human fetal brain.

Results on PAF production are summarised in Table III. When cultures were assayed for PAF production without any challenge, the amount found both in cells and in the supernatant was extremely low, showing the absence of PAF production in unstimulated microglia. On the contrary, when cells were treated with different compounds, only LPS and TNFα significantly increased PAF levels into the cells (respectively $P < 0.02$ and $P < 0.004$). Interestingly, a cytokine such as gamma-interferon, which generally affects microglia (Frei and Fontana, 1989a; Loughlin et al., 1992;), was unable to induce a response, as well as bFGF, which is synthesised in microglia cells (Sawada et al., 1989; Araujo and Cotman, 1992; Presta et al., 1995) or PDGF. Noteworthy is the fact that PAF was not released by microglia, as already shown in other systems (Bussolino et al., 1986; Sogos et al., 1990). Both TNFα and LPS induced PAF production in a concentration- and time-dependent manner. Thus, human microglia are involved in a variety of events related to brain diseases, both through the interaction with the cells of the immune system and via production of molecules of the inflammatory response.

4. MICROGLIA AND THE HISTOGENESIS OF THE NERVOUS SYSTEM

Microglia have been suggested to be involved in the histogenesis of the nervous system via the phagocytic removal of the dead cells occurring through the routine cell death associated with development (Hume et al., 1983; Linden et al., 1986) and for selective elimination of certain axonal projections (Ferrer et al., 1990; Killackew, 1984). Moreover they have been shown to secrete neurotoxic factors (Giulian et al., 1990; Théry et al., 1991). On the other hand, microglia can accumulate nerve growth factor, thus exerting an opposite effect (Gilad and Gilad, 1995). We consequently studied the production of trophic factors and, in particular, of bFGF (Presta et al., 1995). bFGF can stimulate neurite outgrowth (Morrison et al., 1986; Walicke et al., 1986) and exerts trophic functions on neurons in vitro and in vivo (Anderson et al., 1988; Otto et al., 1987; Unsicker et al., 1987). It also induces cell proliferation in astroglial cells (Rogister et al., 1988) and in oligodendrocytes (Saneto and de Vellis, 1985). Thus, the hypothesis can be advanced that bFGF produced by cells of the developing CNS plays a role in normal histogenesis of the human brain. We had previously shown that bFGF was detectable in human fetal brain (Dell'Era et al., 1990) and it was produced by human fetal neurons in vitro (Torelli et al. 1990): Thus, it was of importance to detect whether or not microglia could be a source as well as a target of bFGF.

4.1. Basic FGF Is Present in Human Fetal Microglia

The ability of bFGF to stimulate plasminogen activator (PA) production in endothelial cells has been used as a specific assay for the identification, characterisation and purification of this factor from different tissues and cell types, including adult rat brain, human fetal brain (Dell'Era et al., 1990; Presta et al., 1988) and human fetal neurons (Torelli et al., 1990). Therefore, to evaluate the presence of bFGF-like activity in human fetal microglia cultures, extracts of microglia cells were assayed for their ability to stimulate PA production in endothelial GM 7373 cells. Microglial cell extracts stimulated PA production in a dose-dependent manner, with a half-maximal stimulation at 50–100 μg of protein/ml. Equivalent concentrations of BSA did not affect PA production in GM 7373 cells. No significant difference in the amount of PA-inducing activity present in microglia extracts was observed, regardless the age of the embryo from which cells were obtained (11–15 weeks of gestation) and in vitro passages (from 5 to 40).

To confirm that the PA-inducing activity present in human fetal microglia cultures was due to bFGF, we took advantage of the very high affinity of bFGF for heparin. Indeed, both aFGF and bFGF bind to heparin-Sepharose beads, but a FGF is eluted from the resin with 0.8–1.0 M NaCl, while bFGF is eluted with 1.4–1.6 M NaCl (Lobb et al., 1986). When microglial cell extract was loaded onto a heparin-Sepharose column and the column was washed sequentially with 0,15 M, 0,8 M and 2.0 M NaCl, all the PA-stimulating activity bound to the resin and was eluted with the 2.0 M NaCl wash. Moreover, the PA-inducing activity present in the 2.0 M NaCl eluate was completely quenched by incubation of the sample with neutralising anti-bFGF antibodies, while irrelevant IgG were ineffective. The same anti-bFGF antibodies identified a major Mr 18,000 immunoreactive band when the heparin-binding protein fraction of the microglia extract was probed in a Western blot. A minor Mr 24,000 immunoreactive protein was also observed in this fraction, indicating that trace amounts of high molecular weight bFGF (Florkiewicz et al., 1991) were present in human fetal microglia.

4.2. FGF Receptors Are Present in Human Fetal Microglia

Basic FGF is thought to play an autocrine and/or paracrine role for different cell types (for a review see Basilico and Moscatelli, 1992). This prompted us to investigate for the presence of FGF receptors in bFGF-producing microglia cultures (Presta et al., 1995). Two classes of FGF receptors have been identified. Low affinity binding sites represent heparan sulphate proteoglycans present both on the cell surface and in the extracellular matrix; high affinity binding sites correspond to tyrosine-kinase transmembrane receptors (Basilico and Moscatelli, 1992). Low affinity sites are involved in modulating the interaction of bFGF with its high affinity sites (Yayon et al., 1991) and may play a role in mediating internalisation of bFGF within the cell (Roghani and Moscatelli, 1992: Watkins et al., 1990). Tyrosine- kinase receptors are responsible for intracellular signal transduction and mediate the biological response of the cell to bFGF (Mohammadi et al., 1992; Peters et al., 1992).

The presence of low and high affinity bFGF binding sites in human fetal microglia cultures was evaluated through binding with ^{125}I-bFGF. As already observed for different cell types (Mohammadi et al., 1992; Peters et al., 1992; Watkins et al., 1990) binding of ^{125}I-bFGF to low and high affinity sites was inhibited by addition of soluble heparin to the binding medium. The glycosaminoglycan was more efficient in inhibiting the binding of the growth factor to low affinity sites (ID_{50} = 10 ng/ml) than to high affinity sites (ID_{59} = 3–10 μg/ml). Scatchard plot analysis performed on several preparations of human fetal microglia at different times in culture demonstrated the presence of 10,300 ± 2,500 (mean ± S.E.M.) high affinity sites per cell (Kd = 30 ± 9 pM) and of 380,000 ± 60,000 low affinity sites per cell (Kd = 730 ± 200 nM). Even though some variability was observed among the different microglia preparations, no relationship was observed in the number and/or affinity constant of low and high affinity sites with respect to time in culture or fetal age. Finally, cross-linking of ^{125}I-bFGF to the cell surface of human microglia cultures revealed the presence of a bFGF/receptor complex with molecular weight of 150,000. This indicates a molecular weight for microglial tyrosine kinase FGF receptors of approximately 130,000, in keeping with previous observations on different cell types (Johnson and Williams, 1993) In conclusion, human fetal microglia cultures express both low affinity binding sites (i.e. heparan sulphate proteoglycans) and high affinity binding site (i.e. tyrosine-kinase receptors) for ^{125}I-bFGF.

4.3. Different FGF Receptor Transcripts Are Expressed by Microglia

Four distinct FGF receptor genes have been identified so far (FGFR-1,-2,-3 and -4) and multiple transcripts can be generated by alternative splicing (for a review see Johnson and Williams, 1993). As a preliminary characterisation of the type(s) of FGF receptors present in human fetal microglia, we performed Northern blot analysis of total RNA extracted from microglia cells which was hybridised with cDNA probes for FGFR-1/flg, FGFR-2/bek, FGFR-3 and FGFR-4. Human fetal microglial cells expressed all the four types of FGF receptors, even if to a different extent (Balaci et al., 1994).

4.4. bFGF and FGF Receptor-1/flg Are Co-Localised in Microglial Cells

In order to evaluate the possibility of a coexpression of bFGF and FGF receptor-1/flg, double staining for bFGF and FGFR-1 was performed on microglia. Immunoreactivity for bFGF was widely distributed in the cytoplasm of almost 100% of cells. In keeping with the nuclear localisation of high molecular weight bFGF (Florkiewicz et al., 1991) and its presence in trace amounts in microglia extracts, bFGF staining was also detectable in the nucleus of some cells. Surface staining for FGFR-1/flg was more frequently associated with round-shaped ameboid microglia (Presta et al., 1995).

In conclusion, our data support the hypothesis of a trophic role of microglia during the development as well as the regeneration processes in the human brain.

5. MICROGLIAL AND NEUROTOXICITY

5.1. Microglial Production of Nitric Oxide

Recently, a reactive radical, i.e. nitric oxide (NO), has been added to the list of cytotoxic agents produced by microglia. NO is an unstable nitrogen radical which is generated in different cell types by the concomitant conversion of L-arginine into citrulline through the enzyme NO synthase (NOS) (Moncada et al., 1991). There are at least three distinct isoforms of NOS present in human cells. Two enzymes, the neuronal and the endothelial Ca^{2+}-dependent isoforms (NOS-I and NOS-III, respectively), are constantly expressed and termed constitutive NOS (cNOS). The third enzyme is an inducible Ca^{2+}-independent isoform (NOS-II or iNOS), which is expressed after stimulation with E. coli LPS and/or cytokines, such as interferon-gamma (IFN-gamma), interleukin-1β (IL-β), or tumor necrosis factor-α (TNAα). The induction of human iNOS occurs at the transcriptional level and is mediated by the early activation of some nuclear factors such as NF-kB (Goldring et al., 1995). NO generated at low levels by a cNOS plays an important role in physiological processes, whereas uncontrolled and massive NOS-II-induced NO production can be potentially detrimental to tissue integrity (Suzuki et al., 1995).

Although the induction of an inducible NO pathway in murine microglia has been extensively reported, its presence in human microglial cells is still controversial (Denis, 1994). By using human microglial primary cultures as well as immature human monocytes, some authors observed that cells failed to produce NO in response to various stimuli (Lee et al., 1993b; Peterson et al., 1994). However, NO pathway seems to appear only following maturation of human monocytes into macrophages (Reiling et al., 1994; Martin et al., 1993; Pietraforte et al., 1994). Thus, it may be hypothesised that, despite some differences between microglia and peripheral blood monocyte/macrophages, in human microglia also, NO production is linked to in vitro cell maturation.

Using ramified cells, we have recently examined the presence of L-arginine-NO pathway in human microglia cells after incubation with cytokines and/or LPS. Stimulation of ramified cells for 24 h with LPS and/or TNFα caused a marked enhancement of NO, as measured by nitrite release (Fig. 5). Using reverse transcriptase-polymerase chain reaction (RT-PCR) and Southern blot analysis, we observed that LPS/TNFα produced an increase of NOS-II mRNA expression, the maximal effect being achieved after an 8-h treatment (Colasanti et al., 1995a).

5.2. Regulation of NO Synthesis in Human Ramified Microglial Cells

NO release by microglia plays a pathogenetic role in neuronal cell death during neurodegenerative disorders, including Alzheimer's and Huntington's diseases, multiple sclerosis and AIDS dementia (Zielasek et al., 1996). Thus, preserving NOS-II gene from its undesirable induction may be important for neuronal survival. In fact, NO-induced neurotoxicity can be attenuated by treatment with peptide growth factors, such as basic fibroblast growth factor (bFGF) (Maiese et al., 1993). Recently, we have investigated the effect of bFGF on NO synthesis in human ramified microglial cells. By measuring nitrite,

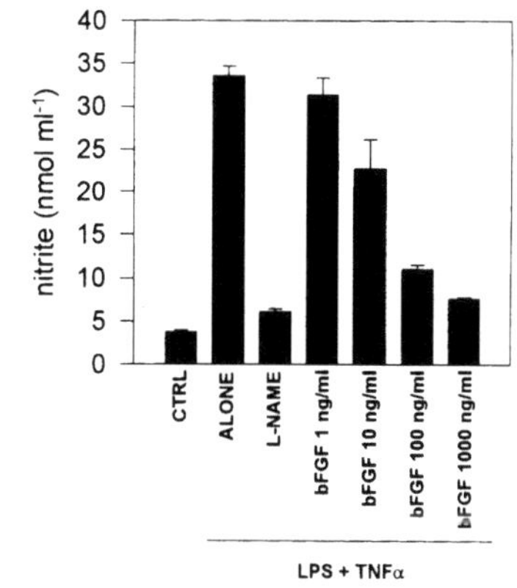

Figure 5. Determination of nitrite (NO) levels in supernatant of 5 x 10^5 human microglial cells. NO was determined by using the Griess reagent (1% sulfanilamide/0.1% naphthylenediamine dihydrochloride/2.5% H_3PO_4. Absorbance was meaured at 540 nm and NO concentration was determined using sodium nitrite as a standard. Treatment with LPS (1 mg ml^{-1}) + TNFα (500 U ml^{-1}) for 24 h increased no levels in microglial cell supernatants. This effect was abolished by coincubation with the NOS inhibitor N-nitro-L-arginine methyl ester (L-NAME 500 mM). bFGF (0.1–1000 ng/ml) was able to prevent dose-dependent LPS/TNFα-induced NO production. The maximal effect was observed at 100 ng/ml of bFGF. Bars represent the mean ± s.e.m. of 6 experiments.

the breakdown product of NO, we found that bFGF dose-dependently inhibited NO release in LPS/TNFα-treated cell supernatants (Fig. 5). The early presence of bFGF during LPS/TNFα induction was essential for inhibition of NO production, suggesting a transcriptional regulation of NOS-II gene expression. Using RT-PCR and Southern blot analysis, we verified that bFGF prevented the NOS-II gene expression as induced by LPS/TNFα (Colasanti et al., 1995b).

Thus, human ramified microglia cells failed to produce NO in the presence of bFGF. It is very intriguing to note the strong correlation between NO pathway and cell maturation state. In fact, bFGF was also able to transform human ramified microglia into ameboid cells. In particular, when treated with bFGF, ramified cells showed morphological changes, first, the cytoplasm became rounded and they formed short processes; second, they proliferated and were functionally activated. Removal of bFGF from the culture medium caused microglia to revert gradually to a ramified phenotype (Di Pucchio et al., 1996).

Down-regulation of NOS-II expression was reported to be achieved by many factors, including NO itself. However, little is known about the regulatory effects on the mechanism by the variable low concentrations of the available NO before iNOS induction. Recently, we have report that pretreatment of human ramified microglia cells with nearly physiological levels of exogenous NO prevented LPS/TNFα-inducible NO synthesis, because it inhibited NO-II mRNA expression by affecting NF-kB activation. Using RT-PCR, we observed that both NO donor sodium nitroprusside (SNP) and authentic NO solution was able to inhibit LPS/TNFα-inducible NOS-II gene expression, this effect being reversed by reduced haemoglobin, a trapping agent for NO. The early presence of SNP during LPS/TNFα induction was essential for inhibition of iNOS mRNA expression (Fig. 6A). Furthermore, SNP is capable of inhibiting LPS/TNFα-inducible nitrite release, as determined by Griess reaction. Finally, using electrophoretic mobility shift assay, we found that SNP inhibits LPS/TNFα-elicited NF-kB activation (Fig. 6B). This suggests that inhibition of iNOS gene expression by exogenous NO may be ascribed to a decreased NF-kB availability (Colasanti et al., 1995c).

In conclusion, our data show that microglia can have multiple, thus pivotal, functions both in the developing and the mature human brain. The complete elucidation of the role they play still requires further investigation as well as their relationship with the other brain cells such as neurons, astrocytes and oligodendrocytes.

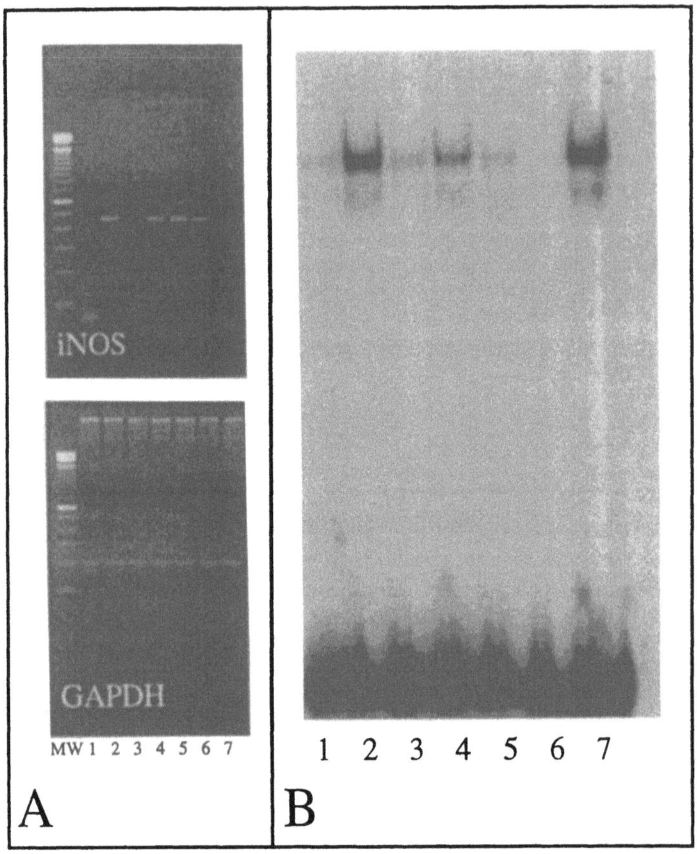

Figure 6. As shown in A, LPS/TNFα increased iNOS mRNA expression in human ramified microglial cells (lane 2), when compared with untreated cells (lane 1), as analysed by RT-PCR. Pretreatment of cells with a NO donor SNP (lane 3) inhibited LPS/TNFα-induced mRNA expression, this effect being reversed by reduced haemoglobin (lane 4). Lane 5: sample 2 pretreated with NO-deprived SNP solution. Lane 6: sample 2 plus SNP added 4h after iNOS upregulation. Lane 7: sample 2 pre-treated with authentic NO solution. PCR products for the GAPDH gene were taken as the references cellular transcript. MW is 100 bp DNA ladder. B Effect of SNP on DNA binding activity of NF-kB transcription factor (EMSA). Nuclear extract of 5 x 10^6 human ramified microglial cells were incubated with [^{32}P]-labelled double stranded oligonucleotide containing the consensus NF-kB DNA binding site (5' GATCCAGAG*GGGACTTTCC*GAGTAC 3') in a reaction mixture. In the competition assay, NF-kB or non specific oligonucleotide competitor was added before addition of labelled probe. The intensity of the retarded band was measured by Phosphor Imager. Treatment of cells with LPS/TNF-α for 0.5 h induced NF-kB activation (lane 2), when compared with control cells (lane 1). A 0.5-h pre-treatment with SNP was able to inhibit this effect (lane 3). Lane 4: LPS/TNF-α-treated cells for 4 h. Lane 5: sample 4 pre-treated with SNP. Lane 6: sample 2 with 100 x cold NF-kB as specific competitor. Lane 7: sample 2 with 100 x cold non specific competitor.

ACKNOWLEDGMENTS

This work was supported by MURST Grants (40–60%) and CNR Grant number 9.02714.CTO4 and 95.02291.CTO4; Regione Autonoma Sardegna, Assessorato Igiene e Sanità, Protocollo d'intesa to F.G. We are grateful to Prof. A. Riva for providing SEM micrograph and to Prof. D. Cocchia for TEM observations.

We thank Mr. Alessandro Cadau for his skillful technical help and Ms. Franca Fadda for carefully typying the manuscript.

REFERENCES

Akiyama H. (1994) Inflammatory response in Alzheimer's disease. Tohoku J. Exp. Med. 174, 295–303

Anderson K.J., Dam D., Lee S., Cotman C.W. (1988) Basic fibroblast growth factor prevents death of lesioned cholinergic neurons in vivo. Nature 332, 360–361

Araujo D.M., Cotman C.W. (1992) Basic FGF in astroglial, microglial, and neuronal cultures: characterisation of binding sites and modulation of release by lymphokines and trophic factors. J. Neurosci. 12, 1668–1678

Balaci L., Presta M., Ennas M.G., Dell'Era P., Sogos V., Lauro G., Gremo F. (1994) Differential expression of the fibro blast-growth factor receptors by human neurons, astrocytes and microglia. NeuroReport 6, 197–200

Basilico C., Moscatelli D. (1992) The FGF family of growth factors and oncogenes. Adv. Cancer Res. 59, 115–165

Bito H., Nakamura M., Honda Z., Izumi T., Iwatsubo T., Seyama Y., Ogura A., Kudo Y., Shimizumi T. (1992) Platelet-activating factor (PAF) receptor in rat brain: PAF mobilizes intracellular Ca^{2+} in hippocampal neurons. Neuron 9, 285–294

Braquet P., Tougui L., Shern T.-Y., Vargaftig B.B. (1987) Perspectives in platelet-activating factor research. Pharmacol. Rev. 39, 97–145

Brochet B., Orogogozo J.M., Dartigues J.F., Henry P., Loiseau P. (1992) Pilot study of Ginkogolide B, a PAF-acether specific inhibitor in the treatment of acute outbreaks of multiple sclerosis. Rev. Neurol. Paris 148, 299–301

Bussolino F., Gremo F., Tetta C., Pescarmona G.P., Camuss G. (1986) Production of platelet-activating factor by chick retina. J. Biol. Chem. 261, 16502–6508

Bussolino F., Soldi R., Arese M., Jaranowska A., Sogos V., Gremo F. (1994) Multiple roles of platelet activating factor in the nervous system. Neurochem. Res., in press

Cabellos C., Macintyre D.E., Forrest M., Burroughs M., Prasad S., Tuomanen E. (1992) Differring roles for platelet activating factor during inflammation of the lung and subarachnoid space. The special case of Streptococcus pneumoniae.J. Clin. Invest. 90, 612–618

Carpenter A.F., Carpenter P.W., Markesbery W.R. (1993) Morphometric analysis of microglia in Alzheimer's disease. J. Neuropathol. Exp. Neurol. 52, 601–608

Chao C.C., Hu S., Sheng W.S., Peterson P.K. (1995) Tumor necrosis factor-alpha production by human fetal microglial regulation by other cytokines. Dev. Neurosci. 17, 97–105

Colasanti M., Persichini T., Di Pucchio T., Gremo F., Lauro G.M. (1995a) Human ramified microglial cells produce nitric oxide upon Escherichia coli lipopolysaccharide and tumor necrosis factor α stimulation. Neurosci. Lett. 200, 144–146

Colandanti N., Di Pucchio T., Persichini T., Sogos V., Presta M., Lauro G.M. (1995b) Inhibition of inducible nitric oxide synthase mRNA expression by basic fibroblast growth factor in human microglial cells. Neurosci. Lett. 196, 45–48

Colasanti N., Persichini T., Menegazzi M., Mariotto S., Giordano E., Caldarera C.M., Sogos V., Lauro G.M., Suzuki H. (1995c) Induction of nitric oxide synthase mRNA expression. Suppression by exogenous nitric oxide. J. Biol. Chem. 270, 26731–26733

Dell'Era P., Ennas M.G., Torelli S., Gremo F., Ragnotti G., Presta M. (1990) Basic fibroblast growth factor in human fetal brain. In Neurology and Neurobiology in Regulation of Gene Expression in the Nervous System (eds Giuffrida Stella A.M., de Vellis J., Perez-Polo J.R.) New York, Wiley-Liss, pp. 425–427

del Rio Hortega P. (1919) El tercer elemento de los centros nerviosos. I La microglia en estado normal. II Intervencion de la microglia en los procesos patologicos. III Maturaleza propable de la microglia. Biol. Soc. Esp. Biol. 9, 69–120

del Rio Hortega P. (1932) Microglia. In Penfield's Cytology and Cellular Pathology of the Nervous System, (ed. Penfield W). pp. 483–534, Harbert and Row, New York

Denis M. (1994) Human monocytes/macrophagtes: NO or no NO? J. Leukoc. Biol. 55, 682–684

Dhib-Jalbut S., Hoffman P.M., Yamabe T., Sun D., Xia J., Eisenber Bergery G., Ruscetti F.W. (1994) Extracellular human T-cell lymphotropic virus type I Tax protein induces cytokine production in adult human microglial cells. Ann. Neurol. 36, 787–790

Di Pucchio T., Ennas M.G., Presta M., Lauro G.M. (1996) Basic fibroblast growth factor modulates in vitro differentiation of human fetal microglia. Neuroreport 7, 2813–1817

Dolman C.L. (1985) Microglia. In Textbook of Neurophatology (eds Davis R.L. and Robertson), Baltimora, William and Wilkins

Ennas M.G., Cocchia D., Silvetti E., Sogos V., Riva A.. Torelli S., Gremo F. (1992) Immunocompetent cell markers in human fetal astrocytes and neurons in culture. J. Neurosci. Res. 32, 424–436

Fedoroff S., Hao C. (1991) Origin of microglia and their regulation by astroglia. In Plasticity and Regeneration of the Nervous System (ed. Timiras P.S. et al.) Plenum Press New York, pp. 135–142

Ferrer I., Bernet E., Soriano E., Del Rio T., Fonseca M. (1990) Naturally occurring cell death in the cerebral cortex of the rat and removal of dead cells by transitory phagocytes. Neuroscience 39, 451–458

Florkiewicz R.Z., Baird A., Gonzales A.M. (1991) Multiple forms of bFGF: differential nuclear and cell surface localization. Growth Factors 4, 265–275

Frei K., Siepl C., Groscurth P., Bodmer S., Scwerdel C., Fontana A. (1987) Antigen presentation and tutor cytotoxicity by interferon-gamma-treated microglia cells. Eur. J. Immunol. 17, 1271- 1278

Frei K., Fontana A. (1989a) Immunoregulatory functions of astrocytes and microglial cells within the Central Nervous System, in Neuroimmune Networks: Physiology and Diseases, pp.127–130, Alan R. Liss Inc., New York

Frei K., Malipiero U., Leist T., Sinkernagel R., Schwab M., Fontana A. (1989b) On the cellular source and function of interleukin-6 produced in the Central nervous system in viral diseases. Eur. J. Immunol. 19, 689–694

Frerichs K.U., Feuerstein G.Z. (1990) Platelet-activating factor- a key mediator in neuroinjury? Cerebrovasc. Brain. Metab. Rev. 2, 148–160

Fujita S., Tsuchihashi Y., Kitamura T. (1981) Origin, morphology and functions of the microglia. Progr. Clin. Biol. Res. 59, 141–169

Gehrmann J., Banati R.B., Kreutzberg G.W. (1993) Microglia in the immune surveillance of the brain: human microglia constitutively express HLA-DR molecules. J. Neuroimmunol. 48, 189–198

Gilad G.M., Gilad V.H. (1995) Chemotaxis and accumulation of nerve growth factor by microglia macrophages. J. Neurosci. Res. 41, 594–602

Giulian D., Baker T.J. (1986) Characterisation of ameboid microglia isolated from developing mammalian brain. J. Neurosci. 6, 2163–2178

Giulian D. (1987) Ameboid microglia as effectors of inflammation in the central nervous system. J. Neurosci. Res. 18, 155–171

Hanahan D.J. (1986) Platelet activating-factor. A biologically active phosphoglyceride. Annu. Rev. Biochem. 55, 483–509

Giulian D., Vaca K., Noonan C.A. (1990) Secretion of enurotoxins by mononuclear phagocytes infected with HIV-I. Science 250: 1593–1596

Giulian D., Li J., Bartel S., Broker J., Li X., and Kirkpatrick J.B. (1995) Cell surface morphology identifies microglia as a distinct class of mononuclear phagocyte. J. Neurosci. 15, 7712–7726

Goldring C.E.P., Narayanan R., Lagadec P. and Jeannin J.F. (1995) Transcriptional inhibition of the inducible nitric oxide synthase gene by competitive binding of NF-kappa B/Rel proteins. Biochem. Biophis. Res. Commun. 209, 73–79

Hassan N.F., Campbell D.E., Rifat S., Douglas S.D. (1991) Isolation and characterisation of human fetal brain-derived microglia in "in vitro" culture. Neuroscience 41, 149–158

Hetier E., Ayala J., Denefle P., Brousseau A., Rouget P., Mallat M., Prochiantz A. (1988) Brain macrophages synthesise interleukin-1 and interleukin-1 mRNA in vitro. J. Neurosci. Res. 21, 391–397

Hetier E., Ayala J., Bousseau A., Denefle P. and Prochiantz A. (1990) Amoeboid microglial cells and not astrocytes synthesize TNF-α in swiss mouse brain cell cultures. Eur. J. Neurosci. 2, 762–768

Hume D.A., Perry W.H., Gordon S. (1983) Immunohistochemical localisation of the macrophage-specific antigen in developing mouse retina: phagocytosis of dying neurons and differentiation of microglia cells to form of regular array in the plexiform layers. J. Cell Biol. 97, 253–257

Hutchins K.D., Dicksons D.W., Rashbaum W.K., Lyman W.D. (1990) Localisation of morphologically distinct microglial populations in the developing human fetal brain: implications for ontogeny. Dev. Brain Res. 55, 95–102

Johnson D.E., Williams L.T. (1993) Structural and functional diversity in FGF receptor multigene family. Adv. Cancer Res. 60, 1–41

Junier M.P., Tiberghien C., Rougeot C., Fafeur V., Dray F. (1988) Inhibitory effect of platelet-activating factor (PAF) on luteinizong hormone-releasing hormone and somatostatin release form rat median eminence in vitro correlated with the characterization of specific PAF receptor sites in rat hypothalamus. Endocrinology 123, 72–80

Killackey H.P. (1984) Glia and the elimination of transient cortical projections. Trends Neurosci. 7, 225–226

Kornechi E., Ehrlich Y.H. (1988) Neuroregulatory and neuropathological actions of the ether-phospholipid platelet-activating factor. Science 240, 1792–1794

Kochanek P.M., Nemoto E.M., Melick J.A., Evans R.W. and Burke (1988) Cerebrovascular and cerebrometabolic effects of intracarotid infused platelet-activating factor in rats. J. Cereb. Blood Flow Metab. 8, 546–551

Lauro G.M., Babiloni D., Apollony C., Buttarelli F., Ennas M.G., Torelli S., Gremo F. (1992) Antigen processing and presenting cells in neuroimmunological network activation and maintenance. In "Recent advances in

Cellular and Molecular Biology", Wegmann and Wegmann Eds., Peeters Press, Lovanio, vol. I, pp. 107–114

Lauro G.M., Babiloni D., Buttarelli F.R., Stace G., Cocchia D., Ennas M.G., Sogos V. and Greno F. (1995) Human microglia cultures: a powerful model to study their origin and immunoreacrive capacity. Int. J. Dev. Neurosci 13, 739–752

Lee S.C., Liu W., Brosnan C.F., Dickson D.W. (1992) Characterisation of primary human fetal dissociated central nervous system cultures with an emphasis on microglia. Lab. Invest. 67, 465–476

Lee S.C., Liu W., Dickson D.W., Brosnan C.F., and Berman J.W. (1993a) Cytokine production by human fetal microglia and astrocytes. J. Immunol. 150, 2659–2678

Lee S.C., Dickson D.W., Liu W. and Brosnam C.F. (1993b) Induction of nitric oxide synthase activity in human astrocytes by interleukin-1β and interferon-gamma. J. Neuroimmunol. 46, 19–24

Lee S.C., Liu W., Brosnan C.F., Dickson D.W. (1994) GM-CSF promotes proliferation of human fetal and adult microglia primary cultures. Glia 12, 309–318

Linden R., Cavalcante L.A., Barradas P.C. (1986) Mononuclear phagocytes in the retina of developing rats. Histochemistry 85,335–339

Ling E.A. (1981) The origin and nature of microglia. In Advances in cellular neurobiology (eds Fedoroff S., Hertz L.), Academic Press, pp. 33–82

Liu W., Brosnan C.F., Dickson D.W., Lee S.C. (1994) Macrophage colony-stimulating factore mediates astrocyte-induced microglial ramification in human fetal central nervous system. Am. J. Pathol. 145, 48–53

Lobb R.R. and Harper J.W. (1986) Purification of heparin-binding growth factors. Anal Biochem. 154, 1–14

Loughlin A.T., Woodroodw M.N., Cuzner M.I.(1992) Regulation of Fc receptor and major histocompatibility complex antigen expression on isolated rat microglia by tumour necrosis factor, interleukin-1 and lipopolysaccharide: effects on interferon-gamma induced activation. Immunology 75, 170–5

Maiese K., Boniece I., De Meo D. and Wagner J.A. (1993) Peptide growth factors protect against Ischemia in culture by preventing nitric oxide toxicity. J. Neurosci 13, 3034–3040

Marcheselli V.L., Rossowska M., Domingo M.T., Braquet P. and Bazan N. (1990) Distinct platelet-activating factor binding sites in synaptic endings and in intracellular membranes of rat cerebral cortex. J. Biol. Chem. 265, 9140–9145

Martin J.H.J. and Edwards S.W. (1993) Changes in mechanisms of monocyte/macrophage-mediated cytotoxicity during culture. J. Immunol. 150, 3478–3486.

McGeer P.L., Klegeris A., Walker D.G., Yasuhara O., McGeer E.G. (1994) Pathological proteins in senile plaques. Tohoku J. Exp. Med. 174, 269–277

Mennerick S., Benz A., Zorumski C.F. (1994) Ultrastructural identification of Ricinus communis agglutinin-I positive cells in primary dissociated cell cultures of human embryonic brain. Arch. Histol. Cytol. 57, 481–491

Mohammadi M., Dione C.A., Li W., Li N., Spivak T., Honegger A.M., Jaye M., Schlessinger J. (1992) Point mutation in FGF receptor eliminates phosphatidynositol hydrolysis without affecting mitogenesis. Nature 358, 681–684.

Moncada S., Palmer R.M.J., Higgs E.A. (1991) Physiology, pathophysiology and pharmacology. Pharmacol. Rev. 43, 109–142

Morrison R.S., Sharm A., de Vellis J., Bradshaw R.A. (1986) Basic fibroblast growth factor supports the survival of cerebral cortical neurons primary culture. Proc. Natl. Acad. Sci. USA 83, 7537–7541

Murabe Y., Sano Y. (1983) Morphological Studies on neuroglia. VII Distribution of brain macrophages in brains of neonatal and adult rats, as determined by means of immunohistochemistry. Cell. Tiss. Res. 229, 85–9527

Oehmichen M. (1983) Inflammatory cells in the central nervous system. Prog. Neuropathol. 5, 277–325

Otto D., Unsicker K., Grothe C. (1987) Pharmachological effects of nerve growth factor and fibroblast growth factor applied to the transectioned sciatic nerve on neuron death in adult rat dorsal root ganglia. Neurosci. Lett. 83, 156–160

Peters K.G., Marie J., Wilson E., Ives H.E., Escobedo J., Del Rosario M., Mirda D., Williams L.T. (1992) Point mutation of an FGF receptor abolishes phosphatidynositol turnover and CA^{2+} flux but not mitogenesis. Nature 358, 678–681

Peterson P.K., Hu S., Anderson W.R., Chao C.C. (1994) Nitric oxide production and neurotoxicity mediated by activated microglia form human versus mouse brain. J. Infect. Dis. 179, 457–460

Pietraforte D., Tritarelli E., Testa U., Minetti M. (1994) gp120 HIV envelope glycoprotein increase the production of nitric oxide in human monocyte-derived macrophages. J. Leukoc. Biol. 55, 175–182

Presta M., Rusnati M., Maier J.A.M., Ragnotti G. (1988) Purification of basic fibroblast growth factor from rat brain: Identification of a Mr 22.000 immunoreactive form. Biochem. Biophys. Res. Commun. 155, 1161–1172

Presta M., Ennas M.G., Torelli S., Gremo F. (1990) Synthesis of urokinase-type plasminogen activator and of type 1-plasminogen activator inhibitor in neuronal cultures of human fetal brain: stimulation by phorbol ester. J. Neurochem. 55, 1647–1654
Presta M., Urbinati C., Dell'Era P., Lauro G., Sogos V., Balaci L., Ennas M.G., Gremo F. (1995) Expression of basic fibroblast growth factor and its receptors in human fetal microglia cells. J. Int. Devl. Neurosci. 13, 29–39
Reiling N., Ulmer A.J., Duchrow M., Ernst M., Flad H.D., Hauschild S. (1994) Nitric oxide synthase: mRNA expression of different isoforms in human monocytes-macrophages. Eur. J. Immunol. 24, 1941–1944
Roghani M., Moscatelli D. (1992) Basic fibroblast growth factor is internalized through both receptor-mediated and heparan sulphate-mediated mechanisms. J. Biol. Chem. 267, 22156–22162
Rogister B., Leprince P., Pettmann B., Labourdette G., Sensenbrenner M., Moonen G. (1988) Brain fibroblast growth factor stimulates the release of plasminogen activator by newborn rat cultured astroglial cells. Neurosci. Lett. 91, 321–326
Saneto R.P., de Vellis J. (1985) Characterisation of cultured rat oligodendrocytes proliferating in a serum-free, chemically defined medium. Proc. Natl. Acad. Sci. USA 82, 3509–3513
Sawada M., Kondo N.,Suzumura A., Marunouchi T. (1989) Production of tumour necrosis factor-α by microglia and astrocytes in culture. Brain Research 491, 394–397
Sheng J.G., Boop F.A., Mrak R.E., Griffin W.S. (1994) Increased neuronal beta-amyloid precursor protein expression in temporal lobe epilepsy: association with interleukin-1 alpha immunoreactivity. J. Neurochem. 63, 1872–1879
Sheng W.S., Hu S., Kravitz F.H., Peterson P.K., Chao C.C. (1995) Tumor necrosis factor alpha upregulates human microglial cell production of interleukin-10 in vitro. Clin. Diagn. Lab. Immunol. 2, 604–608
Sievers J., Parwaresch R., Wottge H.U. (1994) Blood monocytes and spleen macrophages differentiate into microglia-like cells on monolayers of astrocytes: morphology. Glia 12, 245–258
Snyder F. (1985) Chemical and biochemical aspects of platelet activating factor: a novel class of acetylated etherlinked choline phospholipids. Med. Res. Rev. 5, 107–140
Sogos V., Bussolino F., Pilia E., Torelli S., Gremo F. (1990) Acetylcholine-induced production of platelet-activating factor by human fetal brain cells in culture. J. Neurosci. Res. 27, 706–711
Suzumura A., Mezitis S.G.E., Gonatas N.K., Silberberg D.H. (1987) MHC antigen expression on bulk isolated macrophage microglia from newborn mouse brain: induction of Ia antigen expression by gamma interferon. J. Neuroimmun. 15, 263–278
Suzuki H., Menegazzi M., Carcereri de Prati A., Mariotto S., Armato U. (1995) Nitric oxide in the liver: physiopathological roles. Adv. Neuroimmunol. 5, 379–410
Théry C., Chamak B., Mallat M. (1991) Cytotoxic effect of brain macrophages on developing neurons. Eur. J. Neurosci. 3, 1155–1164
Thomas W.E. (1992) Brain macrophages: evaluation of microglia and their functions. Brain Res. Rev. 17, 61–74
Torelli S., Dell'Era P., Ennas M.G., Sogos V., Gremo F., Ragnotti G., Presta M. (1990) Basic fibroblast growth factor in neuronal cultures of human fetal brain. J. Neurosci. Res. 27,78–83
Torelli S., Sogos V., Ennas M.G., Marcello C., Cocchia D., Gremo F. (1991) Human fetal brain cultures: a model to study neuronal proliferation,differentiation and immunocompetence. In "Plasticity and Regeneration of the Nervous System", (eds Timiras P.S. et al.) New York: Plenum Press, pp. 121–134
Ulvestad E., Williams K., Mork S., Antel J., Nyland H. (1994a) Phenotypic differences between human monocytes/macrophages and microglial cells studied in situ and in vitro. J. Neuropathol. Exp. Neurol. 53, 492–501
Ulvestad E., Williams K., Bo L., Trapp B., Antel J., Mork S. (1994b) HLA class II molecules (HLA-DR, -DP, -DQ) on cells in the human studied in situ and in vitro. Immunology 82, 535–541
Unsicker K., Reichert-Preibsch H., Schmidt R., Pettmann B., Labourdette G., Sensenbrenner M. (1987) Astroglial and fibroblast growth factor have neurotrophic functions for cultured peripheral and central nervous system neurons. Proc. Natl. Acad. Sci. USA 84, 5459–5463
Walicke P., Cowan W.M., Ueno N., Baird A., Guillemin R. (1986) Fibroblast growth factor promotes survival of dissociated hyppocampal neurons and enhances neurite extension. Proc. Natl. Acad. Sci. USA 83, 3012–3016
Watkins B.A., Dorn H.H., Kelly W.B., Armstrong R.C., Potts B., Michaels F., Kufta C.V., Dubois-Dalcq M. (1990) Specific tropism of HIV-1 for microglial cells in primary human brain cultures. Science 249, 549–553
Weis S., Neuhaus B., Mehraein P. (1994) Activation of microglia in HIV-1 infected brains is not dependent on the presence of HIV-1 antigens. Neuroreport 21, 1514–1516
Woodroofe M.N., Hayes G.M., Cuzner M.L. (1989) Fc receptor density, MHC antigen expression and superoxide production are increased in interferon gamma-treated microglia isolated from adult rat brain. Immunol. 68, 421–426

Yayon A., Klangsbrun M., Esko J.D., Lader P., Ornitz D.M. (1991) Cell surface, heparin-like molecules are required for binding of basic fibroblast growth factor to its high affinity receptor. Cell 64, 841–848.

Yue T-L., Lysko P.G., Feuerstein G. (1990) Production of platelet-activating factor from rat cerebellar granule cells in culture. J. Neurochem. 54, 1809–1811

Yue T-L., Feuerstein G.Z. (1994) Platelet-activating factor:a putative neuromodulator and mediator in the pathophysiology of brain injury. Critical Review in Neurobiology 8, 11–24

Zielasek J., Hartung H.P. (1996) Molecular of microglia activation. Adv. Neuroimmunol. 6, 191–222

7

MIGRATION AND PROLIFERATION OF MONONUCLEAR PHAGOCYTES IN THE CENTRAL NERVOUS SYSTEM

Michel Mallat, Charles-Félix Calvo, and Alexandre Dobbertin

INSERM U.114, Chaire de Neuropharmacologie
Collège de France, 11 Place Marcelin Berthelot
75231 Paris Cedex 05, France

1. MICROGLIA: FUNCTIONAL ASPECTS

Microglia are resident CNS cells expressing markers of mononuclear phagocytes (Perry, 1994a). Transformation of the microglial phenotype is a hallmark of CNS injuries. This process defined as microglial activation was described in virtually all human neuropathologies, including neurodegenerative diseases, multiple sclerosis and HIV infection (Dickson et al., 1993; McGeer et al., 1993; Kreutzberg 1996).

Microglial activation was extensively characterized in the context of experimental lesions mostly applied to rodent and encompassing penetrating wound, Wallerian and retrograde degeneration, excitatory neurotoxin-induced neuronal death, inflammation, infection and tumor (Gehrmann and Kreutzberg, 1995a). Despite differences related to the type, the localisation or the extent of lesions, microglial reactions share common features. The expression of macrophages markers is upregulated together with modifications of the cell shape. This can lead to phenotypes called ameboid microglial cells or brain macrophages (BM) which differ from resting microglia of the normal CNS by their proliferative, motile and phagocytic behaviours (Perry et al., 1994b; Kreutzberg, 1996).

BM have the ability to influence survival, growth or regeneration of others CNS cells (Mallat and Chamak, 1994) and to promote inflammatory reaction by presenting antigens to T lymphocytes (Gehrmann et al., 1995b). For instance, neuronal degeneration induced by BM was documented in ischemic (Giulian et al., 1990) and mechanical injuries (Thanos et al.,1993), and studies based on in vitro cultures together with neuropathological observations suggest that these cells could be responsible for neuronal degeneration in HIV infection or Alzheimer diseases (Epstein and Gendelman, 1993; Giulian et al., 1996). Despite a clear neurotoxic potency, BM can also produce growth factors such as neurotrophins, and extracellular matrix proteins which support neuronal survival and growth or regeneration of their processes (Mallat and Chamak, 1994). BM also secrete cytokines stimulating astrocyte proliferation and they favour the formation of astrocytic scars (Giulian et al., 1989; Merrill, 1992).

Brain Plasticity, edited by Filogamo *et al.*
Plenum Press, New York, 1997

Beyond lesional contexts, the capacity of BM to promote neuronal and astrocyte growth is likely to operate during ontogenesis (Giulian et al., 1988b; Chamak et al., 1995). In fact, BM are also present in the normal developing brain and these cells contribute to the clearance of cell debris generated by regressive events of neurogenesis (Ling and Wong 1993).

2. CELLULAR ORIGINS OF BM

Similar to other tissue macrophages, compelling evidence indicates that BM stem from CNS invasion by exogenous mesodermal precursors such as blood-derived monocytes (Perry et al., 1994ab). These precursors are generated in different hematopoietic organs, according to the developmental stage. It has been shown that BM detected in embryonic avian brain originate from yolk sack (Cuadros et al., 1993). Embryonic and adult mouse CNS contain progenitors of undefined phenotype which can generate macrophage colonies. (Alliot et al., 1991). In adulthood, monocytes infiltrating the lesioned CNS are derived from bone marrow (Hickey and Kimura, 1988).

During development, infiltrating monocytes give rise to motile BM which spread within different CNS regions. Amplification of this cell population is further achieved by BM proliferation. The progressive maturation of CNS is associated with a transformation of BM into resting ramified microglia and cell death also contributes to the disappearance of BM. Eventually, BM do not occur in normal adult neuropil except for specific regions such as circumventricular organs or neurohypophysis which lack a blood brain barrier (Ling and Wong, 1993; Perry et al., 1994b).

In injured CNS, elicited BM are derived from transformation of resting resident microglial cells and influx of blood-derived monocytes. The respective contributions of these two sources depend on the type of lesion and the time course of the tissue reactions. Injection of excitotoxins into the rat CNS leads to local neuronal death and an early ameboid transformation of resident microglia which is followed by infiltration of monocytes in the lesioned parenchyma (Coffey et al 1990; Marty et al. 1991). During experimental allergic encephalomyelitis (EAE), resident microglia provide a relatively small source of BM compared to the large influx of monocytes (Rinner et al., 1995). In contrast, BM elicited in the facial nucleus after injection of neurotoxic ricin in the facial nerve, appear exclusively derived from resident microglia (Streit et al., 1988). Proliferation of BM contribute to the local expansion of reactive microglia (Perry et al., 1994b).

The mechanisms triggering activation of resting microglia remain poorly defined. In particular, the molecular compounds elicited in lesioned tissue and responsible for early phenotype changes by acting directly on resting cells have not been fully characterized. In contrast, different extracellular agents were identified as factors able to promote BM recruitment by acting on migration or proliferation of both BM and monocytes.

3. GLIAL PRODUCTION OF CHEMOTACTIC AGENTS ACTING ON PHAGOCYTES

The infiltration of monocytes within the developing or injured brain suggests the occurrence of intraparenchymal agents which could increase or polarised the motility of the phagocytes reaching perivascular spaces. Evidence indicates that glial cells including BM release chemotactic compounds which act on mononuclear phagocytes.

3.1. ß-Chemokines

Detection of chemotactic agents using a Boyden chamber assay have allowed to show that purified normal rat BM release factors attracting macrophage precursors recovered from cultures of adult rat bone marrow. Molecular characterization of the chemotactic activity revealed the involvement of monocyte chemotactic protein 1 (MCP1) (Calvo et al. 1996a). This protein belongs to the group of ß-chemokines (chemoattractant cytokines) which share a conserved four cysteine motif linked by intramolecular disulphide bridges and which display the ability to attract mononuclear phagocytes (Baggiolini et al., 1994). Despite spontaneous production of MCP1 by purified BM, expression of MCP1 gene was upregulated when the cells were stimulated with lipopolysaccharide (LPS, extracted from bacteria) or inflammatory cytokines which are produced in injured brain such as interleukin (IL)-1, IL-6 and tumor necrosis factor-α (TNF-α) (Calvo et al., 1996a). Production of MCP1 by a BM cell line was also stimulated by a fragment (25–35) of the ß-amyloid protein, a component of the fibrillar deposit which accumulates together with activated microglia in senile plaques of Alzheimer's disease (Meda et al., 1996) .

At cellular level, the BM synthesis of MCP1 can be visualized by immunocytochemical detection of the protein (Fig1). This approach was used to confirm in situ expression of MCP1 by rat BM during EAE and following penetrating wound of the cerebral cortex or induction of neuronal death by intracerebral injection of kainic acid (Berman et al., 1996; Calvo et al., 1996a). However, BM are not the only glial source of MCP1. This peptide was initially purified from a human glioblastoma cell line (Kuratsu et al., 1989), and the production of MCP1 was induced in cultured astrocytes by LPS and inflammatory cytokines (Hayashi et al., 1995; Hurwitz et al.,1995; Calvo and Mallat, 1996b) (Fig 1). Similar to BM, in situ reactive astrocytes were found to express MCP1 in the contexts of EAE, cerebro-cortical wound, or kainic acid-induced neuronal death (Berman et al., 1996; Calvo et al., 1996a). The respective contributions of BM and astrocytes to the local production of MCP1 seem to vary according to the type of lesion or the species. Indeed, and in contrast to the observations performed in rat brain, the detection of MCP1 synthesis was restricted to astrocytes in mouse models of CNS wound or EAE (Ransohoff et al., 1993; Glabinski et al., 1996).

Upon stimulation with LPS, cultured murine BM and astrocytes secrete macrophage inflammatory protein-1 alpha (MIP-1α), another member of the ß-chemokine family (Hayashi et al., 1995; Murphy et al., 1995). MIP-1α transcripts were observed in sections of human brain infected with HIV, and the cellular sources displayed distribution and morphologies expected of astroglial and microglial cells (Schmidtmayerova et al.,1996). MIP-1α transcripts and proteins were also detected together with MCP1 gene product in rodent brain, following induction of EAE or ischemia (Godiska et al., 1995; Kim et al., 1994; Karpus et al., 1995).

Although expression of ß-chemokine correlates with recruitment of BM in different neuropathological models, the actual role of ß chemokines in monocyte infiltration of CNS remains an open question and is likely to vary according to the type of lesions. This issue was addressed in a mouse model of adoptive relapsing-EAE. It was observed that the mononuclear cell infiltration was reduced if mice were treated by intraperitoneal injections of antibodies neutralizing MIP-1α. However, despite demonstration of MIP-1α depletion within the brain, it was not fully ruled out that the reduction of EAE lesions stemmed from neutralization of MIP-1α acting at the periphery rather than into the CNS (Karpus et al., 1995). Similar administrations of antibodies neutralizing MCP1 failed to reduce EAE lesions (Karpus et al., 1995), but the ability of MCP1 to promote intracerebral

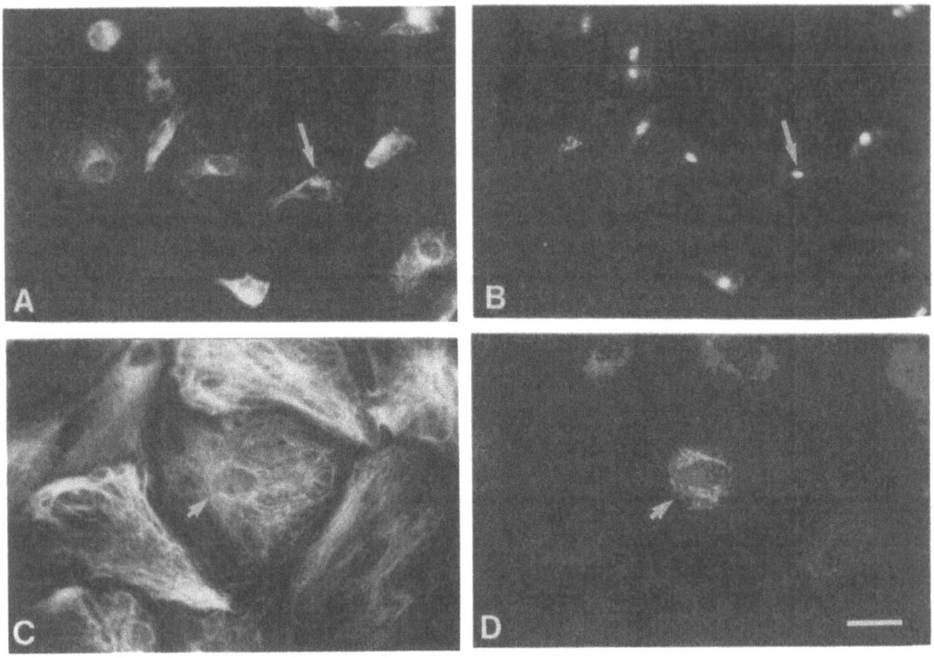

Figure 1. Immunocytochemical detection of MCP1 in glial culture. Astrocytes and BM cultures were derived from the cerebral cortex of 17-day-old rat embryos. Cultures were treated for 12h with LPS (0.5 μg/ml) before immunofluorescent staining. Cultures and immunostaining were performed as described previously (Calvo et al., 1996a). (A,B): BM culture. (A) Surface fluorescein staining with isolectin B4 (Sigma) which is known to label microglia (Streit and Kreutzberg, 1987). (B) Same field as in A, intracellular rhodamine staining using mouse monoclonal anti-MCP1 antibodies (Sakanashi et al., 1994). (C,D) Astrocyte culture. (C) Intracellular fluorescein staining using rabbit polyclonal anti-GFAP antibodies (Dakopatt). (D) Same field as in C, rhodamine staining with anti-MCP1 antibodies. Arrows in each field point to a same cell. MCP1 accumulates in an intracellular compartment which corresponds to the Golgi apparatus. Scale bar: 20 μm.

penetration of monocytes was shown by targeting MCP1 gene expression to oligodendrocytes in transgenic mice. In these animals, perivascular infiltration of phagocytes was observed in close proximity to myelin rich area (Fuentes et al., 1995). Injection of recombinant MCP1 in the hippocampus of normal mice also led to the local infiltration of monocytes (Bell et al., 1996).

3.2. Transforming Growth Factor-ß and Thrombospondin

Production of ß-chemokines is clearly not the unique mechanism whereby reactive glial cell could attract invading monocytes in injured CNS. Both cultured astrocyte and BM can produce transforming growth factor (TGF)-ß1, a cytokine which regulates survival, growth or functions of mesenchymal and epithelial derivatives including neurons and glial cells (Constam et al., 1992; Krieglstein et al., 1995). TGF-ß1 is also known as a potent chemotactic agent acting on monocytes and BM (Wahl et al. 1987, Yao et al., 1990). In the normal CNS, TGF-ß1 seems to be mostly confined to meninges and choroid plexus, but astrocytic or microglial expressions of TGFß1 gene are turned on in human neuropathologies such as HIV infection (Wahl et al., 1991) or in rodent following mechanical lesions or administration of kainic acid (Morgan et al., 1993, Logan et al., 1994).

Most cell types including glia, secrete TGFß1 in a biologically inactive form resulting from the non-covalent association of the mature dimeric TGF-ß to a homodimer of the N-terminal cleavage product of the TGFß precursor (Constam et al., 1992; Flaumenhaft et al., 1993). Thus, a critical step in the regulation of TGFß activity is the release of the active cytokine from this latent precursor. It is not fully understood how activation of latent TGFß is achieved in vivo but different mechanisms were demonstrated in vitro (Flaumenhaft et al., 1993). Relevant to CNS injuries, activation of TGFß can occur through binding of the latent form to thrombospondin-1 (TSP), an extracellular matrix protein which was shown to be secreted by rat BM (Schultz-Cherry et al., 1993; Chamak et al., 1994). Interestingly TSP on its own, regulates adhesion and migration of different cell types and display chemotactic and aptotactic activities on monocytes (Mansfield and Suchard, 1994). Expression of TSP by BM is not restricted to injured CNS as it was also detected in section of developing rat brain. Accumulation of TSP was prominent in clusters of BM located in developing axonal fibre tracts such as Corpus Callosum, at the time when invasion of blood monocytes contribute to the local expansion of microglia (Chamak et al., 1995). Thus, the expression of thrombospondin is compatible with a role in the regulation of intracerebral migration of phagocytes during development.

4. GLIAL-DERIVED FACTORS STIMULATING PROLIFERATION OF PHAGOCYTES

Similar to other mononuclear phagocytes, proliferation of cultured BM can be triggered by three hematopoietic growth factors encoded by distinct genes: colony stimulating factor (CSF)-1 (also called Macrophage-CSF, M-CSF); Granulocyte/macrophage-CSF (GM-CSF) and IL-3 (Giulian and Ingeman, 1988a; Metcalf, 1989; Roth and Stanley, 1992). Increasing evidence indicates that CSF-1 contributes to the development and pathological responses of microglia. Both CSF-1 transcripts and proteins were detected in extracts of the developing mouse brain from embryonic stages up until adulthood (Théry et al., 1990; Chang et al., 1994; Roth & Stanley, 1996). The distribution of CSF-1 transcripts in the developing CNS is heterogeneous. Expression is most prominent in the cerebellum but also occurs in the cerebral cortex (Michaelson et al., 1996). CSF-1 transcripts and protein were also detected in the spinal cord of adult rats and their levels were increased together with BM recruitment during EAE (Hulkower et al., 1993). Expression of CSF-1 in human pathologies remains seldom investigated. Analyses of CSF-1 contents in human cerebrospinal fluid showed increased amounts of this growth factors during encephalopathies associated with HIV infection (Gallo et al., 1994). The cellular source of CSF-1 was studied in cultures derived from developing rodent or human CNS. Cultured astrocytes appeared to produce CSF-1 constitutively. Consistent with the upregulations observed in inflamed CNS, the levels of secreted proteins were significantly increased by stimulating astrocytes with IL-1ß or TNF-α (Hao et al., 1990; Théry et al., 1990; 1992; Lee et al. 1993). Expression of CSF-1 gene was also observed in cultured BM upon stimulation with LPS (Théry et al., 1990; Lee et al., 1993). Contrasting with glia, neurons derived from cerebral cortex do not synthesize CSF-1, but this growth factor was recovered from cultures of mice cerebellar granule cells (Théry et al., 1990; Nohava et al., 1992). However, cultured cerebro-cortical neurons produce a factor which markedly increase the mitogenic response of BM stimulated with CSF-1. This factor was recently identified as TGF-ß2 (Dobbertin et al., submitted).

Investigation of CSF-1 functions has benefited from osteopetrotic (op/op) mice which are characterized by an autosomal recessive mutation in the CSF-1 gene resulting in the absence of CSF-1 (Wiktor-Jedrzejczak et al., 1990). As a consequence, the mutant suffers from severe osteopetrosis due to a deficiency in osteoclasts formation. Other populations of phagocytes are also depleted such as macrophages of peritoneal cavity, spleen marginal zone, liver, and kidney, whereas other populations such as lymph node macrophages are relatively intact (Witmer-Pack et al., 1993). In these CSF-1 deficient mice, microglial cells were observed in the adult brain but their proliferation was dramatically reduced following mechanical or ischemic injuries, compared with normal mice (Witmer-Pack et al., 1993; Raivich et al., 1994; Berezovskaya et al., 1995). Transection of facial nerve induced a 50% increase in microglial cell number within one week, in the facial nucleus of the mutant, while such a lesion led to a 350% increase in normal mice (Raivich et al., 1994). Despite evidence for reduction of microglial growth during development in osteopetrotic mice (Cecchini et al., 1994), the presence of microglia in the adult mutant together with limited proliferation triggered by lesions indicate that other factors different from CSF-1 also support microglial proliferation. A possible contribution of GM-CSF is suggested by the fact that this cytokine can be produced by cultured astrocytes and BM upon stimulation with TNF-α or LPS (Malipiero et al., 1990; Aloisi et al., 1992). IL-3 was also proposed as an autocrine microglial growth factor when it was observed that cultured BM produce this protein (Gebicke-Haerter et al., 1994). Induction of IL-3 receptors mRNA in rat microglial cells was found in vivo as an early marker of microglial activation (Appel et al., 1995). In addition to the CSFs, neurotrophin 3, a factor supporting growth and survival of neuronal subpopulations, was recently shown to stimulate proliferation of cultured rat BM. It is not ruled out that this factor acts by triggering BM production of mitogenic CSFs which would in turn stimulate BM proliferation in an autocrine manner (Elkabes et al., 1996).

5. CONCLUSION

A growing body of evidence indicate that BM display a pivotal role in various pathological remodelling of CNS tissue. However any function of BM depends primarily of their recruitment involving migration, proliferation and differentiation of mononuclear phagocyte in the CNS. Recent investigations support the notion that the recruitment of BM is promoted by cytokines produced by microglial and astrocytic cells during development or in response to CNS injuries. Understanding the molecular mechanism regulating the recruitment of BM may provide new therapeutical approach to control adverse effects of microglia in human neuropathologies.

ACKNOWLEDGMENTS

We thank Pr. J. Glowinski for his constant help and Dr. T. Yoshimura for providing anti-MCP1 antibodies. The work performed in the author's laboratory was supported by INSERM and by Agence Nationale de Recherche sur le SIDA.

REFERENCES

Alliot F, Lecain E, Grima B & Pessac B (1991). Microglial progenitors with a high proliferative potential in the embryonic and adult mouse brain. *Proc Natl Acad Sci USA* 88: 1541–1545.

Aloisi F, Care A, Borsellino G, Gallo P, Rosa S, Bassani A, Cabibbo A, Testa U, Levi G & Peschle C (1992). Production of hemolymphopoietic cytokines (IL-6, IL-8, colony-stimulating factors) by normal human astrocytes in response to IL-1ß and tumor necrosis factor-α. *J Immunol* 149: 2358–2366.

Appel K, Buttini M, Sauter A & Gebicke-Haerter J (1995). Cloning of rat interleukin-3 receptor ß-subunit from cultured microglia and its mRNA expression in vivo. *J Neurosci* 15: 5800–5809.

Baggiolini M, Dewalt B & Mosser (1994). Interleukin-8 and related chemotactic cytokines CXC and CC chemokines. *Adv Immunol* 55: 97–179.

Bell MD, Taub DD & Perry VH (1996). Overriding the brain's intrinsic resistance to leukocyte recruitment with intraparenchymal injections of recombinant chemokines. *Neuroscience* 74: 283–292.

Berezovskaya O., Maysinger D & Fedoroff S (1995). The hematopoietic cytokine colony-stimulating factor in the CNS: Congenital absence of CSF-1 in mice results in abnormal microglial response and increased neuron vulnerability to injury. *Int J Devl Neurosci* 13: 285–299.

Berman JW, Guida MP, Warren J, Amat J & Brosnan CF (1996). Localization of monocyte chemoattractant peptide-1 expression in the central nervous system in experimental autoimmune encephalomyelitis and trauma in the rat. *J Immunol* 156: 3017–3023.

Calvo CF, Yoshimura T, Gelman M & Mallat M (1996a). Production of monocyte chemotactic peptide-1 by rat brain macrophages. *Eur J Neurosci* 8: 1725–1734.

Calvo CF & Mallat M (1996b). Expression of macrophage chemotactic protein-1 in rat glial cells. In *Biology and Physiology of the Blood-Brain Barrier*, eds P Couraud and D Sherman , Plenum Press, New York pp 271–277.

Cecchini MG, Dominguez MG, Mocci S, Wetterwald A, Félix R, Fleisch H, Chilsholm O, Hofstetter W, Pollard JW & Stanley ER (1994). Role of colony stimulating factor-1 in the establishment and regulation of tissue macrophages during postnatal development of the mouse. *Development* 120: 1357–1372.

Chamak B, Dobbertin A & Mallat M (1995). Immunohistochemical detection of thrombospondin in microglia in the developing rat brain. *Neuroscience* 69: 177–187.

Chamak B, Morandi V & Mallat M (1994). Brain macrophages stimulate neurite growth and regeneration by secreting thrombospondin. *J Neurosci Res* 38: 221–233.

Chang Y, Albright S and Lee F (1994). Cytokines in the central nervous system: expression of macrophage colony stimulating factor and its receptor during development. *J Neuroimmunol* 52: 9–17.

Coffey PJ, Perry VH & Rawlins JNP (1990). An investigation into the early stages of the inflammatory response folloing ibotenic acid-induced neuronal degeneration. *Neuroscience* 35: 121–132.

Constam DB, Philipp J, Malipiero UV, Ten Dijke P, Schachner M & Fontana A (1992). Differential expression of transforming growth factor-ß1, -ß2, and -ß3 by glioblastoma cells, astrocytes, and microglia. *J Immunol* 148: 1404–1410.

Cuadros MA, Martin C, Coltey MC, Almendros A & Navascuez J (1993) First appearance distribution and origin of macrophages in the early avian development of the avian central nervous system. *J. Comp Neurol* 330: 113–129.

Dickson DW, Lee SC, Mattiace LA, Yen SC & Brosnan C (1993) Microglia and cytokines in neurological disease, with special reference to AIDS and Alzheimer's disease. *GLIA* 7: 75–83.

Dobbertin A., Schmid P., Gelman M., Glowinski J. & Mallat M. Neurons promote macrophage growth by producing TGF-ß2. *Submitted*

Elkabes S, DiCicco-Bloom M & Black IB (1996). Brain microglia/macrophages express neurotrophins that selectively regulate microglial proliferation and fuction. *J Neurosci* 16: 2508–2521.

Epstein LG & Gendelman HE (1993). Human immunodeficiency virus type 1 infection of the nervous system: pathogenetic mechanisms. *Ann Neurol* 33: 429–436.

Flaumenhaft R, Kojima S, Abe M & Rifkin DB (1993) Activation of latent transforming growth factor ß. *Adv Pharmacol* 24: 51–76.

Fuentes ME, Durham SK, Swerdel MR, Lewin AC, Barton DS, Megill JR, Bravo R & Lira SA. (1995). Controlled recruitment of monocytes and macrophages to specific organs through transgenic expression of monocyte chemoattractant protein-1. *J Immunol* 155: 5769–5776.

Gallo P, De Rossi A, Sivieri s, Chieco-Bianchi L & Tavolato B (1994) M-CSF production by HIV-1-infected monocytes and its intrathecal synthesis. Implications for nerological HIV-1-related disease. *J Neuroimmunol* 51: 193–198.

Gebicke-Haerter PJ, Appel K, Taylor GD, Schobert A, Rich IN, Northoff H & Berger M (1994). Rat microglial interleukin-3. *J Neuroimmunol* 50: 203–214.

Gehrmann J. & Kreutzberg G.W. (1995a) Microglia in experimental neuropathology. In "Neuroglia" eds H. Kettenmann & Ransom B.R. Oxford University Press pp 883–904.

Gehrmann J, Matsumoto Y, Kreutzberg GW (1995b) Microglia: intrinsic immuneffector cell of the brain. *Brain Res Rev* 20: 269–287.

Giulian D., Chen J., Ingeman J.E, George J.K. and Noponen M. (1989) The role of mononuclear phagocytes in wound healing after traumatic injury to adult mammalian brain. *J Neurosci* 9, 4416–4429.

Giulian D, Haverkamp LJ, Yu JH, Karshin W, Tom D, Li J, Kirkpatrick J, Kuo YM & Roher AE (1996). Specific domains of ß-amyloid from Alzheimer plaque elicit neuron killing in human microglia. *J Neurosci* 16, 6021–6037.

Giulian D & Ingeman J (1988a) Colony-stimulating factors as promoters of ameboid microglia. J Neurosci 8: 4707–4717.

Giulian, D T & Robertson C (1990) Inhibition of mononuclear phagocytes reduces ischemic injury in the spinal cord. *Ann. Neurol.* 27: 33–42.

Giulian D, Young DG, Woodward J, Brown DC & Lachman L.B. (1988b) Interleukin-1 is an astroglial growth factor in the developing brain. *J. Neurosci.* 8: 709–714.

Glabinski AR, Balasingam V, Tani M, Kunkel SL, Strieter RM, Yong VW & Ransohoff-RM (1996). Chemokine monocyte chemoattractant protein-1 is expressed by astrocytes after mechanical injury to the brain. *J Immunol* 156: 4363–4368.

Godiska R, Chantry D, Dietsch GN & Gray PW (1995) Chemokine expression in murine experimental allergic encephalomyelitis. *J Neuroimmunol* 58: 167–168.

Hao C, Guilbert LJ & Fedoroff S (1990) Production of colony-stimulating factor-1 (CSF-1) by mouse astroglia in vitro. *J Neurosci Res* 27: 314–323.

Hayashi M, Luo Y, Laning J, Strieter BM & Dorf ME (1995). Production and function of monocyte chemoattractant protein-1 and other b-chemokines in murine glial cells. *J Neuroimmunol* 60: 143–150.

Hickey W, Kimura H (1988) Perivascular microglial cells of the CNS are bone-derived and present antigen in vivo. *Science* 239: 290–292.

Hulkower K, Brosnan CF, Aquino DA, Cammer W, Kulshrestha S, Guida MP, Rapoport DA & Berman JW (1993). Expression of CSF-1, c-fms, and MCP-1 in the central nervous system of rats with experimental allergic encephalomyelitis. *J Immunol* 6: 2525–2533.

Hurwitz AA, Lyman WD & Berman JW (1995). Tumor necrosis factor-alpha and transforming growth factor-b upregulate astrocyte expression of monocyte chemoattractant protein-1. *J Neuroimmunol* 57: 193–198.

Karpus WJ, Lukacs NW, McRae BL, Strieter RM, Kunkel SL & Miller SD (1995). An important role for the chemokine macrophage inflammatory protein-1a in the pathogenesis of T cell-mediated autoimmune disease, experimental autoimmune encephalomyelitis. *J Immunol* 155: 5003–5010.

Kim JS, Gautam SC, Chopp M, Zaloga C, Jones ML, Ward PA & Welch KMA. (1995). Expression of monocyte chemoattractant protein-1 and macrophage inflammatory protein-1 after focal cerebral ischemia in the rat. *J Neuroimmunol* 56: 127–134.

Kreutzberg GW (1996). Microglia: a sensor for pathological events in the CNS *TINS* 9: 312–331.

Krieglstein K, Rufer M, Suter-crazzolara C & Unsicker K (1995) Neural functions of the transforming growth factors ß. *Int J Devl Neurosci* 13: 301–315.

Kuratsu JI, Leonard EJ & Yoshimura T (1989). Production and characterization of human gioma-cell derived monocyte chemotactic factor. *J Natl Cancer Inst* 81:347–351.

Lee SC, Liu W, Roth P, Dickson DW, Berman JW, Brosnan CF (1993). Macrophage colony-stimulating factor in fetal astrocytes and microglia: differential regulation by cytokines and lipopolysaccharide and modulation of class II MHC on microglia *J Immunol* 150: 594–604.

Ling EA, Wong WC (1993) The origin and nature of ramified and amoeboid microglia: a historical review and current concepts. *GLIA* 7: 9–18.

Logan A, Berry M, Gonzalez AM, Frautschy A, Sporn MB & Baird A (1994) Effects of transforming growth factor ß1 on scar production in the injured central nervous system of the rat. *Eur J Neurosci* 6: 355–363.

Malipiero UV, Frei K & Fontana A (1990) Production of hemopoietic colony-stimulating factors by astrocytes. *J Immunol* 144: 3816–3821.

Mallat M & Chamak B (1994) Brain macrophages: neurotoxic or neurotrophic effector cells ?. *J Leukoc Biol* 56: 416–422.

Mansfield PJ & Suchard SJ (1994). Thrombospondin promotes chemotaxis and haptotaxis of human peripheral blood monocytes. *J Immunol* 153: 4219–4229.

Marty S, Dusart I & Peschanski M (1991) Glial changes following an excitotoxic lesion in the CNS-I. Microglia/ Macrophages. *Neuroscience* 45: 529–539.

McGeer PL, Kawamata T, Walker DG, Akiyama H, Tooyama I, McGeer EG (1993) Microglia in degenerative neurological disease. *GLIA* 7: 84–92.

Meda L, Bernasconi S, Bonaiuto C, Sozzani S, Zhou D, Otvos L, Mantovani A, Rossi F, Cassatella MA. (1996) ß-amyloid (25–35) peptide and IFN-g synergistically induce the production of chemeotactic cytokine MCP-1/JE in monocytes and microglial cells. *J Immunol* 157: 1213–1218.

Merrill J.E. (1992). Tumor necrosis factor alpha, interleukin 1 and related cytokines in brain development: normal and pathological. *Devl Neurosci* 14: 1–10.

Metcalf D (1989) The molecular control of cell division, differentiation commitment and maturation in haemopoietic cells. *Nature* 339: 27–30.

Michaelson M.D., Bieri P.L., Mehler M.F., Xu H., Arezzo J.C., Pollard J.W and Kessler J.A. (1996) CSF-1 deficiency in mice results in abnormal brain development. *Development* 122, 2661–2672.

Morgan TE, Nichols NR, Pasinetti GM & Finch CE (1993) TGF-ß1 mRNA increases in macrophages/ microglial cells of the hippocampus in response to deafferentation and kainic acid-induced neurodegeneration. *Exp Neurol* 120: 291–301.

Murphy GM, Jia JX, Song Y, Ong E, Shrivastava R, Bocchini V, Lee YL & Eng LF (1995). Macrophage inflammatory protein 1-α mRNA expression in a immortalized microglial cell line and cortical astrocyte cultures. *J Neurosci res* 40: 755–763.

Nohava K, Malipiero U, Frei K & Fontana A (1992) Neurons and neuroblastoma as a source of macrophage colony-stimulating factor. *Eur J Immunol* 22: 2539–2545.

Perry V.H. (1994a). Microglia, resident macrophages of the CNS. In "*Macrophages and the nervous system*" ed MBIT, RG Landes Company, CRC Press. pp 6–27.

Perry VH, Lawson LJ, Reid DM (1994b). Biology of the mononuclear phagocyte system of the central nervous system and HIV infection. *J Leukoc Biol* 56: 399–406.

Raivich G, Moreno-Flores MT, Möller JC & Kreutzberg GW (1994). Inhibition of posttraumatic microglial proliferation in a genetic model of macrophage colony-stimulating factor defiency in the mouse. *Eur J Neurosci* 6: 1615–1618.

Ransohoff RM, Hamilton TA, Tanie M, Stoler MH, Schick HE, Major JA, Estes ML, Thomas DM & Tuohy VK (1993). Astrocyte expression of mRNA encoding IP-10 and JE/MCP-1 in experimental autoimmune encephalomyelitis. *FASEB J* 7: 592–600.

Rinner WA, Bauer J, Schmidts M, Lassmann H & Hickey WF (1995). Resident microglia and hematogenous macrophages as phagocytes in adoptively transferred experimental autoimmune encephalomyelitis: an investigation using rat radiation bone marrow chimeras. *GLIA* 14: 257–266.

Roth P & Stanley ER (1992). The biology of CSF-1 and its receptor. *Curr Top Microbiol Immunol* 181: 141–167.

Roth P & Stanley ER (1996). Colony stimulating factor-1 expression is developmentally regulated in the mouse. *J Leukoc Biol* 59: 817–823.

Sakanashi Y, Takeya M, Yoshimura T, Feng L, Morioka T & Takahashi K (1994). Kinetics of macrophage subpopulations and expression of monocyte chemoattractant protein-1 (MCP-1) in bleomycin-induced lung injury of rats studied by a novel monoclonal antibody against rat MCP-1. *J Leukoc Biol* 56: 741–750.

Schmidtmayerova H, Nottet HS, Nuovo G, Raabe T, Flanagan CR, Dubrovsky L, Gendelman HE, Cerami A, Bukrinsky M & Sherry B (1996). Human immunodeficiency virus type 1 infections alters chemokine beta peptide expression in human monocyte: implications for recruitment of leukocytes into brain and lymph nodes. *Proc Natl Acad Sci USA* 93: 700–704.

Schultz-Cherry S & Murphy-Ullrich JE (1993). Thrombospondin causes activation of latent transforming growth factor-ß secreted by endothelial cells by a novel mechanism. *J Cell Biol* 122, 923–932.

Streit WJ & Kreutzberg GW (1987). Lectin binding by resting and reactive microglia. *J Neurocytol* 16: 249–260.

Streit WJ & Kreutzberg GW (1988). Response of endogenous glial cells to motor neuron degeneration induced by toxin ricin. *J Comp Neurol* 268: 248–263.

Thanos, S., Mey, J. & Wild, M. (1993) Treatment of the adult retina with microglia-suppressing factors retards axotomy-induces neuronal degradation and enhances axonal regeneration *in vivo* and *in vitro*. *J Neurosci* 13, 455–466.

Théry C, Hétier E, Evrard C & Mallat M (1990) Expression of macrophage colony-stimulating factor gene in the mouse brain during development. *J Neurosci Res* 26:129–133.

Théry C, Stanley ER, Mallat M (1992) Interleukin 1 and tumor necrosis factor-α stimulate the production of colony-stimulating factor 1 by murine astrocytes. *J Neurochem* 59: 1183–1186.

Wahl SM, Allen JB, McCartney-Francis N, Morganati-Kossmann MC, Kossmann T, Ellingsworth L, Ma UEH, Mergenhagen SE & Orenstein JM (1991). Macrophage-and astrocyte-derived transforming growth factor-ß as a mediator of central nervous system dysfunction in acquired immune deficiency syndrome *J Exp Med* 173: 981–991.

Wahl SM, Hunt DA, Wakefield L, Mc Cartney -Francis N, Wahl LM, Roberts AB, Sporn MB (1987) Transforming growth factor beta (TGF-ß) induces monocyte chemotaxis and growth factor production. *Proc Natl Acad Sci USA* 84: 5788–5792.

Wiktor-Jedrzejczak, W, Bartocci A, Ferrante AW, Ahmed-Ansari A, Sell KW, Pollard JW & Stanley ER (1990) Total absence of colony-stimulating factor 1 in the macrophage deficient osteopetrotic (op/op) mouse. *Proc Natl Acad Sci USA* 87: 4828–4832.

Witmer-Pack MD, Hughes DA, Schuler G, Lawson L, McWilliam A, Inaba K, Steinman RM & Gordon S (1993). Identification of macrophages and dendritic cells in the osteopetrotic (op/op) mouse. *J Cell Sci* 104: 1021–1029.

Yao J, Harvath L, Guilbert DL, Colton CA (1990) Chemotaxis by a CNS macrophage, the microglia. *J. Neurosci Res* 27: 36–42.

8

K^+-CHANNELS AND CYTOKINES AS MARKERS FOR MICROGLIAL ACTIVATION

Britta Küst,[1] Manuel Buttini,[2] Andre Sauter,[2] Hendrik W. G. M. Boddeke,[2] and Peter J. Gebicke-Haerter[1*]

[1]Department of Psychiatry
University of Freiburg, F.R.G.
[2]Novartis Pharma Ltd.
Basel, Switzerland

1. INTRODUCTION

Activation of brain microglial cells is a key event in neurodegenerative diseases of the Central Nervous System. Microglia are the macrophages of the brain displaying full immunocompetent functions in injury or disease. Although they express many features typical for and indistinguishable from peripheral blood macrophages, there is an increasing body of evidence accentuating special characteristics of microglia that are distinct from peripheral mononuclear cells. Most likely, these features have been imprinted by the special environmental conditions in brain parenchyma, which are e.g. extremely immunosuppressive. Microglial activation stages have been characterized by a variety of morphological and molecular markers, such as retraction of processes, migratory activities, proliferation, cytokine and oxygen radical production and phagocytosis. Evidently, it depends on the nature and strength of stimulus as to which stage is attained and whether or not the cells develop into full-blown macrophages or fall back into quiescence. This report describes some markers which are both useful to characterize earlier and later stages of microglial activation and may be targeted by therapeutic measures in degenerating diseases of the nervous system.

2. K^+-CHANNELS

We hypothesized, that disturbances in the K^+-homeostasis may be early signals in the activation process of microglia. Very high K^+-concentrations may occur locally when

* Address correspondence to: Peter J. Gebicke-Haerter. Department of Psychiatry, Hauptstr. 8, D-79104 Freiburg i.Br. Phone: -49(761)270–6835; Fax : -49(761)270–6619; e-mail : geha®psylab.ukl.uni-freiburg.de.

Brain Plasticity, edited by Filogamo *et al.*
Plenum Press, New York, 1997

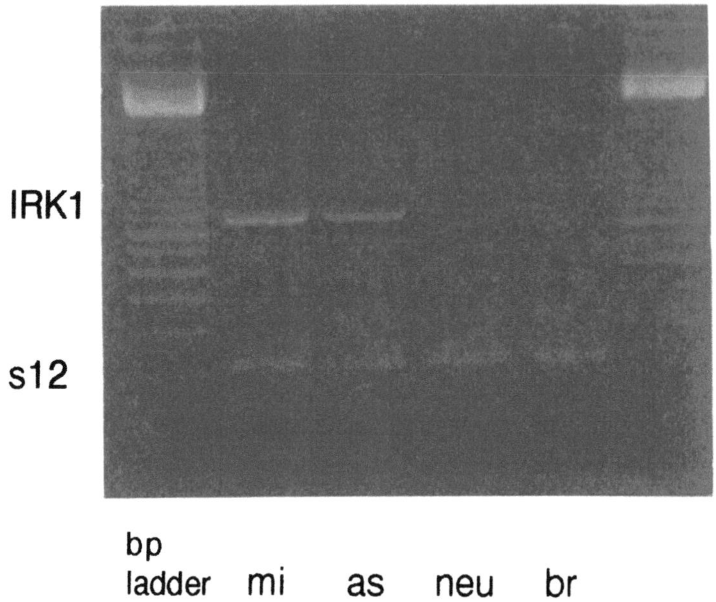

Figure 1. Occurrence of inward rectifying K^+ channel IRK1 mRNA in cultured brain cells. mRNA was reverse transcribed and amplified by PCR using the oligonucleotide primers listed in Table 1. Ribosomal protein s12 mRNA was used as internal control. mi microglia; as astroglia; neu hippocampal neurons; br whole brain. The results indicate predominant glial localization of this channel.

brain cells are damaged or die. Patch-clamp experiments performed by Kettenmann et al. (1990) revealed that microglia possess a rather unique inward rectifying K^+-channel, that is absent in other brain cell types and in peripheral macrophages (Banati et al., 1991). Very recently, our studies on mRNA expression of this channel in microglia, primary astrocytes and hippocampal neurons confirmed its strong expression in microglia. Some mRNA was observed in astrocytes, whereas the neurons were devoid of this mRNA species (Fig. 1). Since no outward K^+-channels were found in microglia, this type of channel was believed to be an important means for the cells to "sense" changes of the extracellular K^+-concentration and to react in an adequate way. Support for this notion came from our studies, using lipopolysaccharide (LPS) to differentiate microglia into more macrophage-like cells (Nörenberg et al., 1992). In this situation, the cells begin to express an outward rectifying K^+-channel (Kv1.3), which can be viewed as compensatory mechanism and, hence, as a kind of "desensitization" (Nörenberg et al., 1994). Our studies on microglial K^+-channels were subsequently pursued on the molecular biological level. In proliferating microglia Kv1.3 mRNA expression was low . Upon addition of LPS, however, this mRNA markedly increased within 2–3 h using quantitative RT-PCR (Fig. 2). These studies also showed the occurrence of additional K^+-channels in microglia. The previous electrophysiological data had already given rise to the suspicion that there was more than one inward rectifier channel, since we found two components of the current, one being Na^+-dependent and rapidly inactivating, the other Na^+-independent and not inactivating (Nörenberg et al., 1994). RT-PCR investigations on cultured microglia revealed mRNA expression of the inward K^+-channels IRK1 and ROMK1 and of the outward voltage-gated rectifier Kv1.6 in addition to Kv1.3 (primers and PCR conditions are summarized in Table 1). In subsequent time-course studies with LPS, we found a sharp decline and then a steady rise of IRK1

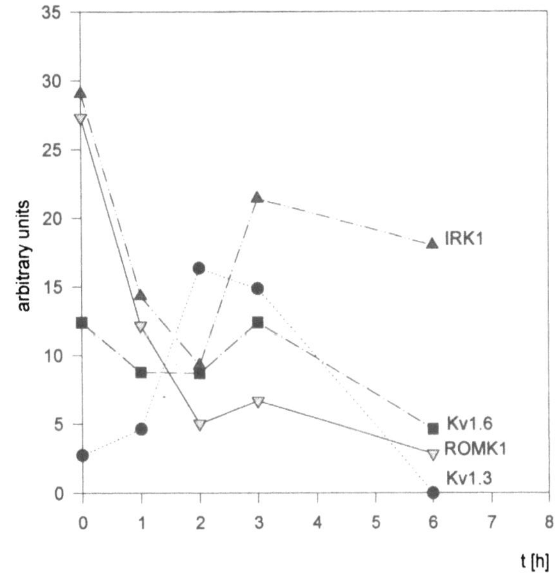

Figure 2. Time-course of two inward (IRK1, ROMK1) and two outward (Kv1.3, Kv1.6) rectifying K^+ channels in cultured microglia treated with LPS (100 ng/ml). Note the transiently increased expression of Kv1.3 mRNA, the marked inhibition of ROMK1, and the sustained high expression of IRK1 mRNAs. PCR-amplified cDNAs were normalized using s12 cDNA as internal standard in a computer-assisted quanitification program.

mRNA expression. ROMK1 mRNA exhibited a similar exponential reduction within the first two hours of LPS treatment but then kept stalling at this low level (Fig. 2). There was no dramatic effect on Kv1.6 mRNA expression. Kv1.3 mRNA, however, was markedly upregulated after addition of LPS reaching peak levels around 3–4 h, which subsequently dropped to baseline within a few hours (Fig. 2).

Subsequently, the occurrence of these four channels in rat brain was reevaluated by RT-PCR. The mRNAs were found in a number of selected brain areas, but the effect of peripherally administered LPS was less pronounced than in culture (Fig. 3). Kv1.3 mRNA was low but detectable in the four brain regions studied. It was increased, however, by LPS in cortex, diencephalon, and -much less- in pons. RT-PCR studies also showed increases of Kv1.6 mRNAs in each brain region, the most marked increases occurring in pons, whereas ROMK1 mRNA was significantly reduced except in cortical tissue. IRK1 mRNA was barely detectable in pons and in cerebellum. Evidently, it was not affected much by LPS treatment in cortex and diencephalon but moderately upregulated in cerebellum.

Table 1. Oligonucleotide primers used for PCR amplification of channel cDNAs

	5`-primer	3`-primer	**Size of PCR product**	**PCR condition**
Kv1.3	5`-GACGAGAAGGACTATCCCCGC	5`-CCCGGGGATACTGTTAAAACCCG	516 bp	36 cycles 64°C
Kv1.6	5`-TCAGTTCCGTGCAGATGGGC	5`-TAGGGAAGAGCGAGTCAACGT	634 bp	35 cycles 62°C
IRK1	5`-AAAGTCCATACCCGACAAC	5`-AGATACGAACTCCGGCATT	843 bp	10 cycles 56°C 23 cycles 54°C
ROMK1	5`-CCATAGGTTACGGATTCAG	5`-CAACTCGGTATTTCCCTTC	595 bp	10 cycles 56°C 29 cycles 54°C

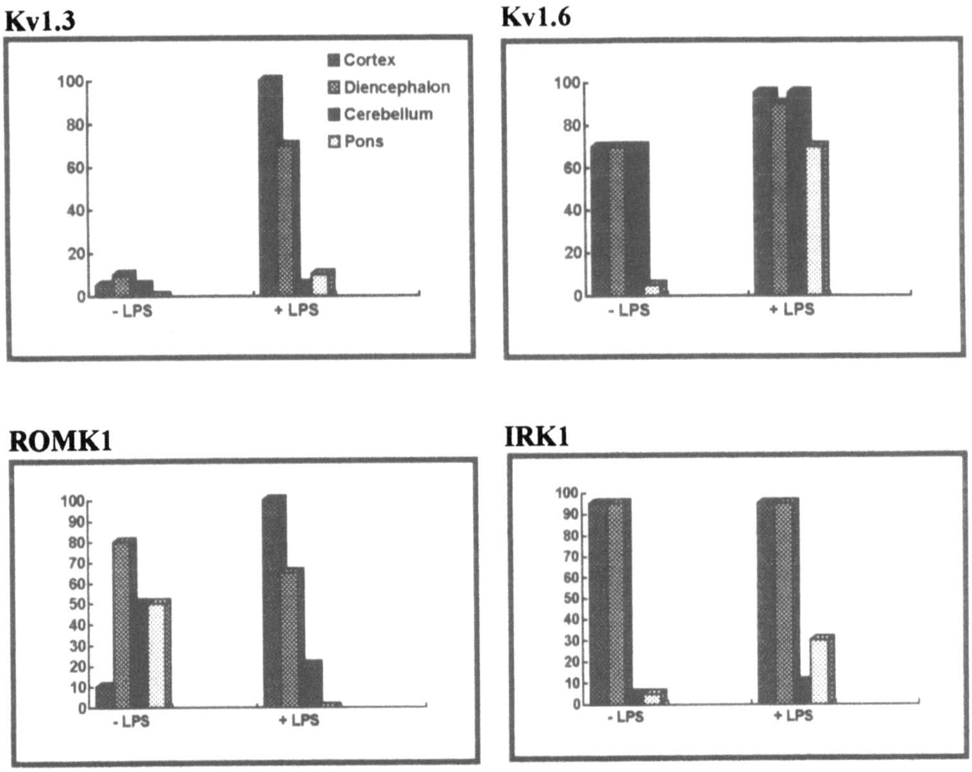

Figure 3. Expression of two inward (IRK1, ROMK1) and two outward (Kv1.3, Kv1.6) rectifying K^+ channels in some rat brain regions. Total RNA was used for RT-PCR and resultant cDNAs were quantified after electrophoretic separation and scanning using s12 cDNA as internal standard.

In situ-hybridizations on brain slices revealed the strongest expression of IRK1 in hippocampus and in midbrain indicating localization in neurons in these areas. Similar results have been reported by others (Karschin et al., 1996; Horio et al., 1996). Additionally, there was diffuse and much weaker staining in other brain regions which would be in favor of a glial localization. This staining appeared to increase upon LPS-treatment. Additionally, there was a significant induction in pons (Fig. 4a). Kv1.6 mRNA was strongly expressed in neurons, as judged by its pattern of appearance (Fig. 4b). The distribution of this channel on the protein level has been investigated by others (Scott et al., 1994; Veh et al., 1995). This does not, of course, preclude expression of this mRNA in glia, if one assumes that its expression in glia is much lower. LPS injections markedly increased this mRNA in the brain areas mentioned above. Kv1.3 mRNA expression was low throughout the brain, which is in agreement with results published elsewhere (Beckh and Pongs, 1990). Besides a diffuse distribution of label suggesting localization in glia, there was distinct labeling of dentate gyrus neurons, whereas no signal was found in cerebellum (Fig. 4c, compare with IRK1, Fig. 4a). The diffuse staining apparently increased upon LPS treatment. These data support our results obtained by RT-PCR (Fig. 3). Finally, ROMK1 mRNA was not traceable by in situ-hybridization in any brain region (not shown). Altogether, one can conclude that ion channels found in cultured microglia are not necessarily restricted to this cell type in vivo. Also, susceptibility of mRNA expression to LPS is not strictly comparable between the cultured cells and whole brain, which is

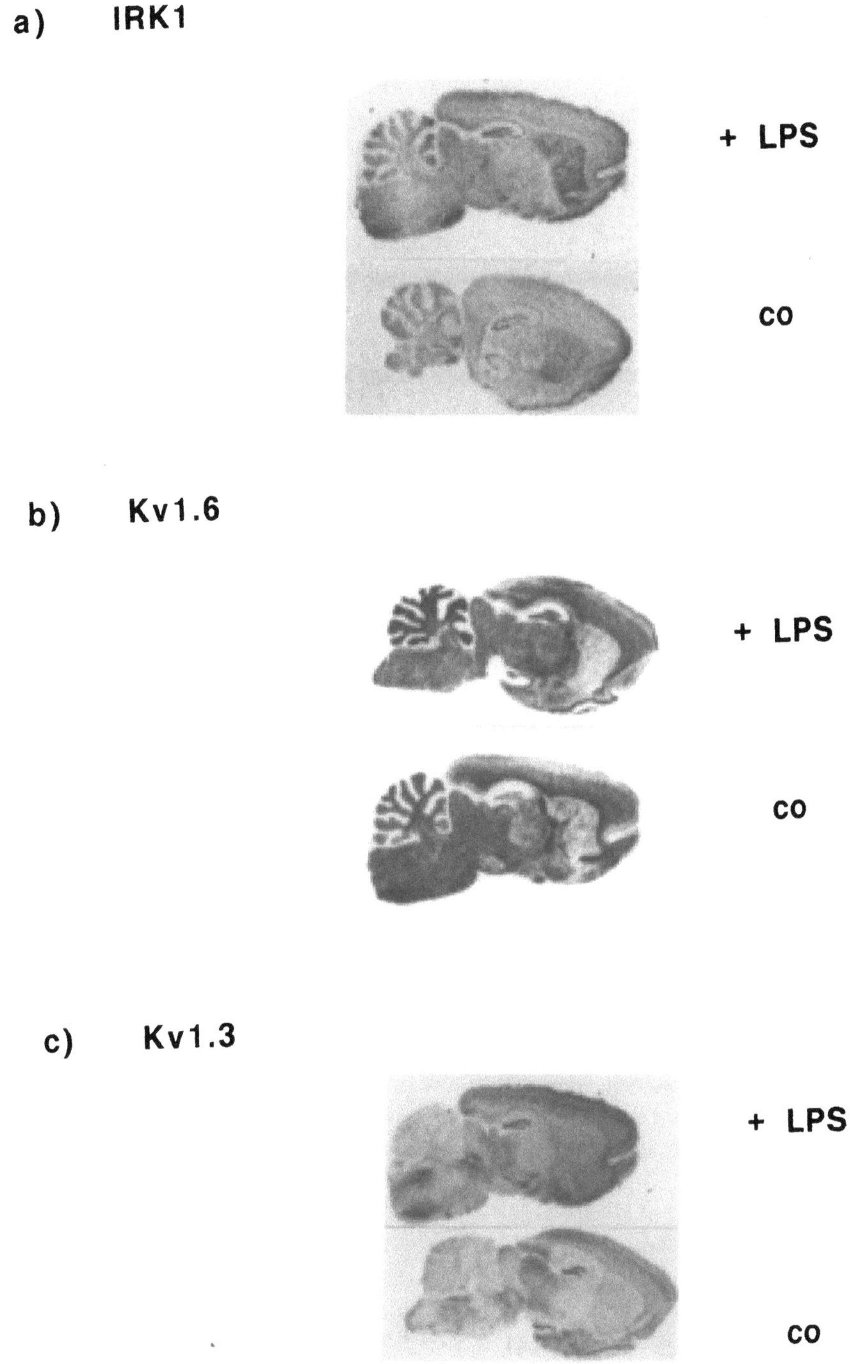

Figure 4. In situ hybridizations of inward rectifying K^+ channel IRK1 and outward rectifying K^+ channels Kv1.3 and Kv1.6. Brain slices from control (co) and LPS-injected rats (+LPS) were subjected to hybridizations using specific ^{32}P-labeled oligonucleotides. (For methods see Buttini and Boddeke, 1995).

not unexpected. However, it is interesting to note, that LPS has differential effects on each channel mRNA. Whether or not this has consequences for microglial activation cannot be predicted at present. In terms of a glial—or even microglial—localization in situ, Kv1.3 may be the best candidate, followed by ROMK1 and IRK1. The sensitivity of resting microglia towards fluctuations of extracellular K^+-concentrations, as hypothesized above, may hence be based on the expression of a special combination of K^+-channels, e.g. IRK1, and simultaneous lack of outward channels. Finally, it is interesting to note that we found an age-dependent delayed mRNA expression of Kv1.3 in cultured microglia upon LPS treatment (Gebicke-Haerter et al., 1996a). This may imply a delayed "desensitization" response and prolonged activation status. Consequently, the risk of overreactions and cell damaging events may be increased in aged microglial cells.

3. CYTOKINES

A number of inflammatory cytokines have been investigated in cultured microglia by us and others. We have shown that IL-1, IL-6, and TNF-α are low or not expressed at all in untreated, freshly isolated microglia (Ganter et al., 1992). Upon treatment of LPS, these cytokines are markedly increased, as shown on the protein level (Ganter et al., 1992) and for IL-6 (unpubl.) and TNF-α also on the mRNA level (Appel et al., 1995a).

Moreover, IL-1 polypeptide can be traced associated with microglia in brain tissue by immunocytochemistry (Giulian et al., 1986; van Dam et al., 1992; 1995). Specific induction of its mRNA in rat brain microglia after peripheral administration of LPS was shown by in situ hybridization (Buttini and Boddeke, 1995). The identity of the cells was confirmed by double labeling with OX42 antibodies which colocalized with the specific oligonucleotide probe (person. commun.).

Tumor necrosis factor-α mRNA was constitutively expressed in cultured microglia as revealed by RT-PCR, but was not found in normal rat brain. Lipopolysaccharide administration, however, induced this mRNA in brain microglia and also increased it in the isolated cells in culture (Appel et al., 1995a; Buttini et al., 1997).

TNF-ß mRNA could not be found in normal rat brain as well and also freshly isolated microglia showed no expression of this mRNA. Exposure of the cultured cells to LPS led to a short-term induction of this message after approximately 4 h and subsiding to undetectable levels beyond 16 h (Appel et al., 1995a). We were unable to detect this mRNA species by in situ hybridization, however, in rat brain after peripheral injection of LPS. Since TNF-ß mRNA was comparably low in the cultured cells already, it may have escaped in situ detection. In view of TNF-ß as shown to be associated with Multiple Sclerosis plaques (Navikas et al., 1996), our negative result could also lend support to the notion that there is a differential reaction of microglia depending on the nature of injury or disease (Gebicke-Haerter et al., 1996b).

Interleukin-3 polypeptide and mRNA were found in cultured microglia as shown by WESTERN blotting and in situ hybridization (Gebicke-Haerter et al., 1994), and mRNA increased upon LPS treatment as shown by quantitative RT-PCR (Appel et al., 1995a). A variety of stimuli were tested for their effects on IL-3 mRNA expression. TNF-α, IL-1, vitamin A, dbcAMP, and phorbol ester as well as phytohemagglutinin led to moderate to large elevations of IL-3 mRNA. In situ hybridizations on rat brain slices, however, did not reveal any positive signals even though some rats were injected with LPS (pers.commun.). Since IL-3 cDNA was amplified from rat brain by RT-PCR, it is feasible again that the sensitivity of in situ hybridization was too low to detect the specific signal.

The situation was different with IL-3 receptor ß-subunit, which was cloned from rat microglia in our laboratory (Appel et al., 1995b). Lipopolysaccharide upregulated this receptor mRNA in the cultured cells. This mRNA stayed at elevated levels for at least one week. In the normal rat brain no mRNA was found. A number of different brain areas were studied by RT-PCR after peripheral LPS injections. Already after three hours, receptor mRNA was detectable in each brain region. mRNA concentrations peaked at six hours and declined thereafter to almost baseline at 24 h. In situ hybridizations revealed occurrence of this mRNA exclusively in microglia upon LPS-treatment of rats. Receptor mRNA was also investigated in these animals using the paradigm of middle cerebral artery occlusion (MCAO). Here, this receptor mRNA stayed at peak level until 24 h and was still found after two days in the infarcted hemisphere. In situ hybridizations, again confirmed its sole localization in microglial cells, although in the penumbra region its expression by invaded blood macrophages could not be excluded.

4. DISCUSSION

Evidently, K^+-channels serve a number of distinct functions in brain, including neuronal activities. Outward rectifying K^+-channels are well known for their compensatory opening upon Na^+ depolarization of nerve cells. Kv1.6 may be a candidate channel involved in these events. It remains - presently - unclear if Kv1.6 is expressed in microglia in situ or in any other glial cell type at all, or if it is induced in one of these cell types by any injury or disease. Our in situ hybridization studies (Fig. 4b) are unable to answer this question. It is more surprising, that IRK1 is expressed also in some special neuronal populations in hippocampus and midbrain (Fig. 4a), since results from previous studies led us and others (Kettenmann et al., 1990; Nörenberg et al., 1992) to assume, that it may be a unique microglial marker. When expression patterns obtained from Kv1.6 are compared with those obtained from IRK1 (Karschin et al., 1996), it is obvious, that IRK1 is predominantly occurring in glial cells. Unfortunately, there are no high resolution pictures or even double labelings available identifying the glial cell types positive for IRK1 probe. Since, in culture, we did not find this channel in neurons (Fig. 1), and it reportedly was absent in astrocytes and oligodendrocytes as determined by patch-clamp experiments (Kettenmann et al., 1990), there is a good chance that the majority of signals seen in in situ hybridizations are associated with microglial cells.

Another good candidate K^+ channel for microglial localization is ROMK1. Apparently, its mRNA is expressed in brain and influenced by LPS. As already reported by others (Kenna et al., 1994), its mRNA is present only in hippocampus and, at low level, in cortex. ROMK1, moreover, is interesting because its mRNA, in contrast to the other channel mRNAs, is down-regulated by LPS both in cultured microglia and in situ. The decline of ROMK1 mRNA may precede the upregulation of Kv1.3. It possibly marks a transition stage between proliferating and non-proliferating microglia, i.e. ROMK1 being an earlier marker for microglia activation than Kv1.3. Evidently, Kv1.3 mRNA expression prevails in the glial compartment, except for its striking, constitutive occurrence in dentate gyrus neurons. Here, it apparently serves a specialized function inherent to these cells.

The molecular mechanisms involved in microglial activation triggered by peripheral administration of LPS remain both fascinating and unresolved issues. Evidently, it is not a direct action of LPS, since it is very unlikely that this hydrophilic substance can cross the blood-brain barrier. Even more importantly, immediate contact of LPS with brain cells in situ, e.g. by injection into brain parenchyma, turned out to be a very poor stimulus for microglia (V.H. Perry, pers. commun.). The best guess, therefore, would be the involvement of some unidentified messenger pathway transforming the LPS signal at the capillary endothelia into a brain-specific message. This message may activate G-protein coupled receptors which are directly linked to K^+ channels or which transmit incoming signals to the cell nucleus and in this way interfere with transcriptional regulation of channel genes.

Some cytokine receptors are coupled to small G-proteins, others to protein tyrosine kinases. In each case, signaling proceeds via protein kinase cascades to the effect of transcriptional activation or inhibition. The IL-3 receptor belongs to a receptor family using the latter signaling pathway, which leads to cell proliferation. For this reason, the IL-3 receptor system has been studied in microglia. Microglia proliferation is believed to be the first stage of activation. Although it is not clear, which ligand induces microglial proliferation in vivo (IL-3, IL-5, and GM-CSF can bind to the IL-3 receptor), the rapid induction of receptor mRNA upon peripheral LPS administration probably marks a crucial event initializing transition of microglia from quiescence to cell division. IL-1 and TNF-α, which are involved in inflammatory processes, have also been shown to support microglial proliferation (Ganter et al., 1992). Their appearance, like IL-6 expression, may be located at the end of proliferation and at the beginning of differentiation into macrophages. In this dual function, these cytokines may then maintain recruitment of additional microglia and keep fueling inflammation, which eventually results in progressive exacerbation of disease.

ACKNOWLEDGMENTS

We are grateful to Prof. M. Berger for generous support of microglial research. Supported by DFG grants Ge 486/6–2;6–3;9–1, and Tropon-Werke, Köln.

REFERENCES

Appel, K., Honegger, P., Gebicke-Haerter, P.J. 1995a. Expression of interleukin-3 and tumor necrosis factor-ß mRNAs in cultured microglia. J.Neuroimmunol. 60:83–91.

Appel, K., Buttini, M., Sauter, A., Gebicke-Haerter, P.J. 1995b. Cloning of rat interleukin-3 receptor ß-subunit from cultured microglia and its mRNA expression in vivo. J.Neurosci. 15:5800–5809.

Banati, R. B., Hoppe, D., Gottmann, K., Kreutzberg, G.W., Kettenmann, H. 1991. A subpopulation of bone marrow-derived macrophage-like cells shares a unique ion channel pattern with microglia. J.Neurosci.Res. 30:593–600.

Beckh, S., Pongs, O. 1990. Members of the RCK potassium channel family are differentially expressed in the rat nervous system. EMBO J. 9: 777–782.

Buttini, M., Boddeke, H.W.G.M. 1995. Peripheral lipopolysaccharide stimulation induces IL-1ß messenger RNA in rat brain microglial cells. Neuroscience. 65:523–530.

Buttini, M., Appel, K., Gebicke-Haerter, P.J., Limonta, S., Boddeke, H.W.G.M. 1997. Peripheral stimulation with lipopolysaccharide induces tumor necrosis factor-α mRNA and protein in rat brain microglial cells. Br.J.Pharmacol.:subm.

Ganter, S., Northoff, H., Männel, D., Gebicke-Haerter, P.J. 1992. Growth control of cultured microglia. J.Neurosci.Res. 33:218–230.

Gebicke-Haerter, P. J., Appel, K., Taylor, G.D., Schobert, A., Rich, I.N., Northoff, H., Berger, M. 1994. Rat microglial interleukin-3. J.Neuroimmunol. 50:203–214.

Gebicke-Haerter, P. J., van Calker, D., Nörenberg, W., Illes, P. 1996a. Ionic and molecular events in the course of microglial activation. Fair or foul for health or disease ? In Topical issues in microglia research. E. A. Ling and C.-K. Tan, editors. Singapore Neurosci. Association, Singapore. 143–164.

Gebicke-Haerter, P., van Calker, D., Illes, P. 1996b. Molecular mechanisms of microglial activation. A. Implications for regeneration and neurodegenerative diseases. Neurochem.Int. 29:1–12.

Giulian, D., Baker, T.J., Shih, N., Lachman, L.B. 1986. Interleukin-1 of the central nervous system is produced by ameboid microglia. J.exp.Med. 164:594–604.

Horio, Y., Morishige, K.I., Takahashi, N., Kurachi, Y. 1996. Differential distribution of classical inwardly rectifying potassium channel mRNAs in the brain: comparison of IRK2 with IRK1 and IRK3. FEBS Lett. 379: 239–243.

Karschin, C., Dimann, E., Stühmer, W., Karschin, A. 1996. IRK(1–3) and GIRK(1–4) inwardly rectifying K^{+} channel mRNAs are differentially expressed in the adult rat brain. J.Neurosci. 16:3559–3570.

Kenna, S., Roeper, J., Ho, K., Hebert, S., Ashcroft, S.J.H., Ashcroft, F.M. 1994. Differential expression of the inwardly-rectifying K-channel ROMK1 in rat brain. Molec.Brain Res. 24: 353–356.

Kettenmann, H., Hoppe, D., Gottmann, K., Banati, R., Kreutzberg, G.W. 1990. Cultured microglial cells have a distinct pattern of membrane channels different from peritoneal macrophages. J.Neurosci.Res. 26:278–287.

Navikas, V., He, B., Link, J.,Haglund, M., Soderstrom, M., Fredrikson, S., Ljungdahl, A., Hojeberg, J., Qiao, J., Olsson, T., Link, H. 1996. Augmented expression of tumor necrosis factor-α and lymphotoxin in mononuclear cells in multiple sclerosis and optic neuritis. Brain. 119:213–223.

Nörenberg, W., Gebicke-Haerter, P.J., Illes, P. 1992. Inflammatory stimuli induce a new K-outward current in cultured rat microglia. Neurosci. Lett. 147:171–174.

Nörenberg, W., Gebicke-Haerter, P.J., Illes, P. 1994. Voltage-dependent potassium channels in activated rat microglia. J.Physiol. (London). 475:15–32.

Scott, V.E.S., Muniz, Z.M., Sewing, S., Lichtinghagen, R., Parcej, D.N., Pongs, O., Dolly, J.O. 1994. Antibodies specific for distinct Kv subunits unveil a heterooligomeric basis for subtypes of α-dendrotoxin-sensitive K^{+} channels in bovine brain. Biochemistry 33:1617–1623.

van Dam, A. M., Brouns, M., Louisse, S., Berkenbosch, F. 1992. Appearance of interleukin-1 in macrophages and in ramified microglia in the brain of endotoxin-treated rats : a pathway for the induction of non-specific symptoms of sickness ? Brain Res. 588:291–296.

van Dam, A. M., Bauer, J., Tilders, F.J.H., Berkenbosch, F. 1995. Endotoxin-induced appearance of immunoreactive interleukin-1ß in ramified microglia in rat brain : a light and electron microscopic study. Neuroscience. 65:815–826.

Veh, R.W., Lichtinghagen, R., Sewing, S., Wunder, F., Grumbach, I.M., Pongs, O. 1995. Immunohistochemical localization of five members of the Kv1 channel subunits: contrasting subcellular locations and neuron-specific co-localizations in rat brain. Eur.J.Neurosci. 7:2189–2205.

Part II

Neuronal Circuitry, Synaptic Plasticity, Regeneration

9

ANATOMICAL AND FUNCTIONAL CHARACTERISTICS OF TRANSPLANTED MONOAMINERGIC NEURONS IN PARAPLEGIC RATS

Minerva Giménez y Ribotta, Christelle Roudet, and Alain Privat*

INSERM U. 336, Developpement
Plasticité et Vieillissement du Système Nerveux
Université Montpellier II. Montpellier, France

1. INTRODUCTION

The transplantation of embryonic tissue in the CNS has been used as a strategy for reducing deficits associated with degenerative diseases and for promoting functional recovery after brain or spinal cord injury (Fisher and Gage, 1993; Freed, 1993; Goldberger et al., 1993a; Nishino 1993).

In spinal cord injury, the general aim has been directed at the development of alternative pathways that facilitate a partial functional connection (Das, 1983; Goldberger et al., 1993b; Nornes et al., 1984; Nothias and Peschanski, 1990; Reier et al, 1986, 1988; Tessler, 1991; Tessler et al., 1992). One approach was to transplant fetal spinal cord into the injury site to restore a partial communication between proximal and distal stumps of the cord (Bregman et al., 1991, 1993; Kunkel-Bagden and Bregman, 1990; Nothias et al., 1990; Reier et al., 1988; Stokes and Reier, 1992). The transplantation, below the lesion, of brainstem monoaminergic neurons constitutes an alternative approach, since these descending systems are involved in the intrinsic spinal cord circuitry that mediates locomotor and autonomic functions (Forssberg and Grillner, 1973; Mas et al., 1985; Rossignol et al., 1986). Thus, the noradrenergic (NA) system regulates these physiological functions involving the adrenoceptors (Mason and Fibiger, 1979; Connor et al., 1981; Shi et al., 1988; Rawlow and Gorka, 1986; Tanabe et al., 1990). The presence of α1- and α2-adrenoceptors in the rat spinal cord has been well characterized (Giron et al., 1985; Jones et al., 1982; Simmons and Jones, 1988; Unnerstall et al., 1984) and a precise topography has been determined by radioautography (Young and Kunhar, 1980; Roudet et al., 1993, 1994). Moreover, it is well established that monoaminergic neurons, following transplan-

* Correspondence to: Alain Privat. INSERM U. 336, Developpement, Plasticité et Vieillissement du Système Nerveux, Université Montpellier II. Montpellier. France, Tel. (33) 04 67 14 33 86; Fax. (33) 04 67 14 33 18.

Brain Plasticity, edited by Filogamo *et al.*
Plenum Press, New York, 1997

tation into the damaged spinal cord of adult rats, can survive and integrate within the host spinal cord (Björklund et al., 1983; Commissiong, 1983, 1984; Foster et al., 1985; König et al., 1989; Nygren et al., 1977; Privat et al., 1986, 1988, 1989; Rajaofetra et al., 1991, 1992a; Reier et al., 1992; Yakovleff et al, 1989, 1995), and specifically reinnervate target regions, re-establishing a pattern comparable to the normal intact innervation (Privat et al., 1988, 1989; Rajaofetra et al., 1992a,b; Yakovleff, 1989, 1995). Indeed, grafted locus coeruleus NA-neurons are also able to restore levels of neurotransmitters in the depleted cord (Björklund et al., 1986; Moorman et al., 1990), to contribute to the recovery of hindlimb flexion and withdrawal reflexes in animals with the chemically depleted cord (Buchanan and Nornes, 1986; Moorman et al., 1990), and to increase the excitability of the spinal stepping generator in animals with the transected spinal cord (Yakovleff et al., 1989, 1995).

In a previous radioautographic study we have shown a significant increase of the α-1 and α-2 adrenoceptors density in two models of spinal cord injury: (1) complete cord transection and (2) chemical denervation by 6OHDA (Roudet et al., 1993, 1994).

In the present report we have used these two models in order to investigate whether embryonic NA-neurons transplanted following a lesion could normalize the denervation-induced increase in spinal α1- or α2-adrenoceptors densities. In addition, the differential restorative effect exerted by transplanted NA-neurons on α-1 or α-2 adrenoceptors densities in the two experimental paradigms was correlated with an ultrastructural immunocytochemical study of functional graft-host connections.

2. MATERIALS AND METHODS

For this study, adult male Sprague-Dawley rats (IFFA-CREDO, France) were divided in five groups: intact animals (n=8), transected animals (n=8), transected-grafted animals (n=14), 6-OH DA lesioned animals (n=8), and 6-OH DA lesioned-grafted animals (n=15).

Procedures of spinal cord transection at thoracic level (T8-T9) and selective NA-lesion by intracisternal injection of 6-OH DA solution (200μg 6-OH DA hydrobromide in 20 μl 0.9% NaCl containing 1% ascorbic acid) were performed as described in previous reports (Rajaofetra et al., 1992a; Roudet et al., 1993).

The cell suspension was prepared from embryos obtained by laparotomy from pregnant rats at embryonic day 13. The microdissection of the locus coeruleus region has been described in detail by König et al. (1988) and Yakovleff et al. (1989). The tissue was mechanically dissociated in calcium- and magnesium-free Puck's solution and the suspension was then centrifuged at 80g for 10 min, resuspended in culture medium, and adjusted to a final concentration of 30,000–50,000 cells/μl.

In the transection model, the optimal period for grafting was one week postlesion whereas in the chemical lesion model, the maximal NA-denervation was obtained only after one month. Thus, after one week or after one month, animals were grafted by the procedure described by Rajaofetra et al. (1992a). Briefly, the animals were placed in a stereotaxic frame and a new laminectomy was performed at T10-T11 level. Four microlitres of the cell suspension were injected into the spinal cord at a depth of 1mm below the pial surface with the help of a metallic needle (0.4 mm diameter) connected to a Hamilton microsyringe. The speed of injection was 1μl/min and the needle was withdrawn after 3 min of total injection.

After a survival time of one month, all animals were sacrificed: (1) by intracardiac perfusion with 5% glutaraldehyde in 50 mM sodium metabisulfite (SMB)- 50 mM cacodylate buffer for the immunocytochemical study, and (2) by decapitation followed by a rapid freezing in isopentane for the quantitative radioautographic study. For immunocytochemical study, transverse vibratome sections (50μm thick), taken above the section (T6 level) and below the section (>3 mm rostral to the graft, at the level of the graft, and at lumbosacral level, >3 mm caudal to graft) were processed for NA immunodetection. For electron microscopy, selected sections from the level of the graft and at 3 to 5 mm caudal to the graft, were collected and processed as previously detailed (Rajaofetra et al., 1992b). For quantitation of varicosity profiles, a minimum of 15 grids per animal were used, each grid carrying at least three sections. Criteria for the identification of synaptic boutons have been also detailed (Rajaofetra et al., 1992b).

For quantitative radioautographic study, transverse cryostat sections (10μm thick) taken above the section (T6 level) and below the section (at lumbar and sacral level), were processed for radioautography as described previously (Roudet et al., 1993, 1994). Optical densities of autoradiograms were quantitated with a computer-based image analysis system (SAMBA 2005, Alcatel) and converted in fmol of radioligand specifically bound per mg of protein using calibrated tritiated microscales as previously reported (Peretti-Renucci et al., 1991). For quantitative analysis, the spinal grey matter was divided into six areas according to the cytoarchitectonic organization of the rat spinal cord (Molander et al., 1984, 1989). Masks of these regions were drawn on the basis of the picture obtained with [^{3}H]-prazosin or [^{3}H]-Rauwolscine binding with the help of a computer mouse. A schematic representation is shown in Fig.1. The masks were modified and rotated to conform individual appearances to ensure adequate comparison between homologous anatomical regions. Three to five sections per level of spinal cord were analyzed for each animal. Autoradiographic data were statistically analyzed using parametric tests as an analysis of variance (Anova) and Bonferroni's post-hoc test.

3. RESULTS

3.1. Model of Spinal Cord Transection

3.1.1. Ultrastructural Immunocytochemical Study. In transected animals, one week after surgery, the spinal cord above the section showed a NA-immunoreactivity distributed in the three main target regions, qualitatively similar to that described in intact animals

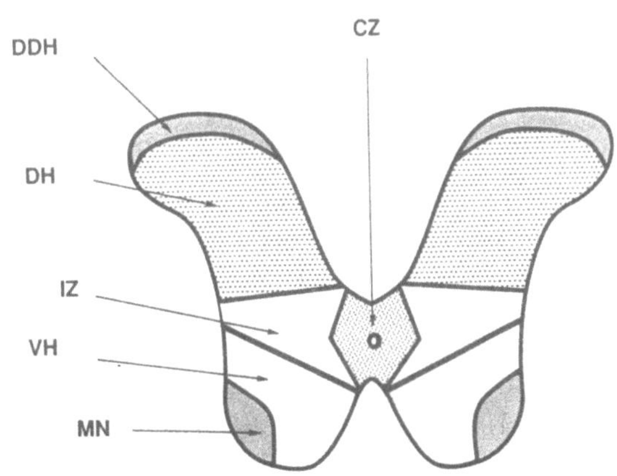

Figure 1. Masks used for the quantitative analysis of [^{3}H]-Prazosin or [^{3}H]-Rauwolscine specific binding sites on a schematic transverse tissue section of the rat spinal cord. The grey matter was divided into six areas: **DDH**, distal dorsal horn, **DH**, dorsal horn, **CZ**, central zone, **IZ**, intermediate zone, **VH**, ventral horn, and **MN**, motoneuron area.

(Björklund and Skagerberg, 1982; Westlund et al., 1982, 1983; Rajaofetra et al., 1992b). The spinal cord below the section was totally devoid of NA-immunoreactivity, consequently to a total loss of descending inputs. No NA-immunoreactive neuronal perikarya were ever detected.

In all transected-grafted animals, one month postgrafting, the spinal cord exhibited a well-developed transplant with many NA-immunoreactive perikarya (Fig. 2A). Most transplants appeared laterally located, although sometimes they were hardly distinguishable from host tissue. Within the transplant, the ovoid, fusiform or multipolar NA-immunoreactive neurons were generally clumped, and gave rise to dense and extensive networks of immunoreactive varicose processes, which extended out in all directions over varying distances, developing at one month postgrafting, a rich plexus of varicose fibres some of which crossed the midline dorsally to the central canal, towards the contralateral side with a clear predilection for the three main target regions of the NA system in the spinal cord: The dorsal horn, the intermediolateral column, and the ventral horn where NA-varicose fibres were distributed around the motoneurons' somata. This specific innervation was visualized up to at least 20 mm rostrocaudally within the spinal cord (Fig. 2B).

Ultrastructural examination of the graft area evidenced numerous immunoreactive perikarya with morphological features of mature NA-neurons, even at this relatively early time postgrafting. Medium-sized perikarya contained a notched nucleus with pale and homogeneous chromatin, and well-developed nucleolus were observed. Their cytoplasm displayed abundant granular endoplasmic reticulum and many mitochondria. Golgi zones were also observed disposed in dictyosomes and giving rise to dense core vesicles. The presence of glial processes was noted in the neuropil around NA-perikarya, dendrites and varicosities. Unlabelled boutons and, occasionally, NA-immunoreactive varicosities were observed contacting NA-immunoreactive perikarya and dendritic processes. These immunoreactive dendrites containing microtubules, mitochondria and multivesicular bodies, exhibited characteristic postsynaptic densities at the contact zone with the afferent boutons which contained predominantly clusters of small spherical vesicles. The synaptic contacts observed at the surface of NA-immunoreactive dendritic profiles were mainly of the asymmetric type (Gray's type I) (Fig. 3A).

NA-immunoreactive axonal fibres were largely distributed in the host tissue, in the vicinity of the transplant. In the ventral horn, at some distance from the transplant (3 to 5 mm), NA-immunoreactive profiles generally appeared as thin and long fibres, with small boutons forming synaptic contacts with unlabelled perikarya and dendritic profiles. These NA-immunoreactive varicosities contained round, small, clear vesicles and formed type II synapses. NA-immunoreactive boutons were also seen to make synapses with dendritic spines (Fig. 3B). Most of the NA-synapses appeared to be of type II. About 58% of the NA-immunoreactive varicosity profiles containing small, spherical vesicles did not display a synaptic junction in single thin sections (Table 1). Among synaptic contacts, 5% were axosomatic, 29% axodendritic and 7% axospinous.

3.1.2. Quantitative Autoradiographic Study (Fig. 4). The density of [^{3}H]-Prazosin binding on α1-adrenoceptors was different according to the level of spinal cord analyzed.

In intact, transected, and transected-grafted animals, above the section, the labeling was homogeneous in the grey matter. Densities ranged from 219 ± 50 to 324 ± 20 fmol/mg prot. and no significant variations were observed in the six areas analyzed at thoracic level among the three groups of animals.

In transected animals, at lumbar and sacral levels, the density of [^{3}H]-Prazosin binding sites was significantly increased by 27% ($p<0.01$) to 66% ($p<0.001$) compared to val-

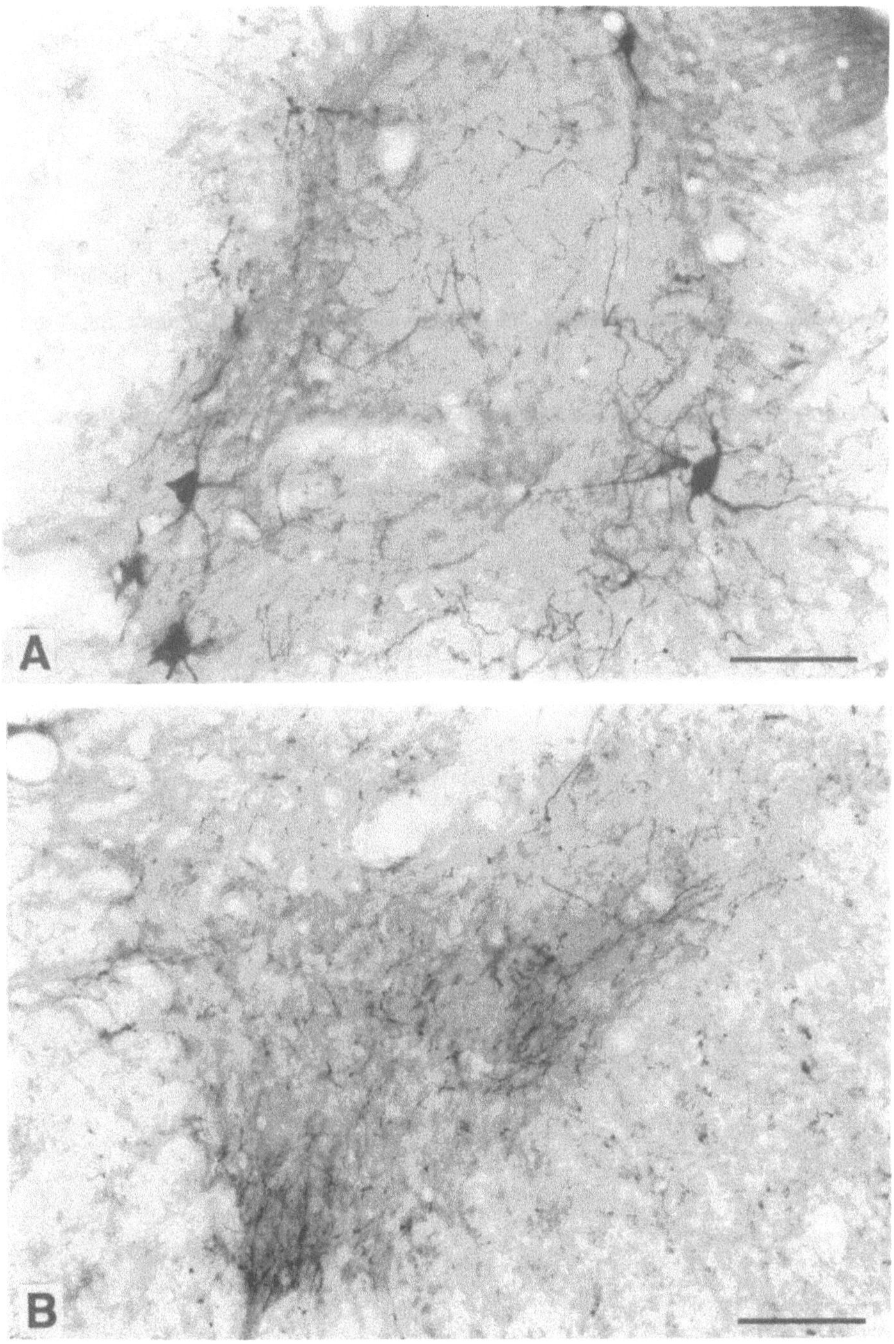

Figure 2. Immunohistochemical detection of NA in the spinal cord transection model. **A.** At one month postgrafting, a transverse section at the graft level shows ovoid and multipolar NA-immunoreactive neurons and their well-developed varicose processes, distributed unilaterally in the grey matter. Bar: 100 μm. **B.** At distance from the graft (>3mm) a dense meshwork of varicose processes is largely distributed around motoneuron somata. Bar: 100 μm.

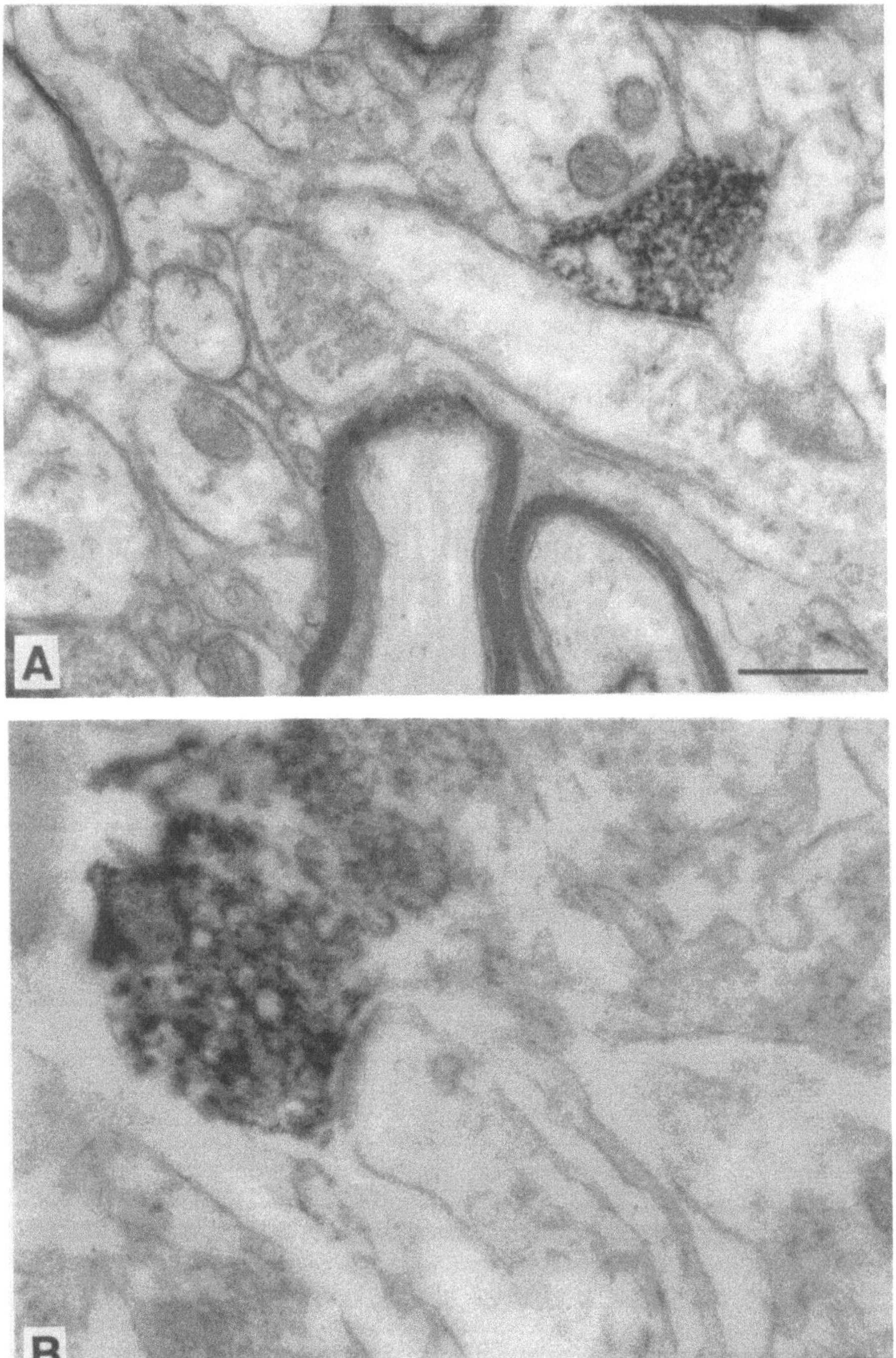

Figure 3. Ultrastructural organization in the spinal cord transection model. **A.** In the vicinity of the graft, a NA-immunoreactive bouton making type I synapse on a dendritic profile. Bar: 0.5 μm. **B.** At distance from the graft (3 to 5 mm) a NA-immunoreactive bouton makes a type II synaptic contact on a dendritic spine. Bar: 0.5 μm.

Table 1. Development of synaptic connectivity (%)

	Control*	Transection n=3	Lesion by 6-OH DA n=3
NS	61.8	58.2	60.5
ADS	38.2 (ADS + ASPS)	29.0	15.1
ASPS		7.3	22.1
ASS		5.5	2.3

An average of 100 varicosity profiles has been counted in each model.

NS, non synaptic varicosities. **ADS**, axodendritic synapses.
ASPS, axospinous synapses. **ASS**, axosomatic synapses.

*From Rajaofetra et al., 1992b.

ues detected in intact animals (359 ± 20 to 485 ± 32 fmol/mg prot. vs 282 ± 19 to 293 ± 27 fmol/mg prot.).

In transected-grafted animals, the density of [^{3}H]-Prazosin binding sites detected at lumbar level was similar to that found in intact animals (260 ± 29 to 410 ± 31 fmol/mg prot. vs 246 ± 25 to 338 ± 37). At variance, at sacral level, the number of binding sites observed remained significantly increased by 33% to 54% and was comparable to that found at the same level in transected animals.

The density of [^{3}H]-Rauwolscine binding on α2-adrenoceptors, above the section, appeared to be homogeneous in the three groups of animals studied, except in the superficial dorsal horn where a significant increase by 33% ($p<0.05$) was observed in transected-grafted animals.

In transected animals, below the section, α2-adrenoceptors densities were increased within the whole grey matter and in particular in the dorsal horn (DDH) and DH areas, confirming our previous observations. (Roudet et al., 1994).

In transected-grafted animals, the density of [^{3}H]-Rauwolscine binding sites detected at lumbar level was not significantly different from that observed in transected animals, except in the superficial dorsal horn (DDH) as well as in DDH and DH at sacral level.

3.2. Model of Spinal Cord Denervation by 6-OH DA

3.2.1. Ultrastructural Immunocytochemical Study. One month after 6-OH DA lesion, animals exhibited an almost total loss of NA-immunoreactive fibres from grey matter of the spinal cord. Only a few isolated NA-immunoreactive processes were occasionally detected in the intermediolateral column and in the ventral horn, as previously shown (Roudet et al., 1993, 1994). These strongly immunoreactive residual fibres showed coarse profiles with large varicosities, an aspect very different from that found in intact animals.

Ultrastructural examination in ungrafted-lesioned animals showed some occasional densely immunoreactive profiles containing numerous vacuoles suggesting a degenerated state of the terminals. Similar varicosities were also observed in grafted spinal cords.

One month postgrafting, in lesioned-grafted animals, as in the first model, a large number of scattered NA-immunostained cell bodies and a dense plexus of immunoreactive

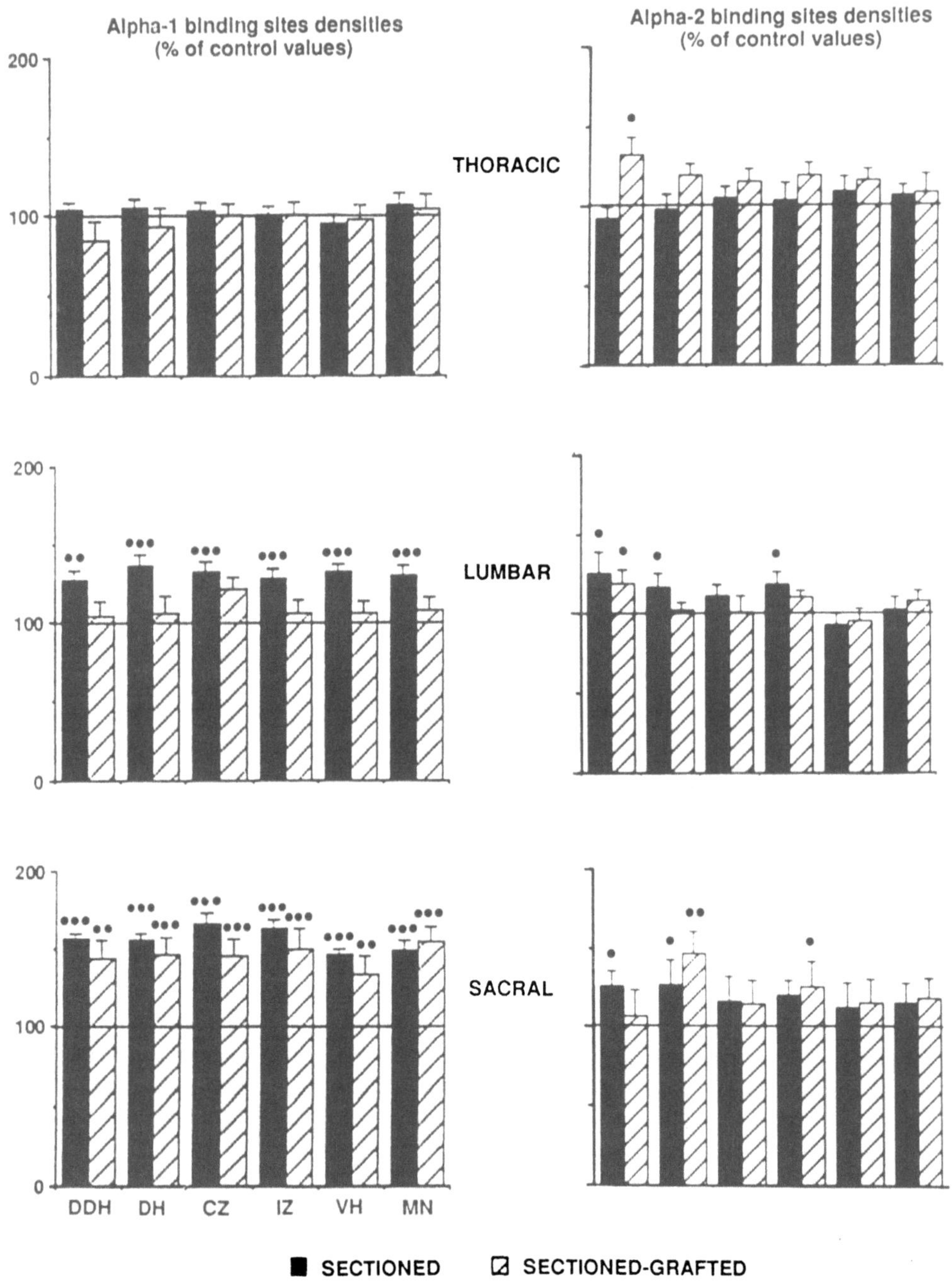

Figure 4. Model of spinal cord transection. Quantitative analysis of α1- and α2-adrenoceptors densities as measured by using [3H]-Prazosin or [3H]-Rauwolscine. Histograms represent the total densities of α-adrenoceptors expressed as the ratio in per cent (mean ± sem) of homologous subregions of the grey matter as defined in Fig. 1 (**DDH, DH, CZ, IZ, VH,** and **MN**) between transected (n=8), and transected-grafted (n=14) and intact (n=8) animals. 100% corresponded to values found in intact group: 244 ± 30 to 315 ± 27, 246 ± 25 to 338 ± 37, and 234 ± 26 to 293 ± 27 fmol/mg prot. at thoracic, lumbar, and sacral levels of the spinal cord respectively. * Comparison with values found in intact animals: * $p<0.05$, ** $p<0.01$, *** $p<0.001$. + Comparison with values found in transected animals: + $p<0.05$, ++ $p<0.01$, +++ $p<0.001$. Note the increase of α-adrenoceptors densities below the section (lumbar and sacral levels) and the effect of the graft, particularly at the lumbar level.

fibres were detected within the transplant. The immunoreactive perikarya gave rise to fine, straight and poorly varicose processes. The extension and the distribution of NA-fibres were similar to that of the first model: Small immunoreactive profiles were observed in the dorsal horn and a dense innervation was detected specifically in the intermediolateral column and in the motoneuron area. These thin processes with relatively few varicosities were intermingled with presumably residual intrinsic NA-fibres. In this model, the innervation was also observed to extend from the transplant into the host spinal cord for a distance of up to about 20mm.

As in the first model, the ultrastructural examination showed immunoreactive cell bodies, largely distributed in the graft area, with evident signs of differentiation. A slightly immunoreactive perikaryon rich in organelles with a nucleus often indented was observed in most of the grafted NA-neurons. Some synaptic boutons were observed to form contacts with NA-immunoreactive perikarya or dendritic profiles. At some distance from the transplant (3 to 5 mm), in the ventral horn, we detected long NA-immunoreactive profiles and some small NA-immunoreactive varicosities exhibiting intrinsic features similar to those already described in the first model. These NA-immunoreactive varicosities were usually forming type II synapses. A few type I synapses were observed with unlabelled dendrites and rarely with perikarya. Interestingly, the majority of these type I synapses involved unlabelled dendritic spines. The frequency of non-synaptic NA-varicosities in single thin sections was similar to that of the first model (about 60%). The axosomatic synaptic contacts were a small minority representing a mean incidence of 2%. The percentage of axodendritic synapses was particularly low (15%), whereas the percentage of axospinous synapses was higher than in the first model (22%). However, the total percentage of axodendritic plus axospinous synapses was similar in the two models studied (Table I).

3.2.2. Quantitative Autoradiographic Study (Fig. 5). After 6-OH DA lesion, all areas of the grey matter at all levels of the spinal cord showed a significant increase by 21% to 68% of [^{3}H]-Prazosin binding sites densities, compared to values found in intact animals (380 ± 24 to 416 ± 14 fmol/mg prot. vs 248 ± 28 to 315 ± 27 fmol/mg prot.).

In lesioned-grafted animals, at thoracic level (which corresponded to non-reinnervated regions according to NA-immunocytochemical results), the density of [^{3}H]-Prazosin binding sites remained significantly increased by 24% to 73% as compared to values found in intact animals (367 ± 54 to 429 ± 58 fmol/mg prot. vs 248 ± 35 to 297 ± 30 fmol/mg prot.) and were similar to those found at the same levels in lesioned animals.

At lumbar level, however, a significant decrease of densities was only observed in three areas of the grey matter analyzed: CZ, VH and MN, corresponding therefore to a partial restorative effect towards values found in intact animals.

At sacral level, which was not systematically reinnervated by transplanted NA-neurons, the densities observed remained significantly increased by 32% to 65% as compared to intact animals (331 ± 31 to 386 ± 47 fmol/mg prot. vs 234 ± 26 to 251 ± 24 fmol/ mg prot.) and were comparable to those of the same level in lesioned animals.

After 6-OHDA lesion, the density of [^{3}H]-Rauwolscine binding showed a significant increase by 32% ($p<0.05$) to 50% ($p<0.01$) in the DDH at all levels of the spinal cord and in all remaining areas at lumbar level as well as in DH at sacral level, as compared to values found in intact animals.

In lesioned-grafted animals, α2-adrenoceptors densities were largely higher than values detected in lesioned animals at all the levels of the spinal cord except at sacral level where the density in CZ and MN remained comparable.

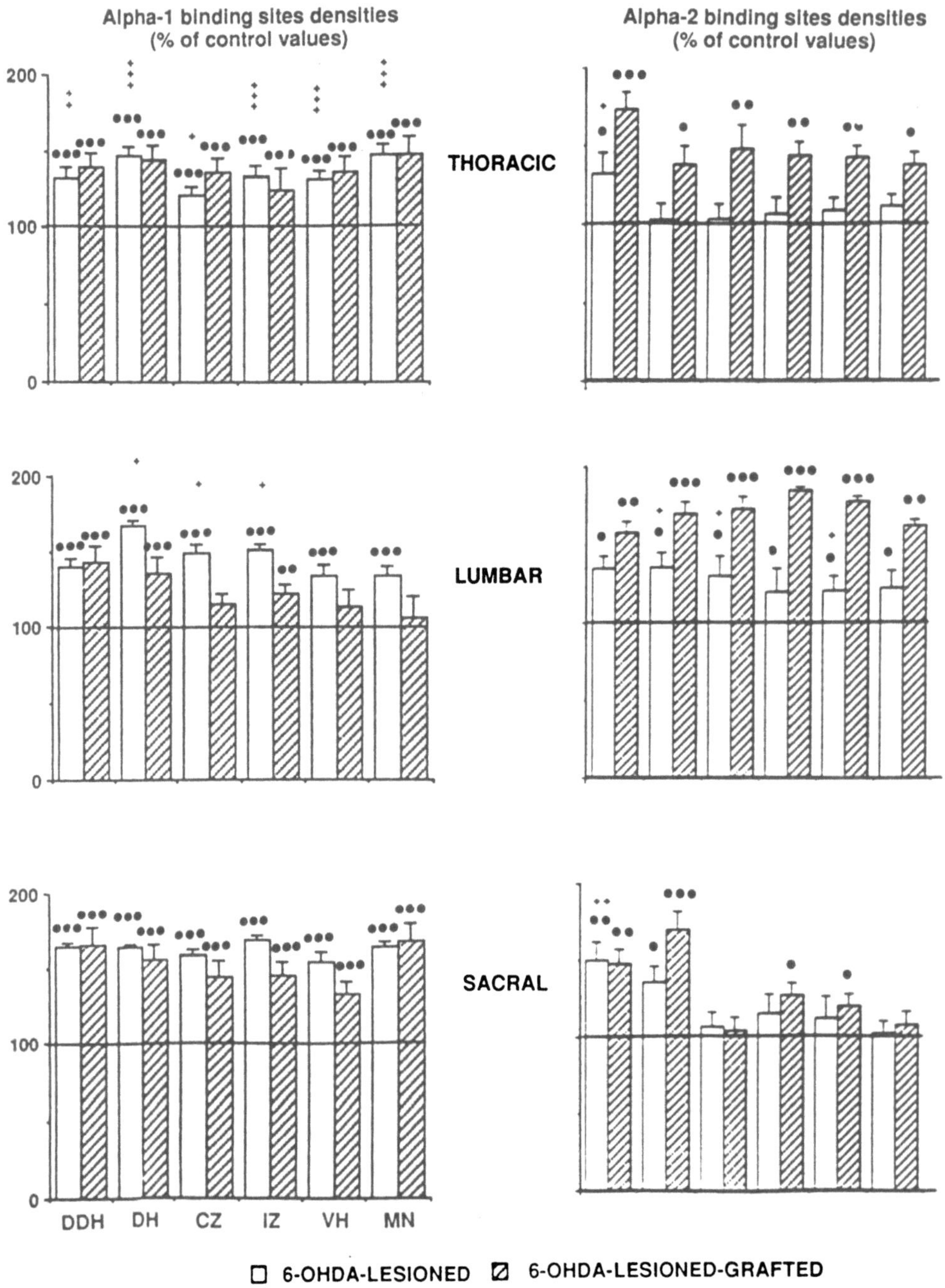

Figure 5. Model of spinal cord denervation by 6-OH DA. Quantitative analysis of α1- and α2-adrenoceptors densities as measured by using [3H]-Prazosin or [3H]-Rauwolscine. Histograms represent the total densities of α-adrenoceptors expressed as the ratio in per cent (mean ± sem) of homologous subregions of the grey matter as defined in Fig. 1 (**DDH**, **DH**, **CZ**, **IZ**, **VH**, and **MN**) between 6-OHDA lesioned (n=8), and 6-OH DA lesioned-grafted (n=14) and intact (n=8) animals. 100% corresponded to values found in intact group: 64 ± 3 to 142 ± 7, 58 ± 2 to 130 ± 8, and 69 ± 4 to 131 ± 11 fmol/mg prot. at thoracic, lumbar, and sacral levels of the spinal cord respectively. * Comparison with values found in intact animals: * $p<0.05$, ** $p<0.01$, *** $p<0.001$. + Comparison with values found in 6-OH DA lesioned animals: + $p<0.05$, ++ $p<0.01$, +++ $p<0.001$.

4. DISCUSSION

Transplants of NA-neurons into the lesioned spinal cord have previously been shown to survive, to reinnervate the host specific target regions, and to induce recovery from some neurochemical and functional deficits induced by the initial lesion (Björklund et al., 1986; Buchanan an Nornes, 1986; Commissiong, 1983, 1984; Moorman et al. 1990; Nornes et al., 1983; Nygren et al., 1977). However, little attention had been given to the ultrastructural features of NA-neurons after transplantation.

The present study, using two experimental models of spinal cord denervation, shows that transplanted locus coeruleus neurons are able to develop an axonal outgrowth in the deafferented areas, reestablishing at one month postgrafting the normal terminal innervation pattern, as already reported in our earlier anatomical studies (Yakovleff et al., 1989, 1995). Moreover, grafted neurons establish synaptic contacts with the host spinal cord. Counts of immunoreactive varicosities suggest some quantitative differences between the two models and those of the normal animals (Rajaofetra et al., 1992b), as to the types of targets contacted. These differences have been correlated with the ability of transplanted NA-neurons to reverse the lesion-induced increase α-adrenoceptors density in the two models of spinal cord denervation.

In the first model of spinal cord transection, the total loss of supraspinal descending inputs provides an environment containing many vacated specific synaptic sites, which could be a major determinant for the extent of neurite outgrowth from the transplant and the formation of synapses (Cotman et al., 1981). At variance, in the second model, obtained by selective destruction of catecholaminergic inputs, the development and the synaptic connectivity of grafted NA-neurons is confronted to a more selective lesion, with, in addition, a residual intrinsic NA-innervation (Roudet et al., 1993) This situation is of interest as it corresponds to a partial lesion model where a competition for vacated postsynaptic sites occurs in the denervated area between the intrinsic residual NA-fibres and the transplant-derived fibres.

In the two lesion models, the increase of α1- or α2-adrenoceptors densities concerned all areas of the grey matter and all levels of the spinal cord (except above the section in the first model). This result could suggest that there is no preferential area affected by denervation and that receptors become sensitive to denervation indicating probably a postsynaptic location to NA-afferents. However, a presynaptic location of some α-adrenoceptors on NA-fibres cannot be excluded.

Whatever the model used, long NA-immunoreactive processes emanating from the grafted neurons ramify in the host spinal cord in a radius of up to 10 mm away from the graft site. As in the intact animals (Mouchet et al., 1992; Rajaofetra, 1992b), the reinnervation in the grafted animals (transected or 6-OHDA lesioned) at one month postgrafting appears to be distributed in the three target regions: superficial layers of the dorsal horn, the intermediolateral column and the ventral horn specifically the area of motoneurons. This specific reinnervation pattern has also been observed when embryonic serotonergic cells from raphe primordia were transplanted following spinal cord injury (König et al., 1989; Privat et al., 1988, 1989; Rajaofetra et al., 1992a). The mechanism by which axons reach their specific target regions is still poorly understood. In our study, results are compatible with an axonal growth guided by diffusible factors specifically produced in the denervated target areas as initially proposed by Ramón y Cajal (1892) for axon guidance during development (Hankin and Lund, 1991; Tessier-Lavigne et al.,1988; Tessier-Lavigne and Placzek, 1991; Tessier-Lavigne and Goodman, 1996). Whatever the lesion model, the expression of the intrinsic programme of the grafted neurons could be present in conjunction with local epigenetic factors that would remain in the adult spinal cord or would be expressed after denervation (Daszuta et al., 1991; Onifer et al., 1993).

Interestingly enough, most of the morphological features of NA-immunoreactive synapses are similar in the two models. Our limited quantification suggests that there exist a difference as to the relative proportion of postsynaptic structures. In the chemical lesion model, axospinous synapses appeared twice as frequent as in the section model. In intact animals (Rajaofetra et al., 1992b) such axospinous synapses are rare. The higher percentage of axospinous synapses in the second model, which was also observed in other models of chemical lesion (Vuillet et al., 1994; Wictorin et al., 1992; Xu et al., 1992) could be a consequence of the presence of fewer synaptic sites available due to only partial denervation (Nieto-Sampedro et al., 1987).

We can then hypothesize that an increased competition for synaptic sites in the chemical lesion model would be able either to trigger the outgrowth of dendritic spines or to conduct to the occupancy of aberrant synaptic sites (Bernstein and Bernstein, 1971, 1973). Our static study does not allow a decision to be made between these two possibilities. However, we cannot exclude other explanations related, for instance, to the timing of the transplantation, or to a nonspecific effect of the neurotoxin on other neuronal populations.

In any case, whatever the model and the corresponding ultrastructural situation, we have shown that the up-regulation of α1-adrenoceptors induced by the lesion returns to normal levels, detected in intact animals, in those areas innervated by grafted NA-neurons. This could explain the slighter normalization of α1-adrenoceptors densities at sacral level in transected-grafted animals consecutively to a large individual variability in the NA-re-innervation. The restorative effect induced by the transplant in the second model of lesion was, however, more restricted to precise areas, concerning only CZ, VH, and MN of the grey matter. The question remains as whether grafted NA-neurons could have specific effects on the ventral grey matter.

The transplant showed, a limited effect on cellular and molecular mechanisms underlying the synthesis rate of spinal α2-adrenoceptors proteins in the transected model. Such differential effects may reflect the modulation of different target neuronal populations. The results in the second model do not allow to draw any conclusion on the mechanisms of a up-regulation after grafting, whether it is due to a competition between NA-fibres from the transplant and those residual intrinsic NA-fibres or to a non specific lesion-related reaction. We cannot exclude that a transplant-induced sprouting of non NA-fibres could up-regulate α2-adrenoceptors.

Although nonspecific mechanisms cannot be excluded, these results suggest that the regulation of postsynaptic sites is not closely dependent on a specific synaptic pattern (which is influenced by the host), but rather on the presence and the release of the appropriate transmitter. Functional restoration of monoaminergic innervation is likely to correspond to the presence of adequate quantities of neurotransmitter in a target area rather to point-to-point connectivity. This gives some hope for the future of clinical applications of such transplant after a traumatic lesion of the spinal cord.

ACKNOWLEDGMENTS

We thank J. R. Teilhac for photography and art work. This study was supported by grants from I.R.M.E., A.F.M., F.R.M.F., D.R.E.T.

REFERENCES

Bernstein, J.J., and Bernstein, M.E. (1971). Axonal regeneration and formation of synapses proximal to the site of the lesion following hemisection of the spinal cord. *Exp. Neurol.*, 30: 336–351.

Bernstein, J.J., and Bernstein, M.E. (1973). Neuronal alteration and reinnervation following axonal regeneration and sprouting in mammalian spinal cord. *Brain Behav. Evol.*, 8: 135–161.

Björklund, A., Nornes, H., and Gage, F.H.. (1986). Cell suspension grafts of noradrenergic locus coeruleus neurons in rat hippocampus and spinal cord: Reinnervation and transmitter turnover. *Neuroscience*, 18: 685–698.

Björklund, A., and Skagerberg, G. (1982). Descending monoaminergic projections to the spinal cord. In B. Sjölund and A. Björklund (Eds), *Brainstem Control of Spinal Mechanisms*, Elsevier, Amsterdam. pp. 55–85.

Björklund, A., Stenevi, U., and Dunnett, S.B. (1983). Transplantation of brainstem monoaminergic 'command' systems: models for functional reactivation of damaged CNS circuitries. In C.C. Kao, R.P. Bunge, and P.J. Reier (Eds), *Spinal Cord Reconstruction*, Raven Press, New York. pp. 397–413.

Bregman, B.S., Bernstein-Goral, H., and Kunkel-Bagden, E. (1991). CNS transplants promote anatomical plasticity and recovery of function after spinal cord injury. *Rest. Neurol. Neurosci.*, 2: 327–338.

Bregman, B.S., Kunkel-Bagden, E., Reier, P.J., Dai, H.N., Mc Atee, M., and Gao, D. (1993). Recovery of function after spinal cord injury: Mechanisms underlying transplant-mediated recovery of function differ after spinal cord injury in newborn and adult rats. *Exp. Neurol.*, 123: 3–16.

Buchanan, J.T., and Nornes, H.O. (1986). Transplants of embryonic brainstem containing the locus coeruleus into spinal cord enhance the hindlimb flexion reflex in adult rats. *Brain Res.*, 381: 225–236.

Commissiong, J.W. (1983). Fetal locus coeruleus transplanted into the transected spinal cord of the adult rat. *Brain Res.*, 271: 174–179.

Commissiong, J.W. (1984). Fetal locus coeruleus transplanted into the transected spinal cord of the adult rat: some observations and implications. *Neuroscience*, 12: 839–853.

Connor, H.E., Drew, G.M., Finch, L., and Hicks, P.E. (1981). Pharmacological characteristics of spinal a-adrenoceptors in rats. *J. Auton. Pharmacol.* 1: 149–156.

Cotman, C.W., Nieto-Sampedro, M., and Harris, E.W. (1981). Synapse replacement in the nervous system of adult vertebrates. *Physiol. Rev.*, 61: 684–784.

Das, G.D. (1983). Neural Transplantation in the spinal cord of the adult mammals. In C.C. Kao, R.P. Bunge, and P.J. Reier (Eds), *Spinal Cord Reconstruction*, Raven Press, New York. pp. 367–397.

Daszuta, A., Marocco C., and Bosler, O. (1991). Serotonin reinnervation of the suprachiasmatic nucleus by intrahypothalamic fetal raphe transplants, with special reference to possible influences of the target. *Eur. J. Neurosci.*, 3: 1330–1337.

Fisher, L.J., and Gage, F.H. (1993). Grafting in the mammalian central nervous system. *Physiol. Rev.*, 73: 583–616.

Forssberg, H., and Grillner, S. (1973). The locomotion of the acute spinal cat injected with clonidine i.v., *Brain Res.*, 50: 184–186.

Foster, G.A., Schultzberg, M., Gage, F.H., Björklund, A., Hökfelt, T., Nornes, H., Cuello, A.C., Verhofstad, A.A.J., and Visser, T.J. (1985). Transmitter expression and morphological development of embryonic medullary and mesencephalic raphe neurons after transplantation to the adult rat central nervous system. I. Grafts to the spinal cord. *Exp. Brain. Res.*, 60: 427–444.

Freed, W.J. (1993). Neural transplantation: Prospects for clinical use. *Cell Transplantation*, 2: 13–31.

Fritschy, J.M., Lyons, W.E., Mullen, C.A., Kosofsky, B.E., Molliver, M.E., and Grzanna, R. (1987). Distribution of locus coeruleus axons in the rat spinal cord: a combined anterograde transport and immunohistochemical study. *Brain Res.*, 437: 176–180.

Giron, L.T., Maccann, S.A., and Crist-Orlando, S.G. (1985). Pharmacological characterization and regional distribution of alpha-noradrenergic binding sites of rat spinal cord. *Eur. J. Pharmacol.*, 115: 285–290.

Goldberger, M.E., Murray, M., and Tessler, A. (1993a). Sprouting and regeneration in the spinal cord - Their roles in recovery of function after spinal injury. In A. Gorio (Ed), *Neuroregeneration*, Raven Press, New York. pp. 241–264.

Goldberger, M.E., Murray, M., and Tessler, A. (1993b). Grafts and functional recuperation. *Rest. Neurol. Neurosci.*, 5: 69–75.

Hankin, M., and Lund, R. (1991). How do retinal axons find their targets in the developing brain? *Trends Neurosci.*, 14: 224–228.

Jones, D.J., Kendall, D.E., and Enna, S.J. (1982). Adrenergic receptors in rat spinal cord. *Neuropharmacology*, 21: 191–195.

König, N., Rajaofetra, N., Sandillon, F., Drian, M.J., Fuentes, C., Favier, F., and Privat, A. (1989). Serotonin-expressing cells from different microregions of the embryonic rat rhombencephalon: Behaviour in cell culture and in transplantation to the adult spinal cord. In F. Gage, A. Privat, and Y. Christen (Eds), *Neuronal grafting and Alzheimer's disease*, Springer-Verlag, Berlin. pp. 150–166.

König, N., Wilkie, M.B., and Lauder, J. (1988). Tyrosine hydroxylase and serotonin containing cells in embryonic rat rhombencephalon: a whole-mount immunocytochemical study. *J. Neurosci. Res.*, 20: 212–223.

Kunkel-Bagden, E., and Bregman, B.S. (1990). Spinal cord transplants enhance the recovery of locomotor function after spinal cord injury at birth. *Exp. Brain Res.*, 81: 25–34.

Mas, M., Zahradnik, M.A., Martino, V., and Davidson, J.M. (1985). Stimulation of spinal serotonergic receptors facilitates seminal emission and suppresses penile erectile reflexes. *Brain Res.*, 342: 128–134.

Mason, S.T., and Fibiger, H.C. (1979). Physiological function of descending noradrenaline projections to the spinal cord: Rôle in post decapitation convulsions. *Eur. J. Pharmacol.*, 57: 29–34.

Molander, C., Xu, Q., and Grant, G. (1984). The cytoarchitectonic organitation of the spinal cord in the rat. I. The lower thoracic and lumbosacral cord. *J. Comp. Neurol.*, 230: 133–141.

Molander, C., Xu, Q., Rivermelian, C., and Grant, G. (1989). The cytoarchitectonic organization of the spinal cord in the rat. II. The cervical and upper thoracic lcord. *J. Comp. Neurol.*, 289: 375–385.

Moorman, S.J., Whalen, L.R., and Nornes, H.O. (1990). A neurotransmitter specific functional recovery mediated by fetal implants in the lesioned spinal cord of the rat. *Brain Res.*, 508: 194–198.

Mouchet, P., Manier, M., and Feuerstein, C. (1992). Immunohistochemical study of the catecholaminergic innervation of the spinal cord of the rat using specific antibodies against dopamine and noradrenaline. *J. Chem. Neuroanat.*, 5: 427–440.

Nieto-Sampedro, M., Kesslak, J.P., Gibbs, R., and Cotman, C.W. (1987). Effects of conditionng lesions on transplant survival, connectivity, and function. In E.C. Azmitia, and A. Björklund (Eds), *Cell and Tissue Transplantation into the Adult Brain*. The New York Academy of Sciences, New York. pp. 108–119.

Nishino, H. (1993). Intracerebral grafting of catecholamine producing cells and reconstruction of disturbed brain function. *Neurosci. Res.*, 16: 157–172.

Nornes, H., Björklund, A., and Stenevi, U. (1983). Reinnervation of the denervated adult spinal cord of rats by intraspinal transplant of embryonic brain stem neurons. *Cell Tissue Res.*, 230: 15–35.

Nornes, H., Björklund, A., and Stenevi, U. (1984). Transplant strategies in spinal cord regeneration. In J.R. Sladek, and D.M. Gash (Eds), Neural Transplants: Development and Function, Plenum Press, New York. pp. 407–421.

Nothias, F., Horvat, J.C., Mira, J.C., Pecot-Dechavassine, M., and Peschanski, M. (1990). Double step neural transplants to replace degenerated motoneurons. Neural Transplantation: From molecular basis to clinical applications. *Prog. Brain Res.*, 82: 239–246.

Nothias, F., and Peschanski, M. (1990). Homotypic fetal transplants into an experimental model of spinal cord neurodegeneration. *J. Comp. Neurol.*, 301: 520–534.

Nygren, L.G., Olson, L., and Seiger, A. (1977). Monoaminergic reinnervation of the transected spinal cord by homologous fetal brain grafts. *Brain Res.*, 129: 225–235.

Onifer, S.M., Whittemore, S.R., and Holets, V.R. (1993). Variable morphological differentiation of a raphé-derived neuronal cell line following transplantation into the adult rat CNS. *Exp. Neurol.* 122: 130–142.

Peretti-Renucci, R., Feuerstein, C., Manier, M., Lorimier, P., Savasta, M., Thibault, J., Mons, N., and Geffard, M. (1991). Quantitative image analysis with densitometry for immunohistochemistry and autoradiography of receptor binding sites. *J. Neurosci. Res.*, 28: 583–600.

Privat, A., Mansour, H., and Geffard, M. (1988). Transplantation of fetal serotonin neurons into the transected spinal cord of adult rats: morphological development and functional influence. In D.M. Gash, and J.R. Sladek (Eds), Transplantation into the Mammalian Central Nervous System, *Progress in Brain Research, Vol. 78*, Elsevier, Amsterdam, pp. 155–166.

Privat, A., Mansour, H., Pavy, A., Geffard, M., and Sandillon, F. (1986). Transplantation of dissociated foetal serotonin neurons into the transected spinal cord of adult rats, *Neurosci. Lett.*, 66: 61–66.

Privat, A., Mansour, H., Rajaofetra, N., and Geffard, M. (1989). Intraspinal transplant of serotonergic neurons in the adult rat. *Brain Res. Bull.* 22: 123–129.

Rajaofetra, N., Kachidian, P., Marlier, L., Poulat, P., Konig, N., Geffard, M., and Privat, A. (1991). Electronmicroscopic detection of the axonal coexistence of serotonin and substance-P in B1-B2 raphe cells transplanted into the transected spinal cord of adult rats. *Brain Res.*, 542: 159–162.

Rajaofetra, N., König, N., Poulat, P., Marlier, L., Sandillon, F., Drian, M.J., Geffard, M., and Privat, A. (1992a). Fate of B1-B2 and B3 rhombencephalic cells transplanted into the transected spinal cord of adult rats: Light and electron microscopic studies. *Exp. Neurol.*, 117: 59–70.

Rajaofetra, N., Ridet, J.L., Poulat, P., Marlier, L., Sandillon, F., Geffard, M., and Privat, A.. (1992b). Immunocytochemical mapping of noradrenergic projections to the rat spinal cord with an antiserum against noradrenaline. *J. Neurocytol.*, 21: 481–494.

Ramón y Cajal, S. (1892). La rétine de vertebrés. In *La Cellule*, Vol IX,.pp. 119–158.

Rawlow, A., and Gorka, Z. (1986). Involvement of post-synaptic α1 and α2-adrenoceptors in the flexor reflex activity in the spinal rats. *J. Neural Trans.*, 66: 93–105.

Reier, P.J., Bregman, B.S., Wujek, J.R., and Tessler, A. (1986). Intraspinal transplantation of fetal spinal cord tissue: An approach toward functional repair of the injured spinal cord. In M.E. Goldberger, A. Gorio, and M. Murray (Eds), *Development and Plasticity of the Mammalian Spinal Cord*, Fidia Research Series, vol. III, Liviana Press, Padova. pp. 251–269.

Reier P.J., Houle, J.D., Jakeman, L., Winialski, D., and Tessler, A.(1988). Transplantation of fetal spinal cord tissue into acute and chronic hemisection and contusion lesions of the adult rat spinal cord. In D.M. Gash, and J.R. Sladek. Jr. (Eds), *Transplantation in the Mammalian CNS*, Elsevier, Amsterdam. pp. 173–179.

Reier, P.J., Stokes, B.T., Thompson, F.J., and Anderson, D.K. (1992). Fetal cell grafts into resection and contusion/compression injuries of the rat and cat spinal cord. *Exp. Neurol.*, 115: 177–188.

Rossignol S., Barbeau, H., and Julien, C. (1986). Locomotion of the adult chronic spinal cat and its modification by monoaminergic agonists and antagonists. In M.E. Goldberger, A. Gorio, and M. Murray (Eds), *Development and Plasticity of the Mammalian Spinal Cord*, Fidia Research Series, vol III, Liviana Press, Padova. pp. 323–345.

Roudet, C., Savasta M., and Feuerstein, C. (1993). Normal distribution of alpha-1-adrenoceptors in the rat spinal cord and its modification after noradrenergic denervation: A quantitative autoradiographic study. *J. Neurosci. Res.*, 34: 44–53.

Roudet, C., Mouchet, M., Feuerstein, C., and Savasta M. (1994). Normal distribution of of alpha-2 adrenoceptors in the rat spinal cord and its modification after noradrenergic denervation: a quantitative autoradiographic study. *J. Neurosci. Res.*, 39: 319–329.

Shi, H., Lewis, D.I., and Coote, J.H. (1988). Effects of activating spinal α-adrenoceptors on sypathetic nerve activity in the rat. *J. Auton. Nerv. Syst.*, 23: 69–78.

Simmons, R.M., and Jones, D.J. (1988). Binding of [^{3}H]-Prazosin and [^{3}H]-p-aminoclonidine to α-adrenoceptors in rat spinal cord. *Brain Res.*, 23: 338–343.

Stokes, B. T., and Reier, P.J. (1992). Fetal grafts alter chronic behavioral outcome after contusion damage to the adult rat spinal cord. *Exp. Neurol.*, 116: 1–12.

Tanabe, M., Ono, H., and Fukuda, H. (1990). Spinal alpha 1- and alpha 2-adrenoceptors mediate facilitation and inhibition of spinal motor transmission respectively. *Jpn. J. Pharmacol.*, 54: 69–77.

Tessier-Lavigne, M., Placzek, M., Lumsden, A.G.S., Dodd, J., and Jessell, T.M. (1988). Chemotropic guidance of developing axons in the mammalian central nervous system. *Nature*, 336: 775–778.

Tessier-Lavigne, M., and Placzek, M. (1991). Target attraction: are developing axons guided by chemotropism? *Trends Neurosci.*,14: 303–310.

Tessier-Lavigne M., and Goodman C.S. (1996). The molecular biology of axon guidance. Science, 274: 1123–1133.

Tessler, A. (1991). Intraspinal Transplants. *Ann. Neurol.*, 29: 115–123.

Tessler, A., Goldberger, M.E., Himes, B.T., Howland, D.R., Itoh, Y., and Pinter, M.J. (1992). Transplant mediated mechanisms of locomotor recovery. *Rest. Neurol. Neurosci.*, 5: 64–65.

Unnerstall, J.R., Kopajtic, T.A., and Kuhar, M.J. (1984). Distribution of α2 agonist binding sites in the rat and human central nervous system: Analysis of some functional, anatomic correlates of the pharmacologic effects of clonidine and related adrenergic agents, *Brain Res. Rev.*, 7: 69–101.

Vuillet, J., Moukhles, H., Nieoullon, A., and Daszuta, A. (1994). Ultrastructural analysis of graft-to-host connections, with special reference to dopamine-neuropeptide Y interactions in the rat striatum, after transplantation of fetal mesencephalon cells. *Exp. Brain Res.*, 98: 84–96.

Westlund, K.N., Bowker, R.M., Ziegler, M.G., and Coulter, J.D.(1982). Descending noradrenergic projections and their spinal terminations. In H.G.J.M. Kuypers, and G.F. Martin (Eds), *Descending Pathways to the Spinal Cord, Progress in Brain Research, Vol. 57*, Elsevier, Amsterdam. 219–238.

Westlund, K.N., Bowker, R.M., Ziegler, M.G., and Coulter, J.D. (1983). Noradrenergic projections to the spinal cord of the rat. *Brain Res.*, 263: 15–31.

Wictorin, K., Simerly, R.B., Isacson, O., Swanson, L.W., and Björklund, A. (1992). Connectivity of striatal grafts implanted into the ibotenic acid lesioned striatum. III. Efferent projecting graft neurons and their relation to host afferents within the graft. *Neuroscience*, 30: 313–330.

Xu, Z.C., Wilson, C.J., and Emson, P.C. (1992). Morphology of intracellularly stained spiny neurons in rat striatal grafts. *Neuroscience*, 48: 95–110.

Yakovleff, A., Roby-Brami, A., Guezard, B., Mansour, H., Bussel, B., and Privat, A. (1989). Locomotion in rats transplanted with noradrenergic neurons. *Brain Res. Bull.*, 22: 115–121.

Yakovleff, A., Cabelguen J.M., Orsal, D., Giménez y Ribotta, M., Rajaofetra, N., Drian M.J., Bussel, B., and Privat A. (1995). Fictive motor activities in adult chronic spinal rats transplanted with embryonic brainstem neurons. *Exp. Brain Res.*, 106: 69–78.

Young, W.S., and Kunhar, M.J. (1980). Noradrenergic a1 and a2 receptors: light microscopic autoradiographic localization. *Proc. Natl. Acad. Sci. U.S.A.*, 77: 1696–1700.

10

REGULATION BY NEUROTRANSMITTERS OF GLIAL ENERGY METABOLISM

Pierre J. Magistretti and Luc Pellerin

Laboratoire de Recherche Neurologique
Institut de Physiologie et Service de Neurologie du CHUV
Faculté de Médecine
Université de Lausanne
7, rue du Bugnon
CH-1005 Lausanne, Switzerland

1. BACKGROUND

Astrocytes are ideally positioned to couple neuronal activity with energy metabolism. Thus particular astrocytic profiles, the end-feet, surround intraparenchymal capillaries, implying that they form the first cellular barrier that energy substrates entering the brain parenchyma, in particular glucose, encounter. In addition, astrocytes possess receptors and reuptake sites for neurotransmitters, and astrocytic processes ensheath synaptic contacts: these features imply that astrocytes are ideally positioned to sense increases in synaptic activity and to couple them with energy metabolism. We have characterized three metabolic processes regulated by neurotransmitters in primary astrocyte cultures prepared from neonatal mouse cerebral cortex: *glycogenolysis, glycogen resynthesis* and *glycolysis*.

2. GLYCOGEN METABOLISM

2.1. Glycogenolysis

Glycogen is the single largest energy reserve of the brain; it is mainly localized in astrocytes, although ependymal and choroid plexus cells, as well as certain large neurons in the brain stem contain glycogen (Magistretti et al., 1993). Glycogen levels are tightly regulated by various neurotransmitters (Table 1). Thus we have shown that Vasoactive Intestinal Peptide (VIP), a neurotransmitter contained in a homogeneous population of bipolar, radially oriented neurons (Magistretti and Morrison, 1988) could promote a cAMP-dependent glycogenolysis in mouse cerebral cortical slices (Magistretti et al.,

Brain Plasticity, edited by Filogamo *et al.*
Plenum Press, New York, 1997

Table 1. Induction of glycogenolysis by neurotransmitters in primary astrocyte cultures of mouse cerebral cortex

Neurotransmitter	EC_{50} (nM)
VIP	3
PHI	6
Secretin	0.5
PACAP	0.08
Noradrenaline	20
Adenosine	800
ATP	1300

PHI, peptide histidine isoleucine.
PACAP, pituitary adenylate cyclase activating peptide.

1981). In view of the morphology and arborization pattern of VIP-containing neurons (Magistretti and Morrison, 1988), we proposed that these cells could regulate the availability of energy substrates locally, within cortical columns (Magistretti et al., 1981; Magistretti and Morrison, 1988). A similar effect had been previously described for NA (Quach et al., 1978), serotonin and histamine (Quach et al., 1980; Quach et al., 1982). The noradrenergic system is organized according to principles strikingly different from those of VIP neurons: the cell bodies of NA-containing neurons are localized in the locus coeruleus in the brain stem from where axons project to various brain areas including the cerebral cortex: here, they enter the rostral pole and progress caudally with a predominantly horizontal trajectory, across a vast rostro-caudal expanse of cortex (Morrison et al., 1978). Given these morphological features we suggested that, in contrast to VIP-containing intracortical neurons, the noradrenergic system could regulate energy homeostasis globally, spanning across functionally-distinct cortical areas (Magistretti et al., 1981; Magistretti and Morrison, 1988).

VIP and NA promote a concentration- and time-dependent glycogenolysis in astrocytes, with EC_{50} of 3 and 20 nM respectively (Magistretti et al., 1983; Sorg and Magistretti, 1991). The effect of NA is mediated by both ß and α_1 receptors. In addition to VIP and NA, adenosine and ATP are also glycogenolytic in astrocytes (Table 1). The initial rate of glycogenolysis activated by VIP and NA is between 5 and 10 nmol/mg prot/min (Sorg and Magistretti, 1991), a value that is remarkably close to glucose utilization of the grey matter as determined by the 2-DG autoradiographic method (Sokoloff et al., 1977). This correlation indicates that the glycosyl units mobilized in response to the two glycogenolytic neurotransmitters can provide quantitatively adequate substrates for the energy demands of the brain parenchyma.

At this stage, it is not yet clear whether the glycosyl units mobilized through glycogenolysis are used by astrocytes to face their energy demands during activation, or whether they are metabolized to a substrate such as lactate (Dringen et al., 1993), which is then released for the use of neurons. A well established fact is that glucose is not released by astrocytes, at least in vitro (Dringen et al., 1993), supporting the view that the activity of glucose-6-phosphatase in astrocytes is very low (Sokoloff et al., 1977; Fishman and Karnovsky, 1986).

2.2. Glycogen Resynthesis

Following this rapid (within 1–5 min) glycogenolysis, VIP and NA induce a massive glycogen resynthesis maximally expressed after 8 hours, bringing glycogen to levels 10x higher than before VIP or NA application (Sorg and Magistretti, 1992). This glycogen resynthesis is mediated by cAMP-dependent induction of *gene expression*. Searching for the genes induced we found both early and late genes related to energy metabolism. Thus, VIP and NA induce C/EBP β and δ (Cardinaux and Magistretti, 1996), which are members of an immediate-early gene family acting as transcription factors for genes involved in energy metabolism regulation in liver and adipose tissue. In addition, transfection of cultured astrocytes with cDNA encoding for C/EBP β amplifies the resynthesis of glycogen evoked by NA (Cardinaux and Magistretti, 1996). We have also observed by Northern blot analysis that the mRNA encoding for glycogen synthase is massively induced within 4 to 6 hours by VIP and NA (Pellegri et al., 1996b). These results strongly suggest that VIP and NA can regulate the expression of early and late genes encoding for gene-products involved in their metabolic effects (Figure 1).

3. GLYCOLYSIS

Astrocytes form the first cellular barrier encountered by glucose, the major fuel of the brain, as it leaves the circulation to enter the brain parenchyma; thus astrocytes constitute a likely site of prevalent uptake of glucose. Up to now, however, functional evidences have been scarce supporting such a role. Another, more recently uncovered, characteristic of astrocytes is the existence of astrocytic processes ensheating synapses. Together with the presence of receptors and transporters for several neurotransmitters on them, astrocytes appear then perfectly adapted to sense neuronal activity. From the foregoing, we postulated that astrocytes may play a prominent role in coupling neuronal activity to local increases in glucose utilization during activation as observed in animals and humans. To

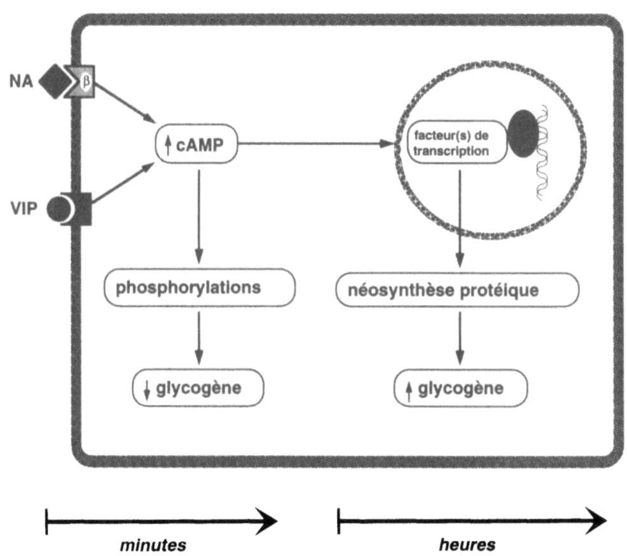

Figure 1. Bidirectional effects of VIP or NA on glycogen in astrocytes.

explore this hypothesis we have studied glucose utilization by mouse cerebral cortex astrocytes in culture using the ^{3}H-2-deoxyglucose (2-DG) as a marker of glucose utilization. We have shown that excitatory amino acids (EAAs) like L-glutamate (L-Glu) and L-aspartate (L-Asp) increase 2-DG uptake and phosphorylation in astrocytes in a concentration-dependent manner (EC_{50} ~ 80 μM for L-Glu) (Pellerin and Magistretti, 1994; Pellerin and Magistretti, 1996). This effect is not receptor-mediated since it can not be prevented nor mimicked by glutamate receptor antagonists and agonists, respectively (Pellerin and Magistretti, 1994; Magistretti and Pellerin, 1996a). Rather, the involvement of a Na^+-dependent glutamate transporter is suggested by *1)* the stereospecificity of the EAA-mediated effect (L-Glu, L-Asp and D-Asp but not D-Glu are active), *2)* the absence of effect when Na^+ is omitted, and *3)* the inhibition by known transporter inhibitors such as threo-hydroxyaspartate (THA). Combining kinetic and Northern blot analyses, we have identified the presence of at least one saturable transport system for EAAs on astrocytes and have detected the presence of the two cloned Na^+-dependent glutamate transporters specific for glial cells, GLAST and GLT1 (Pellerin et al, 1995; Pellegri et al, 1996a). Experiments with ouabain suggest that activation of the Na^+/K^+ ATPase is likely. Using ^{86}Rubidium uptake to directly monitor the activity of the pump, we have demonstrated that EAAs activate the Na^+/K^+ ATPase, most likely via a massive Na^+ influx occurring with EAA uptake (Pellerin and Magistretti, 1996). Moreover, we have shown that EAAs increase lactate formation from extracellular glucose in a concentration-dependent manner (EC_{50} ~ 80 μM for L-Glu) (Magistretti and Pellerin, 1996a). These data thus indicate that EAAs stimulate aerobic glycolysis (i.e. the transformation of glucose into lactate in the presence of sufficient oxygen) in astrocytes by a mechanism involving an activation of the Na^+/K^+ ATPase.

Since lactate can be used as an energy fuel by neurons, we considered the possibility that a selective distribution of lactate dehydrogenase (LDH) isoenzymes could exist among lactate-producing and lactate-consuming cells. We have raised polyclonal antibodies against the LDH_1 (heart-type) and the LDH_5 (muscle-type) subunits and used them for immunohistochemistry of human hippocampus and visual cortex. The results show that the immunoreactivity against LDH_5 subunits is restricted to a population of astrocytes while neurons are stained only by an antibody directed against LDH_1 subunits (Bittar et al., 1996). These data support the idea that some astrocytes would preferentially process glucose glycolytically into lactate which, once released, could be transformed by neurons into pyruvate and enter the TCA cycle to serve as an energy fuel.

In conclusion, since EAA release occurs following the modality-specific activation of a brain region, these data and the proposed model are consistent with the view that during activation, EAA uptake into astrocytes leads to increased glucose utilization and lactate production, which can be subsequently used by neurons to meet their energy needs (Magistretti and Pellerin, 1996b; Tsacopoulos and Magistretti, 1996). Moreover, this mechanism provides a molecular explanation for the unexpected observations made with PET and ^{1}H-MRS for lactate, of an uncoupling between glucose utilization and oxygen consumption, which have led to the notion of activation-induced glycolysis (Fox et al., 1988; Prichard et al., 1991).

4. IMPLICATIONS FOR FUNCTIONAL BRAIN IMAGING

In summary, results obtained in purified preparations of astrocytes and neurons are consistent with the model illustrated in Figure 2, whereby during activation, glutamate re-

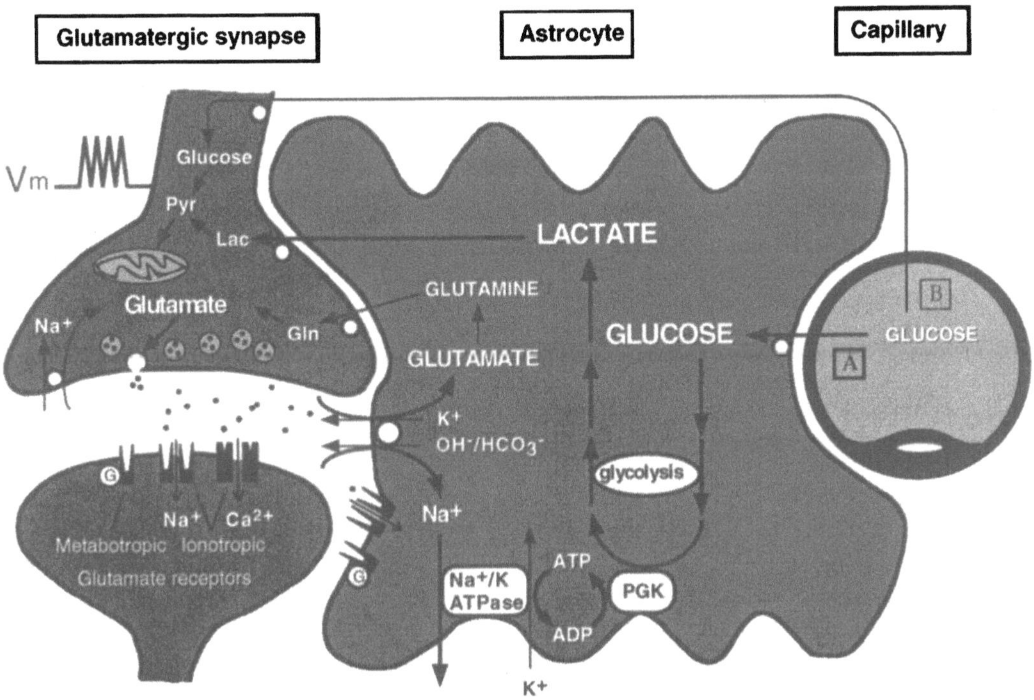

Figure 2. Mechanism for coupling neuronal activity to glucose utilization.

leased from activated synapses and taken up in a sodium-dependent manner by astrocytes activates the Na^+/K^+-ATPase which is preferentially fueled by glycolysis-derived ATP; the ensuing decrease in ATP activates glucose uptake, phosphorylation and processing to lactate which is released into the extracellular space providing an energy substrate for neurons. Indeed, a variety of biochemical and electrophysiological lines of evidence indicate that lactate is an adequate energy substrate for neurons (McIlwain, 1953; Schurr et al., 1988; Larrabee, 1995; Poitry-Yamate et al., 1995). In addition, in vivo microdialysis studies in rodents and ^{1}H-Magnetic Resonance Spectroscopy (MRS) analyses in humans have clearly documented an increased lactate formation during physiological activation (Prichard et al., 1991; Sappey-Marinier et al., 1992; Fray et al., 1996). This is consistent with the existence of an activity-linked glycolysis which is further illustrated by the previously mentioned uncoupling between glucose utilization and oxygen consumption demonstrated in PET activation studies in humans (Fox and Raichle, 1986; Fox et al., 1988).

Thus the glutamate-activated glucose uptake and lactate production in astrocytes provide a simple and straightforward mechanism to couple neuronal activity with glucose utilization during activation (Figure 2). Since lactate can support the energy requirements of neurons, it can be postulated that the activation-linked glycolysis is transient and that indeed lactate is fully oxidized by neurons. Recent in vivo ^{1}H-MRS and functional Magnetic Resonance Imaging studies indicating a “recoupling”, i.e. a delayed increase in oxygen utilization and a consequent decrease in the lactate signal support this view (Frahm et al., 1996). These in vitro and in vivo observations support the notion of an activity-dependent astrocyte-neuron lactate shuttle.

REFERENCES

Bittar PG, Charnay Y, Pellerin L, Bouras C, Magistretti PJ (1996) Selective distribution of lactate dehydrogenase isoenzymes in neurons and astrocytes of human brain. J. Cereb. Blood Flow Metab. 16:1079–1089.

Cardinaux J-R, Magistretti PJ (1996) Vasoactive intestinal peptide, pituitary adenylate cyclase-activating peptide, and noradrenaline induce the transcription factors CCAAT/enhancer binding protein (C/EBP)-ß and C/EBPδ in mouse cortical astrocytes: involvement in cAMP-regulated glycogen metabolism. J. Neurosci. 16:919–929.

Dringen R, Gebhardt R, Hamprecht B (1993) Glycogen in astrocytes: possible function as lactate supply for neighboring cells. Brain Res. 623:208–214.

Fellows LK, Boutelle MG, Fillenz M (1993) Physiological stimulation increases nonoxidative glucose metabolism in the brain of the freely moving rat. J. Neurochem. 60:1258–1263.

Fishman RS, Karnovsky ML (1986) Apparent absence of a translocase in the cerebral glucose-6-phosphatase system. J. Neurochem. 46:371–378l.

Fox PT, Raichle ME (1986) Focal physiological uncoupling of cerebral blood flow and oxidative metabolism during somatosensory stimulation in human subjects. Proc. Natl. Acad. Sci. USA 83:1140–1144.

Fox PT, Raichle ME, Mintun MA, Dence C (1988) Nonoxidative glucose consumption during focal physiologic neural activity. Science 241:462–464.

Frahm J, Krüger G, Merboldt K-D, Kleinschmidt A (1996) Dynamic uncoupling and recoupling of perfusion and oxidative metabolism during focal brain activation in man. Magn. Reson. Med. 35:143–148.

Fray AE, Forsyth RJ, Boutelle MG, Fillenz M (1996) The mechanisms controlling physiologically stimulated changes in rat brain glucose and lactate: a microdialysis study. J. Physiol. 496.1:49–57.

Larrabee MG (1995) Lactate metabolism and its effects on glucose metabolism in an excised neural tissue. J. Neurochem. 64:1734–1741.

Magistretti PJ, Manthorpe M, Bloom FE, Varon S (1983) Functional receptors for vasoactive intestinal polypeptide in cultured astroglia from neonatal rat brain. Regul. Pept. 6:71–80.

Magistretti PJ, Morrison JH (1988) Noradrenaline- and vasoactive intestinal peptide-containing neuronal systems in neocortex: functional convergence with contrasting morphology. Neurosci. 24:367–378.

Magistretti PJ, Morrison JH, Shoemaker WJ, Sapin V, Bloom FE (1981) Vasoactive intestinal polypeptide induces glycogenolysis in mouse cortical slices: a possible regulatory mechanism for the local control of energy metabolism. Proc. Natl. Acad. Sci. 78:6535–6539.

Magistretti PJ, Pellerin L (1996a) Cellular mechanisms of brain energy metabolism: relevance to functional brain imaging and to neurodegenerative disorders. Ann. NY Acad. Sci. 777:380–387.

Magistretti PJ, Pellerin L (1996b) Cellular bases of brain energy metabolism and their relevance to functional brain imaging: evidence for a prominent role of astrocytes. Cerebral Cortex 6:50–61.

Magistretti PJ, Sorg O, Martin J-L (1993) Regulation of glycogen metabolism in astrocytes: physiological, pharmacological, and pathological aspects. In: Astrocytes: pharmacology and function (Murphy S, ed), p. 243. San Diego: Academic Press.

McIlwain H (1953) Substances which support respiration and metabolic response to electrical impulses in human cerebral tissues. J. Neurol. Neurosurg. Psychiatry 16:257–266.

Morrison JH, Grzanna R, Molliver M, Coyle JT (1978) The distribution and orientation of noradrenergic fibers in neocortex of the rat: An immunofluorescence study. J. Comp. Neurol. 181:17–40.

Pellegri G, Pellerin L, Stella N, Martin J-L, Magistretti PJ (1996a) Cellular localization and characterization of glutamate transporters in cortical neurons and astrocytes. Experientia 52:A82.

Pellegri G, Rossier C, Magistretti PJ, Martin J-L (1996b) Cloning, localization and induction of mouse brain glycogen synthase. Mol. Brain Res. 38:191–199.

Pellerin L, Magistretti PJ (1994) Glutamate uptake into astrocytes stimulates aerobic glycolysis: a mechanism coupling neuronal activity to glucose utilization. Proc. Natl. Acad. Sci. USA 91:10625–10629.

Pellerin L, Magistretti PJ (1996) Excitatory amino acids stimulate aerobic glycolysis in astrocytes via an activation of the Na^+/K^+ ATPase. Dev. Neurosci. 18:336–342.

Pellerin L, Martin J-L, Magistretti PJ (1995) Deoxyglucose uptake and c-fos expression, two markers of neural activation, are differentially stimulated by glutamate in astrocytes and neurons in vitro. Soc. Neurosci. Abstr. 21:580.

Poitry-Yamate CL, Poitry S, Tsacopoulos M (1995) Lactate released by Müller glial cells is metabolized by photoreceptors from mammalian retina. J. Neurosci. 15:5179–5191.

Prichard J, Rothman D, Novotny E, Petroff O, Kuwabara T, Avison M, Howseman A, Hanstock C, Shulman R (1991) Lactate rise detected by ^1H NMR in human visual cortex during physiologic stimulation. Proc. Natl. Acad. Sci. USA 88:5829–5831.

Quach TT, Duchemin AM, Rose C, Schwartz JC (1980) [^{3}H]Glycogen hydrolysis elicited by histamine in mouse brain slices: selective involvement of H_1-receptors. Mol. Pharmacol. 17:301–308.

Quach TT, Rose C, Duchemin AM, Schwartz JC (1982) Glycogenolysis induced by serotonin in brain: identification of a new class of receptors. Nature 298:373–375.

Quach TT, Rose C, Schwartz JC (1978) [^{3}H]Glycogen hydrolysis in brain slices: responses to neurotransmitters and modulation of noradrenaline receptors. J. Neurochem. 30:1335–1341.

Sappey-Marinier D, Calabrese G, Fein G, Hugg JW, Biggins C, Weiner MW (1992) Effect of photic stimulation on human visual cortex lactate and phosphates using ^{1}H and ^{31}P magnetic resonance spectroscopy. J. Cereb. Blood Flow Metab. 12:584–592.

Schurr A, West CA, Rigor BM (1988) Lactate-supported synaptic function in the rat hippocampal slice preparation. Science 240:1326–1328.

Sokoloff L, Reivich M, Kennedy C, Des Rosiers MH, Patlak CS, Pettigrew KD, Sakurada O, Shinohara M (1977) The [^{14}C]deoxyglucose method for the measurement of local cerebral glucose utilization: theory, procedure, and normal values in the conscious and anesthetized albino rat. J. Neurochem. 28:897–916.

Sorg O, Magistretti PJ (1991) Characterization of the glycogenolysis elicited by vasoactive intestinal peptide, noradrenaline and adenosine in primary cultures of mouse cerebral cortical astrocytes. Brain Res. 563:227–233.

Sorg O, Magistretti PJ (1992) Vasoactive intestinal peptide and noradrenaline exert long-term control on glycogen levels in astrocytes: blockade by protein synthesis inhibition. J. Neurosci. 12:4923–4931.

Tsacopoulos M, Magistretti PJ (1996) Metabolic coupling between glia and neurons. J. Neurosci. 16:877–885.

11

DENDRITIC DEVELOPMENT OF VISUAL CALLOSAL NEURONS

A. Vercelli,[1] F. Assal,[2] and G. M. Innocenti[3]

[1]Department of Anatomy, Pharmacology and Forensic Medicine
University of Torino, I
[2]Clinique de Neurologie
Hôpital Cantonal Universitaire de Genève, CH
[3]Departement de Biologie cellulaire et de morphologie
Université de Lausanne, CH

Neurons in the cerebral cortex differ from each other in terms of size, dendritic morphology, location, pattern of projection, membrane and response properties. In principle, these features could derive from an interplay between developmental programs intrinsic to the neuron and various extrinsic factors acting upon them. Interestingly, certain features of cortical neurons, such as dendritic arbor and projection pattern, appear to be mutually dependent: in layer V of cerebral cortex of rats and kittens (Figure 1), pyramidal neurons with callosal or corticocortical axons have short apical dendrites, while neurons projecting to the spinal cord or tectum have longer apical dendrites, which reach layer I (Hallman et al., 1988; Hübener et al., 1990; Koester and O'Leary, 1992; Kasper et al., 1993 and 1994). This morphological specificity of the different classes of projection neurons is maintained even in organotypic co-cultures (Bolz et al., 1990, 1991).

Callosally-projecting neurons in adult rodents are distributed in supragranular and infragranular layers. In the supragranular layers, callosally-projecting neurons are restricted to certain regions. For instance, in the visual system they are restricted to the border between areas 17 and 18, whereas their distribution is more uniform in infragranular layers in areas 17 and 18 (Olavarria and Van Sluyters, 1985). In cats, on the other hand, visual callosal connections also originate from cells located in two subzones, one corresponding to the supragranular layers (II/III and upper IV) and the other corresponding to the infragranular layers, particularly layer VI; however, both subzones are restricted to the border between areas 17 and 18 (Innocenti, 1980). Supragranular neurons are the primary source of callosal axons. Both in rodents and in cats, the tangential distribution of callosally-projecting neurons is achieved during development by selective elimination of callosal axons originating from cells located at sites which project through the corpus callosum (CC) only in newborn animals (Innocenti et al., 1977; Innocenti and Caminiti, 1980; Innocenti, 1981; O'Leary et al., 1981; see also Innocenti, 1991, for a review).

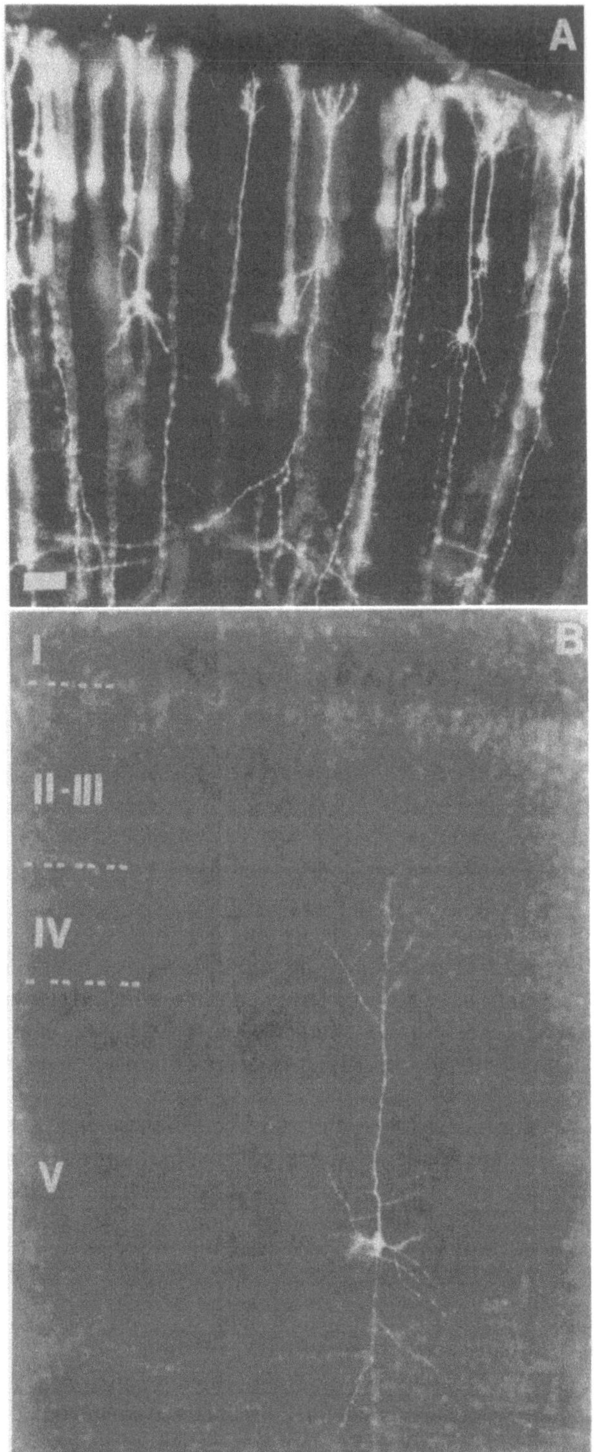

Figure 1. (A) DiI-labeled callosally-projecting neurons in the visual cortex of a P1 rat: all neurons bear apical dendrites reaching the pial surface with their terminal tufts. (B) DiI-labeled callosally-projecting, layer V pyramidal neuron in visual cortex of a P9 rat. The apical dendrite ends in layer II/III. Counterstained with bisbenzimide to show cortical layers (dotted lines). Scale bar = 50 μm.

Most callosally-projecting neurons may be classified as pyramidal cells. They include small or medium-sized pyramids in layer II/III, medium-sized pyramids in layer V (Innocenti, 1980; Diao and So, 1991; Olavarria and Van Sluyters, 1985) and transformed pyramids in layer VI. While the majority of these neurons located in layer V of rodents bear a short apical dendrite, gradually tapering and ending at the layer II/III border, those in the upper layer IV of the cat visual cortex are spiny stellate neurons (Innocenti and Fiore, 1976; Innocenti, 1980; Martin and Whitteridge, 1984). Cells in layer VI of cat visual cortex that send their axons to the contralateral hemisphere have heterogeneous morphologies, and include inverted pyramids as well as fusiform and triangular neurons (see below).

Callosal neurons are known to use the excitatory neurotransmitters aspartate or glutamate (Barbaresi et al., 1987; Conti et al., 1988), and to form asymmetrical synapses (Voigt et al., 1988). Nevertheless, it has been hypothesized that some of them are inhibitory in nature (Peters et al., 1990). In studies that combined use of retrograde transport of fluorescent beads and intracellular injections of Lucifer yellow, we found only one example, in the extrastriate area posteromedia lateral suprasylvian area (PMLS) of the cat, of a callosally-projecting neuron that could be classified as a smooth stellate cell, a morphology which has been previously associated with inhibitory neurotransmitters (Peters et al., 1990). Similarly, Bühl and Singer (1989) reported only one smooth stellate cell among callosally-projecting neurons in areas 17/18.

Until recently, most of the data collected on the development of cortical neurons derived from Golgi impregnations, which allowed detailed studies of dendritic morphology but gave little information about the site of projection of labeled neurons. The introduction of sensitive retrograde tracers such as fluorescent latex microspheres (combined with intracellular injections of Lucifer Yellow) or biotinylated dextran, as well as the lipophilic dyes DiI and DiA permitted visualization of the dendritic arbors of neurons projecting to a given target. Such technical advances, in combination with computer-aided reconstruction of these dendritic arbors, make it feasible to follow the development of neurons with known projections, both qualitatively and quantitatively.

Here we review our results on the development of callosal neurons in the kitten. We give evidence that spiny stellate callosally-projecting neurons originate from pyramidal neurons losing their apical dendrite during development. Moreover, we demonstrate that qualitative and quantitative morphological aspects of callosally-projecting neurons in the kitten depend more on their tangential and laminar location than on the target of their axons.

1. POSTNATAL DEVELOPMENT OF CALLOSALLY-PROJECTING SPINY STELLATE NEURONS

We have studied the development of callosally-projecting neurons in the visual cortex in postnatal day 3 P3, P6 and P10 kittens, using the retrograde diffusion of DiI placed in the CC of paraformaldehyde-fixed brains. In older kittens, callosally-projecting neurons were labeled by a combination of in vivo retrograde transport of green or red beads (respectively FITC- and TRITC-labeled fluorescent latex microspheres; Katz et al., 1984) followed by intracellular injections of Lucifer Yellow (LY) in bead-labeled cells seen in fixed slices (Vercelli et al., 1992).

We found that at P3, all callosally-projecting neurons in the supragranular layers have a pyramidal morphology, with an apical dendrite extending to the pial surface (Figure 2). By P6, some DiI-labeled cells, in the deepest part of the supragranular subzone of

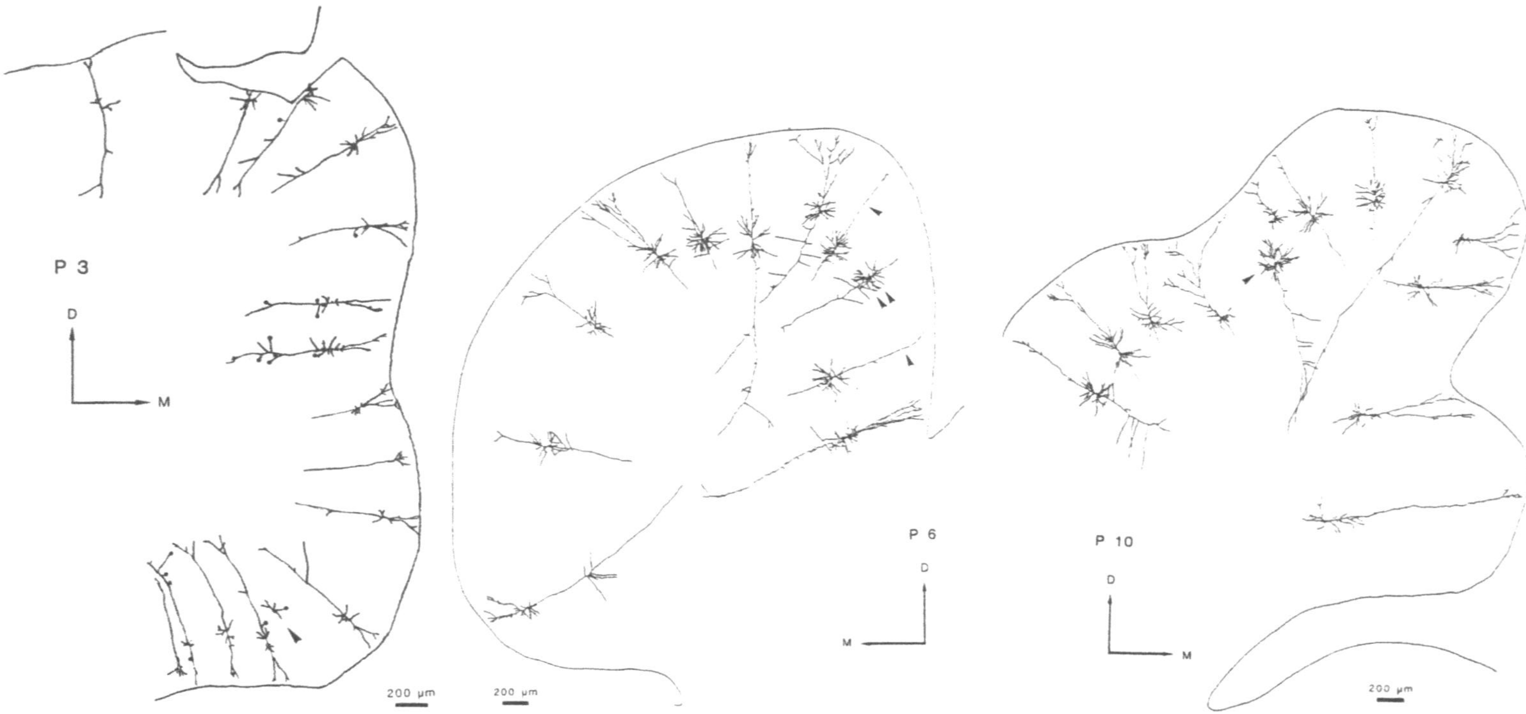

Figure 2. Camera lucida drawings of DiI-labeled callosally-projecting neurons in areas 17 and 18 of P3 (A), P6 (B) and P10 (C) kittens. At P3 all callosally-projecting neurons bear apical dendrites reaching the pial surface, excepted for a neuron (arrowhead) whose apical dendrite is out of the section. At P6, some callosally-projecting neurons (arrowheads) lose their terminal tufts in layer I. At P10, some callosally-projecting neurons (arrowhead), located in deep layer III/upper layer IV, have a spiny stellate morphology. From Vercelli et al., 1992.

callosally-projecting neurons, had apical dendrites which revealed a paucity of spines and of distal branches, and were now beginning to withdraw from the pial surface (Figures 2 and 3). In P10 kitten we could clearly identify spiny stellate neurons that had emitted axons across the callosum—these cell types comprised close to 20% of all contralaterally projecting neurons in the supragranular layers, very similar to the numbers obtained for older, 2 month-old, kittens (Figures 2, 4, and Table 1).

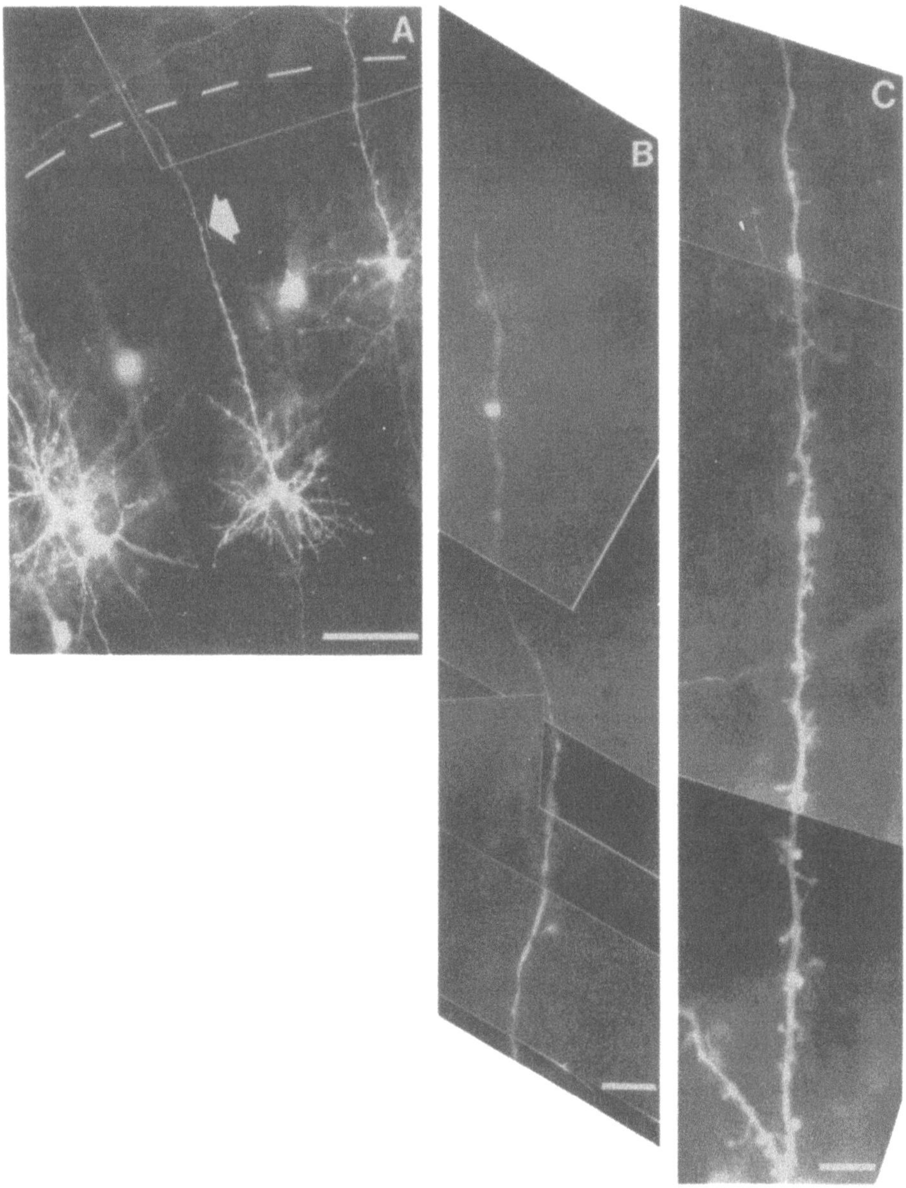

Figure 3. DiI-labeled callosally-projecting neuron (A) in a P6 kitten and details of its apical dendrite (B and C). The distal portion of the apical dendrite (B) has lost its terminal tuft. Scale bars = 100 μm in A and 10 μm in B and C. From Vercelli et al., 1992.

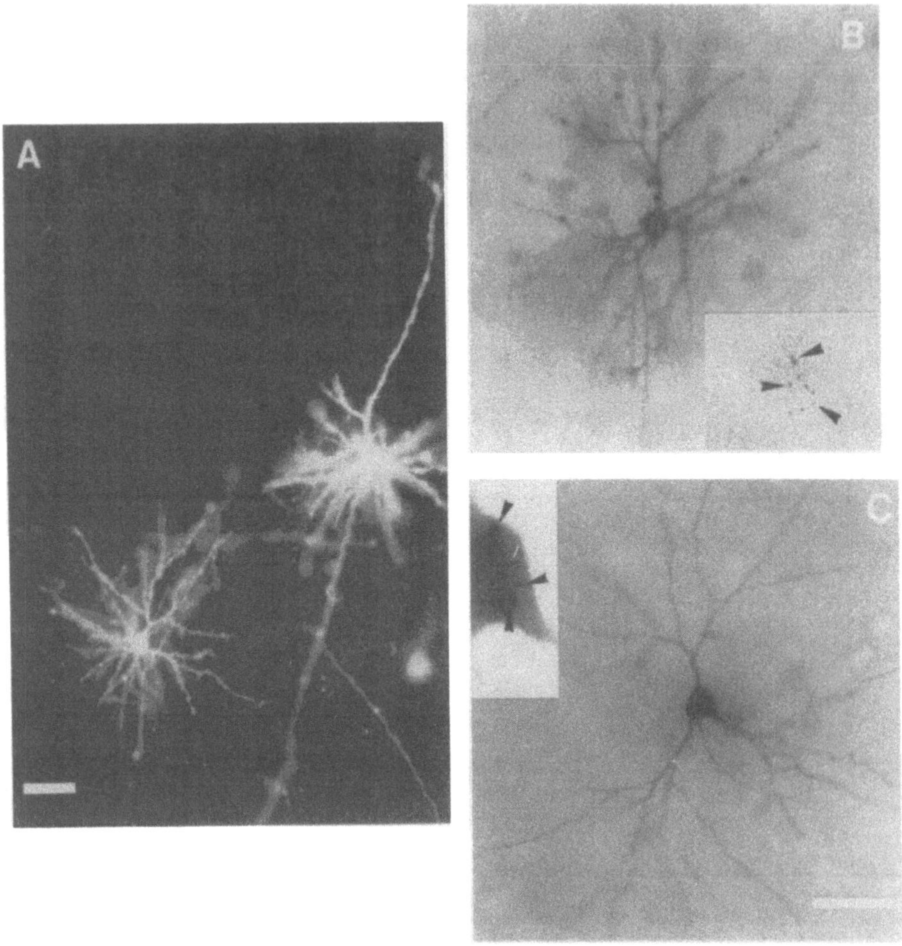

Figure 4. DiI-labeled callosally-projecting neurons in kitten visual cortex at P10 (A and B) and LY-injected callosally-projecting neuron at P57, with retrogradely transported beads in the insert (C). From Vercelli et al., 1992. In A, upper layer IV spiny stellate neuron and layer III pyramid. In B and C spiny stellate neurons.

In order to rule out the possibility that these spiny stellate neurons are identified in development later than are the pyramids because their axons enter the CC at a later time point, we injected green beads in the visual cortex of a P1 kitten; the animal was perfused at P10, and received DiI crystals in the CC. Assuming that the beads remain available for retrograde transport for a limited period of time, we estimate that by P6-P7, axons of

Table 1. Number of pyramidal and stellate callosally-projecting neurons identified at different ages in our study. Labeled neurons at P10 are neurons labeled both by intracallosal DiI and controlaterally-injected green beads

	Pyramids	Stellates	Labeled pyramids	Labeled stellates
P6	239	2		
P10	145	42	12	6
P57	285	63		

bead-labeled neurons had already reached the contralateral visual cortex. The ratio of spiny stellate neurons to pyramidal cells was similar amongst the double (beads and DiI)-labeled neurons (that had sent their axons to the contralateral hemisphere prior to P7) and single, DiI-labeled neurons (whose axons had reached the CC within P10).

We conclude that during development callosally-projecting spiny stellate neurons derive from pyramidal neurons as a result of a selective loss of their apical dendrite. Such a change could be directed by intrinsic (genetic) determinants, by local factors (e.g. from cortical afferents) or by target-derived influences. There is support for each of these hypotheses; Peinado and Katz (1990) report that after infraorbital nerve cut the percentage of spiny stellate neurons in rat somatosensory cortex decreases; also, there seems to be a strong relationship between the morphology of cortical neurons and their target (Hallman et al., 1988; Hübener et al., 1990; Koester and O'Leary, 1992; Kasper et al., 1993 and 1994).

2. THE INFLUENCE OF LOCAL AND TARGET-DERIVED FACTORS ON DENDRITIC DEVELOPMENT IN PYRAMIDAL NEURONS

Weisskopf and Innocenti (1991) demonstrated that pyramidal neurons which lose their callosal projection after removal of their targets, including the contralateral 17/18 border, do not show significant changes in quantitative aspects of dendritic morphology compared to neurons for which the targets are not damaged. In order to study the influence of local vs. target-derived factors on the dendritic arborization of cortical neurons, in another series of experiments we injected, in vivo, green and red beads in area 17, at the 17/18 border, or in areas 18, 19 and PMLS (Rosenquist, 1985) of P3 kittens (N=6). Two months later, we injected bead-labeled cells with LY, in fixed cortical slices of the contralateral hemisphere (Vercelli and Innocenti, 1993). A total of 1263 cells were injected with LY. Of these, one hundred and twenty neurons were fully reconstructed with the aid of the computer, and were analyzed quantitatively. Reconstructed neurons were divided in two groups, i.e. area 17 neurons and neurons restricted to the 17/18 border. Such grouping allowed us to separate dendritic analysis on the basis of the fate of the cell's axon since area 17 neurons that project to the contralateral hemisphere in the newborn kitten lose their projection during development, while those located near the 17/18 border do not (Innocenti and Clarke, 1983 and 1984). Neurons were also classified according to their projection pattern, i.e. those projecting to areas 17 and 18, or those to PMLS (callosally-projecting neurons do not send bifurcating axons to both sites, Innocenti and Clarke, 1983; Seagraves and Innocenti, 1985).

Bead injections in areas 17, 17/18 border, 19 or PMLS of the kitten each labeled callosally-projecting neurons in several contralateral cortical fields. As a rule, the density of labeled neurons was higher at sites homotopic to the injection than elsewhere. All injections resulted in retrogradely filled cells in area 17, far from the 17/18 border, that is, at locations which lose their callosal projections in the adult. The frequency of different morphological types was was scored in relation to the location of their cell body and the termination of their axon: of 1263 callosally-projecting neurons, 1069 were located in layers II-IV. A variety of morphologies were found in all areas, irrespective of the site of termination of their axon.

In layer II-IV, the predominant morphology of injected cells was pyramidal. In contrast, spiny stellate neurons participated in callosal projections originating only from layers III and IV in areas 17 and 18, irrespective of their area of axon termination. Similarly, fusiform and triangular neurons in layer VI were found to project to the CC only from areas 17 and 18, and not from PMLS. Closer inspection of the morphology of 17/18 border neurons having different projections, revealed a whole range of variation in the density and distribution of "perisomatic dendrites" (basal dendrites and proximal side-branches of the apical dendrite) of callosally-projecting pyramids. Neurons with a different distribution of this portion of the dendritic arbors may receive inputs from different thalamic laminae (Hornung and Garey, 1980). When quantitative parameters (soma size, total dendritic or terminal segment length, number of nodes) were considered, differences emerged that were related to the location of the cell body (Table 2). In general, neurons located in area 17 tended to have smaller somata and shorter basal dendrites than neurons located near the 17/18 border. No systematic differences in cellular morphology could be related to the area of termination of the axon.

Pyramidal and polymorphic neurons were found in layers V and VI of all cortical areas under consideration. Fusiform neurons were found in all areas except PMLS.

These results suggest that either the laminar and tangential position or intrinsic neuronal determinants, rather than signals coming from the specific cortical targets in the contralateral hemisphere, control the elaboration of dendritic patterns. They confirm and extend the conclusions of Weisskopf and Innocenti (1991), who found no changes in the dendritic morphology of callosally-projecting pyramids when their contralateral targets were ablated in kittens.

Table 2. Quantitative data on basal dendrites and soma sizes of supragranular pyramidal neurons, subdivided in groups according to their location and site of projection. The values of T.S. (terminal segment), I.S. (internodal segment) and T.D. (total dendritic) lengths and Nodes/Dend (number of nodes per dendrite) are means across cats of the medians per cat and per cell of the values for each dendrite. In each line, the first number relates to the mean, the second to the standard deviation The soma size is the mean of the means for each cat (μm^2). The number (#) of cats, cells and dendrites included in the analysis is indicated for each group. From Vercelli and Innocenti, 1993

			TERMINATION	
			17-18	**PMLS**
ORIGIN	**17**	# Cats, cells	3,16	3, 7
		Soma (μm2)	228 ± 59	245 ± 45
		# Dendrites	68	30
		T.S.(μm)	101.98 ± 13.83	95.21 ± 17.42
		I.S. (μm)	12.64 ± 1.54	15.21 ± 3.09
		T.D.(μm)	680.16±152.32	631.53±251.68
		Nodes/Dend	4.3 ± 0.5	4.3 ± 0.5
	17/18	# Cats, cells	5,35	4, 17
		Soma (μm2)	288 ± 96	344 ± 34
		# Dendrites	156	72
		T.S.(μm)	96.56 ± 15.75	105.59 ± 6.09
		I.S. (μm)	17.62 ± 0.83	17.92 ± 2.33
		T.D.(μm)	713.41±156.68	808.04±246.53
		Nodes/Dend	4.5 ± 0.8	4.3 ± 0.9

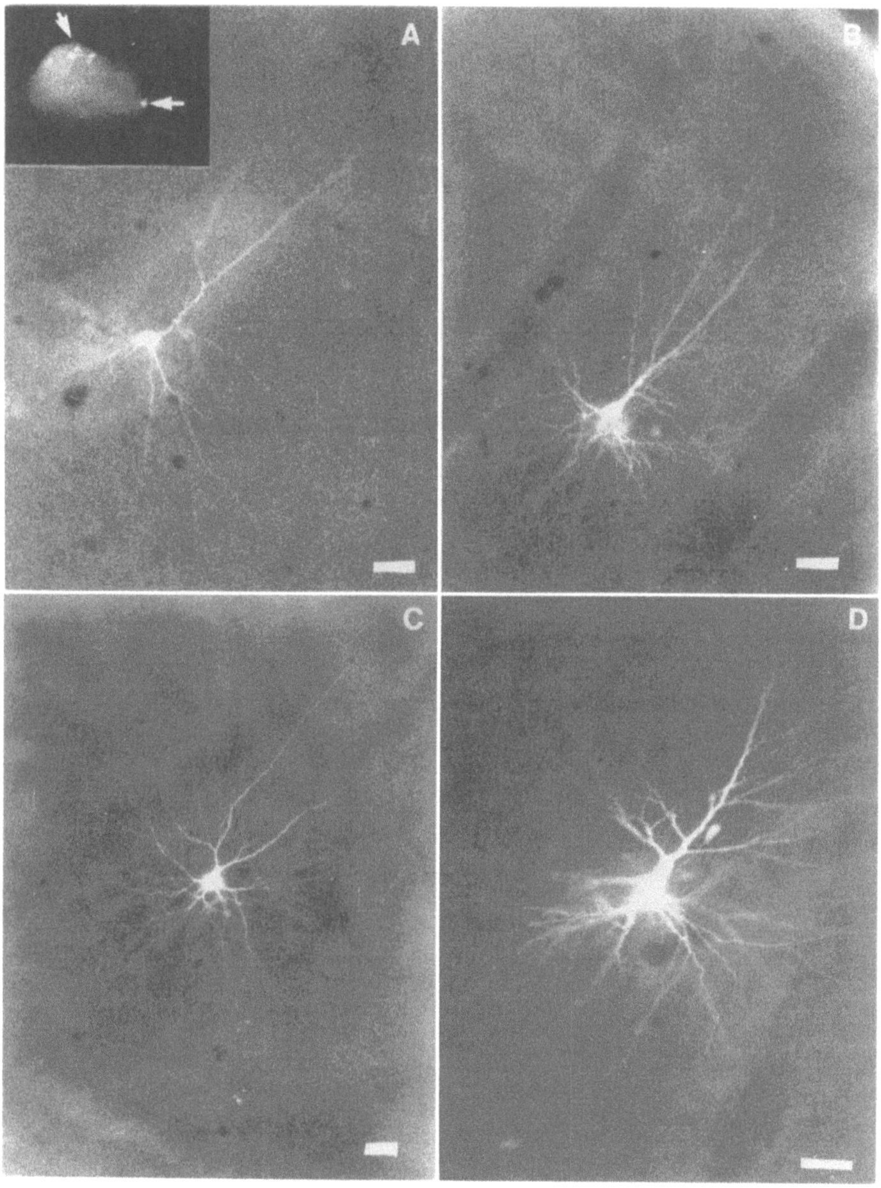

Figure 5. Pyramidal neurons in layer III of area 17/18 border of P57 kittens projecting to contralateral PMLS area, retrogradely labeled by beads (insert in A) and intracellularly-injected with LY. P57 kittens. Scale bar = 50 μm.

3. CONCLUSIONS

Taken together with other studies, we can conclude that the dendritic development of cortical neurons is the result of a combination of an intrinsic developmental program, and of local interactions, most likely with afferent fibers (Fifkova, 1970; Borges and Berry, 1978; Steffen and Van der Loos, 1980; Harris and Woolsey, 1981; Bliss Tieman and Hirsch, 1982).

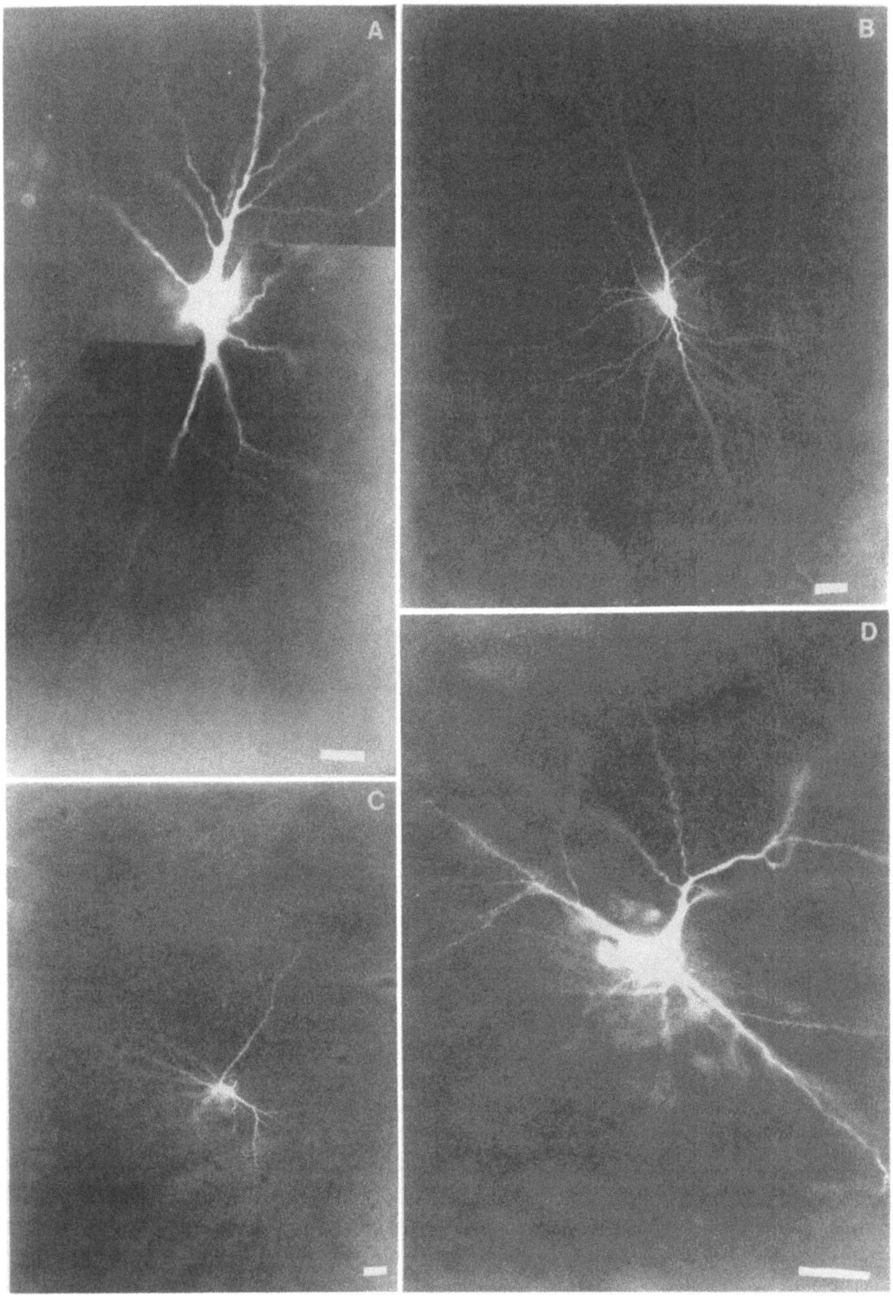

Figure 6. LY-injected, callosally-projecting neurons located in layer VI of areas 17/18 border of P57 kittens. Bypolar neurons in A and B, pyramidal neuron in C and triangular neuron in D. Scale bar = 50 μm.

Additional factors including hormones (Eayrs, 1955), gender (Seymoure and Juraska, 1992), age (Vaughan, 1977), nutrition (Salas et al., 1974) and environmental conditions (Davies and Katz, 1983; Ferrer et al., 1991) have also been implicated in this process.

Our results suggest also that the differences in morphology of cells that project to non-cortical targets, as compared to cortical targets (Hallman et al., 1988), most likely also derive from intrinsic rather than target-derived factors.

In our experiments on kitten visual cortex, we have shown that upper layer IV stellate neurons projecting to the corpus callosum bear an apical dendrite at early stages of development. Moreover, we have demonstrated that the development of several qualitative and quantitative aspects of dendritic morphology of callosally-projecting neurons is influenced by local factors, rather than by factors derived from distal targets. This does not exclude the possibility that retrograde signals conveyed by local axon collaterals can also influence dendritic development. However, other studies have suggested that cortical neurons are influenced in their morphology by local factors, such as sensory experience and more generally, by activity (Globus and Scheibel, 1967; Coleman and Riesen 1968; Valverde, 1968; Winfield, 1981; Juraska, 1982; Wong et al., 1991; Katz and Shatz, 1996). Activity might operate through neurotrophins (Castren et al., 1992; Allendoerfer et al., 1994; Cabelli et al., 1995). Recent studies have documented that neurotrophins can influence the development of dendritic arbors of pyramidal neurons in organotypic cortical cultures; each neurotrophin has a specific influence on dendritic growth, and is effective only during particular developmental periods, depending on the cellular layer, or on whether apical and basal dendrites are considered (McAllister et al., 1995). It will be important, therefore, to assess which neurotrophins are expressed and secreted in each cortical layer, by which cells or terminals, and whether different cortical neurons and/or different compartments of single neurons express subsets of neurotrophin receptors. In turn, neurotrophin-mediated signals could induce a cascade of events including cytoskeletal changes responsible for dendritic and axonal morphology (Riederer et al., 1990; Riederer and Innocenti, 1991; Riederer, 1995). Moreover, it has been recently demonstrated by intravitreal TTX blockade experiments that activity may influence the expression of MAP2, which is a characteristic dendritic protein (Hendry and Bhandari, 1992).

These local influences on the dendritic morphology of cortical neurons probably act in parallel with constraints which establish a specific match between the different cell types and their targets. Genetic programs intrinsic to the neuron could influence its laminar location as well the morphology of its somato-dendritic complex, and the molecular features which control, in rather general ways, the growth and pathfinding of its axon. After encounter with the target and the establishment of synaptic contacts, target-derived factors might influence the maintenance of the connections rather than the development of the dendritic arbor of the cortical neuron. In this respect, cortical neurons appear to be much less sensitive than brainstem and spinal motoneurons to the elimination of distal targets, possibly due to the frequent occurrence of axon collaterals close to their cell bodies (Innocenti, 1980) and in ipsilateral cortical areas (Innocenti et al., 1986).

ACKNOWLEDGMENTS

Supported by MURST and COTRAO grants to AV and by Swiss National Science Foundation grant # 3100–039707.93. to GMI. We are grateful to Dr. Sonal Jhaveri (Department of Brain and Cognitive Sciences, MIT, Cambridge, MA) for reading the manuscript and for her helpful suggestions.

REFERENCES

Allendoerfer K.L., Cabelli R.J., Escandon E., Kaplan D.R., Nikolics K. and Shatz C.J., Regulation of neurotrophin receptors during the maturation of the mammalian visual system. J. Neurosci. 14: 1795–1811 (1994).

Barbaresi P., Conti F., Fabri M. and Manzoni T., D-[^{3}H]aspartate retrograde labeling of callosal and association neurons of somatosensory areas I and II of cats. J. Comp. Neurol. 263: 159–178 (1987).

Bliss Tieman S. and Hirsch H.V.B., Exposure to lines of only one orientation modifies dendritic morphology of cells in the visual cortex of the cat. J. Comp. Neurol. 211: 353–362 (1982).

Bolz J., Novak N., Götz M. and Bonhoefferr T., Formation of target-specific, neuronal projections in organotypic slice cultures from rat visual cortex. Nature 346: 359–362 (1990).

Bolz J., Hübener M., Kehrer I. and Novak N., Structural organization and development of identified projection neurons in primary visual cortex. In: Bagnoli P., Hodos W. (eds) *The Changing Visual System: Maturation and Aging in the Central Nervous System.* Nato ASI Series, Plenum Press pp 233–246 (1991).

Borges S. and Berry M., The Effects of Dark Rearing on the Development of the Visual Cortex of the Rat. J. Comp. Neurol. 180: 277–300 (1978).

Bühl E.H. and Singer W., The callosal projection in cat visual cortex as revealed by a combination of retrograde tracing and intracellular injection. Exp. Brain Res. 75: 470–476 (1989).

Cabelli R.J., Hohn A. and Shatz C., Inhibition of ocular dominance column formation by infusion of NT-4/5 of BDNF. Science 267: 1662–1666 (1995).

Castren E., Zafra F., Thoenen H. and Lindholm D., Light regulates expression of brain-derived neurotrophic factor mRNA in rat visual cortex. Proc. Natl. Acad. Sci. USA 89: 9444–9448 (1992).

Coleman P.D. and Riesen A.H., Environmental effects on cortical dendritic field. I. Rearing in the dark. J. Anat. 102: 363–374 (1968).

Conti F., Fabri M. and Manzoni T., Glutamate-positive corticocortical neurons in the somatic sensory areas I and II of cats. J. Neurosci. 8: 2948–2960 (1988).

Davies C.A. and Katz H.B., The comparative effects of early-life undernutrition and subsequent differential environment on the dendritic branching of pyramidal cells in rat visual cortex. J. Comp. Neurol. 218: 345–350 (1983).

Diao Y.-C. and So K.-F., Dendritic morphology of visual callosal neurons in the golden hamster Brain, Behavior & Evolution 37: 1–9 (1991).

Eayrs J.T., The cerebral cortex of normal and hypothyroid rats. Acta Anat. 25: 160–183 (1955).

Ferrer I., Soriano E., Marti E., Digon E., Reyners H. and Gianfelici de Reyners E., Development of dendritic spines in the cerebral cortex of the microencephalic rat following prenatal x-irradiation. Neurosci. Lett. 125: 183–186 (1991).

Fifkova E., The effect of unilateral deprivation on visual centers in rats. J. Comp. Neurol. 140: 431–438 (1970).

Globus A. and Scheibel A., The effect of visual deprivation on cortical neurons: A Golgi study. Exp. Neurol. 19: 331–345 (1967).

Hallman L.E., Schofield B.R. and Lin C.-S., Dendritic morphology and axon collaterals of corticotectal, corticopontine, and callosal neurons in layer V of primary visual cortex of the hooded rat. J. Comp. Neurol. 272: 149–160 (1988).

Harris R.M. and Woolsey T.A., Dendritic plasticity in mouse barrel cortex following postnatal vibrissa follicle damage. J. Comp. Neurol. 196: 357–376 (1981).

Hendry S.H.C. and Bhandari M.A., Neuronal organization and plasticity in adult monkey visual cortex - Immunoreactivity for Microtubule-Associated Protein-2. Visual Neurosci. 9: 445–459 (1992).

Hornung J.P. and Garey L.J., A direct pathway from thalamus to visual callosal neurons in cat. Exp Brain Res 38:121–123 (1980).

Hübener M.,. Schwarz C. and Bolz J., Morphological types of projection neurons in layer 5 of cat visual cortex. J. Comp. Neurol. 301: 655–674 (1990).

Innocenti G.M., Growth and reshaping of axons in the establishment of visual callosal connections. Science 212: 824–827 (1981).

Innocenti G.M., The primary visual pathway through the corpus callosum: morphological and functional aspects in the cat. Arch. Ital. Biol. 118: 124–188 (1980).

Innocenti G.M., The development of projections from cerebral cortex. Prog. Sens. Physiol. 12: 65–114 (1991).

Innocenti G.M. and Fiore L., Morphological correlates of visual field transformation in the corpus callosum. Neurosci. Lett. 21: 245–252 (1976).

Innocenti G.M. and Caminiti R., Postnatal shaping of callosal connections from sensory areas. Exp. Brain Res. 38: 381–394 (1980).

Innocenti G.M. and Clarke S., Multiple sets of visual cortical neurons projecting transitorily through the corpus callosum. Neurosci. Lett. 41: 27–32 (1983).

Innocenti G.M. and Clarke S., The organization of immature callosal connections. J. Comp. Neurol. 230: 287–309 (1984).

Innocenti G.M., Clarke S. and Kraftsik R., Interchange of callosal and association projections in the developing visual cortex. J. Neurosci. 6: 1384–1409 (1986).

Innocenti G.M., Fiore L. and Caminiti R., Exuberant projection into the corpus callosum from the visual cortex of newborn cats. Neurosci. Lett. 4: 237–242 (1977).

Innocenti G.M., Manzoni T. and Spidalieri G., Patterns of the somesthetic messages transferred through the corpus callosum. Exp. Brain Res. 19: 447–466 (1974).

Juraska J.M., The development of pyramidal neurons after eye opening in the visual cortex of hooded rats: a quantitative study. J. Comp. Neurol. 212: 208–213 (1982).

Kasper E.M., Larkman A.U., Lübke J. and Blakemore C., Pyramidal neurons in layer V of the rat visual cortex. I. Correlation between cell morphology, intrinsic electrophysiological properties and axon targets. J. Comp. Neurol. 339: 459–474 (1993).

Kasper E.M., Lübke J., Larkman A.U. and Blakemore C., Pyramidal neurons in layer V of the rat visual cortex. III. Differential maturation of axon targeting, dendritic morphology, and electrophysiological properties. J. Comp. Neurol. 339: 495–518 (1994).

Katz L.C., Burkhalter A. and Dreyer W.J., Fluorescent latex microspheres as a retrograde neuronal marker for in vivo and in vitro studies of visual cortex. Nature 310: 498–500 (1984).

Katz L.C. and Shatz C.J., Synaptic activity and the construction of cortical circuits. Science 274: 1133–1138 (1996).

Koester S.E. and O'Leary D.D.M., Functional classes of cortical projection neurons develop dendritic distinct ions by class-specific sculpting of an early common pattern. J. Neurosci. 12: 1382–1393 (1992).

Martin K.A.C. and Whitteridge D. Form, function and intracortical projections of spiny neurones in the striate visual cortex of the cat. J. Physiol. (London) 353: 463–504 (1984).

McAllister A.K., Lo D.C. and Katz L.C., Neurotrophins regulate dendritic growth in developing visual cortex. Neuron 15: 791–803 (1995).

Olavarria J. and Van Sluyters R.C. Organization and postnatal development of callosal connections in the visual cortex of the rat J. Comp. Neurol. 239:1–26 1985

O'Leary D.D., Stanfield B.B. and Cowan W.M., Evidence that the early postnatal restriction of the cells of origin of the callosal projection is due to the elimination of axon collaterals rather than. Dev. Brain Res. 1: 607–617 (1981).

Peinado A. and Katz L., Development of cortical spiny stellate cells: retraction of a transient apical dendrite. Soc. Neurosci. Abst. 16: 1127 (1990).

Peters A., Payne B.R. and Josephson K., Transcallosal non-pyramidal cell projections from visual cortex in the cat. J. Comp. Neurol. 302: 124–142 (1990).

Riederer B.M., Guadano-Ferraz A. and Innocenti G.M., Differences in distribution of microtubule-associated protein 5a and 5b during cat cerebral cortex and corpus callosum development: Dependence on phosphorylation. Devl. Brain Res. 56: 235–243 (1990).

Riederer B.M. and Innocenti G.M., Differential distribution of tau proteins in developing cat cerebral cortex and corpus callosum. Eur. J. Neurosci. 3: 1134–1145 (1991).

Riederer B.M., Development of the axonal and dendritic cytoskeleton. Adv. Mol. Cell Biol. 12: 107–142 (1995).

Rosenquist A.C., Connections of visual cortical areas in the cat. In: Peters A, Jones EG (eds). *Cerebral Cortex*. Vol 3, Plenum Publishing Corporation pp 81–117 (1985).

Salas M., Diaz S. and Nieto A., Effects of neonatal food deprivation on cortical spines and dendritic development of the rat. Brain Res. 73: 139–144 (1974).

Segraves M.A. and Innocenti G.M., Comparison of the distributions of ipsilaterally and contralaterally projecting corticocortical neurons in cat visual cortex using two fluorescent tracers. J. Neurosci. 5: 2107–2118 (1985).

Seymoure P. and Juraska J.M., Sex differences in cortical thickness and the dendritic tree in the monocular and binocular subfields of the rat visual cortex at weaning age. Dev. Brain Res. 69: 185–189 (1992).

Steffen H. and Van der Loos H., Early lesions of mouse vibrissal follicles: their influence on dendrite orientation in the cortical barrelfield. Exp. Brain Res. 40: 419–431 (1980).

Valverde F., Structural changes in the area striata of the mouse after enucleation. Exp. Brain Res. 5: 274–292 (1968).

Van der Loos H., The "improperly" oriented pyramidal cell in the cerebral cortex and its possible bearing on problems of neuronal growth and cell orientation. Bull. Johns Hopkins Hosp. 117: 228–250 (1965).

Vaughan D.W., Age-related deterioration of pyramidal cell basal dendrites in rat auditory cortex. J. Comp. Neurol. 171: 501–516 (1977).

Vercelli A., Assal F. and Innocenti G.M., Emergence of callosally-projecting neurons with stellate morphology in the visual cortex of the kitten. Exp. Brain Res. 90: 346–358 (1992).
Vercelli A. and Innocenti G.M., Morphology of visual callosal neurons with different locations, contralateral targets or patterns of development. Exp. Brain Res. 94: 393–404 (1993).
Voigt T., LeVay S., Stamnes M.A., Morphological and immunocytochemical observations on the visual callosal projections in the cat. J. Comp. Neurol. 272: 450–460 (1988).
Weisskopf M. and Innocenti G.M., Neurons with callosal projections in visual areas of newborn kittens: an analysis of their dendritic phenotype with respect to the fate of the callosal axon and of its target. Exp. Brain Res. 86: 151–158 (1991).
Winfield D.A., The postnatal development of synapses in the visual cortex of the cat and the effects of eyelid closure. Brain Res. 206: 166–171 (1981).
Wong R.Ol., Herrmann K. and Shatz C.J., Remodeling of retinal ganglion cell dendrites in the absence of action potential activity. J. Neurobiol 22: 685–697 (1991).

12

NEURONAL AND NON-NEURONAL PLASTICITY IN THE RAT FOLLOWING MYENTERIC DENERVATION

C. Cracco[1,2] and G. Filogamo[1]

[1]Department of Anatomy, Pharmacology, and Forensic Medicine
University of Torino
corso Massimo D'Azeglio 52, I-10126 Torino, Italy
[2]Urological Clinic (Director Prof. S. Rocca Rossetti)
University of Torino
corso Dogliotti 14, I-10126 Torino, Italy

1. INTRODUCTION

The plasticity of the enteric nervous system in adulthood has always been a field of research of the greatest interest. Many experimental models have been developed, allowing to investigate the importance of the peripheral field of innervation in the control of growth and differentiation of enteric neurons during postnatal life (Filogamo, 1987; Filogamo and Cracco, 1995). Among those, we were particularly interested in the chemical ablation of the myenteric plexus by means of a cationic surfactant, benzalkonium chloride (BAC).

2. BENZALKONIUM CHLORIDE (BAC) TREATMENT

Benzalkonium chloride (benzyldimethyltetradecylammonium chloride, BAC) is a cationic surfactant, also used as an antiseptic in the operating rooms. The polar group of the quaternary ammonium salt induces an irreversible depolarization of membranous structures. Subsequent phenomena altering cellular functions are due to the progressive degeneration of membrane structures. The depolarizing effect is more marked on cell membranes with a higher negative charge, such as in the nervous tissue if compared to the smooth muscle (Sato et al., 1978).

A selective, chronic myenteric denervation of the intestine can be produced by local application of a BAC solution (Sato et al., 1978; Sakata et al., 1979). The treatment includes surgical exposure of a definite intestinal loop and serosal application of a 2 mM solution of BAC in saline for 30 minutes (Sato et al., 1978; See et al., 1988, 1990; Cracco

Brain Plasticity, edited by Filogamo *et al.*
Plenum Press, New York, 1997

and Filogamo, 1993). BAC treatment has been performed on various intestinal segments in the rat, including duodenum (Zucoloto et al., 1985), jejunum (See et al., 1988; Zucoloto et al., 1991), ileum (Cracco and Filogamo, 1993), colon (Sato et al., 1978, Sakata et al., 1979) and anorectum (Sato et al., 1978).

Serosal application of BAC to any segment of intestine: i) eliminates greater than 90% of myenteric neurons (Fox et al., 1986); ii) destroys the cells of the longitudinal muscle layer; iii) destroys the outer portion of the circular muscle layer; iv) does not affect the submucosal plexus (See et al., 1988); v) deprives muscle, mucosa and submucosal plexus of sympathetic innervation, by destroying the extrinsic nerves (See at al., 1990). Selectivity of the treatment for the myenteric neurons is: i) due to the serosal application of the BAC solution, gaining access to the myenteric innervation by an inward diffusion through the serosa and the longitudinal muscle layer; ii) the result of poorer repairing ability of nervous tissues, irreversibly damaged at the chronic stage, as compared with other tissues such as smooth muscle, fully recovering after the treatment (Fig.1).

Thus, BAC treatment represents an additional experimental model for studying the contribution of the myenteric plexus to intestinal functions, through its chronic elimination. In the literature, many models of intestinal aganglionosis have been described, including human pathological conditions, such as Hirschprung's disease (Ehrenpreis, 1971) and Chagas' disease (Ferreira-Santos, 1964), and experimentally manipulated animal models, such as the piebald mouse (Webster, 1973) and the lethal spotted mutant mouse (Nagahama et al., 1985; Tennyson et al., 1986; Payette et al., 1987). BAC treatment displays several advantages, if compared to these models. In fact, this technique: i) selectively eliminates the myenteric plexus, instead of producing a complete enteric aganglionosis; ii) can be applied to definite intestinal segments; iii) can be performed in genetically-intact animals; iv) can be used in adulthood, after a normal embryonic and postnatal development; v) allows to investigate dynamically the changes induced by BAC application on the various cellular populations to study, already few hours after treatment, and not only the already established, long-term effects. Finally, many drugs possess surfactant activity, thus, surfactant-induced intestinal denervation may be pharmacologically

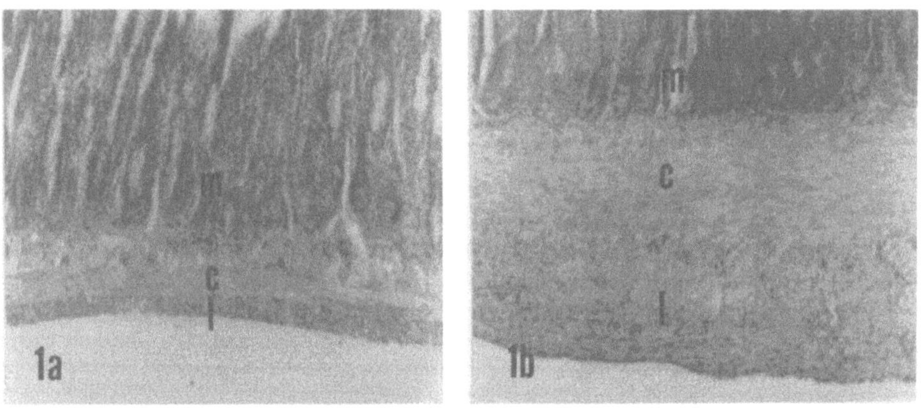

Figure 1. Control (a) and BAC-treated rat ileum (b) one month after BAC application, same magnification. m = mucosa; c = circular smooth muscle; l = longitudinal smooth muscle. Hematoxylin and eosin staining of transverse paraffin sections; 10x. Both smooth muscle layers are significantly increased in thickness after BAC treatment, in absence of the myenteric plexus.

important. For instance, damage to the myenteric and submucosal plexuses has occurred in people chronically abusing laxatives (Smith, 1968; Riemann et al., 1980).

3. MACROSCOPIC CHANGES AFTER BAC TREATMENT

We observed that immediately after the application of the BAC solution, the treated intestinal loop appears edematous and ischaemic. This event does not take place when the gut is sham-operated with saline; therefore, it should not be due to the surgical procedure alone, but to an immediate, inflammatory reaction to BAC.

Already 5 days after BAC treatment, and later on, the intestinal segments are distinguishable among the other loops, being slightly dilated and enveloped in a white, thickened and inflamed serosa. The gut oral to the treated segment appears even more dilated, whereas more proximally it displays a thickened, hypertrophic wall, assuring a preserved flow of the ingesta in spite of the aperistaltic intestinal segment (Cracco and Filogamo, 1993; Filogamo and Cracco, 1995).

4. EFFECTS OF BAC TREATMENT ON THE MUCOSA

About 15 days after BAC treatment the mucosa in a jejunal segment notably increases in thickness, with increased villus length, crypt depth and mitotic rate of the crypt-epithelial cells (See et al., 1990). Since extrinsic denervation alone has no similar effects, it can be concluded that the myenteric plexus normally exerts an inhibitory influence on cell proliferation in the mucosa, the submucosal plexus a chronic, stimulatory one. Myenteric influence might be exerted directly on mucosal elements (myenteric neurons projecting to the mucosa have been demonstrated in the guinea-pig small intestine and canine stomach, Song et al., 1991; Furness et al., 1991), or indirectly, by acting through the submucosal neurons. The final effect, i.e. the stimulated division of the crypt stem cells, has been ascribed to a greater release of VIP (vasoactive intestinal peptide) from the hypertrophic, VIP-immunoreactive submucosal neurons. This neurotransmitter has been recently shown to stimulate mitotic activity in cultured keratinocytes (See et al., 1990).

Other authors (Zucoloto et al., 1988, 1991) demonstrated epithelial hyperplasia in the duodenum and descending colon 5 months after BAC treatment, but failed to confirm any change in epithelial cell proliferation in the jejunum at any time after BAC treatment. According to our observations in the rat ileum, the mucosa displays a slight increase in thickness, especially at longer time intervals after BAC treatment, but morphometric analysis never evidentiates statistically significant differences in comparison with the control values.

5. EFFECTS OF BAC TREATMENT ON THE SUBMUCOSAL PLEXUS

In our experiments we examined the submucosal plexus on whole-mount preparations of both control and BAC-treated ileal segments, peeling away the smooth muscle layers and the mucosa and dissecting laminae of submucosa. The staining techniques employed were: i) AChE (acetylcholinesterase) histochemistry; ii) NADH (ß-nicotinamide adenine dinucleotide)-diaphorase histochemistry; iii) DiI fluorescent tracing (Fig.2); iv)

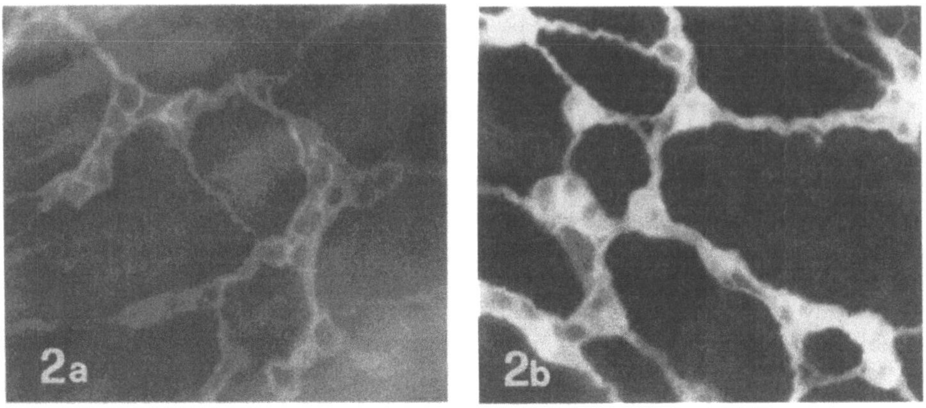

Figure 2. Control (a) and BAC-treated rat ileum (b) 20 days after BAC application, same magnification. DiI tracing of submucosal neurons and internodal strands on laminar preparations; 20x. The submucosal plexus appears hypertrophic after BAC treatment, if compared to the control one.

immunofluorescence for the antibody anti-S100 protein (Honig and Hume, 1989; Vercelli et al., 1992; Cracco and Filogamo, 1993, 1994; Filogamo and Cracco, 1995). In the cases i) and ii) the network of the submucosal plexus is already visible with the naked eye before dissection. On the laminae the stainings employed reveal the regular pattern of ganglia and connecting strands. With all the staining techniques we provided evidence for qualitatively larger neuronal and glial cell bodies, and thicker connecting strands in the BAC-treated ileal segments. Morphometry performed on NADH-diaphorase-stained, stretched submucosal preparations reveal a mean neuronal somatic size of 283 $\mu m^2 \pm 76$ (n=40) in the control rats, 495 $\mu m^2 \pm 134$ (n=32) in the BAC-treated ones. The difference is statistically significant ($t=8.377$, $p<0.001$).

See et al. (1990) also describe enlarged cell bodies of the VIP-immunoreactive neurons of the submucosal plexus, with a mean somatic area similar to ours.

Hypertrophy of the submucosal neurons in the BAC-treated segment may be explained: i) by the lack of inhibitory influence exerted by the neural input (both myenteric and extrinsic) on their growth; ii) by the increased mass of the target tissue innervated (Gabella, 1984), i.e. increased volume of the mucosa (See et al., 1990) and partial reinnervation of the thickened circular muscle layer (Luck et al., 1993). Sprouting of nerve fibers originating from the submucosal plexus is probably mediated by an increased production of trophic factors in the hypertrophic and hyperplastic smooth muscle. A contribution in the production of growth factors might also come from the immune cells (particularly mast cells) present in the inflammatory infiltrate, very prominent already after BAC application, as well as after long time intervals (See et al., 1990).

6. EFFECTS OF BAC TREATMENT ON THE NERVE FIBERS

BAC application eliminates both myenteric and extrinsic innervation to the treated portion (Fox et al., 1983; See et al., 1990). According to our observations on laminar preparations, the elimination is always incomplete (Cracco and Filogamo, 1993; Fig.3), even though affecting extensive areas (Fig.3). Additionally, the amount of nerves detected

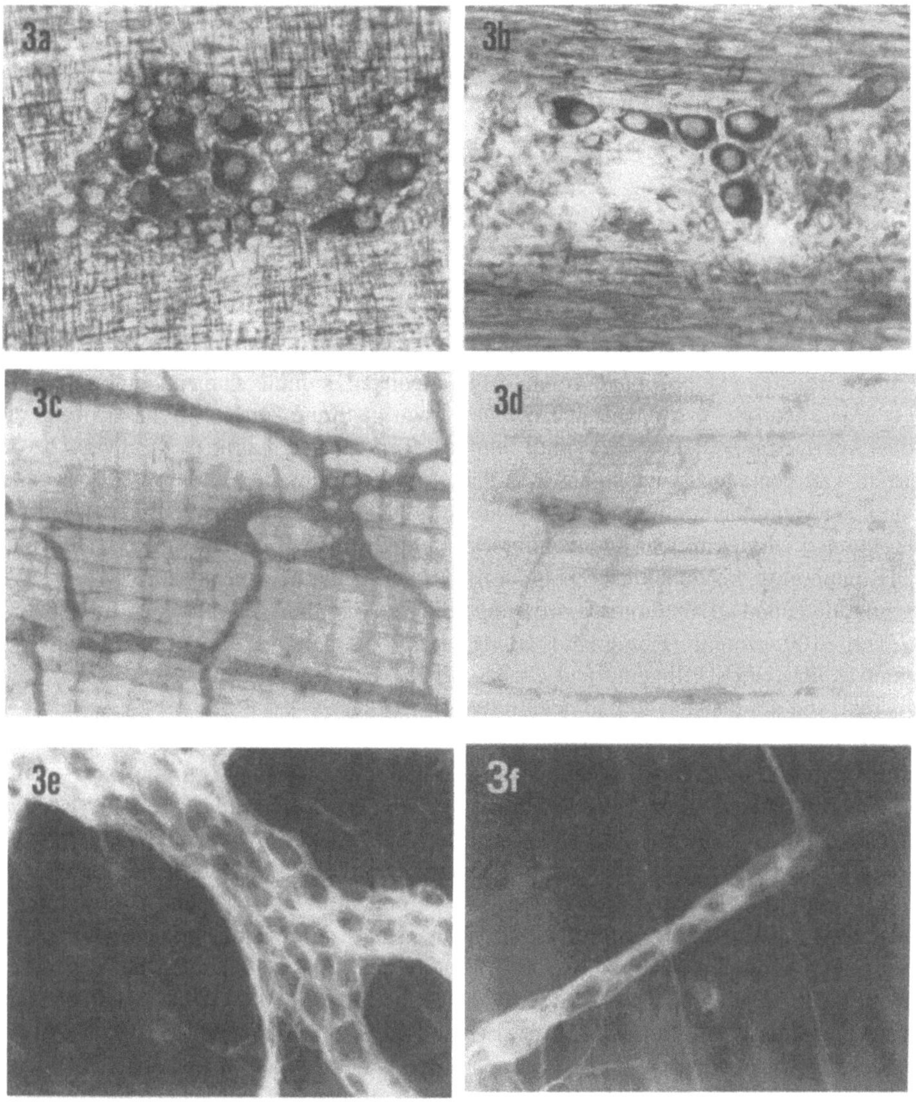

Figure 3. Control (a, c, e) and BAC-treated rat ileum (b, d, f) 20 days after BAC application. NADH-diaphorase staining(a, b; 40x), AChE histochemistry (c, d; 10x) and DiI tracing (e, f; 40x) of myenteric neurons and connecting nerve bundles on laminar preparations. The ablation of the myenteric plexus appears to be extensive; the residual neurons are smaller than the controls; the internodal strands are strongly reduced.

with various techniques seems higher after long time intervals than short-term after BAC application. Growth cones on the tip of isolated nerves can be detected. Other authors (Luck et al., 1993) observed, after 15 days, only few fibers projecting parallel to the circular muscle and running within smooth muscle bundles; after 45 days a greater number of nerve fibers within the circular muscle layer.

Sprouting of novel nerve fibers into the denervated smooth muscle layers might well occur from: i) the submucosal plexus (Luck et al., 1993), even though it cannot be ruled out whether they derive from proliferation of preexisting nerve fibers or from new nerve fiber growth; ii) extraenteric sources; iii) immediately adjacent intestinal segments (Tennyson et al., 1986; Payette et al., 1987; Cracco and Filogamo, 1993); iv) the residual myenteric neurons after BAC application; in fact, myenteric neurons are able to reinnervate submucosal arterioles following extrinsic denervation (Jiang and Surprenant, 1992).

7. EFFECTS OF BAC TREATMENT ON THE MYENTERIC PLEXUS

Serosal application of BAC ablates more than 90% of the myenteric neurons in the treated intestinal segment (Fox et al., 1983, 1986).

According to our results on laminar preparations, extensive areas devoid of myenteric innervation can be observed, others display isolated, spindle-shaped ganglia made up of few small neurons, organized in irregular networks. Some zones only have fine nerve bundles interlacing irregularly. A small number of myenteric neurons can always be detected, in a percentage always higher than that reported in the literature, probably because we employed laminae containing the myenteric plexus rather than sections. These results were obtained employing various techniques, including AChE histochemistry (Fig.3c-d), NADH-diaphorase histochemistry (Fig.3a-b), NADPH-diaphorase histochemistry, silver impregnation (Fig.4), DiI fluorescent tracing (Fig.3e-f), immunofluorescence for the antibody anti-S100 protein (Honig and Hume, 1989; Vercelli et al., 1992; Cracco and Fiogamo, 1993, 1994; Filogamo and Cracco, 1995), confirming that the myenteric ablation is real, and not just a switch to negativity of a marker.

The mean somatic neuronal size in the NADH-diaphorase-stained stretched preparations is 460 $\mu m^2 \pm 110$ (n=40) i the control rats, 285 $\mu m^2 \pm 94$ (n=42) in the BAC-treated ones. The difference is statistically significant ($t=7.69$, $p<0.001$). The shape of the residual myenteric neurons in the BAC treated segments was studied on silver impregnated laminae (Fig.4). In the controls all kind of shapes can be detected, multipolar (similar to Dogiel type I and II, Dogiel, 1899) (Fig.4a-b-e), bipolar (Fig.4c-f) and more rarely unipolar (Fig.4d-f) (Hill, 1927; Schofield, 1968; Wood, 1994); after BAC application only few unipolar neurons can be observed (Fig.4g-h). Many investigations indicate that it is meaningful to distinguish enteric nerve cells by shape, and that an association exists between cell shape, chemistry and function (Hodgkiss and Lees, 1983; Furness et al., 1988).

Finally, we observe in the hypertrophic gut wall orally to the aperistaltic, BAC-treated ileal loop the enlargement of those very small, NADPH-diaphorase-positive cell bodies, already described both in the control and hypertrophic ileum of the adult rat (Cracco and Filogamo, 1995) (Fig.5). Such labeled somata are larger in the hypertrophic intestine (Fig.5c-d-e) than in the control (Fig.5a-b), and are absent in the BAC-treated loops. In any case they are extremely small in comparison to the average size of the myenteric neurons; are located within ganglia and internodal strands and at lower magnifications may be confused with varicosities of the nerve fibers. They could represent those differentiating neurons from a reserve pool, undergoing differentiation under conditions of functional hyperactivity (Filogamo and Vigliani, 1954; Filogamo, 1987). The existence of

Figure 4. Control (a-f; 40x) and BAC-treated rat ileum (g, 100x; h, 40x) two weeks after BAC application. Bielchowski silver impregnation on laminar preparations of myenteric plexus. In the controls all kind of neuronal shapes can be detected, multipolar (a, b, e), bipolar (c, f) and unipolar (d, f); after BAC application only unipolar nerve cell bodies can be evidentiated (g, h).

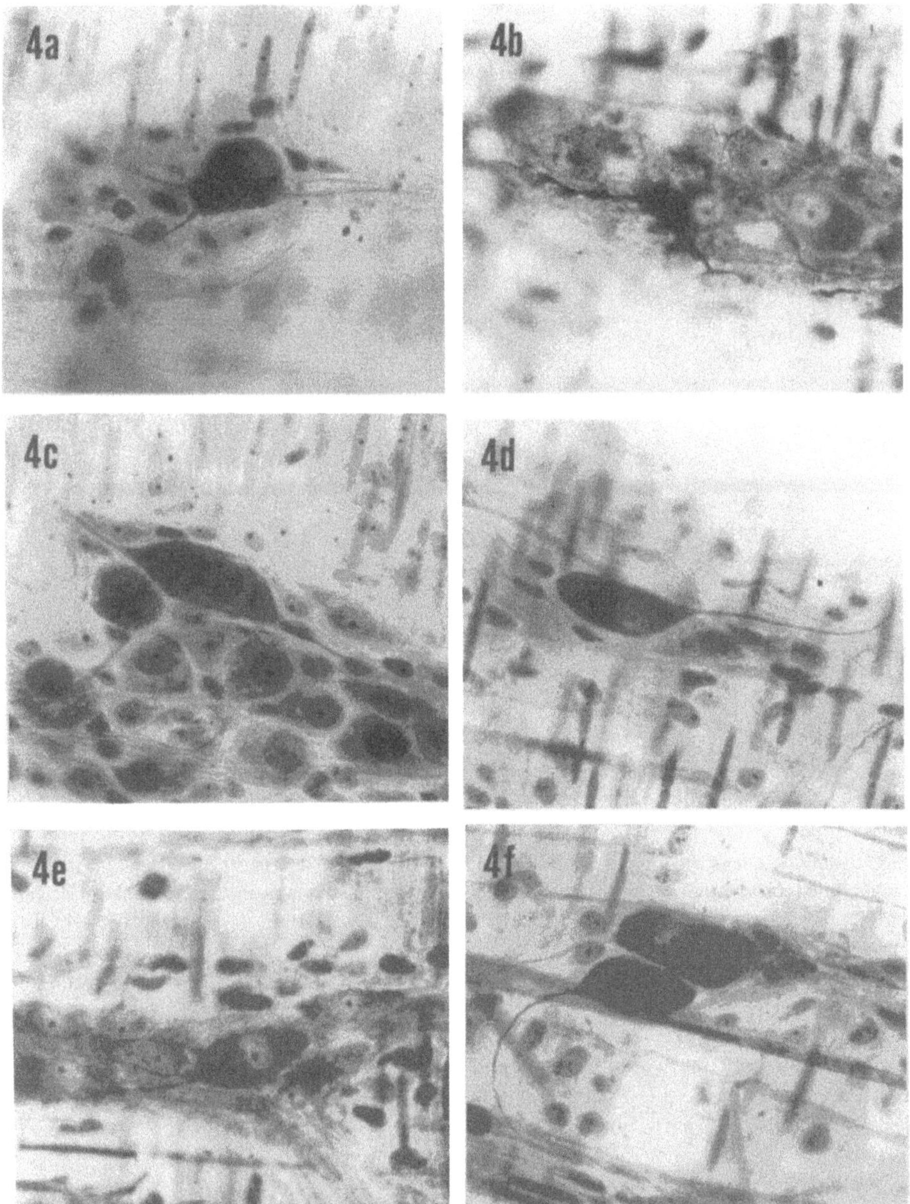
4a
4b
4c
4d
4e
4f
4g
4h

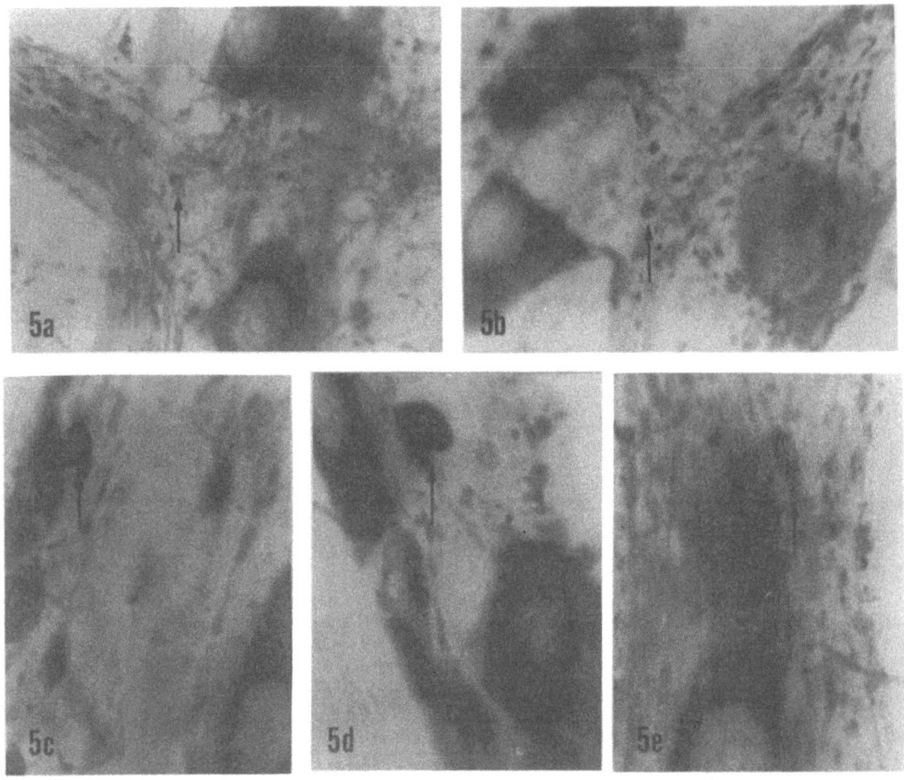

Figure 5. Control (a, b) and hypertrophic rat ileal segment (c, d, e) oral to the BAC-treated loop one month after BAC application, same magnification. NADPH-diaphorase histochemistry on laminar preparations of myenteric plexus; 100x. The very small stained nerve cell bodies seen in the controls (arrows in a and b) are notably enlarged in the hypertrophic intestine (arrows in c, d, e).

newly generated cells with neuronal morphology has already been shown in the adult mammalian nervous system (Reynolds and Weiss, 1992).

8. EFFECTS OF BAC TREATMENT ON THE SMOOTH MUSCLE

BAC treatment eliminates the longitudinal muscle layer and the inner portion of the circular smooth muscle. The surviving smooth muscle rapidly regenerates both smooth muscle layers, notably increased in thickness. In absence of the myenteric plexus, neuronally mediated responses are eliminated, whereas longitudinal muscle contractility is still preserved (Fox and Bass, 1986). No concomitant deficit in smooth muscle contractility occurs, although electric properties seem to be altered (Fox et al., 1983).

The time course and nature of BAC-induced changes have been examined at various time intervals after treatment (See et al., 1988). Within 12 hours cellular disruption can be observed within both smooth muscle layers. After 24 hours the longitudinal muscle layer and the outer part of the circular muscle layer are degenerated and display an abundant inflammatory infiltrate (mainly macrophages). Within 2 days after treatment degeneration and edema disappear, whereas a newly formed longitudinal muscle layer becomes evident, due to the proliferation of the remaining circular muscle cells (even though the possibility cannot be completely ruled out that some longitudinal muscle cells survive treatment and repopulate the layer, as well as the migration from outside the treated area of newly formed longitudinal muscle cells). By 4/5 days after treatment the thickness of both smooth muscle layers closely resembles the control one, while a high rate of cell division can be detected within both smooth muscle layers. Between 5 and 15 days after treatment the thickness of both smooth muscle layers increases, between 15 and 105 days the hyperplasia of the circular muscle layer decreases to control values, whereas it is irreversible for the longitudinal muscle layer. Inflammatory infiltrates can still be detected in the treated tissues, particularly of mast cells.

Our results substantially confirm the data previously reported by other authors. Camera lucida drawings have been performed on transverse, hematoxylin and eosin-stained 10 um-thick paraffin sections, obtained from the day following BAC treatment up to two months afterwards . Measurements reveal that the thickness of both smooth muscle layers is reduced two and three days after BAC application, if compared to the controls, whereas after five days the thickness is again similar to the controls and after ten days its increase is evident. After two weeks the thickness of both smooth muscle layers is nearly doubled; after one month it becomes three-fold the original values (Fig.1a-b); after 45 days it is again about twice the controls. After two months the decrease regards especially the circular muscle layer, which is reduced to no more than 1.5-fold the control values, while the longitudinal muscle layer still remains about twice its original thickness.

Thus, the damage to the smooth muscle layers is repairable. Regeneration of the longitudinal muscle layer occurs with a pattern similar to its prenatal development, in a sort of recapitulation of the embryonic pattern of migration. In fact, development of the circular muscle layer nears completion before the longitudinal muscle layer becomes evident, first between the circular muscle layer and the myenteric plexus.

Hyperplasia of the smooth muscle layers might result from: i) an increased workload (the BAC-treated intestinal loop is an aperistaltic segment, thus acting as an incomplete obstruction to the flow of the ingesta, Filogamo and Vigliani, 1954); ii) intrinsic tissue factors acting as mitogens (possibly derived from the inflammatory infiltrate, or from the residual myenteric neurons, or from the damaged, degenerating cellular populations); iii) the myenteric denervation, through the loss of a stabilizing influence of the nerve-derived, putative trophic factors, maintaining cells in a quiescent state. Similarly, absence of enteric neurons associated with thickening of the muscularis mucosae and of the muscularis externa have been described in the lethal spotted mutant mouse (Tennyson et al., 1986), in Hirschprung's disease (Ehrenpreis, 1971) and in Chagas' disease (Ferreira-Santos and Carril, 1964).

9. EFFECTS OF BAC TREATMENT ON THE MESENTERIC NEURONS

Nerve cells are known to be located along the mesenteric nerve bundles supplying normal intestinal segments in the rat (Kuntz and Jacobs, 1955; Cracco and Filogamo, 1993). In a previous work (Cracco and Filogamo, 1993) we demonstrated that these mesenteric neurons, when present along nerve fascicles supplying myenterically-denervated ileal loops and immediately adjacent intestinal segments, are similar to the controls in position, shape and size, but display striking differences with regard to their arrangement and density per nerve. In the control rats occasional, small aggregates of mesenteric neurons are present; more commonly, isolated nerve cell bodies are scattered along the nerve trunks. In the BAC-treated rats, abundant clusters made up of numerous neurons are located along the nerves, their mean density per nerve displaying a two/five-fold increase over the control values. Such changes have been interpreted as an attempt at reinnervation of the BAC-treated ileal segments from extraenteric sources. The hypothesis of a reserve pool of potential nervous elements present along the mesenteric nerves has been put forward, in order to explain the increase in identifiable neurons per nerve without identifiable mitoses, occurring in adulthood under these conditions of functional hyperactivity (Filogamo and Cracco, 1995). Different amounts of diffusible factors from the hypertrophic and hyperplastic smooth muscle cells in the BAC-treated loop might promote neuronal differentiation of morphologically undifferentiated elements, possibly left behind along the neural crest migratory pathway to the gastrointestinal wall. Again, this event would confirm the capacity of adult nervous cells to divide and differentiate, as observed in the mammalian central nervous system (Reynolds and Weiss, 1992).

10. CONCLUSION

BAC treatment of the rat intestine represents a useful experimental model, in order to study: i) the effects of the selective, chronic elimination of the myenteric plexus; ii) the work overload of the submucosal plexus, due to the absence of the myenteric plexus and to its increased territory of innervation (thickened smooth muscle layers and mucosa); iii) the regenerative ability of the enteric nervous system; iv) the regeneration of smooth muscle.

REFERENCES

Cracco C. and Filogamo G., Mesenteric neurons in the adult rat are responsive to ileal treatment with benzalkonium chloride. Int. J. Devl. Neurosci. 11 (1993), 49–61.

Cracco C. and Filogamo G., Quantitative study of the NADPH-diaphorase-positive myenteric neurons of the rat ileum. Neuroscience 61 (1994), 351–359.

Dogiel A.S., Ueber den Bau der Ganglien in den Geflechten des Darmes und der Gallenblase des Menschen und der Saugetieren. Arch. Anat. Physiol. Leipzig, Anat. Abt. (1899), 130–158.

Ehrenpreis T., Hirschprung's disease. Am. J. Dig. Dis. 16 (1971), 1032–1052.

Ferreira-Santos R. and Carril C.F., Acquired megacolon in Chagas' disease. Dis. Colon Rect. 7 (1964), 353–364.

Filogamo G., Neuronal modulation by number nad volume: a review and critical analysis. In: Model Systems of Development and Ageing of the Nervous System, Vernadakis A., Privat A., Lauder J.M., Timiras P. and Giacobini E. eds, Martinus Nijhoff, Boston, MA (1987), pp. 1–16.

Filogamo G. and Vigliani F., Ricerche perimentali sulla correlazione tra estensione del territorio di innervazione e grandezza e numero delle cellule gangliari del plesso mienterico (di Auerbach) del cane. Riv. Patol. Nerv. Ment. 75 (1954), 441–462.

Filogamo G. and Cracco C., Models of neuronal plasticity and repair in the enteric nervous system. It. J. Anat. Embryol. 100 (Suppl.) (1995), 185–195.

Fox D.A., Epstein M.L. and Bass P., Surfactants selectively ablate enteric neurons of the rat jejunum. J. Pharmacol. Exp. Ther. 227 (1983), 538–544.

Fox D.A. and Bass P., Pharmacological characterization of rat jejunal contractility after chronic ablation of the myenteric plexus. J. Pharmacol. Exp. Ther. 238 (1986), 372–377.

Fox D.A., Herman J.R. and Bass P., Differentiation between myenteric plexus and longitudinal muscle of the rat jejunum as the site of action of putative enteric neurotransmitters. Eur. J. Pharmacol. 131 (1986), 39–47.

Furness J.B., Bornstein J.C. and Trussell D.C., Shapes of nerve cells in the myenteric plexus of the guinea-pig small intestine revealed by intracellular injection of dye. Cell Tissue Res. 254 (1988), 561–571.

Furness J.B., Lloyd K.C.K., Sternini C. and Walsh J.H., Evidence that myenteric neurons of the gastric corpus project to both the mucosa and the external muscle: myectomy operations on the canine stomach. Cell Tissue Res. 266 (1991): 475–481.

Gabella G., Size of neurons and glial cells in the intramural ganglia of the hypertrophic intestine of the guinea-pig. J. Neurocytol. 13 (1984), 73–84.

Hill C.J., A contribution to our knowledge of the enteric plexuses. Phil. Transact. Royal Soc. London 215 (1927), 355–387.

Hodgkiss J.P. and Lees G.M., Morphological studies of electrophysiologically identified myenteric plexus neurons in the guinea-pig ileum. Neuroscience 8 (1983), 593–608.

Honig M.G. and Hume R.I., Dil and DiO: versatile fluorescent dyes for neuronal labelling and pathway tracing. TINS 12 (1989), 333–341.

Jiang M.M. and Surprenant A., Re-innervation of submucosal arterioles by myenteric neurones following extrinsic denervation. J. Aut. Nerv. Syst. 37 (1992), 145–154.

Kuntz A. and Jacobs M.W., Components of the periarterial extensions of celiac and mesenteric plexuses. Anat. Rec. 123 (1955), 509–520.

Luck M.S., Dahl J.L., Boyeson M.G. and Bass P., Neuroplasticity in the smooth muscle of the myenterically and extrinsecally denervated rat jejunum. Cell Tissue Res. 271 (1993), 363–374.

Nagahama M., Ozaki T. and Hama K., A study of the myenteric plexus of the congenital aganglionosis rat (spotting lethal). Anat. Embryol. 171 (1985), 285–296.

Payette R.F., Tennyson V.M., Pham T.D., Mawe G.M., Pomeranz H.D., Rothman T.P. and Gershon M.D., Origin and morphology of nerve fibers in the aganglionic colon of the lethal spotted (ls/ls) mutant mouse. J. Comp. Neurol. 257 (1987), 237–252.

Reynolds B.A. and Weiss S., Generation of neurons and astrocytes from isolated cells of the adult mammalian central nervous system. Science 255 (1992), 1707–1710.

Riemann J.F., Schmidt H. and Zimmerman W., The fine structure of colonic submucosal nerves in patients with chronic laxative abuse. Scand. J. Gastroenterol. 15 (1980), 761–768.

Sakata K., Kunieda T., Furuta T. and Sato A., Selective destruction of intestinal nervous elements by local application of benzalkonium solution in the rat. Experientia 35 (1979), 1611–1613.

Sato A., Yamamoto M., Imamura K., Kashiki Y.. Kunieda T. and Sakata K., Pathophysiology of aganglioic colon and anorectum: an experimental study on aganglionosis produced by a new method in the rat. J. Ped. Surg. 13 (1978), 399–405.

Schofield G.C., Anatomy of muscular and neural tissues in the alimentary canal. In: Handbook of Physiology, Code C.F. ed., Sec.6, Vol. IV, American Physiological Society, Washington, pp. 1597–1627.

See N.A., Epstein M.L., Schultz F., Pienkowski T.P. and Bass P., Hyperplasia of jejunal smooth muscle in the myenterically denervated rat. Cell Tissue Res. 253 (1988), 609–617.

See N.A., Epstein M.L., Dahl J.L. and Bass P., The myenteric plexus regulates cell growth in rat jejunum. J. Aut. Nerv. Syst. 31 (1990), 219–230.

Smith B., Effect of irritant purgatives on the myenteric plexux in man and the mouse. Gut 9 (1968), 139–143.

Song Z.M., Brookes S.J.H. and Costa M., Identification of myenteric neurons which project to the mucosa of the guinea-pig small intestine. Neurosci. Lett. 129 (1991), 294–298.

Tennyson V.M., Pham T.D., Rothman T.P. and Gershon M.D., Abnormalities of smooth muscle, basal laminae, and nerves in the aganglionic segments of the bowel of lethal spotted mutant mice. Anat. Rec. 215 (1986), 267–281.

Vercelli A., Assal F. and Innocenti G.M., Emergence of callosally projecting neurons with stellate morphology in the visual cortex of the kitten. Exp. Brain Res. 90 (1992), 346–358.

Webster W., Embryogenesis of the enteric ganglia in normal mice and in mice that develop congenital aganglionic megacolon. J. Embryol. Exp. Morphol. 30 (1973), 573–585.

Wood J.D., Application of classification schemes to the enteric nervous system. J. Aut. Nerv. Syst. 48 (1994), 17–29.

Zucoloto S., Muccillo G., Wright N.A. and Allison M.R., Chronic effects of alcohol on the epithelium of the small intestine using two experimental models. Virchows Arch. (Cell Pathol.) 49 (1985), 365–371.

Zucoloto S., Diaz J.A., Oliveira J.S.M. et al., Effects of chemical ablation of myenteric neurones on intestinal cell proliferation. Cell Tissue Kin. 21 (1988), 213–218.

Zucoloto S., Silva J.C., Oliveira J.S.M. and Muccillo G., The chronological relationship between the thickening of smooth muscle, epithelial cell proliferation and myenteric neural denervation in the rat jejunum. Cell Prolif. 24 (1991), 15–20.

Part III

Neuronal Cell Death, Neuroprotection

13

NERVE GROWTH FACTOR AND OXIDATIVE STRESS IN THE NERVOUS SYSTEM

Zhaohui Pan, Deepa Sampath, George Jackson, Karin Werrbach-Perez, and Regino Perez-Polo

Department of Human Biological Chemistry and Genetics
University of Texas Medical Branch at Galveston
Galveston, Texas, 77555-0652

1. NERVE GROWTH FACTOR AND ITS FUNCTION

Nerve growth factor is a target-derived neurotrophic factor acting on sympathetic and neural crest-derived sensory neurons in the peripheral nervous system (PNS) and some populations of cholinergic neurons in the central nervous system (CNS; Levi-Montalcini and Hamburger, 1951; 1953; Levi-Montalcini, 1987; Hefti and Weiner, 1986; Thoenen and Barde, 1980). Nerve growth factor isolated from mouse submaxillary glands has a sedimentation coefficient of 7S which lacks biological activity (Varon *et al.*, 1972; Stach and Shooter, 1980) and is made up of two alphas, one beta, and two gamma subunits ($\alpha_2\beta\gamma_2$). The biologically active form is the β-subunit, a homodimer made up of two identical polypeptides. Each chain contains three intrachain disulfide bonds which are crucial for biological activity since the reduction of these bonds abolishes biological activity (Greene and Shooter, 1980; Perez-Polo *et al.*, 1990; Fahnestock, 1991). There are significant homologies in the amino acid sequence of the α- and γ-subunits (Greene *et al.*, 1969; Thomas *et al.*, 1981; Evans and Richards, 1985). The γ-subunit of 7S NGF (γ-NGF) is an arginine- or lysine-specific trypsin-like serine proteinase in the kallikrein gene family (Thomas *et al.*, 1981; Evans and Richards, 1985; Evans *et al.*, 1987). The γ-NGF has been postulated to function during the processing of the β-NGF precursor and may participate in cellular migration or tissue remodeling. The γ-NGF can also cleave recombinant single chain urokinase-type plasminogen activators which might be involved in cellular migration by activating a proteinase cascade (Wolf *et al.*, 1993).The α-subunit also belongs to the kallikrein gene family (Evans and Richards, 1985). Although α-NGF may protect β-NGF from proteolytic degradation or inhibit NGF biological activity via formation of the 7S complex, a definitive biological function for the α-subunit has not been established.

Nerve growth factor is essential for the development and maintenance of neurons in the PNS and the CNS (Misko *et al.*, 1987, Johnson *et al.*, 1988). In the PNS, NGF is synthesized and released from non-neuronal target tissues (Kromer and Cornbrooks, 1986; Taniuchi *et al.*, 1988). In the CNS, both astrocytes (Carman-Krzan *et al.*, 1991; Furukawa

Brain Plasticity, edited by Filogamo *et al.*
Plenum Press, New York, 1997

et al., 1987) and certain neurons (Gonzalez *et al.*, 1990) secrete NGF. During development, NGF is taken up after binding to its receptor at nerve terminals and retrogradely transported through the axon to the neuronal soma. The interaction of NGF with its receptor initiates a series of signal transduction events that start from the binding on the membrane to a receptor, internalization of the NGF receptor complex, and transport along the axon retrogradely to the soma. Neurons may perish when they fail to successfully compete for NGF, which is typically released in limited amounts, or when they fail to transport NGF to the soma at critical times during development, after injury or chronic stress (Thoenen and Barde, 1980; Hamburger *et al.*, 1981; Thoenen *et al.*, 1971; Oppenheim *et al.*, 1982; Hamburger and Yip, 1984; Levi-Montalcini, 1987; Barde, 1989).

For CNS and PNS tissues studied *in vitro*, the main documented functions of NGF have been the maintenance of cell survival, differentiation, and protection and improved recovery from noxious stimuli (Chun and Patterson, 1977a; 1977b; 1977c; Greene, 1977a; 1977b; Levi-Montalcini and Angeletti, 1963; Berg, 1984; Chao, 1992; Jackson, *et al*, 1995). Nerve growth factor is required for the survival of embryonic sympathetic ganglion neurons, dorsal root ganglion cells, and cholinergic neurons of the basal forebrain including the the medial septal nucleus, nucleus basalis, and nucleus of the diagonal band of Broca (Honegger and Lenoir, 1982; Gnahn *et al.*, 1983; Hefti, 1986; Hartikka and Hefti (1988). Moreover, NGF can prevent neuronal degeneration in the PNS and the CNS. Smith *et al.* (1993) has observed that deprivation of NGF causes exaggerated neuronal degeneration in superior cervical ganglia. Nerve growth factor protects hippocampal and cortical neurons against iron-induced degeneration, via generation of free radicals (Zhang *et al.*, 1993).

Nerve growth factor initiates a number of metabolic changes in responsive neurons that include the induction of neurotransmitter biosynthetic enzymes, such as tyrosine hydroxylase and dopamine β-hydroxylase (Thoenen *et al.*, 1971; Chun and Patterson, 1977a; Macdonnell *et al.*, 1977a; Max *et al.*, 1978; Otten *et al.*, 1977), and the stimulation of choline acetyltransferase (ChAT) activity and cholne uptake *in vivo* in sympathetic (Greene and Rein, 1977; Thoenen and Barde, 1980), and cholinergic neurons *in vivo*(Mobley *et al.*, 1986; Johnson *et al.*, 1987a), and in cultured neonatal and fetal basal forebrain and striatal neurons. (Gnahn *et al.*, 1983; Hatanaka and Tsukui, 1986; Martinez *et al.*, 1985; Lorenzi *et al.*, 1993). Nerve growth factor can also increase the activity of phenylethanolamine-N-methyl transferase (Liuzzi *et al.*, 1977), ornithine decarboxylase (MacDonnell *et al.*, 1977b), and acetylcholinesterase (Rieger *et al.*, 1980).

Rat pheochromocytoma (PC12) cells, a neuronal crest-derived cell line (Greene and Tischler, 1976), have been used extensively as a model for the study of NGF action. When treated with NGF the chromaffin-like PC12 cells undergo marked morphologic, physiologic, and biochemical changes consistent with a neuronal phenotype (Levi *et al.*, 1988; Fujita *et al.*, 1989). As part of the signal transduction process responsible for cell rescue from neuronal death by NGF in PC12 cultures, NGF induces expression and activation of transcription factors such as AP-1, NFκBand other gene producst such as *c-myc,* NGFI-A, and NGFI-B (Kruijer *et al.*, 1985; Greenberg *et al.*, 1985; Wu *et al.*, 1989; Milbrandt, 1988; Watson and Milbrandt, 1990; Liqi and Perez-Polo, 1995; Liqi et al, 1996, in press). These transcription factors play critical roles in the induction of critical enzymes such as ornithine decarboxylase and tyrosine hydroxylase (Gizang-Ginsberg and Ziff, 1990; Greene and McGuire, 1978; Guroff *et al.*, 1981; Feinstein *et al.*, 1985; Muller *et al.*, 1993) and also of cytoskeletal proeins such as F-actin (Paves *et al.*, 1988), α-actin and myosin light chain (Henke *et al.*, 1991), neural-specific protein GAP-43 (Federoff *et al.*, 1988; Basi *et al.*, 1987), VGF genes (Salton *et al.*, 1991), microtubules (Drubin *et al.*, 1985), tubulin (Fernyhough and Ishii, 1987), and intermediate filament protein (Leonard *et al.*, 1988).

The expression of NGF can be stimulated by cytokines (Yoshida and Gage, 1992), thyroid hormone (Giordano *et al.*, 1992), steroids (Perez-Polo, *et al*, 1977; Follesa and Mocchetti, 1993), tumor necrosis factor (TNF, Hattori *et al.*, 1993), proteases (Neveu *et al.*, 1993), glutamate depolarization (Thoenen, 1991; Pechan *et al.*, 1993), injury (Pechan *et al.*, 1992; Lu *et al.*, 1991; Heumann *et al.*, 1987; Richardson and Ebendal, 1982), and stress (Foreman *et al*, 1993). For example, interleukin-1β increases NGF expression in cultured hippocampal cells (Friedman *et al.*, 1990); thus, arachidonic acid lipoxygenation stimulation of interleukin-1β can induce secretion of NGF in rat neonatal cortical astrocytic primary cultures (Carman-Krzan and Wise, 1993). Also, interleukin-4 and interleukin-5 can modulate NGF synthesis and secretion in astrocytes (Awatsuji *et al.*, 1993). Giordano (*et al.*, 1992) has reported that thyroid hormone significantly increases NGF mRNA levels in the cortex and hippocampus of young adult rats. Tumor necrosis factor has also been found to stimulate the synthesis and secretion of immunoreactive NGF in quiescent mouse fibroblasts by elevating NGF mRNA levels (Hattori *et al.*, 1993). Neveu *et al.* (1993) found that proteases, such as α-thrombin, collagenase, trypsin, plasmin, α-chymotrypsin, or elastase, enhance NGF mRNA levels in cultured astrocytes. The excitatory amino acid neurostransmitter glutamate also increases NGF mRNA expression in rat astrocytic cultures (Pechan *et al.*, 1993). Nerve growth factor synthesis is regulated during development and altered after nerve injury (Heumann *et al.*, 1987; Richardson and Ebendal, 1982) while lethal doses of H_2O_2 can induce NGF mRNA increases in rat astrocytic cultures (Pechan, 1992). Taken together these findings are consistent with the interpretation that NGF expression is associated with developmental events and stress response mechanisms in the nervous system.

2. FREE RADICALS AND ANTIOXIDANT SYSTEMS

Stressful conditions in the CNS, such as ischemia, hyperoxia, hemorrhage, trauma, and aging, share a free radical component. Specifically there is a relationship between pathophysiology and imbalance in oxidative homeostasis (Floyd, 1990). Most free radicals are reactive and undergo reactions in which unpaired electrons become paired, while some may be relatively stable. In biological systems, oxygen-derived free radicals or reactive oxygen species (ROS) are the most common free radicals (Fisher, 1988; Farber *et al.*, 1990). Reactive oxygen species include singlet oxygen ($^1O_2$), superoxide ions ($HO_2\cdot$), hydroperoxyl (HOO·) and peroxyl radicals (·OOR), hydroxyl radical (OH·), organic peroxide (ROO·), nitric oxide (·NO), and peroxynitrite (ONOO-). Some ROS are not free radicals, but are highly reactive and participate in free radical generation. These ROS include hydrogen peroxide (H_2O_2), ozone (O_3), and hydroxychloride (HOCl). Reactive oxygen species are also by-products of oxidation-reduction reactions in the mitochondrial respiratory chain, in microsomal electron transport, activated phagocytosis, and the cyclooxygenase pathway and are associated with cytochrome P450 oxidase, NADPH oxidase, and xanthine oxidase enzymatic activity (Bandy and Davison, 1990; Farber *et al.*, 1990; Trush and Kensler, 1991). During phagocytosis, NADPH oxidase, a membrane bound enzyme, is activated in neutrophils and macrophages to generate H_2O_2. Xanthine dehydrogenase is activated to xanthine oxidase which generates H_2O_2 in a variety of cell types by exposure to phorbol esters, tumor necrosis factor-α, or ischemia-reperfusion injury (Till *et al.*, 1991; Friedl *et al.*, 1989). The presence of free iron or other divalent cations convert $O_2^-\cdot$ or H_2O_2 to OH·, probably the most reactive ROS and the molecular species most responsible for genotoxic lesions in mammals via the Fenton reaction (Koppenol, 1983; Mossman *et al.*,

1987). There are other radicals in biological systems that are sulphur-centered, carbon-centered, and carbon-oxygen centered. For example, carbon-centered radicals are of the allyl-types and are products of autoxidation of membrane lipids. Carbon-oxygen centered radicals have a wide variety of forms, including ascorbate, quinones, and tyrosine. Thiol radicals generated from thiols and disulfides play important roles in cellular function and malfunction. A definitive differential analysis of the roleof each radical type in cellular dysfunction is lacking although it is known that the downstream most proximal injurious coponent are the hydroxyl radicals.

Free radicals are noxious to molecular targets that include proteins, nucleotides, and lipids (Halliwell and Gutteridge, 1989; Cross *et al.*, 1987; Floyd, 1990). Membrane peroxidation caused by free radicals alters lipid fluidity and alters membrane permeability thus affecting membrane-bound enzyme activity. Exposure to ROS can result in protein aggregation, crosslinking, fragmentation, and oxidation, all events that perturb ion transport, increase calcium influx and alter enzyme activities. Perhaps of the utmost consequence is the ROS-induced damage to the deoxyribose rings and bases of DNA and RNA that result in mutagenesis, translational errors, inhibition of protein synthesis, and the induction of stress response genes which in turn may trigger apoptosis.

Cells have defense systems against free radicals which may be constitutively synthesized *in vivo*, taken up in the diet, or induced by the pesence of ROS. The antioxidants synthesized in mammalian systems include proteins, enzymes, and transition metal-binding proteins. The vitamins, ascorbate, tocopherol, carotene, and lycopene, come from diet. These antioxidants are present in membrane bound, extracellular, or intracellular structures. There are also extracellular antioxidant defenses such as iron-bound transferrin; copper-containing protein ceruloplasmin, albumin; hemoglobin- or heme-bound protein haptoglobin or hemopexin; iron and copper chelator urate, which also reacts with singlet oxygen; the lipid-soluble or lipoprotein-bound reactive oxygen scavengers, α-tocopherol, β-carotene, and lycopene; and the water-soluble antioxidant ascorbate. The major intracellular antioxidant defense systems are superoxide dismutase (SOD), catalase (CAT), and glutathione peroxidase (GSH-Px), and the glutathione S-transferases (GST). There are secondary protection systems, also called repair systems, such as systems for protein, lipid, and DNA repair or degradation, including proteinases, peptidases, phospholipases, acyltransferases, polymerases, ligases, exonuclease, endonucleases, and glucosylases. Repair system activation can also trigger other stress response gene families associated with inflammatory and pro-apoptotic events.

Superoxide dismutase (SOD) catalyzes the conversion of two molecules of $O_2^{-}\cdot$ to O_2 and H_2O_2. There are three forms of SOD in eukaryotic cells: CuZnSOD is found in the peroxisomes and cytoplasm, a different form of CuZnSOD is found in extracellular fluids (also called EC-SOD), MnSOD is localized in the mitochondria. Catalase reduces H_2O_2 to H_2O and O_2 and is found in peroxisomes (80%) and cytosol (20%) in most tissues. While catalase activity is especially high in liver and red blood cells, it is very low in brain and spinal cord, specially so in neurons there. Glutathione peroxidase catalyzes the oxidation of GSH to GSSG at the expense of hydroperoxides that include H_2O_2 and ROO·. There are two forms of GSH-Px, the Se-independent and Se-dependent forms. The Se-dependent GSH-Px is a tetramer of *Mr* 84,000 with high specificity towards both H_2O_2 and organic hydroperoxides (Little and O'Brien, 1968) that exists in cytosol (70%) and mitochondria (30%). The other form of Se-dependent GSH-Px, also called phospholipid hydroperoxide glutathione peroxidase (PLGSH-Px), has a *Mr* 20,000 (Thomas *et al.*, 1990) and is a monomer of GSH-Px found in cytosol and biomembranes that reduces phospholipid hydroperoxides to alcohols, a property not common for GSH-Px. The Se-independent GSH-

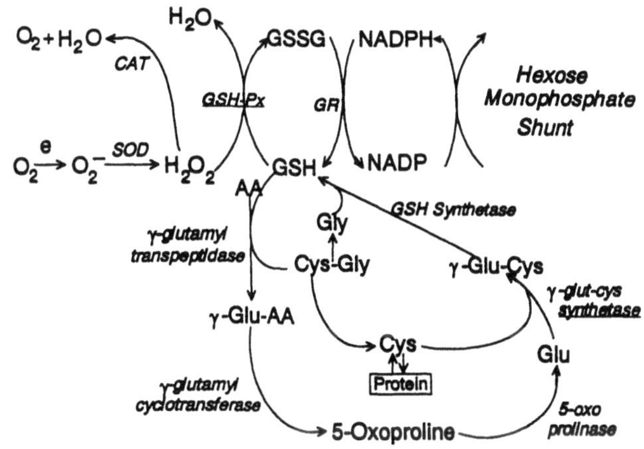

Figure 1. Pathways of glutathione metabolism (modified from Meister, 1983). AA: amino acid.

Px is one class of GST isoenzyme, class α (Lawrence and Burk, 1976) that has two subunits with *Mr*s of approximately 26,000 and 24,000. The GST-α is found in cytosol and mitochondria and acts on organic hydroperoxides (Sun, 1990).

The activity of GSH-Px is regulated by GSH regeneration systems, reequiring glucose-dependent NADPH for the reduction of GSSG, catalyzed by glutathione reductase (GSH-R). Glutathione reductase is an NADPH linked dimmer of *Mr* 120,000, each with FAD at its active site. Glutathione peroxidase regenerates GSH at the expense of NADPH. The GSH-Px system is also regulated by the hexose monophosphate shunt where the rate limiting enzyme is glucose-6-phosphate dehydrogenase, which is activated by GSSG and NADP, but inhibited by NADPH (Fisher, 1985; Doroshow *et al.*, 1990; Farber *et al.*, 1990). A succint description of antioxidant enzyme pathways is shown in Figure 1.

Exposure of cells to ROS changes cellular metabolic events. For example, H_2O_2 inhibits ADP phosphorylation via the inhibition of glycolysis, reduces energy charge (Hyslop *et al.*, 1988; Spragg *et al.*, 1985), and stimulates the Na^+-independent Ca^+ release from mitochondria through the oxidation of pyridine nucleotide oxidation followed by the hydrolysis of NAD^+ to ADP-ribose and nicotinamide (Richer and Kass, 1991). However, not all oxidative responses are catabolic. Increased oxidant levels can stimulate the expression of those genes that code for antioxidant enzymes, regulatory proteins for defense systems, and oncogenes that enhance cell proliferation, differentiation or induce the pro-apoptotic expression of c-*jun*, *c-fos* and *c-myc* (Rao *et al.*, 1993; 1992; Amstad *et al.*, 1990). For example, both H_2O_2 and UV radiation enhance transcription of the heme oxygenase gene, a major stress protein in human cells (Tyrrell *et al.*, 1993) and ozone and zinc oxide stimulate both metallothionine and heme oxygenase expression in rat lung (Cosma *et al.*, 1992). The hypothesis that pathological states are associated with free radical-related phenomena has been made for inflammatory disorders, rheumatoid arthritis, retinopathy of prematurity, toxic liver injury, reperfusion injury, atherosclerosis, lung disorder, tumor promotion, porphyria, Parkinson's disease, Down's syndrome, Alzheimer's disease, and aging (Jesberger and Richardson, 1991; Trush and Kensler, 1991).

For example, β-amyloid, the widely distributed abnormal protein present in plaques in the brains of Alzheimer's patients, is reported to enhance glutamate toxicity via generation of free radicals in cortical neurons (Koh *et al.*, 1990). Lipid peroxidation has also been reported in tissues from patients with Alzheimer's disease (Richardson *et al.*, 1990a; 1990b; Andorn *et al.*, 1990). In brains of patients suffering from Parkinson's disease, which is characterized by a loss of midbrain dopaminergic neurons, there are reports of increased iron concentrations (Riederer *et al.*, 1989; Sofic *et al.*, 1988), which may account for the increased lipid peroxidation (Damier *et al.*, 1993). Finally, it has also been suggested that aging-associated pathology may reflect the deleterious effects of free radicals (Harman, 1956; 1957; 1981; 1984; Pryor, 1977; 1982).

3. GLUTATHIONE IN BIOLOGICAL SYSTEMS

Glutathione is a tripeptide (L-γ-glutamyl-L-cysteinylglycine) where the N-terminal of L-cysteine is linked to the γ-carboxyl group of glutamate.The first step in GSH synthesis is catalyzed by γ-glutamylcysteine synthetase (GCS, EC 6.3.2.2), which is rate limiting and is regulated by negative feedback by GSH. γ-Glutamylcysteine synthetase catalyzes formation of a bond between L-glutamate and L-cysteine to form γ-glutamylcysteine (Figure 1) at the expense of one mole of adenosine triphosphate (ATP). Glutathione synthetase catalyzes a second step, in which glycine is linked to the cysteine carboxyl group of γ-glutamylcysteine to form GSH. Intracellular GSH can be a substrate for membrane-bound γ-glutamyl transpeptidase, which transfers GSH or GSSG γ-glutamyl moieties to amino acid acceptors which in turn are substrates for γ-glutamyl cyclotransferase (EC 2.3.2.2), which cyclizes the glutamyl moiety of γ-glutamyl amino acids to 5-oxoproline, releasing free amino acids as part of a reaction that generates glutamate in a reaction coupled to the cleavage of ATP to adenosine diphosphate (ADP). These reactions are responsible for the cellular turnover of GSH and are commonly identified as the γ-glutamyl cycle (Meister. 1981; 1983; 1988a,b; Meister and Anderson, 1983).

Glutathione reductase catalyzes the reduction of oxidized GSH (GSSG) to the reduced form (GSH) so as to maintain intracellular GSH homeostasis. Glutathione levels are responsible for the detoxification of xenobiotics, carcinogens, free radicals, and lipid peroxides. The reduction of H_2O_2 and other peroxides at the expense of GSH can be catalyzed by selenium-dependent GSH-Px. By maintaining the reduced states of pyridine nucleotides GSH can also maintain the thio state of proteins. Glutathione can complex with metals and reduce metal-mediated oxidation damage (Freedman *et al.*, 1989; Milne *et al.*, 1993). Glutathione serves as reservoir for the storage of cysteine in cells. Since the thiol group of cysteine is not stable (cysteine will auto-oxidize to cystine), cysteine cannot be stored extracellularly. Glutathione deficiency linkage to brain dysfunction has been seen in enlarged and degenerated cerebral cortex mitochondria (Jain *et al.*, 1991), and after cerebral ischemic injury in rats (Mizui *et al.*, 1992).

Glutathione can be effectively depleted by a selective transition state inhibitor of GCS, L-buthionine-SR-sulfoximine (BSO, Huang *et al.*, 1988). Treatment with BSO sensitizes cells to ROS (Naganuma *et al.*, 1990). Alternatively, GSH levels can be increased via the administration of substrates of GCS and GSH synthetase, respectively, and cysteine precursors, such as, L-2-oxothiazolidine-4-carboxylate and *N*-acetyl-L-cysteine (Anderson, at al., 1990).

4. EFFECTS OF NERVE GROWTH FACTOR ON GLUTATHIONE PEROXIDASE CYCLE IN PC12 CELLS

The central nervous system is susceptible to oxidative damage due to its high oxygen consumption, low concentrations of antioxidant enzymes, high concentrations of free iron which are involved in the generation of hydroxyl free radicals, and high concentrations of oxidizable substances, such as catecholamines and unsaturated lipids (Jesberger and Richardson, 1991; Halliwell and Gutteridge, 1989). The reduction of these abundant substrates generates toxic ROS that promote neuronal degeneration. It has been proposed that a loss of antioxidant balance results in increased oxygen radical influx in brain areas, which is in part responsible for neuronal lesions and disease. For example, GSH-Px, SOD, and catalase activities are two or three orders of magnitude lower in brain as compared to liver in the rat (Ciriolo *et al.*, 1991). The Se-dependent GSH-Px system has ben shown to be necessary for the detoxification of hydrogen peroxide in the brain. There are also reports of low levels of GSH-Px-positive glial cells around dopaminergic neurons a, which reneders them more susceptible to oxidative damage (Damier *et al.* 1993). Similarly, Tayarani *et al.* (1989) have reported decreased levels of catalase activity together with increased levels of SOD activity in aged brain capillaries, a combination that is likely to result in increased H_2O_2 levels.

Nerve growth factor regulates some antioxidant enzymes in CNS. Nistico *et al.* (1992) has reported that treatment with exogenous NGF restores catalase activity and increases SOD and GSH-Px activity in different brain areas in aged rats. *In vitro* pretreatment of PC12 cells with NGF reduces H_2O_2 cytotoxicity while increasing catalase activity (Tiffany-Castiglioni and Perez-Polo, 1981; Jackson *et al.*, 1990a; 1990b). Nerve growth factor treatment stimulates GSH-Px, G6P-D, and GCS activities resulting in increased GSH concentrations (Sampath and Perez-Polo, 1994; Pan and Perez-Polo, 1993). Thus, NGF can regulate the GSH redox cycle and the levels of GSH itself in PC12 cells.

The NGF regulation of antioxidant and energy homeostasis is not a simple event. The action of NGF on PC12 cells is multifactorial. Given that NGF treatment stimulates GSH-Px activity after three days and G6P-D after two days while NGF protection of PC12 cells from applied oxidative stress becomes significant as early as twenty four hours, it is clear that there are other signalling pathways involved in the protection of PC12 cells from H_2O_2 via exogenous NGF treatment (Figure 2).

Although G6P-D is a primary generator of NADPH, the slight increase in G6P-D activity observed after NGF treatment is unlikely to account for the effects of NGF on survival. The changes in GSH concentrations and GCS activity observed, on the other hand,

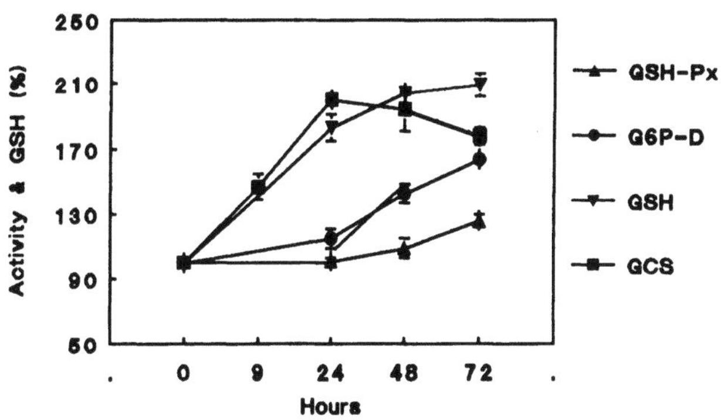

Figure 2. Time course of relative activities of GSH-Px, G6P-D, GCS, and relative GSH concentrations in NGF-pretreated cells. Cells were plated in 35-mm petri dishes overnight and were treated with NGF (20 ng/ml) for 9 to 72 hours. Activity was normalized to the same amount of protein and presented as a percentage of the control cells cultured for the same number of days. Data are the means of triplicate cultures ± SEM. From Pan and Perez-Polo, 1993, with permission.

do correlate in a dose dependent fashion with the extent of NGF protection of PC12 cells from applied oxidative stress and with the NGF concentrations present (Pan et al, 1993). The minimum NGF concentration that elicits a response in GSH concentrations and GCS activity and, also, in cellular protection from serum deprivation is about 1 ng/ml (3.7 (10^{-11} M), a concentration that is similar in magnitude to the measured K_d for the high affinity equilibrium binding dissociation constant of NGF to high affinity receptors on PC12 cells. Glutathione peroxidase and G6P-D activities do not increase as markedly or promptly as GSH and GCS do after NGF treatment, but they are significantly stimulated when treated with higher NGF concentrations for longer times, in a similar manner to that reported for the NGF regulation of catalase in PC12 cells (Jackson *et al.*, 1990a). Thus, it would appear that while NGF does stimulate catalase, GSH-Px, and G6P-D activities in a delayed fashion, the increases in antioxidant enzymes and G6P-D after NGF treatment may not be the events that account for early protection from oxidative stress by NGF. The evidence for two time courses and two NGF dosage effects may reflect the presence of two signal transduction pathways depending on tyrosine phosphorylation and ceramide metabolism respectively as well as protection vs. recuperation regulation.

Glutathione concentrations and GCS activity are increased about two-fold after NGF treatment for twentyfour hours. This increase mimmics the increase in cell survival that is observed after H_2O_2 and NGF treatment. Glutathione concentrations are also influenced by GSH-R activity (Figure 1). Although BDNF has been reported to stimulate GSH-R activity in SY5Y human neuroblastoma cells (Spina *et al.*, 1992), NGF has no effect on GSH-R activity in PC12 cells, regardless of the NGF concentration used, for up to three days after NGF treatment. Since the GSSG concentrations in PC12 cells is below 0.1 nmoles per milligram protein, 100 times lower than the GSH concentrations, GSSG concentrations do not influence the induction of GSH by NGF. Synthesis of GSH by GCS is one component responsible for much of the NGF-induced increase in GSH concentrations. When GSH is depleted by BSO (10µM), NGF has almost no protective effect against subsequent H_2O_2 treatments (Figure 3). GSH is not only a useful antioxidant but also an important substrate of GST for oxidative injury repair. Since the response of GSH to NGF in PC12 cells is significant and prompt, and GSH depletion by BSO abolishes all NGF protection, one can

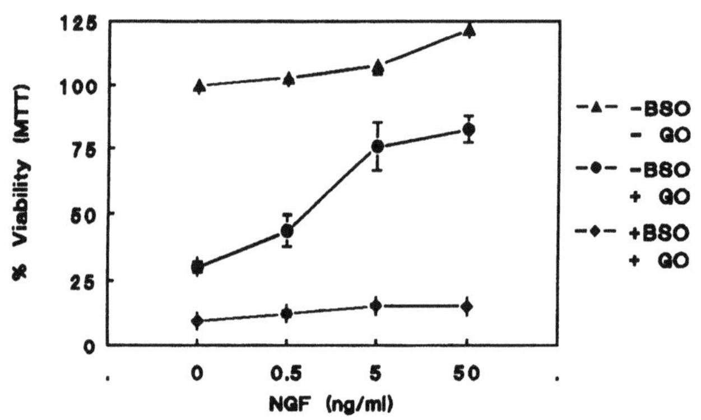

Figure 3. Dose response effect of BSO on GSH concentrations of the control and NGF-pretreated cells. Cells were plated in 35-mm petri dishes. One day after plating, cells were treated with different concentrations of BSO in the presence or absence of NGF (20 ng/ml) for 24 hours. Glutathione concentrations were measured by DTNB assay. Data are the means of triplicate cultures ± SEM.From Pan and Perez-Polo, 1993, with permission.

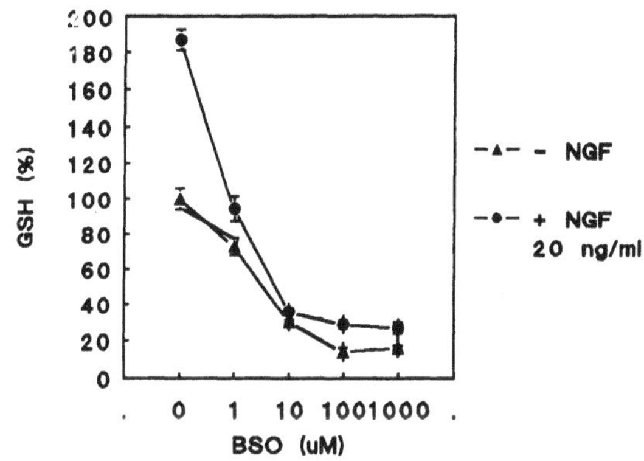

Figure 4. Dose-response effect of BSO on the viability of PC12 cells treated with or without glucose oxidase. Cells were plated in 24-well dishes. One day after plating, cells were treated with different amount of BSO. After 24 hours, half of cells in 24-well plates were subjected to glucose (10 mM)/glucose oxidase (5 mU/ml) in RPMI 1640 serum free medium for 3 hours. Cell viability was quantified by the MTT assay and expressed as a percentage of the control cells without any treatment (100%). Cells for BSO toxicity were directly assayed for viability after 24 hours of BSO treatment. Data are the means of triplicate cultures ± SEM and presented as a percentage of the control.From Pan and Perez-Polo, 1993, with permission.

conclude that the regulation of GSH by NGF is a major component of the protection of PC12 cells by NGF when oxidative homeostasis is perturbed.

At doses above 10 μM, BSO is toxic to PC12 cells in a dose dependent form. About 50% of the cells died after a treatment with 0.1 mM BSO for 24 hours (Figure 3). PC12 cells are more susceptible to BSO treatment when compared to other tumor cell lines, such as the A549 human lung carcinoma and the NRK-49F rat kidney fibroblasts, which continue to proliferate after 0.1 mM BSO treatment (Kang *et al.*, 1991; Kang and Enger, 1992). At 10 μM of BSO, about 80% of the intracellular GSH is depleted in PC12 cells (Figure 4), and these BSO-treated cells are very vulnerable to subsequent oxidative treatment (Figure 5). These results suggest that at a low dose, BSOacts to inhibit GCS while BSO concentrations in the millimolar range affect cellular metabolism beyond GSH synthesis.

When PC12 cells are treated with H_2O_2 there is a depletion of NADH and when cells are treated with both NGF and H_2O_2 the NGF-treated cells recuperate from the H_2O_2 exposure in a more rapid and complete fashion in terms of their return of NADH concentrations to control levels (Jackson *et al.*, 1992; Figure 6). While the effects of NGF on pyridine metabolism have not been fully characterized it appears that NGF may stimulate intracellular ATP concentrations and energy metabolism as it regulates GSH concentrations. γ-Glutamylcysteine synthetase catalyzes the reaction of L-glutamate and L-cysteine to synthesize γ-glutamylcysteine at the expense of one mole of ATP, while GSH synthetase catalyzes a second reaction combining γ-glutamylcysteine and glycine to yield the final product, GSH, also in an ATP-dependent reaction. Thus, to generate one mole of GSH requires two moles of ATP. Nerve growth factor stimulates ATP stores in PC12 cells (Jackson *et al.*, 1992) and may regulate intracellular energy metabolism as part of its regulation of GSH synthesis. Since GSH synthesis also requires glutamate, a side effect may be the reduction of glutamate toxicity due to the effects of NGF on GSH metabolism. It is interesting to observe that increased glutamate stimulates NGF synthesis *inn vivo*, suggesting a more pervasive involvement of NGF in all aspects of GSH metabolism (Zafra *et al.*, 1990).

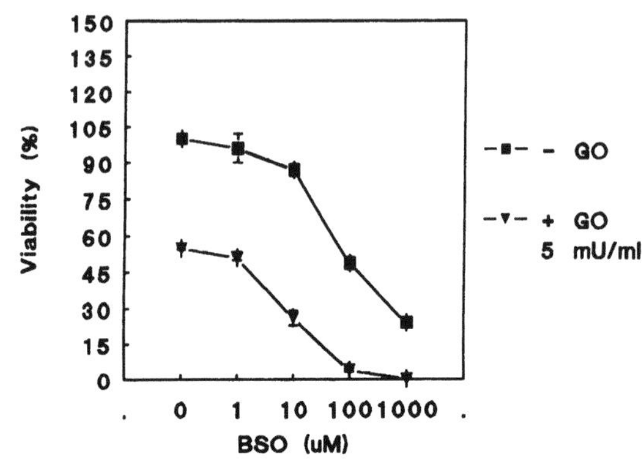

Figure 5. Protection by pretreatment with varying doses of NGF against glucose oxidase in PC12 cells co-treated with or without BSO. Cells were plated in 24-well plates overnight and were treated with different concentrations of NGF in the presence or absence of 10 μM BSO for 24 hours. NGF ± BSO treated cells were subjected to glucose (10 mM)/glucose oxidase (5 mU/ml) for 3 hours in RPMI 1640 serum free medium. Then, the H_2O_2 generating medium was replaced with RPMI 1640 with serum and cells were allowed to recover for one hour. Cell viability was measured with the MTT assay and expressed as a percentage of the control cells without any treatment (100%). Data are the means of triplicate cultures ± SEM. From Pan and Perez-Polo, 1993, with permission.

Increasing GSH concentrations also modulates thiol-disulfide exchange by increasing the concentrations of thiol-disulfide in cells, a reaction catalyzed by the thiol-transferases. Many enzymes in glucose metabolism, for example, glycogen phosphorylase phosphatase, phosphofructokinase, hexokinase, and pyruvate kinase, are regulated by the availability of thiol-disulfide (see review of Ziegler, 1985). Since Ca^{2+}-ATPase activity has been reported to be sensitive to sulfhydryl reagents, GSH may prevent the oxidation of

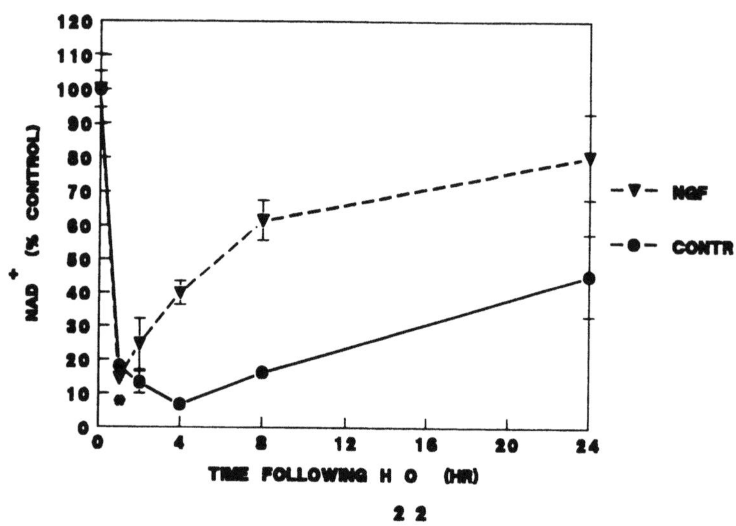

Figure 6. Effect of treatment with 0.5mM H_2O_2 on intracellular NAD^+ in control (●) and NGF-treated (▼) PC12 cells. Analysis of NAD^+ depletion was performed as described previously. From Jackson, *et al*, 1992, with permission.

thiol groups on Ca^{2+}-sequestering cellular systems, thus preserving intracellular Ca^{2+} homeostasis (Orrenius *et al.*, 1983). Therefore, NGF could be expected to affect glycolysis, gluconeogenesis, and intracellular Ca^{2+} concentrations through its regulation of the GSH/GSSG ratio.

5. NERVE GROWTH FACTOR REGULATION OF GAMMA-GLUTAMYLCYSTEINE SYNTHETASE

Gamma-glutamylcysteine synthetase (glutamate-cysteine ligase; EC 6.3.2.2; GCS) consists of two subunits (*Mr* 72,614 and 27,700 respectively). The heavy subunit catalyzes the first step of GSH synthesis, and exhibits nonallosteric feedback inhibition by GSH (Seeling and Meister, 1985). Since Huang *et al.* (1988) found that GCS was inactivated by BSO in the presence of ATP, it is likely that a γ-glutamyl phosphate intermediate is involved in GSH synthesis. The heavy subunit alone has lower activity and affinity for glutamate, and higher sensitivity to inhibition by GSH than the isolated holoenzyme, suggesting that the light subunit may also play a significant regulatory role under physiological conditions (Huang *et al.*, 1993a). Although the light subunit does not have catalytic activity, it may contribute to the activity of the heavy subunit by altering the binding of substrate at the active site. Yan and Meister (1990) isolated the cDNA clones encoding the heavy subunit from rat kidney from a λgt11 cDNA library by immunoscreening with antibody against the isolated enzyme and screening with oligonucleotide probes derived from several peptides. Northern analysis showed a single 4.1 Kb mRNA in rat kidney. The activity and mRNA of GCS are regulated by a number of factors that vary in different cell types. For example, glucagon and phenylephrine decrease hepatocyte GSH by inhibiting GCS (Lu *et al.*, 1991), while insulin and hydrocortisone increase GSH concentrations in hepatocytes by stimulating GCS activity (Lu *et al.*, 1992). The activity of GCS is stimulated by heat shock in K562 erythroid cells and by methyl mercury in kidney cells through a mechanism that increases GCS mRNA levels (Woods *et al.*, 1992; Kondo *et al.*, 1993). It has been reported that GCS mRNA is increased in cisplatin-resistant carcinoma cells (Goldwin *et al.*, 1992) and in melphalan-resistant DU-145 human prostate carcinoma cells, both associated with increased GSH concentrations (Bailey *et al.*, 1992).

NGF increases GSH concentrations in PC12 cells after 24 hours of treatment by stimulating GCS, the rate-limiting enzyme in GSH synthesis. There are many mechanisms by which NGF can influence or regulate GCS activity. For example, NGF may increase the rate of L-cysteine uptake and the increased amounts of L-cysteine may be partly responsible for increased GCS activity. The results of actinomycin D and cycloheximide experiments demonstrate that NGF-induced GCS activity is blocked by these two inhibitors, suggesting that NGF might regulate GCS activity either at the translational or transcriptional level. NGF is not able to increase GCS mRNA in the short term. Rather NGF stabilizes GCS mRNA in actinomycin D-treated cells. Actinomycin D blocks DNA-dependent RNA synthesis and ultimately causes mRNA degradation. After 6 hours, GCS mRNAs are preserved by 70% in cells treated with NGF plus actinomycin D as compared to the untreated controls (Pan et al, 1997). Whereas the half-life of GCS mRNA in actinomycin D-treated cells was 5.5 hours, NGF increased the half-life of GCS mRNA to 10.5 hours in the actinomycin D-treated cells. NGF has an effect on GCS stability only in the short term while stimulating GCS mRNA transcription in the long term (Pan et al, 1997). The mechanisms by which NGF stabilizes GCS mRNA are not known.

6. NGF EFFECT ON L-CYSTEINE OR L-CYSTINE UPTAKE

The synthesis of GSH is dependent upon the availability of precursor amino acids, particularly on L-cysteine. L-Cysteine is a rate-limiting building block in GSH synthesis and thus regulates GSH concentrations in *in vitro* and *in vivo* systems (Bannai and Tateishi, 1986). Intracellular GSH concentrations are strongly affected by extracellular concentrations of L-cysteine or L-cystine. L-Cysteine is usually not stable outside the cell and autooxidizes to L-cystine which can be taken up by cells and rapidly reduced to L-cysteine. L-Cysteine is then released from cells to maintain extracellular L-cysteine concentrations (Bannai and Ishii, 1980; 1982; Bannai and Kitamora, 1980). In addition, L-cysteine can be produced by hydrolyzing GSH, which is secreted from cells by the action of the membrane bound enzyme γ -glutamyltranspeptidase during GSH turnover (Abbott and Meister, 1986; Ballatori *et al.*, 1988).

The increases in GSH concentrations following NGF treatment are also due to increased transport of L-cysteine and L-cystine. γ-Glutamylcysteine synthetase activity is also stimulated by enhanced substrate concentrations of L-cysteine. Based on time course of uptake data, L-cysteine, unlike L-cystine, is taken up more rapidly within the first 5

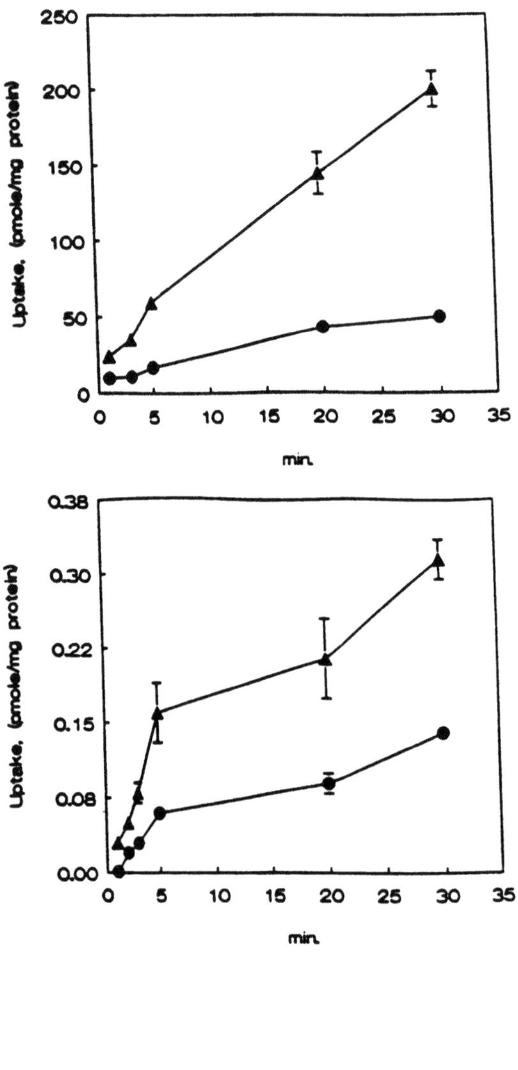

Figure 7. Time course of L-cysteine (top) or L-cystine (bottom) uptake in NGF-treated cells. Cells were plated in 24-well plate in serum containing medium for 2 days. After treatment of NGF, L-[^{35}S]cysteine or L-[^{35}S]cystine uptake at dose of 1 Ci/ml were measured for different periods of time. Each point represents mean of triplicate dishes ± SEM (control: circle; +NGF triangle). From Pan and Perez-Polo, 1996, with permission.

minutes after NGF treatment when compared to uptake rates at later time points (Pan and Perez-Polo, 1996). The L-cysteine transport system may depend upon sodium channels which are activated by NGF in PC12 cells within 1 minute, since L-cysteine transport is Na^+-dependent. L-cystine uptake follows Michaelis-Menten kinetics. The value of Vmax (70 pmole/minute mg protein) in NGF-treated cells is 3–4 fold greater than that (20 pmole/minute mg protein) of controls without treatment. BSO, which is a specific inhibitor of GCS at low doses, does not block NGF-induced L-cysteine uptake and has no effect L-cysteine or L-cystine uptake rates when compared to controls (Figure 7 and 8; Pan and Perez-Polo, 1996).

Increased intracellular L-cysteine not only participates in GSH synthesis, but also serves as an antioxidant itself. In Figure 7 it can be seen that NGF protects PC12 cells from oxidative stress either by increasing intracellular GSH concentrations or by increasing intracellular L-cysteine concentrations. Cells treated with NGF have a greater resistance to oxidants than cells treated with NGF and BSO, which exhibit greater resistance than cells tested with BSO alone. In NGF-treated cells, oxidative protection is due to increases in both GSH concentrations and L-cysteine concentrations. Thus, the mechanisms by which NGF protects PC12 cells from oxidative stress are multifactorial.

L-Cystine transport is dependent on the exchange of glutamate through system x_c-, which is Na^+-independent. The L-cystine transport system x_c- is highly specific for L-cystine and L-glutamate in human fibroblasts (Bannai and Kitamura, 1980), rat hepatoma cells (Makowski and Christensen, 1982), rat hepatocytes (Takada and Bannai, 1984), mouse peritoneal macrophages (Watanabe and Bannai, 1987), cultured rat astrocytes (Cho and Bannai, 1990), rat neuroblastoma (Murphy *et al.*, 1989), and immature cortical neurons (Murphy *et al.*, 1990). L-Cysteine transport, on the other hand, is mediated by a Na^+-

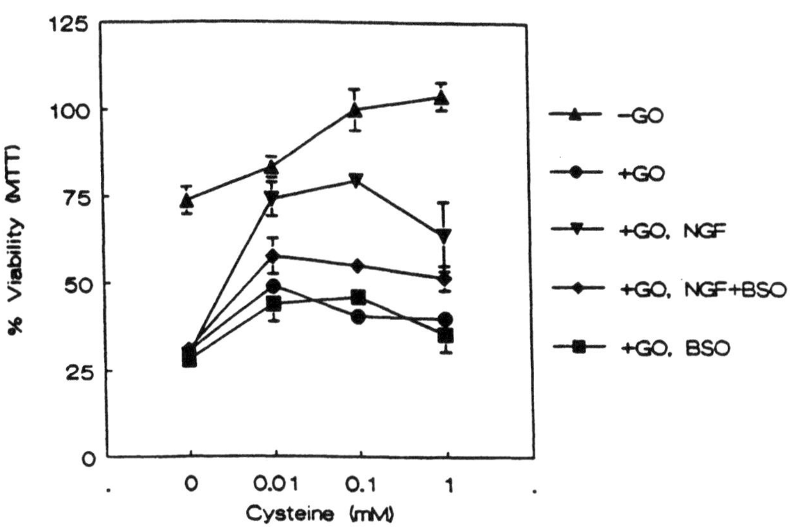

Figure 8. Protection of PC12 cells from oxidative attack in medium containing various doses of L-cysteine. Cells were seeded in 24-well plates overnight in normal medium. Cells were treated with 50 ng/ml NGF and 10 M BSO at different doses of L-cysteine. After 24 hours, cells received 50 mU/ml glucose oxidase for 1 hour in RPMI 1640 L-cystine-free medium. Cell viability was measured with the MTT assay and expressed as a percentage of the control cells without any treatment (100%). Data are the means of triplicate cultures ± SEM. From Pan and Perez-Polo, 1996, with permission.

dependent system expressed in a variety of cell types (Simmons *et al.*, 1993; Bannai and Ishii, 1982; Kilberg *et al.*, 1981, Kilberg *et al.*, 1979).

L-Cysteine is not only required as a precursor for the synthesis of GSH and a number of other biological compounds, but also functions as a cellular mediator. In lymphoid cells, exogenous administration of GSH, L-cysteine, and L-cysteine derivatives, such as N-acetyl-cysteine (Mihm and Droge, 1990; Mihm *et al.*, 1991), inhibits NFκB (nuclear factor κ B) activation whereas H2O2 (Schreck *et al.*, 1991), tumor necrosis factor α (TNFα), IL-1, mitogens, lipopolysaccharides, and muramyl peptides stimulate NFκBactivation (Grilli *et al.*, 1992; Baeuerle, 1991; Rooney *et al.*, 1990). In summary, the regulation of GSH by NGF is a shared pathway that determines cellular outcome in the nervous system in response to stress and injury.

REFERENCES

Abbott, W.A. and Meister, A. (1986): Intrahepatic transport and utilization of biliary glutathione and its metabolism. *Proc. Natl. Acad. Sci. USA* **83,** 1246–1250.

Amstad, P., Crawford, D., Muehlematter, D., Zbinden, I., Larsson, R., and Gerutti, P. (1990): Oxidant stress induces the proto-oncogenes, *C-fos* and *C-myc* in mouse epidermal cells. *Bull. Cancer* **77,** 501–502.

Anderson, M.E. Naganuma, A., and Meister, A. (1990): Protection against cisplatin toxicity by administration of glutathione ester. *FASEB J.* **4,** 3251–3255.

Andorn, A.C., Britton, R.S., and Bacon, R.R. (1990): Evidence that lipid peroxidation and total iron are increased in Alzheimer's brain. *Neurobiol. Aging.* **11,** 316.

Awatsuji, H., Furukawa, Y., Hirota, M., Murakami, Y., Nii, S., Furukawa, S., and Hayashi, K. (1993): Interleukin-4 and -5 as modulators of nerve growth factor synthesis/secretion in astrocytes. *J. Neurosci. Res.* **34,** 539–45.

Baeuerle, P.A. (1991): The inducible transcription activator NF B: regulation by distinct protein subunits. *Biochim. Biophys. Acta* **1072,** 63–80.

Bailey, H.H., Gipp, J.J., Ripple, M., Wilding, G., and Mulcahy, T.R. (1992): Increase in -glutamylcysteine synthetase activity and steady-state messenger RNA levels in melphalan-resistant DU-145 human prostate carcinoma cells expressing elevated glutathione levels. *Cancer Res.* **52,** 5115–5118.

Ballatori, N., Jacob, R., Barrett, C., and Boyer, J.L. (1988): Biliary catabolism of glutathione and differential reabsorption of its amino acid constituents. *Am. J. Physiol.* **254,** G1-G7.

Bandy, B. and Davison, A.J. (1990): Mitochondrial mutations may increase oxidative stress: implications for carcinogenesis and aging. *Free Rad. Biol. Med.* **8,** 523–539.

Bannai, S. and Ishii, T. (1980): Formation of sulfhydryl groups in the culture medium by human diploid fibroblasts. *J. Cell. Physiol.* **104,** 215–223.

Bannai, S. and Ishii, T. (1982): Transport of cystine and cysteine and cell growth in cultured human diploid fibroblasts: effect of glutamate and homocysteate. *J. Cell. Physiol.* **112,** 265–272.

Bannai, S. and Kitamura, E. (1980): Transport interaction of L-cysteine and L-glutamate in human diploid fibroblasts in culture. *J. Biol. Chem.* **255,** 2372–2376.

Bannai, S. and Tateishi, N. (1986): Role of membrane transport in metabolism and function of glutathione in mammals. *J. Membr. Biol.* **89,** 1–8.

Barde, Y-A. (1989): Trophic factors and neuronal survival. *Neuron* **2,** 1525–1534.

Basi, G.S., Jacobson, R.D., Virag, I., Schilling, J., Skene, J.H.P. (1987): Primary structure and transcriptional regulation of GAP-43, a protein associated with nerve growth. *Cell* **49,** 785–791.

Berg, D.K. (1984): New neuronal growth factors. *Annu. Rev. neurosci.* **7,** 149–170.

Bergelson, S., Pinkus, R., and Daniel, V. (1994): Intracellular glutathione levels regulate *Fos/Jun* induction and activation of glutathione S-transferase gene expression. *Cancer Res.* **54,** 36–40.

Carman-Krzan, M. Vige, X., and Wise, B.C. (1991): Regulation by interleukin-1 of nerve growth factor secretion and nerve growth factor mRNA expression in rat primary astroglial cultures. *J. Neurochem.* **56,** 636–643.

Carman-Krzan, M. and Wise, B.C. (1993): Arachidonic acid lipoxygenation may mediate interleukin-1 stimulation of nerve growth factor secretion in astroglia cultures. *J. Neurosci. Res.* **34,** 225–232.

Chao, M.V. (1992): Neurotrophic receptors: a window into neuronal differentiation. *Neuron* **9,** 583–593.

Cho, Y. and Bannai, S. (1990): Uptake of glutamate and cystine in C-6 glioma cells and cultured astrocytes. *J. Neurochem.* **55,** 2091–2097.

Chun, L.L.Y. and Patterson, P.H. (1977a): Role of nerve growth factor in the development of rat sympathetic neurons *in vitro*. I. Survival, growth, and differentiation of catecholamine production. *J. Cell Biol.* **75,** 694–704.

Chun, L.L.Y. and Patterson, P.H. (1977b): Role of nerve growth factor in the development of rat sympathetic neurons *in vitro*. II. Developmental study. *J Cell Biol.* **75,** 705–711.

Chun, L.L.Y. and Patterson, P.H. (1977c): Role of nerve growth factor in the development of rat sympathetic neurons *in vitro*. III. Effect on acetylcholine production. *J Cell Biol.* **75,** 712–718.

Ciriolo, M.R., Fiskin, K., Martino, A.D.E., Corasaniti, M.T., Nistico, G., and Rotilio, G. (1991): Age-related changes in Cu,Zn superoxide dismutase, Se-dependent and -independent glutathione peroxidase and catalase activities in specific areas of rat brain. *Mech. Ageing Dev.* **61,** 287–297.

Cosma, G., Fulton, H., Defeo, T., and Gordon, T. (1992): Rat lung metallothionein and heme oxygenase gene expression following ozone and zinc oxide exposure. *Toxical. Appl. Pharmacol.* **117,** 75–80.

Cross, C.E., Halliwell, B., Borish, E.T., Pryor, W.A., Ames, B.N., Saul, R.L. *et al.*, (1987): Oxygen radicals and human diseases. *Annu. Int. Med.* **107,** 526–545.

Damier, P., Hirsch, E.C., Zhang, P., Agid, Y., and Javoy-Agid, F. (1993): Glutathione peroxidase, glial cells and Parkinson disease. *Neurosci.* **52,** 1–6.

Doroshow, J.H., Akman, S., Chu, F.F., and Esworthy, S. (1990): Role of the glutathione-glutathione peroxidase cycle in the cytotoxicity of the anticancer quinones. *Pharma. Thera.* 47, 359–370.

Drubin, D.G., Feinstein, S.C., Shooter, E.M., and Kirschner, M.W. (1985): Nerve growth factor induced neurite outgrowth in PC12 cells involves the coordinated induction of microtubule assembly and assembly promoting factors. *J. Cell Biol.* **101,** 1799–1807.

Evans, B.A. and Richards, R.I. (1985): Genes for the and subunits of mouse nerve growth factor are contiguous *EMBO J.* **4,** 133–138.

Evans, B.A., Drinkwater, C.C., and Richards, R.I. (1987): Mouse glandular kallikrein genes: structure and partial sequence analysis of the kallikrein gene locus. *J. Biol. Chem.* **262,** 8027–8034.

Fahnestock, M. (1991): Structure and biosynthesis of nerve growth factor. *Curr. Topics microbiol. Immunol.* **165,** 1–26.

Farber, J.L., Kyle, M.E., and Coleman, J.B. (1990): Biology of diseases: Mechanisms of cell injury by activated oxygen species. *Lab. Invest.* **62,** 670–679.

Federoff, H.J., Gradbczyk, E., and Fishman, M.C. (1988): Dual regulation of GAP-43 gene expression by nerve growth factor and glucocorticoids. *J. Biol. Chem.* **263,** 19290–19295.

Feinstein, S.C., Dana, S.L., McConlogue, L., Shooter, E.M., and Coffino, P. (1985): Nerve growth factor rapidly induces ornithine decarboxylase mRNA in PC12 rat pheochromocytoma cells. *Proc. Natl. Acad. Sci. USA.* **82,** 5761–5765.

Fernyhough, P. and Ishii, D.N. (1987): Nerve growth factor modulates tubulin transcript levels in pheochromocytoma PC12 cells. *Neurochem. Res.* **12,** 891–899.

Fisher, A.B. Intracellular production of oxygen-derived free radical. *Oxygen Radicals and Tissue Injury.* Barry Halliwell ed., Upjohn Company. pp34–39.

Floyd, R.A. (1990): Role of oxygen free radicals in carcinogenesis and brain ischemia. *FASEB J.* **4,** 2587–2597.

Follesa, P. and Mocchetti, I. (1993): Regulation of basic fibroblast growth factor and nerve growth factor mRNA by beta-adrenergic receptor activation and adrenal steroids in rat central nervous system. *Mol. Pharmcol.* **43,** 132–138.

Foreman, P.J., Taglialatela, G., Angelucci, L., Turner, C.P. & Perez-Polo, J.R. Nerve growth factor & Nerve growth factor Receptor mRNA change in rodent CNS Following stress activation of the hypothalamo-Pituitary-Adrenocortical axis. J. Neurosci. Res. *36*:10–18, 1993.

Freedman, J.H., Ciriolo, M.R., and Peisach, J. (1989): The role of glutathione in copper metabolism and toxicity. *J. Biol. Chem.* **264,** 5598–5605.

Friedman, W.J., Larkfors, L., Ayer-LeLievre, C., Ebendal, T., Olson, L., and Persson, H. (1990): Regulation of beta-nerve growth factor expression by inflammatory mediators in hippocampal cultures. *J. Neurosci. Res.* **27,** 374–382.

Fridle, H.P., Till, G.O., Ryan, U.S., and Ward, P.A. (1989): Mediator-induced activation of xanthine oxidase in endothelial cells. *FASEB J.* **3,** 2512–2518.

Fujita, K., Lazarovici, P., and Guroff, G. (1989): Regulation of the differentiation of PC12 pheochromocytoma cells. *Environ. Health Perspect.* **80,** 127–142.

Furukawa, S., Furukawa, Y., Satoyoshi, E., and Hayashi, K. (1987): Regulation of nerve growth factor synthesis/secretion by catecholamines in cultured mouse astroglial cells. *Biochem. Biophys. Res. Commum.* **147,** 1048–1054.

Furukawa, K., Onodera, H., Kogure, K., and Akaike, N. (1993): Time-dependent expression of Na and Ca channels in PC12 cells by nerve growth factor and cAMP. *Neurosci. Res.* **16,** 143–147.

Giordano, T., Pan, J.B., Casuto, D., Watanabe, S., and Arneric, S.P. (1992): Thyroid hormone regulation of NGF, NT-3 and BDNF RNA in the adult rat brain. *Mol. Brain Res.* **16,** 239–245.

Gizang-Ginsberg, E. and Ziff, E.B. (1990): Nerve growth factor regulates tyrosine hydroxylase gene transcription through a nucleoprotein complex that contains *c-fos*. *Genes Dev.* **4,** 447–491.

Gnahn, H., Hefti, F., Heumann, R., Schwab, M.E., and Thoenen, H. (1983): NGF-mediated increase in choline acetyltransferase (ChAT) in the neonatal rat forebrain: evidence for physiological role of NGF in the brain. *Dev. Brain Res.* **9,** 45–52.

Goldwin, A.K,, Meister, A., O'Dwyer, P.J., Huang, C.S., Hamilton, T.C., and Anderson, M.E. (1992): High resistance to cisplatin in human ovarian cancer cell lines is associated with marked increase in glutathione synthesis. *Proc. Natl. Acad. Sci. USA* **89,** 3070–3074.

Gonzalez, D. Les Dees, W., Hiney. J.K., Ojeda, S.R., Santo, R.P. (1990): Expression of -nerve growth factor in cultured cells derived from the hypothalamus and cerebral cortex. *Brain Res.* **511,** 249–258.

Greenberg, M.E., Greene, L.A., and Ziff, E.B. (1985): Nerve growth factor and epidermal growth factor induce rapid transient changes in proto-oncogene transcription in PC12 cells. *J Biol. Chem.* **260,** 14101–14110.

Greene, L.A., Shooter, E.M., and Varon, S. (1969): Subunit interaction and enzymatic activity of mouse 7S nerve growth factor. *Biochemistry* **8,** 3735–3741.

Greene, L.A. and Tischler, A.S. (1976): Establishment of a nonadrenergic clonal line of rat adrenal pheochromocytoma cells which respond to nerve growth factor. *Proc. Natl. Acad. Sci. USA* **73,** 2424–2448.

Greene, L.A. (1977a): Quantitative *in vitro* studies on the nerve growth factor requirement of neurons. I. Sympathetic neurons. *Dev. Biol.* **58,** 96–105.

Greene, L.A. (1977b): Quantitative *in vitro* studies on the nerve growth factor requirement of neurons. I. Sensory neurons. *Dev. Biol.* **58,** 106–113.

Greene, L.A. and Rein, G. (1977): Synthesis, storage, and release of acetylcholine by a nerve growth factor responsive line of rat pheochromocytoma cells. *Nature* **268,** 349–351.

Greene, L.A. and McGuire, J.C. (1978): Induction of ornithine decarboxylase by nerve growth factor dissociated from effects on survival and neurite outgrowth. *Nature* **276,** 119–193.

Grilli, M., Chiu, J.J., and Lenardo, M.J. (1992): NF B and Reparticipants in a multiform transcriptional regulatory system. *Annu. Rev. Immunol.*

Guroff, G., Dickens, G., and End, D. (1981): The induction of ornithine decarboxylase by nerve growth factor and epidermal growth factor in PC12 cells. *J. Neurochem.* **37,** 342–349.

Halliwell, B. and Gutteridge, M.C. (1989): Lipid peroxidation: a radical chain reaction, in *Free Radicals in Biology and Medicine* (Halliwell B. 2nd. ed.) pp.188–276, Oxford, Clarendon Press; New York, Oxford University Press.

Hamburger, V. and Yip, (1984): Reduction of experimentally induced neuronal death in spinal ganglia of the chick embryo by nerve growth factor. *J. Neurosci.* **4,** 767–774.

Hamburger, V., Brunso-Bechtold, J.K., and Yip, J.W. (1981): Neuronal death in the spinal ganglia of the chick embryo and its reduction by nerve growth factor. *J Neurosci.* **1,** 60–71.

Harman, D. (1956): A theory based on free radical and radiation chemistry. *J. Gerontol.* **11,** 298–300.

Harman, D. (1957): Prolongation of the normal life span by radiation protection chemistry. *J. Gerontol.* **12,** 257–263.

Harman, D. (1981): The aging process. *Proc. Natl. Acad. Sci. USA.* **78,** 7124–7128.

Harman, D. (1984): Free radical theory of aging: The "free radical" diseases. *Age* **7,** 111–131.

Hartikka, J. and Hefti, F. (1988): Development of septal cholinergic neurons in culture: plating density and glial cells modulate effects of NGF on survival, fiber growth, and expression of transmitter-specific enzymes. *J. Neurosci.* **8,** 2967–2985.

Hatanaka, H. and Tsukui, H. (1986): Differential effects of nerve growth factor and glioma-conditioned medium on neurons cultured from various regions of fetal rat central nervous system. *Dev. Brain Res.* **30,** 47–56.

Hattori, A., Tanaka, E., Murase, K., Ishida, N., Chatani, Y., Tsujimoto, M., Hayashi, K., and Kohno, M. (1993): Tumor necrosis factor stimulates the synthesis and secretion of biologic active nerve growth factor in nonneuronal cells. *J. Biol. Chem.* **268,** 2577–2582.

Hefti, F. (1986): Nerve growth factor promotes survival of septal cholinergic neurons after fimbrial transections. *J. Neurosci.* **6,** 2155–2162.

Hefti, F. and Weiner, W.J. (1986): Nerve growth factor and Alzheimer's disease. *Annu. Neurol.* **20,** 275–281.

Henke, R.C., Tolhurst, O., Sentry, J.W., Gunning, P., and Jeffrey, P.L. (1991): Expression of actin and myosin genes during PC12 differentiation. *Neurochem. Res.* **16,** 675–679.

Heumann, R. (1987): Regulation of the synthesis of nerve growth factor. *J. Exp. Biol.* **132,** 133–150.

Honegger, P. and Lenoir, D. (1982): Nerve growth factor (NGF) stimulation of cholinergic telencephalic neurons in aggregating cell cultures. *Dev. Brain Res.* **3,** 229–238.

Huang, C.S., Moore, W.R., and Meister, A. (1988): On the active site thiol of -glutamylcysteine synthetase: Regulationships to catalysis, inhibition, and regulation. *Proc. Natl. Acad. Sci. USA* **85,** 2464–2468.

Huang, C.S., Chang, L.S., Anderson, M.E., and Meister, A. (1993a): Catalytic and regulatory properties of the heavy subunit of rat kidney -glutamylcysteine synthetase. *J. Biol. Chem.* **268,** 19675–19680.

Huang, C.S., Anderson, M.E., Meister, A. (1993): Amino acid sequence and function of the light subunit of rat kidney -glutamylcystine synthetase. *J. Biol. Chem.* **268,** 20578–20583.

Hyslop, P.A., Hinshaw, D.B., Halsey, Jr., Schraufstatter, I.U., Sauerheber, R.D., Spragg, R.G., Jackson, J.H., and Cochrane, C.G. (1988): Mechanisms of oxidant-mediated cell injury: The glycolytic and mitochondrial pathways of ADP phosphorylation are major intracellular targets inactivated by hydrogen peroxide. *J. Biol. Chem.* **263,** 1665–1675.

Jackson, G.R., Apffel, L., Werrbach-Perez, K., and Perez-Polo, J.R. (1990a): Role of nerve growth factor in oxidant-antioxidant balance and neuronal injury. I. Stimulation of hydrogen peroxide resistance. *J. Neurosci. Res.* **25,** 360–368.

Jackson, G.R., Werrbach-Perez, K., and Perez-Polo, J.R. (1990b): Role of nerve growth factor in oxidant-antioxidant balance and neuronal injury. II. A conditioning lesion paradigm *J. Neurosci. Res.* **25,** 369–374.

Jackson, G.R., Werrbach-Polo, K., Ezell, E.L., Post, J.F.M., and Perez-Polo, J.R. (1992): Nerve growth factor effects on pyridine nucleotides after oxidant injury of rat pheochromocytoma cells. *Brain Res.* **592,** 239–248.

Jackson, G.R., & J.R. Perez-Polo. Neurotrophin Regulation of energy homeostasis in the central nervous system. Developmental Neuroscience 16:285–290, 1994.

Jain, A., Martensson, J., Stole, E., Auld, A., and Meister, A. (1991): Glutathione deficiency leads to mitochondrial damage in brain. *Proc. Natl. Acad. Sci. USA* **88,** 1913–1917.

Jesberger, J.A. and Richardson, J.S. (1991): Oxygen free radicals and brain dysfunction. *Intern. J. Neurosci.* **57,** 1–17.

Johnson, M.V., Rutkowski, J.L., Wainer, B.H., Long, J.B., Mobley, W.C. (1987a): NGF effects on developing forebrain cholinergic neurons are regionally specific. *Neurochem. Res.* **12,** 985–994.

Johnson, Jr., Manning, P.T., and Wilcox, C. (1988): The biology of nerve growth factor *in vivo.* from Neurobiology of Amino Acids, Peptides, and Trophic factor. Ferrendelli, J.A., Collins, R.C.and Johnson, E.M. (eds). Kluwer Academic Publishers. pp101–140.

Kang, Y.J., Emary, D., and Enger, M.D. (1991): Buthionine sulfoximine induced growth inhibition in human lung carcinoma cells does not correlate with glutathione depletion. *Cell Biol. Toxicol.* **7,** 249–261

Kang, Y.J. (1992): Exogenous glutathione decreases cellular cadmium uptake and toxicity. *Drug Metab. Disposi.* **20,** 714–718.

Kang, Y.J. and Enger, M.D. (1992): Buthionine sulfoximine-induced cytostasis does not correlate with glutathione depletion. *Am. J. Physiol.* **262,** C122–127.

Kilberg, M.S., Christensen, H.N., and Handlogten, M.E. (1979): Cysteine as a system-specific substrate for transport system ASC in rat hepatocytes. *Biochem. Biophys. Res. Commun.* **88,** 744–751.

Kilberg, M.S., Handlogten, M.E., and Christensen, H.N. (1981): Characteristics of system ASC for transport of neutral amino acids in the isolated rat hepatocytes. *J. Biol. Chem.* **256,** 3304–3312.

Koh, J.Y., Duffy, L.K., and Kirschner, D.A. (1990): -Amyloid protein increases the vulnerability of cultured cortical neurons to excitotoxic damage. *Brain Res.* **533,** 315–350.

Kondo, T., Yoshida, K., Urata, Y., Goto, S., Gasa, S., and Taniguchi, N. (1993): -glutamylcysteine synthetase and active transport of glutathione S-conjugate are responsive to heat shock in K562 erythroid cells. *J. Biol. Chem.* **268,** 20366–20372.

Koppenol, W.H. (1983): Thermodynamics of the Fenton-Driven Haber-Weiss and related reactions. In: Oxy-radicals and Their Scavenger System: Vol 1 Molecular Aspects. Cohen, G. and Greenwald, R.A. (eds). Elsevier, New York. pp. 84–88.

Kromer, L.F. and Cornbrooks, C.J. (1986): Transplants of Schwann cell cultures promote axonal regeneration in the adult mammalian brain. *Proc. Natl. Acad. Sci. USA.* **82,** 6330–6334.

Kruijer, W., Schubert, D., and Verma, I.M. (1985): Induction of the proto-oncogene *fos* by nerve growth factor. *Proc. Natl. Acad. Sci. USA.* **82,** 7330–7334.

Lawrence, R.A. and Burk, R.F. (1976): Glutathione peroxidase activity in selenium deficient rat liver. *Biochem. Biophys. Res. Comm.* **71,** 952–958.

Leonard, D.G.B., Gorham, J.D., Cole, P., greene, L.A., and Ziff, E.B. (1988): A nerve growth factor-related messenger RNA encodes a new intermediate filament protein. *J. Cell. Biol.* **106,** 181–193.

Levi, A. Biocca, S., Cattaneo, A., and Callissano, P. (1988): The mode of action of nerve growth factor on PC12 cells. *Mol. Neurobiol.* **2,** 201–226.

Levi-Montalcini, R. and Hamburger, V. (1951): Selective growth stimulating effects of mouse sarcoma on the sensory and sympathetic nervous system of the chick embryo. *J. Exp. Zool.* **116,** 321–362.

Levi-Montalcini, R. and Hamburger, V. (1953): A diffusible agent of mouse sarcoma, producing hyperplasia of sympathetic ganglia and hyperneurotization of viscera in the chick embryo. *J. Exp. Zool.* **123,** 233–288.

Levi-Montalcini, R. and Angeletti, P.U. (1963): Essential role of the nerve growth factor on the survival and maintenance of dissociated sensory and sympathetic embryonic nerve cells *in vitro*. *Dev. Biol.* **7,** 653–659.

Levi-Montalcini, R. (1987): The nerve growth factor 35 years later. *Science* **237,** 1154–1162.

Liuzzi, A., Foppen, F.H., and Kopin, I.J. (1977): Stimulation and maintenance by nerve growth factor of phenylethanolamine-N-methyltransferase in superior cervical ganglia of adult rats. *Brain Res.* **138,** 309–315.

Little, C. and O'Brien, (1968): An intracellular GSH-peroxidase with lipid peroxide substrate. *Biochem. Biophys. Res. Comm.* **31,** 145–150.

Lorenzi, M.V., Knusel, B,, Hefti, F., and Strauss, W.L. (1992): Nerve growth factor regulation of choline acetyltransferase gene expression in rat embryo basal forebrain cultures. *Neurosci. Lett.* **140,** 185–188.

Lu, B., Yokoyama, M., Dreyfus, C.F., and Black, I.B. (1991): NGF expression in actively growing brain glia. *J. Neurosci.* **11,** 318–326.

Lu, S.C., Garcia-Ruiz, C., Kuhlenkamp, J,, Ookhtens, M., Salas-Prato, M., and Kaplowitz, N. (1991): Hormonal regulation of GSH efflux. *J. Biol. Chem.* **265,** 16088–16095.

Lu, S.C., Ge, J.L., Kuhlenkamp, J., and Kaplowitz, N. (1992): Insulin and glucocorticoid dependence of hepatic -glutamylcysteine synthetase and glutathione synthesis in the rat. *J. Clin. Invest.* **90,** 524–532.

MacDonnell, P.C., Tolson, N., Yu, M.W., and Guroff, G. (1977): The de novo synthesis of tyrosine hydroxylase *in vitro*: the effects of nerve growth factor. *J. Neurochem.* **28,** 843–849.

MacDonnell, P.C., Nagaich, K., Lashmanan, J., and Guroff, G. (1977): Nerve growth factor increases activity of ornithine decarboxylase in superior cervical ganglion of young rats. *Proc. Natl. Acad. Sci. USA.* **74,** 4681–4685.

Makowski, M. and Christensen, H.N. (1982): Contrasts in transport systems for anionic amino acids in hepatocytes and hepatoma cell line HTC. *J. Biol. Chem.* **257,** 5663–5670.

Martensson, J. Lai, J.C.K., and Meister, A. (1990b): High-affinity transport of glutathione is part of a multicomponent system essential for mitochondrial function. *Proc. Natl. Acad. Sci. USA.* **87,** 7185–7189.

Martinez, H.J., Dreyfus, C.F., Jonakait, G.M., and Black, I.B. (1985): Nerve growth factor promotes cholinergic development in brain striatal cultures. *Proc. Natl. Acad. Sci. U.S.A.* **82,** 7777–7781.

Max, S.R., Rohrer, H., Otten, U., and Thoenen, H. (1978): Nerve growth factor-mediated induction of tyrosine hydroxylase in the rat superior ganglia *in vitro*. *J. Biol. Chem.* **253,** 8013–8015.

Meister, A. (1981): On the cycles of glutathione metabolism and transport. *Curr. Topi. Cellu. Regul.* **18,** 21–58.

Meister, A. (1983): Metabolism and transport of glutathione and other gamma-glutamyl compounds, in *Function of Glutathione: Biochemical, Physiological, Toxicological, and Clinical Aspects* (Larsson A. et al. ed) pp. 1–23. Raven Press, New York.

Meister, A. (1988a): On the discovery of glutathione. *TIBS* **13,** 185–188.

Meister, A. (1988b): Glutathione metabolism and its selective modification. *J. Biol. Chem.* **263,** 17205–17208.

Meister, A. and Anderson, M.E. (1983): Glutathione. *Annu. Rev. Biochem.* **52,** 711–760.

Mihm, S. and Droge, W. (1990): Intracellular glutathione levels controls DNA-binding activity of NF B-like proteins. *Immunobiology* **181,** 245.

Mihm, S., Ennen, J., Pessara, U., Kurth, R., and Droge, W. (1991): Inhibition of HIV-1 replication and NF B activity by cysteine and cysteine derivatives. *AIDS* **5,** 497–503.

Milbrandt, J. (1988): Nerve growth factor induces a gene homologous to the glucocorticoid receptor gene. *Neuron* **1,** 183–188.

Milne, L., Nicotera, P., Orrenius, S., and Burkitt, M.J. (1993): Effects of glutathione and chelating agents on copper-mediated DNA oxidation: pro-oxidant and antioxidant properties of glutathione. *Arch. Biochem. Biophys.* **304,** 102–109.

Misko, T.P., Radeke, M. J., and Shooter, E.M. (1987): Nerve growth factor in neuronal development and maintenance. *J. Exp. Biol.* **132,** 177–190.

Mizui, T., Kinouchi, H., and Chan, P.H. (1992): Depletion of brain glutathione by buthionine sulfoximine enhances cerebral ischemic injury in rats. *Am. J. Physiol.* **262,** H313-H317.

Mobley, W.C., Rutkowski, J.L., Tennekoon, G.I., Gemski, J., Buchanan, K., Johnson, M.V. (1986): Nerve growth factor increases choline acetyltransferase activity in developing basal forebrain neurons. *Mol. Brain Res.* **1,** 53–62.

Moore, W.R., Anderson, M.E., Meister, A., Murata, K., and Kimura, A. (1989): Increased capacity for glutathione synthesis enhances resistance to radiation in *Escherichia coli*, a possible model for mammalian cell protection. *Proc. Natl. Acad. Sci. USA.* **86,** 1461–1464.

Mossman, B.T., Marsh, J.P., Shatos, M.A., Doherty, J., Gilbert, R., and Hill, S. (1987): Implication of active oxygen species as second messengers of asbestos toxicity. *Drug Chem. Toxicol.* **10,** 157–180.

Muller, S.R., Huff, S.Y., Goode, B.L., Marshall, L., Chang, J., and Feinstein, S.C. (1993): Molecular analysis of the nerve growth factor inducible ornithine decarboxylase gene in PC12 cells. *J. Neurosci. Res.* **34,** 304–314.

Murphy, T.H., Miyamoto, M., Sastre, A., Schnaar, R.L., and Coyle, J.T. (1989): Glutamate toxicity in a neuronal cell line involves inhibition of cystine transport leading to oxidative stress. *Neuron* **2,** 1547–1558.

Murphy, T.H., Schnaar, R.L., and Coyle, J.T. (1990): Immature cortical neurons are uniquely sensitive to glutamate toxicity by inhibition of cystine uptake. *FASEB J.* **4,** 1624–1633.

Naganuma, A., Anderson, M.E., Meister, A. (1990): Cellular glutathione as a determinant of sensitivity to mercuric chloride toxicity. *Biochem. Pharmacol.* **40,** 693–697.

Neveu, I., Jehan, F., Handrot-Perrus, M., Wion, D., and Brachet, P. (1993): Enhancement of the synthesis and secretion of nerve growth factor in primary cultures of glial cells by proteases; a possible involvement of thrombin. *J. Neurochem.* **60,** 858–867.

Nistico, G., Ciriolo, M.R., Fiskin, K., Iannone, M., Demartino, A., and Rotilio, G. (1992): NGF restores decrease in catalase activity and increases superoxide dismutase and glutathione peroxidase activity in the brain of aged rats. *Free Rad. Biol. Med.* **12,** 177–181.

Oppenheim, R.W., Maderdrut, J.L., and Wells, D.J. (1982): Cell death of motorneurons in the chick embryo spinal cord. VI. Reduction of naturally occurring cell death in the Thoracolumbar Column of Terni by nerve growth factor. *J. Comp. Neurol.* **210,** 174–189.

Orrenius, S., Jewell, S.A., Bellomo, G., Thor, H., Jones, D.P., and Smith, M.T. (1983): Regulation of calcium compartmentation in the hepatocyte - A critical role of glutathione. in Function of Glutathione: Biochemical, Physiological, Toxicological, and Clinical Aspects (Larsson A. et al., ed.) pp. 261–271, Recen Press. New York.

Otten, U., Schwab, M., Gagnon, C., and Thoenen, H. (1977): Selective induction of tyrosine hydroxylase and dopamine- -hydroxylase by nerve growth factor: comparison between adrenal medulla and sympathetic ganglia of adult and newborn rats. *Brain Res.* **133,** 291–303.

Pan, Z. and Perez-Polo, J.R. (1993): Role of nerve growth factor in oxidant homeostasis: Glutathione metabolism. *J. Neurochem.* **61,** 1713–1721.

Pan, Z., and J.R. Perez-Polo. Regulation of γ-glutamylcysteine synthetase activity by nerve growth factor. Intl. J. Develop. Neurosci. 14:5 559–566, 1996.

Pan, Z., and J.R. Perez-Polo. Increased uptake of l-cysteine and L-cystine by nerve growth factor in rat pheochromocytoma cells. Brain Res., 740:21–26, 1996.

Paves, H., Neuman, T., Metsis, and Saarma, M. (1988): Nerve growth factor (NGF) induces rapid redistribution of F-actin in PC12 cells. *FEBS Lett.* **235,** 141–143.

Pechan, P.A., Chowdhury, K., Gerdes, W., and Seifert, W. (1993): Glutamate induces the growth factors NGF, bFGF, the receptor FGF-R1 and *c-fos* mRNA expression in rat astrocyte culture. *Neurosci. Lett.* **153,** 111–114.

Pechan, P.A., Chowdhury, K., and Seifert, W. (1992): Free radicals induce gene expression of NGF and bFGF in rat astrocyte culture. *Neuroreport* **3,** 469–472.

Perez-Polo, J.R., Hall, K., Livingston, K., and Westlund, K. (1977): Steroid induction of nerve growth factor synthesis in cell culture. *Life Science* **21,** 1535–1544.

Perez-Polo, J.R., Foreman, P.J., Jackson, G.R., Shan, D-E, Taglialatela, G., Thorpe, L.W., Werrbach-Perez, K. (1990): Nerve growth factor and neuronal cell death. *Mol. Neurobiol. 4,* 57–91.

Pryor, W.A. (1977): Free Radicals in Biology. The involvement of radical reactions in aging and carcinogenesis. In Mathieu J. (ed): *"Medicine Chemistry V"* Amsterdam: Elsevier, pp 331–359.

Pryor, W.A. (1982): Free radical biology: Xenobiotics, cancer, and aging. *Annu. N. Y. Acad. Sci.* **393,** 1–30.

Rao, G.N., Lassegue, B., Griendling, K.K., and Alexander, R.W. (1993): Hydrogen peroxide stimulates transcription of c-jun in vascular smooth muscle cell: Role of arachidonic acid. *Oncogene* **8,** 2759–2764.

Rao, G.M. and Berk, B.C. (1992): Active oxygen species stimulate vascular smooth muscle cell growth and proto-oncogene expression. *Circ. Res.* **70,** 593–599.

Richardson, J.S., Subbarao, K.V., and Ang, L.C. (1990a): Biochemical indices of peroxidation in Alzheimer's and control brains. *Trans. Am. Soc. Neurochem.* **21,** 113.

Richardson, J.S., Subbarao, K.V., and Ang, L.C. (1990b): Autopsy Alzheimer's brains show increased peroxidation to an in vitro iron challenge. *Neurobiol. Aging.* **11,** 286.

Richardson, P.M. and Ebendal, T. (1982): Nerve growth factor activities in rat peripheral nerve. *Brain Res.* **246,** 57–64.

Richter, C. and Kass, G.E.N. (1991): Oxidative stress in mitochondria: Its relationship to cellular Ca^{2+} homeostasis, cell death, proliferation, and differentiation. *Chem. Biol. Interact.* **77,** 1–23.

Riederer, P., Sofic, E., Rausch, W.D., Schmidt, B., Reynolds, G.P., Jellinger, K., and Youdim, M.B.H. (1989): Transition metals, ferritin, glutathione, and ascorbic acid in Parkinsonian brains. *J. Neurochem.* **52,** 515–520.

Rieger, F., Shelanski, M.L., and Greene L.A. (1980): The effects of nerve growth factor on acetylcholinesterase and its multiple forms in cultures of rat PC12 pheochromocytoma cells: increased total specific activity and appearance of the 16S molecular form. *Dev. Biol.* **76,** 238–243.

Rooney, J.W., Emery, D.W., and Sibley, C.H. (1990): Slow response variant of the B lymphoma 10Z/3 defective in LPS activation of NF B. *Immunogenetics* **31,** 73.

Sampath, D, G.R. Jackson, K. Werrbach-Perez & J.R. Perez-Polo. Effects of NGF on Glutathione peroxidase and catalase in PC12 cells. J. Neurochemistry *62*:2476–2479, 1994.

Sampath, D., V. Holets, J.R. Perez-Polo. Effect of spinal cord photolesion injury on catalase mRNA levels. IJDN 13: 645–654, 1995.

Salton, S.R.J., Fischberg, D.J., Dong, K-W. (1991): Structure of the gene encoding VGF, a nervous system-specific mRNA that is rapidly and selectively induced by nerve growth factor in PC12 cells. *Mol. Cell Biol.* **11,** 2335–2349.

Schreck, R., Rieber, P., and Baeuerle, P.A. (1991): Reactive oxygen intermediates as apparently widely used messengers in the activation of the NF B transcription factor and HIV-1. *EMBO J.* **10,** 2247–2258.

Seelig, G.F. and Meister, A. (1985): Glutathione biosynthesis; -glutamylcysteine synthetase from rat kidney. *Methods Enzymol.* **113,** 379–390.

Simmons, T.W., Adners, M.W., and Ballatori, N. (1993): Amino acid transport and glutathione homeostasis: What is the mechanism for cysteine uptake from bile? *Hepatology* **18,** 700–702.

Singhal, R.K., Anderson, M.E., and Meister, A. (1987): Glutathione, a first line of defense against cadmium toxicity. *FASEB J.* **1,** 220–223.

Smith, C.J., Johnson, E.M. Jr., Osborne, P., Freeman, R.S., Neveu, I., and Brachet, P. (1993): NGF deprivation and neuronal degeneration trigger altered beta-amyloid precursor protein gene expression in the rat superior cervical ganglia *in vivo* and *in vitro*. *Brain Res. Mol. Brain Res.* **17,** 328–334.

Sofic, E., Riederer, P., Heinsen, H., Bechmann, H., Reynolds, G.P., Hebenstreit, G., Youdim, M.B.H. (1988): Increased iron (III) and total iron content in post mortem substantia nigra of Parkinsonian brain. *J. Neural. Trans.* **74,** 199–205.

Sohal, R.S., Arnold, L., and Orr, W.C. (1990): Effect of age on superoxide dismutase, catalase, glutathione reductase, inorganic peroxides, TBA-reactive material, GSH/GSSG, $NADPH/NADP^+$, and $NADH/NAD^+$ in drosophila melanogaster. *Mech. Aging Dev.* **56,** 223–235.

Spragg, R.G., Hinshaw, D.B., Hyslop, P.A., Schraufstatt, I.U., and Cochrane, C.G. (1985): Alterations in adenosine triphosphate and energy charge in cultured endothelial and $P388D_1$ cells after oxidant injury. *J. Clin. Invest.* **76,** 1471–1476.

Stach, R.W. and Shooter, E.M. (1980): Cross-linked 7S nerve growth factor is biologically inactive. *J. Neurochem.* **34,** 1499–1505.

Sun, Y. (1990): Free radicals, antioxidant enzymes, and carcinogenesis. *Free Rad. Biol. Med.* **8,** 583–599.

Taglialatela, G., M. Gegg, J.R. Perez-Polo, L.R. Williams, G. Rose. Evidence for DNA fragmentation in the CNS of aged Fisher -344 rats. NeuroReport,7:077–980, 1996.

Takada, A. and Bannai, S. (1984): Transport of cystine in isolated rat hepatocytes in primary culture. *J. Biol. Chem.* **259,** 2441–2445.

Taniuchi, M., Clark, H.B., Schweitzer, J.B., and Johnson, E.M. Jr. (1988): Expression of nerve growth factor receptors by Schwann cells of axotomized peripheral nerves: ultrastructural location, suppression by axonal contact, and binding properties. *J. Neurosci.* **8,** 664–681.

Tayarani, I., Cloez, I., Clement, M., Bourre, J.M. (1989): Antioxidant enzymes and related trace elements in aging brain capillaries and choroid plexus. *J. Neurochem.* **53,** 817–824.

Thoenen, H. (1991): The changing scene of neurotrophic factors. *TINS* **14,** 165–170.

Thoenen, H., Angeletti, P.U., Levi-Montalcini, R., and Kettler, R. (1971) Selective induction of tyrosine hydroxylase and dopamine hydroxylase in rat superior cervical ganglia by nerve growth factor. *Proc. Natl. Acad. Sci. USA.* **68,** 1598–1602.

Thoenen, H. and Barde, Y-A. (1980): Physiology of nerve growth factor. *Physiol. Rev.* **60,** 1284–1335.

Thomas, J.P., Maiorino, M.,, Ursini, F., and Girotti, A.W. (1990): Protective action of phospholipid hydroperoixde glutathione peroxidase against membrane-damaging lipid peroxidation: *In situ* reduction of phospholipid and cholesterol hydroperoxides. *J. Biol. Chem.* **265,** 454–461.

Till, G.O., Friedle, H.P., and Ward, P.A. (1991): Lung injury and complement activation: Role of neutrophils and xanthine oxidase. *Free Rad. Biol. Med.* 10, 379–86.

Tong, L., J.R. Perez-Polo. Transcription factor DNA binding activity in PC12 cells undergoing apoptosis after glucose deprivation. Neuroscience Letters 191:137–140, 1995.

Tong, L., & J.R. Perez-Polo. Effect of NGF on AP-1, NFκB, and Oct-1 DNA binding activity in apoptotic PC12 cells: extrinsic and intrinsic elements. J. Neurosci. Res., 45:1–12, 1996.

Trush, M.A. and Kensler, T.W. (1991): An overview of the relationship between oxidative stress and chemical carcinogenesis. *Free Rad. Biol. Med.* **10,** 201–210.

Tyrrell, R.M., Applegate, L.A., and Tromvoukis, Y. (1993): The proximal promoter region of the human heme oxygenase gene contains elements involved in stimulation of transcriptional activity by a variety of agents including oxidants. *Carcinogenesis* **14,** 761–765.

Varon, S., Nomura, J., Perez-Polo, J.R., and Shooter, E.M. (1972): The isolation and assay of the nerve growth factor proteins. In Fried M (ed): "Methods in Neurochemistry." New York: M. Dekker, Inc., Vol 3 p4.

Watanabe, H. and Bannai, S. (1987): Induction of cystine transport activity in mouse peritoneal macrophages. *J. Exp. Med.* **165,** 628–640.

Wolf, B.B., Vasudevan, J., Henkin, J., and Gonias, S.L. (1993): Nerve growth factor-gamma activates soluble and receptor-bound single chain urokinase-type plasminogen activator. *J. Biol. Chem.* **268,** p16327–16331.

Woods, J.S., Davis, H.A., and Baer, R.P. (1992): Enhancement of -glutamylcysteine synthetase mRNA in rat kidney by methyl mercury. *Arch. Biochem. Biophys.* **296,** 350–353.

Wu, B.Y., Fodor, E.J., Edwards, R.H., and Rutter, W.J. (1989): Nerve growth factor induces proto-oncogene c-jun in PC12 cells. *J. Biol. Chem.* **264,** 9000–9003.

Yan, N. and Meister, A. (1990): Amino acid sequence of rat kidney -glutamylcysteine synthetase. *J. Biol. Chem.* **265,** 1588–1593.

Yoshida, K. and Gage, F.H. (1992): Cooperative regulation of nerve growth factor synthesis and secretion in fibroblasts and astrocytes by fibroblast growth factor and other cytokines. *Brain Res.* **569,** 14–25.

Zafra, F., Hengerer, B., Leibrock, J., Thoenen, H., and Lindholm, D. (1990): Activity dependent regulation of BDNF and NGF mRNAs in the rat hippocampus is mediated by non-NMDA glutamate receptors. *EMBO J.* **9,** 3545–3550.

Zhang, Y., Tatsuno, T., Carney, J.M., and Mattson, M.P. (1993): Basic FGF, NGF, and IGFs protect hippocampal and cortical neurons agonist iron-induced degeneration. *J. Cereb. Blood Flow Metab.* **13,** 378–388.

Ziegler, D.M. (1985): Role of reversible oxidation-reduction of enzyme thiol-disulfides in metabolic regulation. *Annu. Rev. Biochem.* **54,** 305–329.

14

ROLE OF ASTROCYTES IN GLUTAMATE HOMEOSTASIS

Implications for Excitotoxicity

Arne Schousboe,[1] Ursula Sonnewald,[2] Gianluca Civenni,[3] and Georgi Gegelashvili[1]

[1]PharmaBiotec Res. Center
Department of Biological Sciences
Royal Danish School of Pharmacy
DK-2100 Copenhagen, Denmark
[2]MR-Center, SINTEF UNIMED
7034 Trondheim, Norway
[3]Present address: Institute of Biochemistry and Molecular Biology
University of Bern
CH-3000 Bern, Switzerland

1. INTRODUCTION

Glutamate which is one of the most abundant amino acids in the central nervous system (CNS) plays two prominent roles in brain function as an important metabolite coupling tricarboxylic acid (TCA) cycle and amino acid metabolism and as the major excitatory neurotransmitter (see Schousboe and Frandsen, 1995). The latter function is fine tuned by homeostatic mechanisms which during pathological conditions such as energy failure may be easily impaired leading to overexposure of glutamate receptors and subsequent neuronal damage, a phenomenon termed excitotoxicity (Lucas and Newhouse, 1957; Olney et al., 1971; Lipton and Rosenberg, 1994; Schousboe and Frandsen, 1995). The mechanisms responsible for the maintenance of extracellular glutamate concentrations within a very narrow physiological range involve control of its release and uptake as well as its intracellular metabolism. The present review shall deal with these aspects with the main emphasis on glutamate uptake and metabolism. It has been generally accepted for a number of years that astrocytes play a very important role in these processes (see, Schousboe, 1981) and therefore emphasis will be placed on a discussion of the role of astrocytes in glutamate homeostasis.

Brain Plasticity, edited by Filogamo *et al.*
Plenum Press, New York, 1997

2. RELEASE PROCESSES

2.1. Physiological Release

Neurotransmitter glutamate is released primarily *via* an exocytotic mechanism triggering fusion of glutamate containing vesicles with the presynaptic plasma membrane during influx of Ca^{++} caused by membrane depolarization (Nicholls and Attwell, 1990). As the influx of Ca^{++} is mediated by a variety of voltage gated Ca^{++} channels, blockers of such channels are efficient inhibitors of release of neurotransmitter glutamate (Wheeler et al., 1994;1996). Varming et al. (1997) have studied the role of different types of voltage gated Ca^{++} channels for this process. Using glutamatergic cerebellar granule neurons in culture they showed that the most efficatious blockers of glutamate release were antagonists of L-type channels (nifedipine) and P-type channels (ω-agatoxin IVA) whereas the Q-type channel blocker ω-conotoxin MVIIC had no effect. The ω-conotoxins GVIA and MVIIA acting on N-channels only partly block K^{+}-stimulated glutamate release from these cells (Elster et al., 1994; Huston et al., 1995; Varming et al., 1997). Nifedipine has, however, not consistently been found to inhibit K^{+}-stimulated glutamate release from these neurons (Belhage et al., 1992; Didier et al., 1993). In addition to the vesicular mechanism, glutamate may be released during depolarization by reversal of the high affinity glutamate carriers (Nicholls and Attwell, 1990). This release mechanism may be particularly important during energy failure (see 2.2).

The depolarization coupled glutamate release can also be reduced by activation of either $GABA_A$ or $GABA_B$ receptors (Meier et al., 1984; Belhage et al., 1991; Kardos et al., 1994). It appears that the subunit composition of the $GABA_A$ receptors plays a role for the ability of these receptors to mediate an inhibitory action on release of transmitter glutamate (Belhage et al., 1991; Schousboe and Redburn, 1995).

2.2. Pathological Release

It is well established that conditions leading to energy failure in the brain such as hypoglycemia or ischemia will result in a pronounced increase in the extracellular concentration of glutamate and aspartate (Benveniste et al., 1984; Hagberg et al., 1985; Sandberg et al., 1986; Globus et al., 1988). Since the exocytotic process is energy dependent (Nicholls and Attwell, 1990) and since the glutamate carriers in order to be fully operational depend on an intact membrane potential (Barbour et al., 1988) it is likely that release of glutamate *via* reversal of the carrier could contribute significantly to this overflow of glutamate (Sánchez-Prieto and González, 1988; Kauppinen et al., 1988). Using the compound phenylsuccinate which has been shown to inhibit glutamate release from the vesicular pool *via* an inhibitory action on its biosynthesis in this pool (Palaiologos et al., 1988), Christensen et al. (1991) have shown that glutamate which is released during ischemia does not originate from the vesicular transmitter pool, thus indirectly confirming the notion that glutamate released during ischemia originates from the cytosolic, metabolic glutamate pool.

3. GLUTAMATE TRANSPORT

3.1. Cloning and Cellular Localization

Almost 50 years ago reports appeared showing that brain slices had a large capacity for glutamate uptake (Stern et al., 1949) and it was subsequently firmly established that a

Table I. Distribution of the cloned Na-dependent, Cl-independent high-affinity glutamate transporters

Transporter/species	Brain region	Cell type
GLAST		
human	cerebellum>>other regions[a,b]	glioma cells[c]
bovine	cerebellum[d]	
rat	forebrain ≈ cerebellum[d]	neurons[e,f]
	cerebellum>>cerebrum[e,g,h]	glia[e,h,i,j,k]
		PC12 cells[l]
mouse		cultured astroglia[m]
GLT1		
human	forebrain>>hindbrain>spinal cord[b]	glioma cells[c]
	cerebellum<<other brain regions[a]	
rat	cerebral cortex>striatum≈brainstem>	glia[e,j,k]
	>cerebellum[h]	neurons[n]
mouse	forebrain≈brainstem>cerebellum[o]	
EAAC1		
human	uniformly among major regions[b]	glioma cells[c]
	cerebral cortex>other brain regions[b]	
rabbit	all major regions[p]	neurons[p]
rat	cerebral cortex>other regions[c]	neurons[p,q]
	uniformly among major regions[p]	C6 glioma cells[c]
EAAT4 human	cerebellum[r]	ND

[a]Arriza et al. (1994); [b]Nakayama et al. (1996); [c]Palos et al. (1996); [d]Tanaka (1993); [e]Rothstein et al. (1994); [f]Li et al. (1994); [g]Storck et al. (1992); [h]Lehre et al. (1995); [i]Otori et al. (1994); [j]Chaudhry et al. (1995); [k]Kondo et al. (1995); [l]Ramachandran et al. (1993); [m]Gegelashvili et al. (1996); [n]Torp et al. (1994); [o]Kirschner et al. (1994); [p]Kanai and Hediger (1992); [q]Bjørås et al. (1996); [r]Fairman et al. (1995).

high affinity transport mechanism was responsible for the efficient glutamate uptake (Balcar and Johnston, 1972a,b; 1973). Such high-affinity uptake of glutamate has been demonstrated in a variety of brain cellular and subcellular preparations (see Schousboe, 1981) and it has been firmly established that uptake into glial elements, particularly astrocytes prevails over that in neurons (see Schousboe and Westergaard, 1995). Recently, four glutamate transporters have been cloned (Kanai and Hediger, 1992; Storck et al., 1992; Pines et al., 1992; Fairman et al., 1995) and their brain regional and cellular localizations have been summarized in Table I. Generally, the two transporters GLAST (glutamate-aspartate-transporter) and GLT-1 (glutamate-transporter-1) have a glial localization whereas EAAC1 (excitatory amino acid carrier-1) cloned from rabbit intestine appears to be primarily located in neurons, albeit it has also been found in glioma cell lines. The newly cloned transporter named EAAT4 (excitatory amino acid transporter-4) is apparently only present in cerebellum but its cellular localization has not been reported. Interestingly, this particular transporter has been shown to possess chloride channel activity thus possibly constituting an alternative ligand gated chloride channel (Fairman et al., 1995).

3.2. Regulation of Expression

Based on studies of glutamate uptake into astrocytes cultured from different parts of the brain it was suggested that astrocytic glutamate uptake might be particularly pronounced in brain areas with high glutamatergic activity (Schousboe and Divac, 1979). This may point to regulatory mechanisms governing expression of glutamate carriers and it has been observed that neuronally released factors may influence activity and expres-

Table II. Expression of GLAST protein and [^{3}H]D-aspartate uptake activity in cerebral cortical astrocytes cultured in the presence of different glutamate receptor agonists and antagonists

Treatment	[^{3}H]D-asp uptake (% of control)	GLAST expression (% of control)
Control (serum-medium)	100 ± 3	100 ± 9
+dBcAMP	182 ± 4$^{***a)}$	197 ± 14
+Glutamate (1mM)	120 ± 5$^{**a)}$	182 ± 6$^{***a)}$
+KA (50μM)	119 ± 7$^{*a)}$	169 ± 6$^{***a)}$
+AMPA (50μM)	101 ± 5	107± 13
+tACPD (1mM)	102 ± 5	92 ± 6
+dBcAMP + glutamate	210 ± 6$^{*b)}$	210±16
+dBcAMP +AMPA	185 ±10	256±15$^{**b)}$
+dBcAMP + tACPD	188 ± 9	228±13
+Glutamate + CNQX (1mM)	101 ± 3	N.D.
+KA + CNQX (1mM)	104 ± 4	N.D.

Primary cultures of cortical astrocytes were prepared from newborn mice essentially as described by Hertz et al. (1989). D-[22,3-^{3}H]aspartate uptake was assayed at 37°C as described by Drejer et al. (1983). Uptake values (pmol x min^{-1} x mg^{-1} x protein) estimated for control cultures were considered as 100%.

Protein dot blot and Western blot analyses were carried out in order to evaluate the expression of glutamate transporters. Cultures were solubilized in 20 mM Na-phosphate buffer, pH 7.4, containing 1% SDS, 1 mM EDTA, 1 mM PMSF. The extracted protein was dot blotted onto nitrocellulose membranes (5μg protein per dot); or subjected to SDS-PAGE and Western blotting (10-100 μg per lane). Subsequent procedures, including incubation with the anti-GLAST polyclonal antibody (A511) were performed essentially as described previously (Lehre et al., 1995; Levy et al., 1995).

At the final stages, dot blots were incubated with [^{125}I]protein G (3000 dpm μl^{-1}) in phosphate buffered saline (PBS) containing 0.2% Tween 20 and 0.25% gelatine, for 90 min. simultaneously, parallel samples were incubated with iodinated protein G only, i.e. in the absence of the primary antibodies. After washing (4 x 10 min) with PBS/0.2% Tween 20, blots were dried, exposed to radiosensitive phosphor screens and scanned (Phosphorimager SI, Molecular dynamics). Optical densities of spots and background were evaluated using ImageQuant software (Molecular Dynamics). Values are expressed relative to the expression in control cultures (100%), statistically significant differences from plain astrocytes (a) or astrocytes grown in the presence of dBcAMP (b) shown by asterisks (*P<0.05; **P<0.01; ***P<0.001; Student's t-test). Values are mean ± S.E.M. of 8-18 separate cultures for each condition. the mean ± S.E.M. value for the D-aspartate uptake in the control 1 (untreated) cultures was 472.5±14 pmol x min^{-1}.

From Gegelashvili et al. (1996).

sion of astrocytic glutamate carriers both *in vitro* and *in vivo* (Drejer et al., 1983; Chaudhry et al., 1995; Levy et al., 1995). As shown in Table II, treatment of astrocytes cultured from mouse cerebral cortex with glutamate or kainate led to an increased expression of GLAST protein as well as capacity for uptake of [^{3}H]D-aspartate, a phenomenon blocked by the AMPA ((RS)-2-amino-3-(3-hydroxy-5-methylisoxazol-4-yl)propionate)/KA (kainate) antagonist CNQX (6-cyano-7-nitro- quinoxaline-2,3-dione) (Gegelashvili et al., 1996). Interestingly, Northern blot analyses revealed that the mRNA for GLAST was not upregulated. On the other hand, tACPD (trans-(1S,3R)-1-amino-1,3-cyclopentanedicarboxylic acid) which activates metabotropic glutamate receptor subtypes (Schoepp and Conn, 1993) in combination with dibutyryl cyclic AMP (dBcAMP) greatly stimulated the level of GLAST mRNA but had no effect on expression of GLAST protein or D-aspartate uptake (Gegelashvili et al., 1996). These findings clearly show that astrocytic glutamate uptake capacity is highly regulated and that glutamate receptors present on the astrocytes may take part in this, functioning as sensors. This notion is clearly in keeping with the proposal that a correlation may exist between glutamatergic activity in discrete brain areas and the capacity for astrocytic glutamate uptake in that particular region.

4. GLUTAMATE METABOLISM

4.1. Oxidative Metabolism versus Glutamine Formation

As a result of detailed studies of compartmentation of glutamate metabolism in the brain (see Berl and Clarke, 1983) it was suggested that glutamate is primarily if not exclusively converted to glutamine through the action of glutamine synthetase (GS). This notion has been further strengthened by the subsequent unequivocal demonstration that GS is exclusively found in glial cells and predominantly in astrocytes (Martinez-Hernandez et al., 1977; Norenberg and Martinez-Hernandez, 1979). However, studies of metabolism of [^{14}C]glutamate in cultured astrocytes have clearly shown that glutamate is not exclusively metabolized to glutamine but additionally to a considerable extent to CO_2 obviously involving TCA cycle activity (Yu et al., 1982; Yudkoff et al., 1986; Waniewski and Martin, 1986; Zielke et al., 1989; Farinelli and Nicklas, 1992). Oxidation of glutamate through the TCA cycle requires convertion to α-ketoglutarate (αKG), a process which can proceed as a transamination most likely catalyzed by aspartate aminotransferase (AAT) or as an oxidative deamination *via* glutamate dehydrogenase. Some controversy exists in the literature as to the quantitative importance of transamination versus oxidation. Using the transaminase inhibitor aminooxyacetic acid (AOAA) it has been found that either transamination plays no role at all (Yu et al., 1982) or accounts for up to 70% of the glutamate being metabolized in the TCA cycle to CO_2 (Farinelli and Nicklas, 1992). Using [^{13}C]glutamate and NMR spectroscopy allowing convertion to lactate and other compounds involving TCA cycle activity to be quantified, Westergaard et al. (1996) have found that AOAA only slightly inhibited these processes. Accordingly, under these conditions glutamate oxidation proceeds primarily *via* oxidative deamination. Interestingly, it was also observed that the opposite reaction, i.e. synthesis of glutamate from αketoglutarate (αKG) requires a transamination probably involving AAT (Westergaard et al., 1996). It should also be mentioned that conversion of glutamate to lactate *via* TCA cycle activity as first reported by Sonnewald et al. (1993) seems to be correlated to the external glutamate concentration. Thus, McKenna et al. (1996) have reported that lactate formation from glutamate plays an increasingly prominent role at higher glutamate concentrations. This may suggest that during glutamatergic activity, astrocytes taking up a large amount of glutamate can utilize this glutamate not only for internal energy production but *via* synthesis of lactate can provide neurons with an important energy source (Schurr et al., 1988; Maran et al., 1994; Takata and Okada, 1995; Schousboe et al., 1997; Sonnewald et al., 1997).

5. EXCITOTOXICITY

5.1. Role of Glutamate Receptors

Numerous studies in the brain *in vivo* as well as in brain tissue preparations including cultured neurons (see reviews by Choi, 1988; Schousboe and Frandsen, 1995) have demonstrated that activation of N-methyl-D-aspartate (NMDA) as well as non-NMDA receptors leads to neuronal degeneration. The corresponding role of metabotropic glutamate receptors is less well defined as reports exist showing either no significant effect or protective or toxic actions of agonists selectively activating these receptors (Koh et al., 1991a,b; Aleppo et al., 1992; Kawai et al., 1992; Thomsen et al., 1993). The mechanisms leading to neuronal damage subsequent to activation of the ionotropic glutamate receptors involve disturbances of Ca^{++} homeostasis, mitochondrial functions, second messenger production (cyclic nucleotides, NO, lipids), free radical formation and activation of early response genes (Didier et al., 1989;1992; Moncada et al., 1991; Lerea et al., 1992; Lerea and McNamara, 1993; Bonfoco et al., 1995; Choi, 1995; Gorman et al., 1995; Ankarcrona et al., 1995; Schousboe and Frandsen, 1995; Schinder et al., 1996; Griffiths et al., 1997; Hansen et al., 1997).

That activation of ionotropic glutamate receptors is the primary event triggering neuronal damage subsequent to elevation of the extracellular glutamate concentration is perhaps best supported by the repeated observations that glutamate receptor antagonists protect against glutamate induced neurotoxicity in the brain as well as in a variety of *in vitro* nervous tissue preparations including cultured neurons (for references see Schousboe and Frandsen, 1995). It should, however, be pointed out that the relative significance of NMDA receptors and non-NMDA receptors may be somewhat controversial as different *in vitro* and *in vivo* models (cultured neurons - ischemic brain damage) have yielded conflicting results depending on the experimental protocols particularly with regard to exposure periods (Choi et al., 1988; Frandsen et al., 1989; Rod and Auer, 1989; Sheardown et al., 1990; Swan and Meldrum, 1990; Buchan et al., 1991a,b).

5.2. Role of Glutamate Transporters

As pointed out above (2.2) energy failure in CNS leads to excessive overflow of glutamate and aspartate increasing the extracellular concentration of these amino acids by at least 10-fold. A major part of this glutamate is likely to have been recruited from the cytoplasmic pool and is released by reversal of the glutamate carriers (Nicholls and Attwell, 1990). Thus, under these conditions activity of the carriers contribute to neuronal degeneration. Under physiological conditions, on the other hand, normal activity of the glutamate carriers is of paramount importance for maintenance of a low extracellular glutamate concentration and thus for prevention of excitotoxic processes. This has been demonstrated *in vivo* by removal of glutamatergic afferents in hippocampus which leads to a reduction of high affinity glutamate uptake and concomittant enhancement of glutamate neurotoxicity (Köhler and Schwarcz, 1981). Moreover, injection of KA e.g. in striatum leads to much more pronounced neuronal damage (Coyle and Schwarcz, 1976; McGeer and McGeer, 1976) than infusion of glutamate which is essentially non-toxic (Mangano and Schwarcz, 1983). This is best explained by the fact that glutamate can be removed by the high affinity transporters whereas kainate cannot (Johnston et al., 1979). This notion is underlined by the finding in cultured neurons that a blocker of glutamate and aspartate transport enhances the cytotoxic potency of these amino acids but not that of NMDA, AMPA and KA (Frandsen and Schousboe, 1990; Robinson et al., 1993) which are not transported by the carriers (Johnston et al., 1979; Drejer et al., 1982). Moreover, glutamate may be less toxic than non-transportable glutamate receptor agonists (Rosenberg et al., 1992). That astrocytic glutamate uptake may be of particular importance in this context is demonstrated by the finding that neuronal cultures containing a large number of astrocytes are less sensitive than pure neuronal cultures to glutamate cytotoxicity (Rosenberg and Aizenman, 1989; Rosenberg et al., 1992). Further compelling evidence that astrocytic glutamate uptake by GLT-1 and GLAST is indispensible to prevent excitotoxic neuronal damage comes from studies on knockout mice. Deletion of each one of the glutamate transporters showed that deletion of GLT-1 and GLAST had more severe consequences than deletion of EAAC 1 for the ability to clear glutamate from the extracellular space and for excitotoxic neuronal damage (Rothstein et al., 1996). Moreover, aberrant glutamate uptake due to dramatic downregulation of the astroglial GLT1 transporter was found to be the critical factor in sporadic forms of amyotrophic lateral sclerosis (ALS), a neurodegenerative disease with a pronounced excitotoxic component (Rothstein et al., 1995).

5.3. Role of Glutamate Metabolism

Although failures in mitochondrial energy metabolism in which glutamate is an essential component are believed to underlie a number of neurodegenerative diseases such as Parkinsonism and Alzheimer's Disease (see Beal et al., 1993) a direct link between the activity of a glutamate metabolizing enzyme and neurodegeneration has only been reported for olivoponto cerebellar atrophy (Plaitakis and Berl, 1983). In this disease there is a significant reduction in the activity of glutamate dehydrogenase. In the brain, this enzyme seems to catalyze primarily oxidation of glutamate as the opposite reaction, reductive amination of αKG does not take place (Cooper et al., 1979). This is contradictory to the thermodynamic equilibrium of the glutamate dehydrogenase reaction which favors glutamate biosynthesis but it should be kept in mind that the concentrations of αKG, NADH and NH_4^+ in the brain are normally quite low altering the prevailing direction of the reaction (Yu et al., 1982). In keeping with this, an elevated NH_4^+ concentration will inhibit glutamate oxidation (Yu et al., 1984). Whether or not astrocytic glutamate dehydrogenase may play a more prominent role than neuronal glutamate dehydrogenase in the development of olivoponto cerebellar atrophy is not known but it should be kept in mind that the activity of this enzyme in astrocytes is similar to that present in neurons (Schousboe et al., 1977; Larsson et al., 1985).

6. CONCLUDING REMARKS

Although the triggering event in excitotoxicity clearly is the activation of glutamate receptors, a prerequisite for this to occur in a manner leading to neuronal damage is a pathological increase in the extracellular glutamate (or aspartate) concentration. As this is governed primarily by the glutamate transporters and secondarily by the glutamate metabolic machinery it is also obvious that a better understanding of regulatory mechanism controlling these processes is instrumental in the design of therapeutic strategies aimed at preventing neuronal degeneration. As astrocytes play a prominent role in this context, studies on these cells are perhaps particularly useful. In our laboratories we have chosen to concentrate on a characterization of regulatory mechanisms for expression of astrocytic glutamate transporters as well as on detailed studies of the metabolic events in astrocytes pertinent to glutamate homeostasis. Preliminary studies in our laboratory as well as results from other investigators (Swanson et al., 1997) clearly suggest that neurons influence the expression of GLAST and GLT-1 in astrocytes.

ACKNOWLEDGMENTS

The expert secretarial assistance of Ms Hanne Danø is cordially acknowledged. The work has been supported by grants from the Danish State Biotechnology Programmes (1991–1995 and 1996–1999), The Lundbeck Foundation and the CEC Programmes B102-CT92-1159 and -CT94-1248.

REFERENCES

Aleppo G, Pisani A, Copani A, Bruno V, Aronica E, D'Agata V, Canonico PL, Nicoletti F (1992) Metabotropic glutamate receptors and neuronal toxicity. *Adv Exp Med Biol* 318:137–145.

Ankarcrona M, Dypbukt JM, Bonfoco E, Zhivotovsky B, Orrenius S, Lipton SA, Nicotera P (1995) Glutamate-induced neuronal death: A succession of necrosis or apoptosis depending on mitochondrial function. *Neuron* 15:961–973.

Arriza JL, Fairman WA, Wadiche JI, Murdoch GH, Kavanaugh MP, Amara SG (1994) Functional comparisons of three glutamate transporter subtypes cloned from human motor cortex. *J Neurosci* 14:5559–5569.

Balcar VJ, Johnston GAR (1972a) The structural specificity of the high affinity uptake of L-glutamate and L-aspartate by rat brain slices. *J Neurochem* 19:2657–2666.

Balcar VJ, Johnston GAR (1972b) Glutamate uptake by brain slices and its relation to the depolarization of neurones by acidic amino acids. *J Neurobiol* 3:295–301.

Balcar VJ, Johnston GAR (1973) High affinity uptake of transmitters: Studies on the uptake of L-aspartate, GABA, L-glutamate and glycine in cat spinal cord. *J Neurochem* 20:529–539.

Barbour B, Brew H, Attwell D (1988) Electrogenic glutamate uptake in glial cells is activated by intracellular potassium. *Nature* 335:433-

Beal MF, Hyman BT, Koroschetz W (1993) Do defects in mitochondrial energy metabolism underlie the pathology of neurodegenerative diseases? *Trends Neurosci* 16:125–131.

Belhage B, Damgaard I, Saederup E, Squires RF, Schousboe A (1991) High- and low-affinity GABA-receptors in cultured cerebellar granule cells regulate transmitter release by different mechanisms. *Neurochem Int* 19:475–482.

Belhage B, Rehder V, Hansen GH, Kater SB, Schousboe A (1992) ^{3}H-D-Aspartate release from cerebellar granule neurons is differentially regulated by glutamate- and K^{+}-stimulation. *J Neurosci Res* 33:436–444.

Benveniste H, Drejer J, Schousboe A, Diemer NH (1984) Elevation of the extracellular concentrations of glutamate and aspartate in rat hippocampus during transient cerebral ischemia monitored by intracerebral microdialysis. *J Neurochem* 43:1369–1374.

Berl S, Clarke DD (1983) The metabolic compartmentation concept. In: *Glutamine, Glutamate and GABA in the Central Nervous System* (Hertz L, Kvamme E, McGeer EG, Schousboe A, eds), Alan R Liss, New York, pp 205–217.

Bjørås M, Gjesdal O, Erickson JD, Torp R, Levy LM, Ottersen OP, Degree M, Storm-Mathiesen J, Seeberg E, Danbolt NC (1996) Cloning and expression of a neuronal rat brain glutamate transporter. *Mol Brain Res* 36:163–168.

Bonfoco E, Krainc D, Ankarcrona M, Nicotera P, Lipton SA (1995) Apoptosis and necrosis: two distinct events induced respectively by mild and intense insults with *N*-methyl-D-aspartate or nitric oxide/superoxide in cortical cell cultures. *Proc Natl Acad Sci USA* 92:7162–7166.

Buchan A, Li H, Cho S-H, Pulsinelli W (1991a) Blockade of the AMPA receptor prevents CA1 hippocampal injury following severe but transient forebrain ischemia in adult rats. *Neurosci Lett* 132:255–258.

Buchan A, Li H, Pulsinelli W (1991b) The N-methyl-D-aspartate antagonist, MK-801, fails to protect against neuronal damage caused by transient, severe forebrain ischemia in adult rats. *J Neurosci* 11:1049–1056.

Choi DW (1988) Glutamate neurotoxicity and diseases of the nervous system. *Neuron* 1:623–634.

Choi DW (1995) Calcium: still center-stage in hypoxic-ischemic neuronal cell death. *Trends Neurosci* 18:58–60.

Choi DW, Koh J-Y, Peters S (1988) Pharmacology of glutamate neurotoxicity in cortical cell culture. Attenuation by NMDA antagonists. *J Neurosci* 8:185–196.

Chaudhry FA, Lehre KP, Van-Lookeren M, Campagne M, Ottersen OP, Danbolt NC, Storm-Mathisen J (1995) Glutamate transporters in glial plasma membranes: highly differentiated localizations revealed by quantitative ultrastructural immunocytochemistry. *Neuron* 15:711–720.

Christensen T, Bruhn T, Diemer NH, Schousboe A (1991) Effect of phenylsuccinate on potassium- and ischemia-induced release of glutamate in rat hippocampus monitored by microdialysis. *Neurosci. Lett* 134:71–74.

Cooper AJL, McDonald JM, Gelbard AS, Gledhill RF, Duffy TE (1979) The metabolic fate of ^{13}N-labeled ammonia in rat brain. *J Biol Chem* 254:4982–4992.

Coyle JT, Schwarcz R (1976) Lesion of striatal neurones with kainic acid provides a model for Huntington's chorea. *Nature* 263:244–246.

Didier M, Roux P, Piechaczyk K, Verrier B, Bockaert J, Pin JP (1989) Cerebellar granule cell survival and maturation induced by K^{+} and NMDA correlate with c-*fos* protooncogene expression. *Neurosci Lett* 107:55–67.

Didier M, Roux P, Piechaczyk M, Mangeat P, Devilliers G, Bockaert J, Pin JP (1992) Longterm expression of the c-*fos* protein during the in vitro differentiation of cerebellar granule cells induced by potassium and NMDA. *Mol Brain Res* 12:249–258.

Didier M, Héaulme M, Gonalons N, Soubrié P, Bockaert J, Pin JP (1993) 35 mM K^+-stimulated $^{45}Ca^{++}$-uptake in cerebellar granule cell cultures mainly results from NMDA receptor activation. *Eur J Pharmacol* 244:57–65.

Drejer J, Larsson OM, Schousboe A (1982) Characterization of glutamate uptake into and release from astrocytes and neurons cultured from different brain regions. *Exp Brain Res* 47:259–269.

Drejer J, Meier E, Schousboe A (1983) Novel neuron-related regulatory mechanisms for astrocytic glutamate and GABA high affinity uptake. *Neurosci Lett* 37:301–306.

Elster L, Saederup E, Schousboe A, Squires RF (1994) ω-Conotoxin binding sites and regulation of transmitter release in cerebellar granule neurons. *J Neurosci Res* 39:424–429.

Fairman WA, Vandenberg RJ, Arriza JL, Kavanaugh MP, Amara SG (1995) An excitatory amino-acid transporter with properties of a ligand-gated chloride channel. *Nature* 375:599–603.

Farinelli SE, Nicklas WJ (1992) Glutamate metabolism in rat cortical astrocyte cultures. *J Neurochem* 58:1905–1915.

Frandsen Aa, Schousboe A (1990) Development of excitatory amino acid induced cytotoxicity in cultured neurons. *Int J Devl Neurosci* 8: 209–216.

Frandsen Aa, Drejer J, Schousboe A (1989) Direct evidence that excitotoxicity in cultured neurons is mediated *via* N-methyl-D-aspartate (NMDA) as well as non-NMDA-receptors. *J Neurochem* 53:297–299.

Gegelashvili G, Civenni G, Racagni G, Danbolt NC, Schousboe I, Schousboe A (1996) Glutamate receptor agonists up-regulate glutamate transporter GLAST in astrocytes. *Neuroreport* 8:261–265.

Globus MYT, Busto R, Dietrich D, Martinez E, Valdes I, Ginsberg MD (1988) Effect of ischemia on the *in vivo* release of striatal dopamine, glutamate and γ-aminobutyric acid studied by intracerebral microdialysis. *J Neurochem* 51:1455–1464.

Gorman AM, Scott MP, Rumsby PC, Meredith C, Griffiths R (1995) Excitatory amino acid-induced cytotoxicity in primary cultures of mouse cerebellar granule cells correlates with elevated, sustained c-*fos* proto-oncogene expression. *Neurosci Lett* 191:116–120.

Griffiths R, Malcolm C, Ritchie L, Frandsen A, Schousboe A, Scott M, Rumsby P, Meredith C (1997) Association of c-*fos* mRNA expression and excitotoxicity in primary cultures of mouse neocortical and cerebellar neurons. *J Neurosci Res*, in press.

Hagberg H, Lehmann A, Sandberg M, Nyström B, Jacobsen I, Hamberger A (1985) Ischemia-induced shift of inhibitory and excitatory amino acids from intra- to extracellular compartments. *J Cereb Blood Flow Metab* 5:413–419.

Hansen HS, Lauritzen L, Strand AM, Vinggaard AM, Frandsen Aa, Schousboe A (1997) Characterization of glutamate-induced formation of *N*-acyl-phosphatidylethanolamine and *N*-acylethanolamine in cultured neocortical neurons. *J Neurochem*, in press.

Hertz L, Juurlink BHJ, Hertz E, Fosmark H, Schousboe A (1989) Preparation of primary cultures of mouse (rat) astrocytes. In: *A Dissection and Tissue Culture Manual for the Nervous System* (Shahar A, Vellis J De, Vernadakis A, Haber B, eds), Alan R Liss, New York, pp 105–108.

Huston E, Cullen GP, Burley JR, Dolphin AC (1995) The involvement of multiple types of calcium channel subtypes in glutamate release from cerebellar granule cells and its modulation by $GABA_B$ receptor activation. *Neuroscience* 68:465–478.

Johnston GAR, Kennedy SME, Twitchin B (1979) Action of the neurotoxin kainic acid on high affinity uptake of L-glutamatic acid in rat brain slices *J Neurochem* 32:121–127.

Kanai Y, Hediger MA (1992) Primary structure and functional characterization of a high-affinity glutamate transporter. *Nature* 360:467–471.

Kardos J, Elster L, Damgaard I, Krogsgaard-Larsen P, Schousboe A (1994) Role of $GABA_B$ receptors in intracellular Ca^{2+} homeostasis and possible interaction between $GABA_A$ and $GABA_B$ receptors in regulation of transmitter release in cerebellar granule neurons. *J Neurosci Res* 39:646–655.

Kauppinen RA, McMahon HT, Nicholls DG (1988) Ca^{++}-Dependent and Ca^{++}-independent glutamate release, energy status and cytosolic free Ca^{++} concentration in isolated nerve terminals following metabolic inhibition: possible relevance to hypoglycaemia and anoxia. *Neuroscience* 27:175–182.

Kawai M, Horikawa Y, Ishihara T, Shimamoto K, Ohfune Y (1992) 2-(Carboxycyclopropyl)glycines: binding, neurotoxicity and induction of intracellular free Ca^{++} increase. *Eur J Pharmacol* 211:195–202.

Kirschner MA, Gopeland NG, Gilbert DJ, Jenkins NA, Amara SG (1994) Mouse excitatory amino acid transporter EAAT2: isolation, characterization, and proximity to neuroexcitability loci on mouse chromosome 2. *Genomics* 24:218–224.

Koh J, Palmer E, Lin A, Cotman CW (1991a) A metabotropic glutamate receptor agonist does not mediate neuronal degeneration in cortical culture. *Brain Res* 561:338–343.

Koh J, Palmer E, Cotman CW (1991b) Activation of the metabotropic glutamate receptor attenuates *N*-methyl-D-aspartate neurotoxicity in cortical cultures. *Proc Natl Acad Sci USA* 88:9431–9535.

Kondo K, Hashimoto H, Kitanaka J, Sawada M, Suzumura A, Marinouchi T, Baba A (1995) Expression of glutamate transporters in cultured glial cells. *Neurosci Lett* 188:140–142.

Köhler C, Schwarcz R (1981) Monosodium glutamate: Increased neurotoxicity after removal of neuronal reuptake sites. *Brain Res* 211:485–491.

Larsson OM, Drejer J, Kvamme E, Svenneby G, Hertz L, Schousboe A (1985) Ontogenetic development of glutamate and GABA metabolizing enzymes in cultured cerebral cortex interneurons and in cerebral cortex *in vivo*. *Int J Devl Neurosci* 3:177–185.

Lehre KP, Levy LM, Ottersen OP, Storm-Mathisen J, Danbolt NC (1995) Differential expression of two glial glutamate transporters in the rat brain: quantitative and immunocytochemical observations. *J Neurosci* 15:1835–1853.

Lerea LS, McNamara JO (1993) Ionotropic glutamate receptor subtypes activate c-*fos* transcription by distinct calcium-requiring intracellular signaling pathways. *Neuron* 10:31–41.

Lerea LS, Butler LS, McNamara JO (1992) NMDA and non-NMDA receptor mediated increase of c-*fos* mRNA in dentate gyrus neurons involves calcium influx via different routes. *J Neurosci* 12:2973–2981.

Levy LM, Lehre KP, Walaas SI, Storm Mathisen J, Danbolt NC (1995) Down-regulation of glial glutamate transporters after glutamatergic denervation in the rat brain. *Eur J Neurosci* 7:2036–2041.

Li HS, Niedzielski AS, Beisel KW, Hiel H, Wenthold RJ, Morley BJ (1994) Identification of a glutamate/aspartate transporter in the rat cochlea. *Hear Res* 8:235–242.

Lipton SA, Rosenberg PA (1994) Mechanisms of disease: excitatory amino acids as a final common pathway for neurologic disorders. *N Engl J Med* 330:613–622.

Lucas DR, Newhouse JP (1957) the toxic effect of sodium-L-glutamate on the inner layers of retina. *AMA Arch Ophthalmol* 58:193–201.

Mangano RM, Schwarcz R (1983) Chronic infusion of endogenous excitatory amino acids into rat striatum and hippocampus. *Brain Res Bull* 10:47–51.

Maran A, Cranston I, Lomas J, Macdonald I, Amiel SA (1994) Protection by lactate of cerebral function during hypoglycaemia. *Lancet* 343:16–20.

Martinez-Hernandez A, Bell KP, Norenberg MD (1977) Glutamine synthetase-glial localization in the brain. *Science* 195:1356–1358.

McGeer EG, McGeer PL (1976) Duplication of biochemical changes of Huntington's chorea by intrastriatal injections of glutamic and kainic acids. *Nature* 263:517–519.

McKenna MC, Sonnewald U, Huang X, Stevenson J, Zielke RH (1996) Exogenous glutamate concentration regulates the metabolic fate of glutamate in astrocytes. *J Neurochem* 66:386–393.

Meier E, Drejer J, Schousboe A (1984) GABA influences functionally active low-affinitive GABA receptors on cultured cerebellar granule cells. *J Neurochem* 43:1737–1744.

Moncada S, Palmer RMJ, Higgs EA (1991) Nitric ocide: Physiology, pathophysiology, and pharmacology. *Pharmacol Rev* 43:109–142.

Nakayama T, Kawakami, Tanaka K, Nakamura S (1996) Expression of three glutamate transporter subtype mRNAs in human brain regions and peripheral tissues. *Mol Brain Res* 36:189–192.

Nicholls D, Attwell D (1990) The release and uptake of excitatory amino acids. *Trends Pharmacol Sci* 11:462–468.

Norenberg MD, Martinez-Hernandez A (1979) Fine structural localization of glutamine synthetase in astrocytes of rat brain. *Brain Res* 161:303–310.

Olney JW, Ho OL, Rhee V (1971) Cytotoxic effects of acidic and sulfur containing amino acids on the infant mouse central nervous system. *Exp Brain Res* 14:61–70.

Otori Y, Shimada S, Tanaka K, Ishimoto I, Tano Y, Tohyama M (1994) Marked increase in glutamate-aspartate transporter (GLAST/GluT-1) mRNA following transient retinal ischemia. *Mol Brain Res* 27:310–314.

Palaiologos G, Hertz L, Schousboe A (1988) Evidence that aspartate amino transferase activity and ketodicarboxylate carrier function are essential for biosynthesis of transmitter glutamate. *J Neurochem* 51:317–320.

Palos T, Ramachandran B, Boado R, Howard B (1996) Rat C6 and human astrocytic tumor cells express a neuronal type of glutamate transporter. *Mol Brain Res* 37:297–303.

Pines G, Danbolt NC, Bjørås M, Zhang Y, Bendahan A, Eide L, Koepsell H, Storm-Mathisen J, Seeberg E, Kanner BI (1992) Cloning and expression of a rat brain L-glutamate transporter. *Nature* 360:464–467.

Plaitakis A, Berl S (1983) Involvement of glutamate dehydrogenase in degenerative neurological disorders. In: *Glutamine, Glutamate and GABA in the Central Nervous System* (Hertz L, Kvamme E, McGeer EG, Schousboe A, eds), Alan R Liss, New York, pp 609–618.

Ramachandran B, Houben K, Rozenberg YY, Haigh JR, Varpetian A, Howard BD (1993) Differential expression of transporters for norepinephrine and glutamate in wild type, variant, and WNT1-expressing PC12 cells. *J Biol Chem* 268:23891–23897.

Robinson MB, Djali S, Buchhalter JR (1993) Inhibition of glutamate uptake with L-trans-pyrrolidine-2,4-dicarboxylate potentiates glutamate toxicity in primary hippocampal cultures. *J Neurochem* 61:2099–2103.

Rod M, Auer, R (1989) Pre- and post-ischemic administration of dizocilpine (MK-801) reduces cerebral necrosis in the rat. *Can J Neurol Sci* 16:340–344.

Rosenberg PA, Aizenman E (1989) Hundred-fold increase in neuronal vulnerability to glutamate toxicity in astrocyte-poor cultures of rat cerebral cortex. *Neurosci Lett* 103:162–168.

Rosenberg PA, Amin S, Leitner M (1992) Glutamate uptake disguises neurotoxic potency of glutamate agonists in cerebral cortex in dissociated cell culture. *J Neurosci* 12:56–61.

Rothstein JD, Martin L, Levey AI, Dykes-Hoberg M, Jin L, Wu D, Nash N, Kunci RW (1994) Localization of neuronal and glial glutamate transporters. *Neuron* 13:713–725.

Rothstein JD, Van Kammen M, Levey AI, Martin LJ, Kuncl RW (1995) Selective loss of glial glutamate transporter GLT-1 in amyotrophic lateral sclerosis. *Ann Neurol* 38:73–84.

Rothstein JD, Dykes-Hoberg M, Pardo CA, Bristol LA, Jin L, Kuncl RW, Kanai Y, Hediger A, Wang Y, Schielke JP, Welty DF (1996) Knockout of glutamate transporters reveals a major role for astroglial transport in excitotoxicity and clearance of glutamate. *Neuron* 16:675–686.

Sánchez-Prieto J, González P (1988) Occurrence of a large Ca^{++}-independent release of glutamate during anoxia in isolated nerve terminals (synaptosomes). *J Neurochem* 50:1322–1324.

Sandberg M, Butcher SP, Hagberg H (1986) Extracellular overflow of neuroactive amino acids during severe insulin-induced hypoglycemia: *in vivo* dialysis of rat hippocampus. *J Neurochem* 47:178–184.

Schinder AF, Olson EC, Spitzer NC, Montal M (1996) Mitochondrial dysfunction is a primary event in glutamate neurotoxicity. *J Neurosci* 16:6125–6133.

Schoepp DD, Conn PJ (1993) Metabotripic glutamate receptors in brain function and pathology. *Trends Pharmacol Sci* 14:13–20.

Schousboe A (1981) Transport and metabolism of glutamate and GABA in neurons and glial cells. *Int Rev Neurobiol* 22:1–45.

Schousboe A, Divac I (1979) Differences in glutamate uptake in astrocytes cultured from different brain regions. *Brain Res* 177: 407–409.

Schousboe A, Frandsen Aa (1995) Glutamate receptors and neurotoxicity. In: *CNS Neurotransmitters and Neuromodulators: Glutamate* (Stone TW, ed), CRC Press, Boca Raton, FL, pp 239–251.

Schousboe A, Redburn DA (1995) Modulatory actions of GABA on $GABA_A$ receptor subunit expression and function. *J Neurosci Res* 41:1–7.

Schousboe A, Westergaard N (1995) Transport of neuroactive amino acids in astrocytes. In: *Neuroglia* (Kettenmann H, Ransom B, eds), Oxford University Press, New York, pp 246–258.

Schousboe A, Svenneby G, Hertz L (1977) Uptake and metabolism of glutamate in astrocytes cultured from dissociated mouse brain hemispheres. *J Neurochem* 29:999–1005.

Schousboe A, Westergaard N, Waagepetersen HS, Larsson OM, Bakken IJ, Sonnewald U (1997) Trafficking between glia and neurons of TCA cycle intermediates and related metabolites. *Glia*, in press.

Schurr A, West C, Rigor BM (1988) Lactate-supported synaptic function in the rat hippocampal slice preparation. *Science* 240:1326–1327.

Sheardown MJ, Nielsen EO, Hansen AJ, Jacobsen P, Honoré T (1990) 2,3-Dihydroxy-6-nitro-7-sulfamoylbenzo(F)quinoxaline: a neuroprotectant for cerebral ischemia. *Science* 247:571–574.

Sonnewald U, Westergaard N, Petersen SB, Unsgård G, Schousboe A (1993) Metabolism of [U-^{13}C]glutamate in astrocytes studied by ^{13}C NMR spectroscopy: Incorporation of more label into lactate than into glutamine demonstrates the importance of the TCA cycle. *J Neurochem* 61:1179–1182.

Sonnewald, U., Westergaard, N., and Schousboe, A. Glutamate transport and metabolism in astrocytes. Glia (1997) in press.

Stern JR, Eggleston LV, Hems R, Krebs HA (1949) Accumulation of glutamic acid in isolated brain tissue. *Biochem J* 44:410–418.

Storck T, Schulte S, Hofmann K, Stoffel W (1992) Structure, expression, and functional analysis of a Na^+-dependent glutamate/aspartate transporter from rat brain. *Proc Natl Acad Sci USA* 89:10955–10959.

Swan J, Meldrum B (1990) Protection by NMDA antagonists against selective cell loss following transient ischemia. *J Cereb Blood Flow Metab* 10:343–351.

Swanson RA, Liu J, Miller JW, rothstein JD, Farrel K, Stein BA, Longuemare MC (1997) Neuronal regulation of glutamate transporter subtype expression in astrocytes. *J Neurosci* 17:932–940.

Takata T, Okada Y (1995) Effects of deprivation of oxygen or glucose on the neural activity in the guinea pig hippocampal slices - intracellular recording study of pyramidal neurons. *Brain Res* 683:109–116.

Tanaka K (1993) Expression cloning of a rat glutamate transporter. *Neurosci Res* 16:149–153.

Thomsen C, Frandsen Aa, Suzdak PD, Andersen CF, Schousboe A (1993) Effects of t-ACPD on neuronal survival and second messengers in cultured cerebral cortical neurons. *Neuroreport* 4:1255–1258.

Torp R, Danbolt NC, Babaie E, Bjørås M, Seeberg E, Storm-Mathisen J, Ottersen OP (1994) Differential expression of two glial glutamate transporters in the rat brain: an in situ hybridization study. *Eur J Neurosci* 6:936–942.

Varming T, Christophersen P, Schousboe A, Drejer J (1997) Pharmacological characterisation of voltage sensitive calcium channels and neurotransmitter release from mouse cerebellar granule cells in culture. *J Neurosci Res* 48:1–10 (1997).

Waniewski RA, Martin DL (1986) Exogenous glutamate is metabolized to glutamine and exported by rat primary astrocyte cultures. *J Neurochem* 47:304–313.

Westergaard N, Drejer J, Schousboe A, Sonnewald U (1996) Evaluation of the importance of transamination versus deamination in astrocytic metabolism of [U-^{13}C]glutamate. *Glia* 17:160–168.

Wheeler DB, Randall A, Tsien RW (1994) Roles of N-type and Q-type Ca^{2+}-channels in supporting hippocampal synaptic transmission. *Science* 264:107–111.

Wheeler DB, Randall A, Tsien RW (1996) Changes in action potential duration alter reliance of excitatory synaptic transmission on multiple types of Ca^{2+} channels in rat hippocampus synaptic transmission. *Science* 264:107–111.

Yu AC, Schousboe A, Hertz L (1982) Metabolic fate of [^{14}C]-labeled glutamate in astrocytes in primary cultures. *J Neurochem* 39:954–960.

Yu ACH, Schousboe A, Hertz L (1984) Influence of pathological concentrations of ammonia on metabolic fate of ^{14}C-labeled glutamate in astrocytes in primary cultures. *J Neurochem* 42:594–597.

Yudkoff M, Nissim I, Hummeler K, Medow M, Pleasure D (1986) Utilization of [^{15}N]glutamate by cultured astrocytes. *Biochem J* 234:185–192.

Zielke HR, Tildon JT, Baab PJ, Hopkins IB (1989) Synthesis of glutamate and glutamine in dibutyryl cyclic AMP-treated astrocytes. *Neurosci Lett* 97:209–214.

15

NERVE CELL DEATH INDUCED BY Ca^{2+} IONOPHORES IN DISSOCIATED HIPPOCAMPAL CULTURES

Protective Action of the NMDA Antagonist MK-801

N. Safran,[1*] R. Haring,[2] A. Shainberg,[3] R. Zisling,[4] A. H. Futerman,[4] and A. Shahar[2]

[1]Koret School of Vet. Med. Hebrew University of Jerusalem 76100 Israel
[2]Israel Institute for Biological Research
Ness-Ziona 70450, Israel
[3]Bar-Ilan University
Ramat-Gan 52900, Israel
[4]Weizmann Institute of Science
Rehovot 76100, Israel

ABBREVIATIONS

AA, arachidonic acid; AOCC's, agonist operated calcium channel; CNQX, 6-cyano-7-nitroquinoxaline-2,3-dione; DMEM, Dulbecco's modified Eagle's medium; DMSO, D-methyl sulfoxide; DIV, days *in-vitro;* EAA's, excitatory amino acids; GFAP, glial fibrillary acidic protein; Ka, Kainate; LDH, lactate dehydrogenase; LH, Locke HEPES buffer; MEM, minimum essential medium; MK-801, (+)-5-methyl-10,11-dihydro-5H-dibenzo [a,d] cyclohepten-5,10-imine maleate; NF, neurofilament; NMDA, N-methyl-D-aspartate; NSE, neuron specific enolase; PBS, phosphate buffer saline; PLA_2, phospholipase A_2; VOCC's, voltage operated calcium channel; XTT, (2,3-bis[2-methoxy-4-nitro-5-sulfophenyl]-2H-tetrazolium-5-carboxanilelide inner salt).

1. INTRODUCTION

Lasalocid (X-537A) is a polyether ionophore compound (Westley et al., 1970) that was isolated from *streptomyces lasaliensis* in 1951 (Berger et al., 1951). It is a lipid-soluble mate-

* Corresponding author: Dr Noam Safran, Koret School of Veterinary Medicine, The Hebrew University of Jerusalem, P. O. Box 12, Rehovot, 76100, Israel. Tel: 972-8-9481021; Fax: 972-8-9467940.

rial which facilitates the passage of divalent and monovalent ions through lipophilic-biological membranes (Westley, 1977; Reed, 1982; Aebi, 1989). Lasalocid is used as a broad-spectrum anticoccidial agent and has been approved for use as a coccidiostat for chickens, and as a growth promoter for cattle (Bergen et al., 1984). However, lasalocid residues in commercial food were found to cause a paralytic syndrome in dogs (Safran et al., 1993a) and overdose in poultry feed caused neuromuscular deficit in chickens (Perelman et al., 1986).

We have developed an *in-vitro* model using brain neurons in culture to study the neurotoxic effects of lasalocid in mouse (Safran et al., 1993b) and rat (Safran et al., 1996). This model contributed to a better understanding of the mechanism of lasalocid neurotoxicity at the cellular level, and may help in search for possible neuroprotective mechanism.

Calcium-related neuronal necrosis is a well documented phenomenon (Schanne et al., 1979) in which the pathogenecity of calcium influx occurs mainly via agonist operated calcium channels (AOCC's) rather than through voltage operated calcium channels (VOCC's) (Leonard and Salpeter, 1979). Most of the neurodegenerative processes which occur through the AOCC's are related to the excitatory amino acids (EAA's) aspartate and glutamate and their receptors (Choi, 1987; 1988). Agonists for all EAA receptor subtypes (e.s n-methyl-D-aspartate, quisqualate and kainate receptors) induce Ca^{2+} influx in neurons (Choi, 1987; Murphy and Miller, 1988,1989; Ogura et al., 1988; Wahl et al., 1993; Frandsen and Schousboe, 1993). This influx is barely reduced by VOCC's blockers (Lasarewicz et al., 1987; Frandsen and Schousboe, 1992,1993). In contrast, the activation of the EAA's receptors and the subsequent Ca^{2+} influx was completely blocked by a combination of NMDA and non-NMDA selective antagonists (Frandsen and Schousboe, 1993). The neurotoxicity induced by NMDA seems to be similar to that induced by glutamate in various neuronal tissues and was completely blocked by the selective NMDA antagonist MK-801, or the omission of external Ca^{2+} (Lehmann, 1987; Ellrem and Lehmann, 1989; Frandsen and Schousboe, 1992, 1993). Stimulation of the NMDA receptor-mediated mechanism, or treatment with Ca^{2+} ionophore, increased the free Ca^{2+} concentration in rat hippocampal neurons (Sanfeliu et al., 1990). Increased levels of intracellular Ca^{2+} may be responsible for the activation of phospholipases and lipases (phospholipase A_2 (PLA_2) etc.), increase lipolysis and proteolysis and the production of free radicals. The lipid peroxidation and damage to membrane proteins may contribute to cell death (Farooqui and Horrocks, 1994). It has been suggested that elevated intracellular concentration of Ca^{2+}, triggered by normal excitation of NMDA receptors or by Ca^{2+} ionophores may participate in activating PLA_2, resulting in the release of arachidonic acid in cultured neurons from striatum (Dumuis et al., 1988), cerebellum (Lazarewicz et al., 1988, 1990), hippocampus (Sanfeliu et al., 1990) and cerebral cortex (Tapia-Arancibia et al., 1992). This release pathway was absent in hippocampal astrocytes (Sanfeliu et al., 1990). In this study we characterize the morphological and biochemical changes caused by the Ca^{2+} ionophore, lasalocid in cultures of hippocampal neurons. We show that lasalocid-mediated neuronal toxicity is attenuated by MK-801 (NMDA receptor/channel antagonist) but not by nimodipine and methoxyverapamil (VOCC's antagonists) or by CNQX, a non-NMDA receptor/channel antagonist (Honor'e et al., 1988).

2. MATERIALS AND METHODS

2.1. Hippocampal Cultures

For biochemical assays cultures in high density were prepared as previously described by Shahar et. al., (1989). Briefly, fetal rat hippocampus (gestation day 18–19) were dissociated mechanically after removal of meninges and seeded in individual 35 mm plastic dishes ($1x10^6$ cells/dish) or 12-well dishes ($0.5x10^6$ cells/well) pre-coated with 0.01% poly-L-lysine (Sigma). The nutrient medium consisted of 5% heat-inactivated horse serum (Gibco), 3% Ultraser (Gibco) and 90% Dulbecco's modified Eagle's medium (DMEM) (Beth-Haemek, Israel). The medium was supplemented with L-glutamine (2 mM), glucose (6 mg/ml) as well as penicillin G (100 IU/ml). Cells were incubated for 18 hr in serum-free DMEM, washed and treated with various ligands for 2–4 hr in magnesium-free Locke-HEPES (LH) medium (pH=7.4), consisting of 154 mM NaCl, 5.6 mM KCl, 3.6 mM $NaHCO_3$, 1.3 mM $CaCl_2$, 5.6 mM glucose and 10 mM HEPES. Cytotoxicity was determined by continuous observation using phase contrast microscopy.

For morphological and $[Ca^{2+}]_i$ assays, cultures were prepared according to the method of Hirschberg et al., (1996) with some modifications. Briefly, fetal rat hippocampus (gestation day 18–19) were dissociated by trypsinization (0.25% for 15 min at 37°C). The tissue was washed in magnesium/calcium-free Hank's balanced salt solution (Gibco) and dissociated by repeated passage through a constricted Pasteur pipette. Cells were plated in minimal essential medium (MEM) with 10% heat inactivated horse serum, at a density of 240,000/24-mm glass coverslip (Chance Proper, Warley, UK) that had been precoated with poly-L-lysine (1 mg/ml). After allowing 2–4 hr for the cells to adhere to the substrate, coverslips were transferred into 24-well multidishes /100 mm-dish (Nunc) containing a monolayer of astroglia (Hirschberg et al., 1996). Neurons were placed with cells facing downward and were separated from the glia by paraffin "feet" as described (Goslin and Banker, 1991). Cultures were maintained in a serum-free medium (MEM) which included the N2 supplements (Goslin and Banker, 1991), ovalbumin (0.1%, w/v), and pyruvate (0.1 mM).

2.2. Morphological Assessment

Cultures were visualized and photographed with a phase contrast Nikon inverted microscope. Viable neurons exhibited pyramidal to oval perikaria and well distinguished nerve fibers. Damaged neurons presented a swollen, granulated and vacuolated soma and the nerve processes appeared retracted and beaded. The degenerated neurons tended to detach from the culture substrate.

Scanning electron microscopy was made on cultures grown on 12-mm round coverglasses. Cultures were fixed in 2.5% gluteraldhyde for 1hr at 4°C, washed 3 times in phosphate buffer saline (PBS) and in distilled water and dehydrated in increments of ethanol. Critical-point drying was performed in a Polaron apparatus, followed by spattering with gold-palladium in a Polaron spattering unit. Observations were made in a Joel 35C at 25 kV.

2.3. Immunocytochemistry

Hippocampal dissociated cultures were grown on 12-mm cover slides pretreated and coated with poly-L-lysine and maintained in culture for 12–14 days (Shahar et al., 1989). After 10 min of preincubation in LH buffer containing the experimental antagonist, the cells were washed and incubated with various ligands for up to 2 hr. Cultures were fixed for 15–20 min in 4°C with 4% paraformaldehyde in PBS pH=7.4. The cells were permeabilized by exposure to 0.3% solution of Triton X-100 in PBS for 5 min, and the non-specific staining was blocked by 10% normal goat serum for 30 min. Subsequently cells were incubated overnight at 4°C in PBS containing the primary antibody. For enzyme-linked immunostaining the cells were incubated with neuron specific enolase (NSE) primary antibodies and were further processed using ABC biotin-avidin kit (Zymed Labs) with 3',3-diaminobenzidine tetrahydrochloride as a substrate. For radio-immuno assay (RIA), a commercial glial fibrillary acidic protein (GFAP) and neurofilament (NF) first antibodies (Boerhinger Menhiem) were used at a 1:200 and 1:50 dilutions (respectively) in PBS containing 0.02% Tween-20. Cells were further processed using secondary ^{125}I-rabbit IgG (1:100) for 2 hr at 37°C, and the cover slides were washed extensively with cold PBS three times and the radioactivity in the attached cells was measured by liquid scintillation counting.

2.4. Measurement of $[Ca^{2+}]_i$

2.4.1. Fura-2 Indicator. Cells were plated on 24 mm round glass cover-slips coated with 0.01 % poly-L-lysine (Hirschberg et al., 1996). On day 14 *in vitro*, cells were washed twice in LH buffer, and loaded with 5 µM Fura-2AM, and 1.5 µM pluronic acid (Molecular Probes, Eugene, OR) in LH buffer at 37°C for 30–45 min, in the dark. Loading was followed by three washes in LH buffer, and the neuronal cells were randomly viewed on an inverted epiflurescence microscope (Ziess, Axiovert 135M; Fluor 63 / 1.3 oil objective) to which an intensified couple-charge device C2400 (ICCD) camera (Hamamatsu, Japan) was mounted. Images were obtained at two excitation wave-lengths alternating between 340 and 380 nm using a frame grabing software (Galai, Israel). The ratio (340:380) of the emission intensities at 520 nm (after background subtraction) is proportional to $[Ca^{2+}]_i$. Because of our primary interest in relative rather than absolute changes in $[Ca^{2+}]_i$ levels, values are expressed at 340:380 nm fluorescence ratios. Calcium concentrations were computed using an equation according to Grynkiewicz et al., (1985), and the values given for $[Ca^{2+}]_i$ should be regarded as quantitated calcium levels. Imaging experiments were conducted at room temperature in a 1 ml static bath containing the ligands. After obtaining baseline measurements the ligands were added and ratio images were obtained at the indicated times. The morphological changes of the selected cells were evaluated by phase contrast images.

2.4.2. Indo-1 Indicator. On day 14 *in-vitro*, cells were washed twice in LH buffer, and loaded with 3 µM Indo-1/AM, and 1.5 µM pluronic acid (Molecular Probes, Eugene, OR) in LH buffer at 37°C for 30–45 min, in the dark. Loading was followed by three washes in LH buffer, and neuronal cells were randomly viewed on an inverted microscope (Ziess, Axiovert 135M) to which two phoyomultipliers were mounted. Exitation of indo-1 was at 355 nm with 75w Xe lamp. Emission was detected at 2 wavelength 405 nm for Ca^{2+}-liganded indo fluorescence and 495 nm for Ca^{2+}-free indo fluorescence and the ratio is presented in the data. Following acquisition of images after 30 min of incubation at room temperature in a 1 ml static bath control media.

2.5. Quantitative Neurotoxicity Studies

2.5.1. [^{3}H]-Arachidonic Acid (AA) Release. The present procedure was based on the methods described previously (Sanfeliu et al., 1990; Lazarewicz et al., 1990). Briefly, dissociated hippocampal neuronal cell cultures (Shahar et al., 1989) (14 DIV) were labeled for 18 hr with 0.1 µCi/ml [5,6,8,9,11,12,14,15 -^{3}H] arachidonic acid [^{3}H]AA (100 Ci/mmol), Dupont / New England Nuclear (Boston, MA). The cells were washed three times with LH buffer containing 0.1% fatty acid-free bovine serum albumin (BSA) for 10 min. After 10 min of preincubation in LH buffer containing the tested antagonists, the cells were washed and incubated with various ligands for up to 2 hr. The medium samples were collected and centrifuged at 6,000 x g for 10 min in 4°C, to remove possible contamination by cell debris. The radioactivity present in the supernatant was measured by liquid scintillation counting.

2.5.2. XTT-Based Assay. The proportion of cultured hippocampal neurons survival was assessed by measuring the extent of mitochondrial activity in living cells using the XTT-based assay. XTT (2,3-bis[2-Methoxy-4-nitro-5-sulfophenyl]-2H-tetrazolium-5-carboxanilelide inner salt) is reduced by mitochondrial dehydrogenase to a soluble colored formazan product. The intensity of color formation (O.D.) is proportional to mitochondrial activity. Dissociated hippocampal cells were plated at a density of 75,000 cells per well of 96 well culture plate and were grown for 14 day *in-vitro* at the same conditions and media described previously. Following exposure to various treatments, cells were incubated with XTT for 4 hr at 37°C and the O.D. at 580 nm was measured by a Elisa reader plate.

2.5.3. Lactate Dehydrogenase (LDH) Release. Cytotoxicity was assessed by measuring LDH activity in 200–500 µl aliquots of culture medium according to the method of Koh and Choi (1987), using an assay kit (Sigma, 340-LD). LDH release was expressed as a percentage of total LDH activity in aliquots of medium taken after freezing/thawing of the same cultures.

2.6. Statistics

All assays were routinely carried out in 3–4 replicates, and the results were expressed as the mean ± SEM. Data presented in the figures are from representative experiments which were repeated at least three times. The statistical comparisons were done using paired and unpaired Student's t-tests (two-tailed) and one-way analysis of variance (ANOVA).

3. RESULTS

3.1. Nerve Cell Insult

After exposure to lasalocid (1 µM) for 1 hr, hippocampal neurons (14 days *in-vitro*) became swollen and vacuolated exhibiting picnotic nuclei and retracted beaded nerve fibers (Fig. 1B,E). Most of the neurons were degenerated after 1–2 hr and detached from the substrate. No significant damage could be observed in the glial and other non-neuronal cells. Exposure of cultures to lasalocid (1.5 µM) induced a decrease (of 40–75%) in the number of neuron specific enolase-positive cells already after 30–60 min. The reduction in the number of NSE-stained neurons was due to detachment of the degenerated neurons from the culture substrate. The concomitant exposure of cultures to both lasalocid (1.5 µM) and MK-801 (10 µM) resulted in protection of nerve cell degeneration as expressed by increase in number of NSE-positive cells in the cultures (Fig. 2A). Further evaluation of the lasalocid-mediated selective toxicity in neuronal cell population was determined by radio-immuno assay. Neurofilament (NF) was used as a specific marker for neurons, and the glial fibrillary acidic protein (GFAP) was used as marker for astrocytic

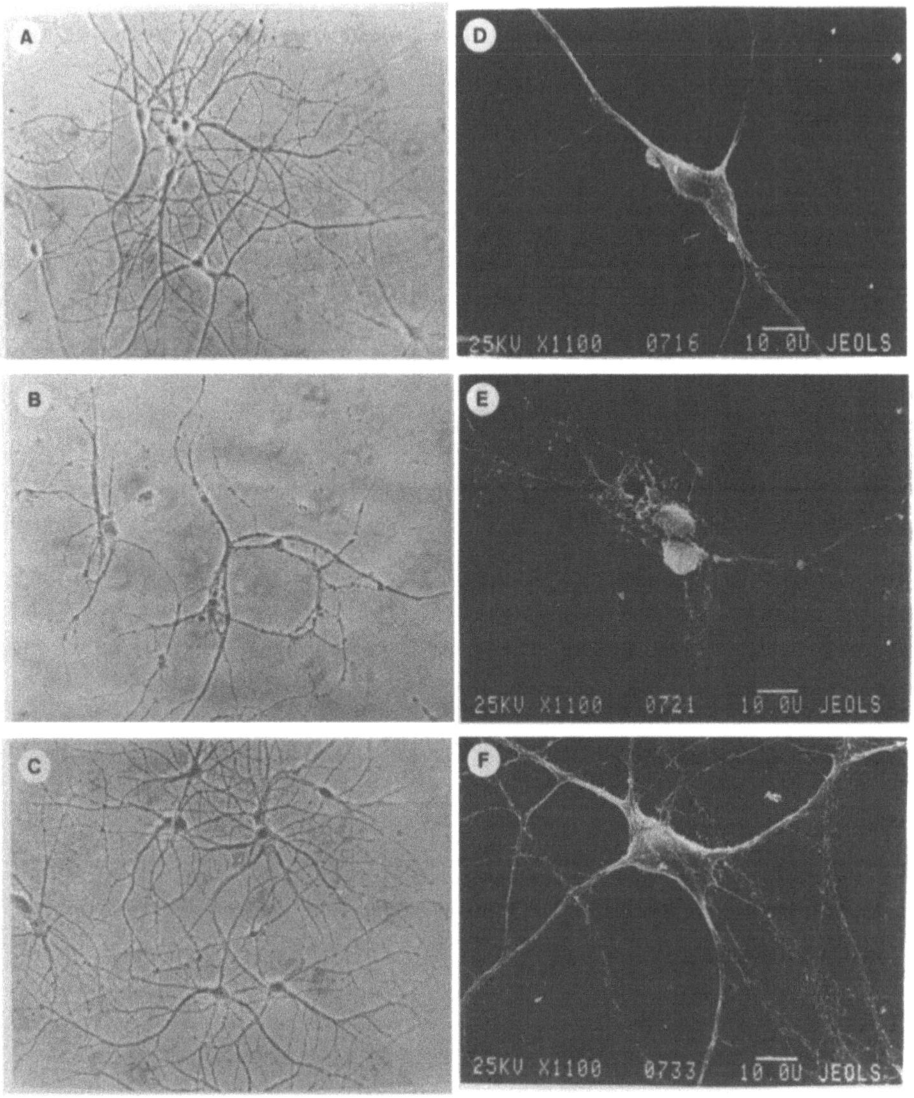

Figure 1. Phase-contrast micrographs (A,B,C) (x 100) and scanning electron micrographs (D,E,F) (x 550) of neuronal hippocampal cells. A,D - Control cultures in 0.01% DMSO; B,E - Lasalocid (1 μM); C,F - Lasalocid (1 μM) and MK-801 (10 μM). Note the neurodegenerative changes in B,E following exposure of 1hr to lasalocid and the neuroprotection by MK-801 in C,F.

cells. After 1 hr exposure to lasalocid (1 μM), a marked decrease of the neuronal population was observed as expressed by decrease in labeling for the neuronal marker NF (fig. 2B). In contrast, the astrocytic glial cells were not damaged following lasalocid treatment as shown by the consistent level of the GFAP stain (Fig. 2B). Treatment of hippocampal cultures with lasalocid (1 μM) for 30 min caused a significant increase in $[Ca^{2+}]_i$ in neurons but not in glial cells. The $[Ca^{2+}]_i$ increase in neuronal cells extended to the maximal level after 15 min of exposure, while there was no change in $[Ca^{2+}]_i$ in the glial cells (Fig. 3A). Furthermore, as can be seen in fig. 3B, the neuronal $[Ca^{2+}]_i$ was not increased by

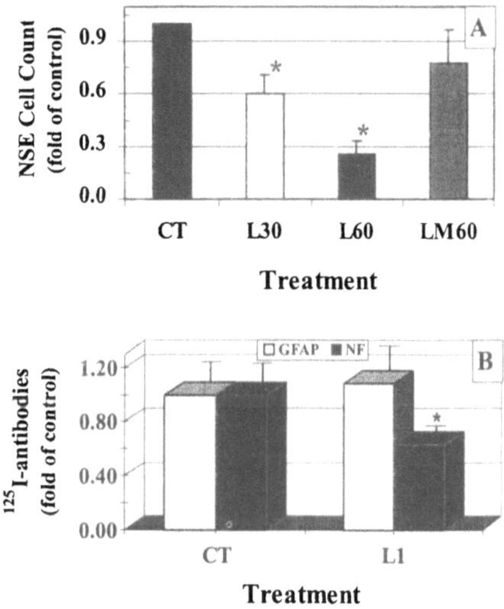

Figure 2. (A) Immunocytochemical staining of hippocampal neurons for neuron specific enolase (NSE). CT - control in 0.01% DMSO; L30, L60 - Cultures exposed to lasalocid (1.5 μM) for 30, 60 min; LM60 - Co-incubation of lasalocid (1.5 μM) and MK-801 (10 μM) for 60 min. Values represent the mean and SEM of determinations made in 5 fields from each of 4 separate cultures from 3 separate experiments. The values were expressed as the fold of control from the same experiment. *$P < 0.01$ compared with the values of the control. (B) Radio-immuno assay was performed using NF (1:50) and GFAP (1:200) antibodies in: CT - control cultures in 0.01% DMSO; L1 - lasalocid 1μM. Values represent the mean and SEM of 4 replicates from three separate experiments following the subtraction of non-specific binding (without first antibodies). * $p < 0.001$ compared with control.

lasalocid incubated in Ca^{2+} free medium. This suggests that the increase of intracellular Ca^{2+} was mediated by Ca^{2+} influx (Fig. 3B).

3.2. Nerve Cell Protection

In order to prevent the lasalocid-mediated neurotoxicity we have studied the ability of VOCC's and AOCC's blockers to attenuate the Ca^{2+} influx and the subsequent neurodegeneration. The lasalocid induced Ca^{2+} influx and elevation of $[Ca^{2+}]_i$ in hippocampal neuronal cells were blocked by the AOCC's antagonist MK-801(10 μM) (an NMDA receptor/channel antagonist). In contrast, the VOCC's antagonist nimodipine (10 μM), was not effective (Fig. 4). CNQX (10 μM) which is another AOCC's antagonist (an non-NMDA receptor/channel antagonist) only slightly inhibited the increase in $[Ca^{2+}]_i$, while co-incubation of CNQX and MK-801 totally blocked the Ca^{2+} influx (Fig. 4). In addition to the morphological neurotoxicity and the selective neuroprotection previously described (Fig. 1, 2), the differences in Ca^{2+} influx induction and $[Ca^{2+}]_i$ increase were correlated with the activation of PLA_2 and release of ^{3}H-AA to the medium (Fig. 5A). The neuronal survival differences were quantitated biochemically by both XTT method (Fig. 5B) and LDH release methods (Fig. 5C). In correlation with the $[Ca^{2+}]_i$ determination assays, the selective neurodegenerative processes caused by lasalocid and NMDA could be protected by the NMDA receptor/channel antagonist MK-801, in contrast to nimodipine that was unable to protect the cells and even exacerbates the lasalocid-induced damage. In agreement with the $[Ca^{2+}]_i$ determination described previously, the non-NMDA receptor/channel antagonist CNQX (10 μM) only partially inhibited the neurotoxic changes caused by lasalocid as can be determined by LDH release activity and ^{3}H-AA release to the medium (Fig. 6). The lasalocid-mediated neuronal damage was entirely blocked by the simultaneous exposure of cultures to both MK-801(10 μM) and CNQX (10 μM) (Fig. 6). The lasalocid-induced selective neurodegeneration and the MK-801-related neuroprotection are not common features of Ca^{2+} ionophores. Exposure of dissociated hippocampal cells to another calcium ionophore - calcimycin (A23187), caused dose-dependent degeneration of neuronal and non-neuronal cells. The glia and other non-neuronal elements, showed

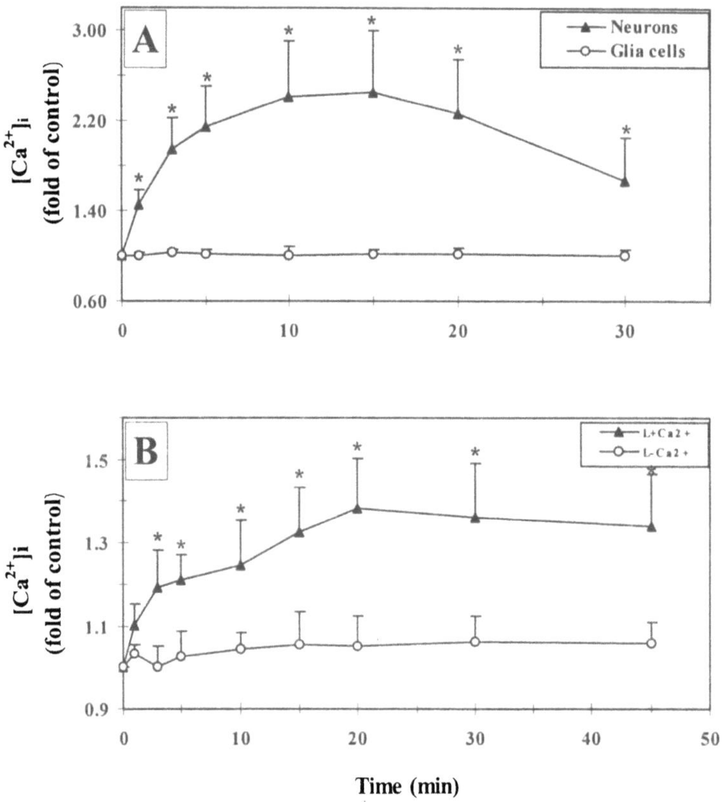

Figure 3. (A) Selective increase of $[Ca^{2+}]_i$ in cultured hippocampal neurons exposed to lasalocid (1 μM), as determined by Fura-2AM indicator. Values at the indicated time are expressed as fold of the $[Ca^{2+}]_i$ measured before lasalocid exposure and represent the mean and SEM of 6 neuronal cells and 7 glia cells from a representative experiment that was repeated at least 3 times. *$p < 0.001$ compared with glial cells. (B) Measurements of $[Ca^{2+}]_i$ in lasalocid-exposed hippocampal neurons incubated in the presence (1 mM) or absence of Ca^{2+} in the medium as determined by Indo-1 indicator. Values at the indicated time are expressed as fold of the $[Ca^{2+}]_i$ measured before lasalocid exposure and represent the mean and SEM of 12 neuronal cells from 4–5 separate fields of a representative experiment that was repeated 3 times. * $p < 0.01$ compared with Ca^{2+} free medium.

marked granulation and vacuolization of cell bodies which resulted in complete cell degeneration. The two ionophores, calcimycin and lasalocid induced Ca^{2+} influx, $[Ca^{2+}]_i$ increase (Fig. 7A), activation of PLA_2 and [^{3}H]-AA release (Fig. 7B, C). However, MK-801 (10 μM) blocked the the lasalocid-mediated $[Ca^{2+}]_i$ increase and [^{3}H]-AA release at high lasalocid concentration (3 μM) (Fig. 7B), but was not potent in protecting the cells even very low calcimycin concentrations (0.05–1μM) (Fig. C).

4. DISCUSSION

In a previous study we have shown that lasalocid caused selective cytotoxicity in primary cultures of cerebral neurons, but not in cultured fibroblasts, striated muscle, myoblasts or dissociated heart myoctes (Safran et al., 1993b). This selective lasalocid-me-

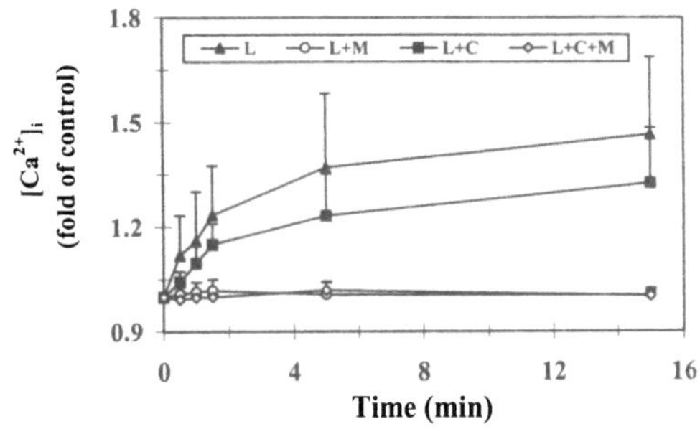

Figure 4. Measurements of $[Ca^{2+}]_i$ as determined by Indo-1 indicator, in hippocampal neurons exposed to: [L]-lasalocid (1 μM); [L+M] - lasalocid (1 μM) and MK-801(10 μM); [L+D] - lasalocid (1 μM) and nimodipine (10 μM); [L+C] - lasalocid (1 μM) and CNQX (10 μM); [L+C+M] - lasalocid (1 μM) and CNQX (10 μM) and MK-801(10 μM). Values are expressed as fold of the $[Ca^{2+}]_i$ measured before each ligand exposure and represent the mean and SEM of 15 neuronal cells from 4–5 separate fields of a representative experiment that was repeated 3 times. * $P < 0.001$ for (L),(L+D), (L+C) compared with (L+M),(L+C+M), ** $P < 0.05$ for (L+C) compared with (L),(L+D).

diated toxicity was also observed in neuroblastoma cell-line cultures. However, cultures of non-neuronal cell lines (CG6, PC12) did not express cellular damage after exposure to lasalocid (Safran et al., 1993b).

In the present study, we used both neuronal (Hirschberg et al., 1996) as well as mixed neuronal-glia (Shahar et al., 1989) cultures of hippocampus in order to answer the following questions: I. Why was lasalocid selectively neurotoxic to the cultured neuronal population and not to the glial cells? II. What is a possible neuroprotection mechanism for

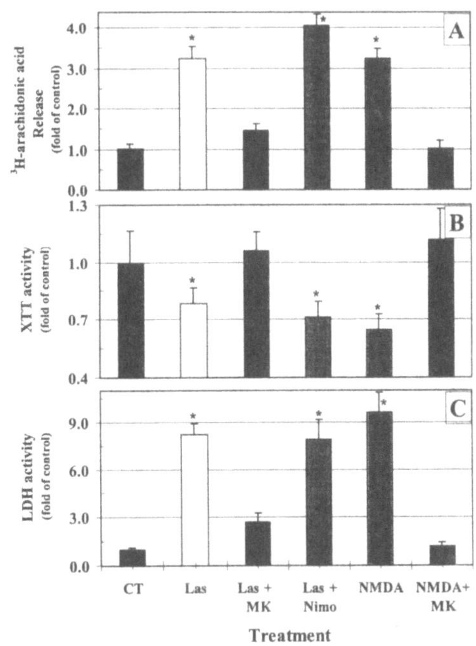

Figure 5. (A) The activation of PLA_2 as measured by ^{3}H-arachidonic acid release. (B) The survival of cells was determined using the XTT-based assay; (C) The neurotoxicity was quantitated by LDH activity released. CT - control in 0.01% DMSO; Las - lasalocid (1 μM); MK - MK-801 (10 μM); Nimo - nimodipine (10 μM); NMDA - N-methyl-D-aspartate (100 μM). For (A,C) Results are expressed as fold of control and represent the means and SEM from at least 3 experiments with 3 replicates. * $P < 0.001$ compared with control; For (B) Results are expressed as fold of control and represent the means and SEM from at least 2 experiments with 8 replicates. * $P < 0.01$ compared with control.

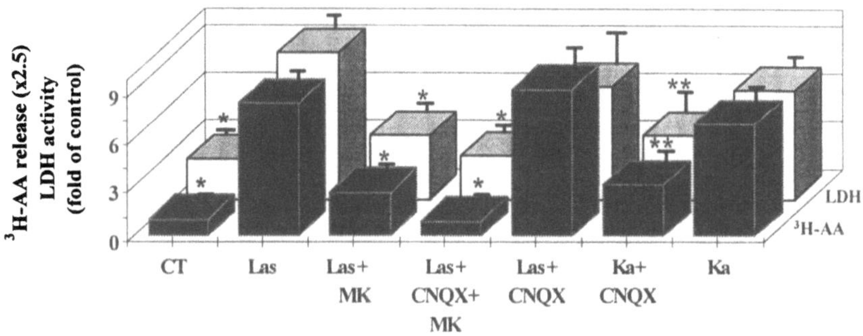

Figure 6. Measurements of LDH activity release and ³H-arachidonic acid release in hippocampal neurons exposed to various ligands: CT - control in 0.01% DMSO; Las - lasalocid (1 μM); MK - MK-801 (10 μM); CNQX - CNQX (10 μM); Ka - kainaic acid (100 μM). Results are expressed as fold of control and represent the means and SEM from at least 3 experiments with 3 replicates. * P < 0.0001 compared with lasalocid (Las) treated cultures. ** P < 0.001 compared with kainaic acid (Ka) treated cultures.

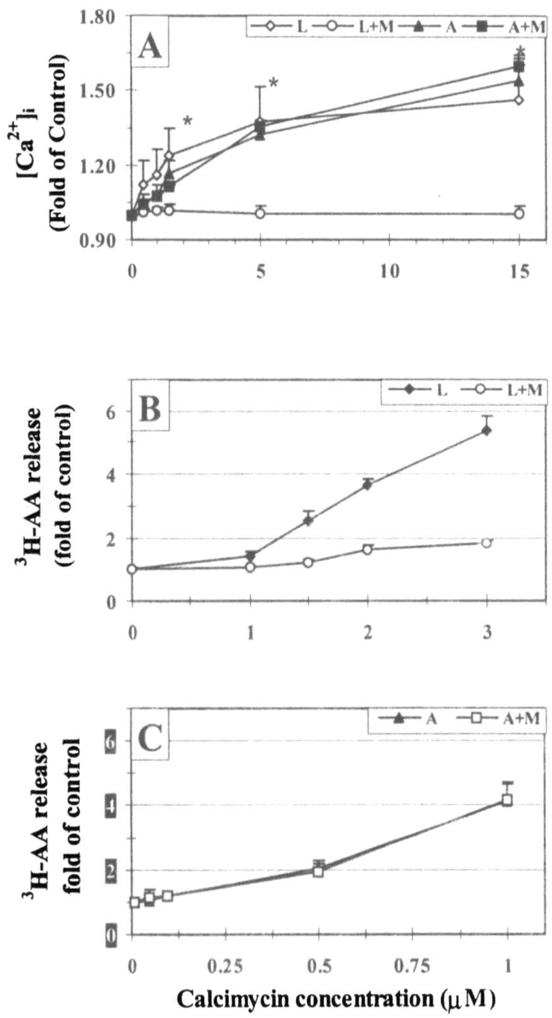

Figure 7. (A) Measurements of $[Ca^{2+}]_i$ increase in cultures loaded with the Indo-1 indicator. Values represent the mean and SEM of 12 to 15 neuronal cells from 4–5 separate fields of a representative experiment that was repeated 3 times. (B,C) Measurements of ³H-arachidonic acid release into the medium. Results are expressed as fold of control and represent the means and SEM from at least 3 experiments with 3 replicates. *P < 0.01 compared with other treatments of the same experiments at the indicated time.

the lasalocid insult? III. Is the mechanism by which lasalocid exerts neurotoxicity, similar to that of other Ca^{2+} ionophores?

Treatment of hippocampal cultures with lasalocid caused a time dependent degenerative process of the neuronal population as expressed by NSE positive cells, while the non-neuronal cells that were NSE negative were not damaged. The use of the neuronal markers (NSE, NF) and the astrocytic marker (GFAP) had also shown a significant decrease in the number of surviving nerve cells due to detachment of most degenerated neurons. The astrocytic population remained intensively labeled and undamaged. Furthermore, selective elevation of $[Ca^{2+}]_i$ and cytotoxicity were observed only in lasalocid-exposed hippocampal neurons but not in glia cells. Previous studies have shown concentration-dependent $^{45}Ca^{2+}$ influx only in nerve cells but not in non-neuronal elements (Safran et al, 1993b). These findings suggest that the mechanism of lasalocid neurotoxicity involves a specific entity that is not present in non neuronal cells.

Lasalocid is a polyether ionophore compound that facilitates the passage of Ca^{2+} through lipophilic-biological membranes. Its inability to increase $[Ca^{2+}]_i$ and neurotoxicity in Ca^{2+} free medium, suggest that the mechanism of lasalocid-induced nerve cell injury involves Ca^{2+} influx. This phenomenon is similar to the NMDA toxicity, a mechanism that is attenuated by the omission of external Ca^{2+}. In contrast, the toxicity induced by quisqualate (a non-NMDA receptor agonist) is independent of the presence of external Ca^{2+} (Garthwaite and Garthwaite, 1989; Frandsen and Schousboe, 1992), and the kainaic acid-induced toxicity (another non-NMDA receptor agonist) is even exacerbated in Ca^{2+} free medium (Lehmann, 1990). In order to prevent the lasalocid-mediated neurotoxicity we have studied the ability of VOCC's and AOCC's blockers to attenuate the Ca^{2+} influx and the subsequent neurodegenerative disorders. Lasalocid-mediated Ca^{2+} influx, $[Ca^{2+}]_i$ increase, ^{3}H-AA release and the following neurotoxicity were not blocked by nimodipine (Ca^{2+} channel antagonist antagonist of VOCC's - which is present in both neuronal and non-neuronal cell types) or by CNQX (an antagonist of the non-NMDA receptors) (Honor'e et al., 1988). In contrast, MK-801, antagonist of the NMDA receptor/channel complex that is present only in neuronal cells, exclusively blocked the lasalocid-induced Ca^{2+} influx, elevation of $[Ca^{2+}]_i$ and completely prevented its neurotoxic insult. Frandsen and Schousboe (1992, 1993) reported that Ca^{2+} influx in neurons can be only slightly reduced by VOCC's blockers, in contrast to AOCC's antagonists that may completely block the influx mediated by EAA receptor agonists. The cytotoxicity induced by lasalocid seemed to be similar to that induced by NMDA in hippocampal neurons (Ellren and Lehmann, 1989). VOCC's blockers had no neuroprotective effect, whereas NMDA receptor antagonists completely blocked the neurotoxicity (Choi, 1985, 1987; Lehmann, 1987; Ellren and Lehmann, 1989; Frandsen and Schousboe, 1992, 1993). Therefore, it is likely that the mechanism of lasalocid-induced neurotoxicity involves Ca^{2+} influx through the NMDA receptor/channel and subsequent $[Ca^{2+}]_i$ increase leading to neurodegeneration. Previous studies showed that glial cells lack NMDA receptor (Teichberg, 1991) and did not show Ca^{2+} uptake responses to NMDA (Usowicz et al., 1989; Ahmed et al., 1990; Holzwarth et al., 1994). These studies together with the selective lasalocid neurotoxicity, strongly support the assumption that lasalocid neurotoxicity is attributed to the involvement of NMDA receptor/channel that is exclusively present in neuronal cells and not in glial cells.

Elevation of intracellular calcium level plays a crucial role in regulating the activity of the NMDA receptor/channel (Markram and Segal, 1991). It has been suggested that elevated intracellular Ca^{2+} ions, triggered by normal excitation of NMDA receptors or by Ca^{2+} ionophores, may participate in activating PLA_2 resulting in the release of arachidonic acid in cultured neurons from striatum (Dumuis et al., 1988), cerebellum (Lazarewicz et al., 1988, 1990), hippocampus (Sanfeliu et al., 1990) and cerebral cortex (Tapia-Arancibia et al., 1992). Unlike lasalocid, exposure of dissociated hippocampal cells to another calcium ionophore - calcimycin (A23187), induced dose-dependent degeneration of all the neuronal and non-neuronal cultured cells. The two ionophores, calcimycin and lasalocid induced Ca^{2+} influx, $[Ca^{2+}]_i$ increase, activation of PLA_2, $[^3H]$-AA release and degeneration of the hippocampal nerve cells. However, while MK-801 attenuated $[Ca^{2+}]_i$ increase, $[^3H]$-AA release and the neurodegenerative process induced by lasalocid, the calcimycin-mediated Ca^{2+} influx, and the subsequent toxicity effects were not blocked by MK-801. In conclusion, the selective neurotoxicity induced by lasalocid is exerted by a mechanism that involves the NMDA receptor/channel complex, a feature that is not shared with other ionophores.

There is considerable evidence that implicates the NMDA receptor/channel activity in acute and chronic neurodegenerative diseases including ischemia, stroke, Alzheimer and Parkinson diseases, AIDS-associated dementia etc. (Meldrum and Garthwaite, 1990, 1991; Murphy et al., 1990, 1991; Frandsen and Schousboe, 1993; Farooqui and Horrocks, 1994). The specific mode of action of lasalocid might serve as research model for further studies on neuronal insult and neuroprotection.

REFERENCES

Aebi R. (1989) Coccidiosis prevention with lasalocid. Zootecnica Int 3, 31–35.

Ahmed Z. Lewis C. A. Faber D. S. (1990) Glutamate stimulates release of Ca^{2+} from internal stores in astroglia. Brain Res 526, 165–169.

Bergen W. G. and Bates D. B. (1984) Ionophores: Their effect on production efficiency and mode of action. J Ani Sci 58, 1465–1483.

Berger J. Rachlin A. I. Scott W. E. Sternbach L. H. and Goldberg M. W. (1951) The isolation of three new crystalline antibiotics from streptomyces. J Am Chem Soc 73, 5295–5298.

Choi D. W. (1985) Glutamate neurotoxicity in cortical cell culture is calcium dependent. Nerosci Lett 58, 293–297.

Choi D. W. (1987) Ionic dependence of glutamate neurotoxicity. J of Neurosci 7, 369–379.

Dumuis A. Sebben M. Haynes L. Pin J. P. Bockaert J. (1988) NMDA receptors activate the arachidonic acid cascade system in striatal neurons. Nature 336, 68–70.

Ellren K. and Lehmann A. (1989) Calcium dependency of N-methyl-D-aspartate toxicity in slices from immature rat hippocampus. Neuroscience 32, 371–379.

Farooqui A. A. and Horrocks L. A. (1994) Involvement of glutamate receptors, lipases, and phospholipases in long-term potentiation and neurodegeneration. J Neurosci Res 38, 6–11.

Frandsen A. and Schousboe A. (1992) Mobilization of dantrolene-sensitive intracellular calcium pools is involved in the cytotoxicity induced by quisqualate and N-methyl-D-aspartate but not by (RS)-2-amino-3-(3-hydroxy-5-methylisoxazol-4-yl) propionate and kainate in cultured cerebral cortical neurons. Proc Natl Acad Sci USA 89, 2590–2594.

Frandsen A. and Schousboe A. (1993) Excitatory amino acid-mediated cytotoxicity and calcium homeostasis in cultured neurons. J Neurochem 60, 1202–1211.

Garthwaite G. and Garthwaute J. (1989) Differential dependence on Ca^{2+} of N-methyl-D-aspartate and quisqualate neurotoxicity in young rat hippocampal slices. Neurosci lett 97, 316–322.

Goslin K. and Banker G. (1991) Culturing nerve cells, pp. 251–281, MIT Press, Cambridge.

Grynkiewicz G. Poenie M. Tsien R. Y. (1985) A new generation of Ca^{2+} indicators with greatly improved fluorescence properties. J Biol Chem 260, 3440–3450.

Hirschberg K. Zisling R. Echten-Deckert G. and Futerman A.H. (1996) Ganglioside synthesis during the development of neuronal polarity. J Biol Chem 271, 14876–14882.

Holzwarth J. A. Gibbons S. J. Brorson J. R. Philipson L. H. Miller R. J. (1994) Glutamate receptor agonists stimulate diverse calcium responses in different types of cultured rat cortical glial cells. J Neurosci 14, 1879–1891.

Honor'e T. Davies J. N. Drejer J. Fletcher G. J. Jacobson P. Lodge D. Neilsen F. E. (1988) Quinoxalinediones: potent competitive non NMDA glutamate receptor antagonist. Science 241, 701–703.

Koh J. Y. and Choi D. W. (1987) Quantitative determination of glutamate mediated cortical neuronal injury in cell culture by lactate dehydrogenase efflux assay. J Neurosci Meth 20, 83–90.

Lazarewicz J. W. Wroblewski J. T. Palmer M. E. Costa E. (1988) Activation of N-methyl-D-aspartate-sensitive glutamate receptors stimulates arachidonic acid release in primary cultures of cerebellar granule cells. J Neuropharmacology 27, 765–769.

Lazarewicz J. W. Wroblewski J. T. Costa E. (1990) N-methyl-D-aspartate-sensitive glutamate receptors induce calcium-mediated arachidonic acid release in primary cultures of cerebellar granule cells. J Neurochem 55, 1875–1881.

Lehmann A. (1987) Pharmacological protection against the toxicity of N-methyl-D-aspartate in immature rat cerebellar slices. Neuropharmacology 26, 1751–1761.

Lehmann A. (1990) Kainaic acid neurtoxicity in slices from the immature rat hippocampus: protection by chloride reduction and exacerbation by calcium omission. Neurosci Res Commun 6, 27–36.

Leonard J. P. and Salpeter M. M. (1979) Agonist-induced myopathy at the neuromuscular junction is mediated by calcium. J Cell Biol 82, 811–819.

Markram H. and Segal M. (1991) Calcium potentiates responses of rat hippocampal neurons to N-methyl-D-aspartate. Brain Res 540, 322–324.

Murphy S. N. and Miller R. J. (1988) A glutamate receptor regulates Ca^{2+} mobilization in hippocampal neurons. Proc Natl Acad Sci USA 85, 8737–8741.

Murphy S. N. and Miller R. J. (1989) Two distinct quisqualate receptors regulate Ca^{2+} homeostasis in hippocampal neurons in vitro. Mol Pharmacol 35, 671–680.

Ogura A. Akita K. and Kudo Y. (1990) Non NMDA receptor mediates cytoplasmic Ca^{2+} elevation in cultured hippocampal neurons. Neurosci Res 9, 103–113.

Perelman B. Abarbanel J.M. Gur-Lavie A. Meller Y. Elad T. (1986) Clinical and Pathological changes caused by the interaction of lasalocid and chloramphenicol in broiler chickens. Avi Path 15, 279–288.

Reed P. W. (1992) Biochemical and biological effect of carboxylic acid ionophores. In: Polyether Antibiotics, Naturally Occurring Acid Ionophores. Edited by J. W. Westley. Vol. I: Biology. pp 185–302. Marcel Dekker, Inc., New York.

Safran A. Aizenberg I. and Bark H. (1993a) Paralytic syndrome in dogs caused by lasalocid residues in a commercial ration. J Am Vet Med Ass 202, 1273–1275.

Safran N. Shainberg A. Haring R. Gurwitz D. Shahar A. (1993b) Selective neurotoxicity induced by lasalocid in dissociated cerebral cultures Toxic *in vitro* 7, 345–352.

Safran N. Haring R. Gurwitz D. Shainberg A. Halili I. Levy A. Bogin E. Shahar A. (1996). Selective neurotoxicity induced by the ionophore lasalocid in rat dissociated cerebral cultures, involvement of the NMDA receptor/channel. NeuroToxic. 17,.

Sanfeliu C. Hunt A. and Patel A. J. (1990) Exposure to N-methyl-D-aspartate increases release of arachidonic acid in primary cultures of rat hippocampal neurons and not in astrocytes. Brain Res 526, 241–248.

Schanne F. A. X. Kane A.B. Young E. E. and Farber J.L. (1979) Calcium dependence of toxic cell death: a final common pathway. Science 206, 700–702.

Shahar A. De Vellis J. Vernadakis A. and Haber B. (1989) A dissection and tissue culture manual of the nervous tissue. Alan R. Liss, Inc., New York.

Shier W. T. Angerhofer C. K. and Dubourdieu D. J. (1987) Role of stress in the initial injury stages of cell killing by altered intracellular calcium. Toxicol Lett 39, 283–293.

Tapia-Arancibia L. Rage F. R'ecasens M. Pin J. P. (1992) NMDA receptor activation stimultaes phospholipase A_2 and somatostatin release from rat cortical neurons in primary cultures. Eur J Pharmacol 225, 253–262.

Teichberg VJ. (1991) Glia glutamate receptors: likely actors in brain signaling. FASEB J 5, 3086–3091.

Usowicz M. M. Gallo V. Cull-Candy S. G. (1988) Multiple conductance channels in type-2 cerebellar astrocytes activated by excitatory amino acids. Nature 339, 380–383.

Westley J. W. Evans R. H. Williams T. and Stempel A. (1970) Structure of antibiotic X-537A. In: Chemical communication, p. 71. Burlington House, London.

Westley J. W. (1977) Polyether antibiotics: versatile carboxylic acid ionophores produced by streptomyces. Adv App Microb 22, 177.

16

IS INCREASED NEUROTOXICITY A BURDEN OF THE AGEING BRAIN?

Andrea Vaccari, PierLuigi Saba, Ignazia Mocci, and Stefania Ruiu

Department of Neuroscience "B. Brodie"
Neurotoxicology Unit
Via Porcell 4, 09124 Cagliari, Italy

1. THE "WOOD-WORM" HYPOTHESIS FOR ENVIRONMENTAL NEUROTOXICITY

Our living environment is a melting pot for more than 4.5 million natural and synthetic chemicals, which an individual may come in contact with at any given time: only 10,100 of them have been described in the *Merck Index (11th Ed.,1989)*. It is now thought that a lifetime's continuous exposure to trace amounts of endogenously formed and/or environmental toxins (such as industrial chemicals, pesticides, food additives, or abused and therapeutic drugs), may provoke neuronal degenerative events such as those occurring in Parkinson's and other diseases (Barbeau et al., 1987; Schoenberg et al., 1987a; Tanner, 1989; Koller et al., 1990; Calne, 1991; see Dawson et al., 1995). A process, in other words, similar to the continuous gnawing of small wood-worms which can lead to the destruction of even huge pieces of furniture. Epidemiologic and experimental evidence supports this hypothesis. In industrialized countries, during the last 100 years, there has been an unequivocal lengthening of the life expectancy of individuals, accompanied by a dramatic increase in the rate of neurodegenerative Parkinson's and Alzheimer's diseases (Lilienfeld et al., 1990; Rajput, 1992). Meanwhile, the threshold age for the onset of Parkinsonism has fallen (Schoenberg, 1987b; Tanner, 1989). On the other hand, in the developing countries less likely to suffer from environmental pollution, the incidence of Parkinson's disease is much lower (Schoenberg, 1987b), and Alzheimer's dementia is actually absent in Nigeria (Osuntokun et al., 1991). It is true that an increase in the average age, and thus the greater number of vulnerable individuals, may play an important role in increasing the incidence of neurodegenerative diseases in developed countries. While the existence of a clear correlation has been demonstrated between neuropathies and exposure to selected environmental toxicants (Barbeau et al., 1987; Rajput et al., 1987; Schoenberg, 1987a: Tanner, 1989; Calne, 1991; Semchuk et al., 1992; Dawson et al., 1995), there is only controversial evidence that genetic factors may help neurotoxicity (Golbe, 1990; Rajput, 1992; Vieregge, 1994). Thus, the *wood-worm* theory for neurotoxicity implies that

Brain Plasticity, edited by Filogamo *et al.*
Plenum Press, New York, 1997

the neuronal damage underlying Parkinson's, Alzheimer's and additional neurodegenerative diseases may be triggered by a number of environmentally-diffused organic and inorganic neurotoxins that display an unusually high affinity for selected brain regions and neuronal populations.

2. THE SYNAPTIC VESICLES AS A TARGET AND PUTATIVE CONTAINER FOR NEUROTOXINS

According to the *wood-worm* hypothesis for environmental neurotoxicity, the chronic exposure to very small and *per se* harmless concentrations of toxicants would also lead to the accumulation of toxically significant amounts of the agents in selected brain stores. Should the pools, thereafter, be suddenly depleted, or should their toxic contents even leak out slowly, a number of neuronal targets such as mitochondria might be affected, thus triggering the cascade of neurodegenerative events. Vesicle organelles in presynaptic nerve endings, the physiological sink for legitimate transmitters, have all the requisites needed to take up and release xenobiotics and toxicants, in at least two ways: by means of a carrier-dependent, energy-activated transport, or by carrier-independent, lipophilic diffusion (Johnson, 1988). Thus, monoamine plasma and vesicular membrane transporters can either vehiculate synthetic or natural analogues of the respective neurotransmitters, or be affected in their function by non-transported molecules (Johnson, 1988; Simantov, 1995). This imperfect selectivity implies that putative neurotoxins may "abuse" the transporter in order to enter the cell. Several established catecholaminergic toxins such as 6-hydroxydopamine and other hydroxylated derivatives of dopamine (Breese & Taylor, 1970; Slivka & Cohen, 1985), MPTP (1-methyl-4-phenyl-1,2,3,6-tetrahydro-pyridine) and MPP^+ (1-methyl-4-phenylpyridinium ion) (for a review see Tipton & Singer, 1993), and N-(2-chloroethyl)-N-ethyl-2-bromobenzylamine (DSP-4) (Ross, 1976), work in this way. It must be stressed, however, that the vesicular storage of toxins is not synonymous with neurotoxicity. On the contrary, it has been suggested that the internalization of toxins in organelles such as the chromaffin granules of Parkinson-resistant adrenals and, perhaps, in synaptic vesicles, in the absence of any consistent leakage, may act as a neuroprotective mechanism (Reinhard et al., 1990; Edwards, 1993).

We have here focused our attention on dopamine-rich striatal synaptic vesicles, considering them as a possible storage box for selected, transported toxins such as MPP^+, the extremely potent Parkinson-provoking oxidative metabolite of the meperidine analogue MPTP. Both toxins are known to destroy presynaptic dopamine terminals and dopaminergic cells (see Tipton & Singer, 1993) in the brain of rodents, non-human primates, and also humans. This disruption of dopaminergic pathways is reflected by dramatic reductions in striatal dopamine levels, tyrosine hydroxylase activity, and in the number of $[^3H]$mazindol-labelled, neuronal uptake sites for dopamine (see Ali et al., 1994). We have also identified the vesicular transporter for dopamine as a functionally relevant target for chemically heterogeneous, environmentally-diffused agents, which do not necessarily undergo vesicular accumulation.

3. THE EXPERIMENTAL MODEL

At the very beginning of the experiments summarized herein, the availability of a rapid, easy method helped us to screen putative neurotoxins for their interference with the membrane transport system for dopamine in striatal synaptic vesicles. We used the crude synaptosomal fraction as a source of broken and resealed synaptic vesicles, where the biogenic amine [^{3}H]tyramine, under appropriate assay conditions, mainly associates (85%) with the vesicular transporter for dopamine (Vaccari, 1986; 1993), and partially (15%) with the neuronal transporter and additional sites (Vaccari & Gessa, 1989). This presynaptically-located binding process is highly sensitive (K_i =6–30 nM) to reserpine (Vaccari, 1986) and tetrabenazine (Vaccari et al., 1991), two established markers of the vesicular transporter for monoamines, and poorly sensitive to cocaine, an inhibitor of the neuronal transporter (Vaccari et al., 1991). The monoamine vesicular transporter is known to exist in two different conformations, which preferentially bind reserpine and/or tetrabenazine (Darchen et al., 1989), and to which the ligand dihydrotetrabenazine has differential access (Naudon et al., 1996). Recently the two vesicular transporters vesicular monoamine transporter ($VMAT_1$) and $VMAT_2$ have been cloned (see Borowsky & Hoffman, 1995), the former being expressed in rat adrenals, and the latter in the brain only, where it is associated with small synaptic vesicles. Reserpine and ketanserin potently inhibit both transporters, and define a high-affinity monoamine uptake recognition site, and a lower affinity monoamine binding site that may release amines into the vesicle lumen, respectively (see Erickson et al., 1996). [^{3}H]Tyramine would bind at both conformational sites (of $VMAT_2$?) as indicated by downward curvilinear Scatchard plots of equilibrium binding assays, within a wide range of ligand concentrations. Consistently, both reserpine and tetrabenazine displace specifically bound [^{3}H]tyramine with two-site kinetics, and compete with the ligand for both high and low-affinity sites. [^{3}H]Dihydrotetrabenazine (Scherman et al., 1988; Henry & Scherman, 1989) was used to label the monoamine vesicular transporter in frozen cerebellar membranes.

4. THE [^{3}H]TYRAMINE-LABELLED VESICULAR TRANSPORTER FOR DOPAMINE IS SENSITIVE TO HETEROGENEOUS XENOBIOTICS

4.1. Parkinsogenic Toxins

The Parkinson-provoking toxin MPP^+ and the vesicular marker tetrabenazine are equipotent ($K_D \approx 30$ nM) at inhibiting competitively the striatal binding of [^{3}H]tyramine (Table 1). The interaction of MPP^+ analogues with the vesicular transporter obeys structural requirements which make the protoxin MPTP an approximately 120-fold less potent inhibitor of specifically-bound [^{3}H]tyramine, and the two bipyridyl herbicides paraquat (a "double" MPP^+) and diquat even less active compounds (Table 1). Several amino-aldehyde adducts putatively sharing neurotoxic properties in alcoholism (Collins, 1982; Myers, 1989), and the excitotoxins kainate and glutamate, display low affinity for the transporter. On the other hand, the established catecholaminergic toxins 6-hydroxydopamine and DSP-4, and the MPTP/reserpine analogue harmine, potently displace specifically bound [^{3}H]tyramine (Table 1). In more general terms, xenobiotics less eager to attach to the vesicular carrier can hardly be expected to accumulate in the synaptic organelles, though this does not hinder them from damaging the vesicles and additional targets. This appears to be particularly true for tetrahydroisoquinoline adducts originating from foods such as wine and cheese, whose suggested (Makino et al., 1988) putatively parkinsogenic accumulation in the central nervous system is not supported here. The demon-

Table 1. The [^{3}H]tyramine-labelled vesicular transporter for dopamine as a target of recognized and potential neurotoxins*

Compound	Affinity K_i (μM)
Caffeine	>100
β-Carboline	5.1
DSP-4	3.0
Harmine	1.2
6-Hydroxydopamine	1.3
Kainic acid	>50
Glutamic acid	>50
MPP^+	0.035
MPTP	4.2
Picrotoxin	>100
Salsolinol	>100
Tetrahydro-β-carboline	>50
1,4-Tetrahydroisoquinoline	>100
5,8-Tetrahydroisoquinoline	>100
Tetrahydropapaveroline	>100
Tryptophol	>100

*Data taken from Vaccari and Saba, 1995. Adult rat striatal membranes.

strated interaction of MPP^+ with site(s) located on the vesicular carrier for dopamine triggers two functional events, i.e. the energy-dependent accumulation of the toxin within striatal vesicles (Del Zompo et al., 1993), and the dramatic, mostly vesicular release of dopamine evoked by the toxin and tyramine *in vivo* (Vaccari et al., 1991). As a last consideration, both MPP^+ and tyramine also appear to label noradrenaline vesicles, with kinetic parameters differing from those in typically dopaminergic organelles (Vaccari et al., 1993; Vaccari, 1993). This finding strongly suggests that the vesicular carriers for dopamine and noradrenaline are different entities. In other words, the $VMAT_2$ transporter would be expressed in slightly different subtypes, which may underlie a regional specificity in the rate of the endovesicular sequestration of MPP^+, with either neurodegenerative or neuroprotective consequences, depending on the neurotransmitter characterization of the brain area involved (Vaccari et al., 1993).

4.2. Pesticides and Industrial Chemicals

A large number of miscellaneous environmental chemicals have been shown to interfere with the tyramine binding process, thus becoming candidates for disrupting either the vesicular transport function of neurotransmitters, or for entering the vesicles and constituting potentially toxic deposits. Among pesticides, several insecticide, herbicide and fungicide products display a wide range of affinities for the vesicular transporter for dopamine (Table 2). Organochlorine- and pyrethroid-insecticides, as well as dithiocarbamate fungicides, represent three chemical families endowed with overall high affinity (in the low micromolar or nanomolar range) for the vesicular transporter. Organophosphate- and carbamate insecticides, and heterogeneous herbicides, as a whole, are poorly active. The few organo-halogenated solvents tested, which have extensive applications such as soil fumigation, production of anti-knocking gasolines, dry-cleaning fluids, plastics etc., have K_i values exceeding 100 μM (Table 3), which denotes low affinity for the vesicular transporter. The only exception is the polychlorobiphenyl Arochlor 1254: since

Table 2. The [^{3}H]tyramine-labelled vesicular transporter for dopamine as a target of miscellaneous pesticides*

Compounds	Affinity K_i(μM)	Compounds	Affinity K_i (μM)
Insecticides		Fungicides	
(*pyrethroids*)		(*dithiocarbamates*)**	
Allethrin	5.1	Ferbam	6.8
Deltamethrin	0.15	Mancozeb	0.84
Fenvalerate	0.44	Maneb	9.1
Permethrin	1.9	Propineb	9.9
Pyrethrins	2.1	Thiram	3.1
Resmethrin	2.7	Zineb	0.45
		Ziram	0.87
(*organochlorines*)			
Chlordecone	1.1	(*miscellaneous*)**	
DDT	2.6	Biphenyl	38
Dieldrin	2.9	Copper oxychloride	0.83
Lindane	>10	Copper sulphate	1.0
Methoxychlor	2.9	Dichloran	1.2
		PCNB	6.2
(*organophosphates*)**			
Chlorpyriphos	7.8	Herbicides	
Fenithrothion	38	(*miscellaneous*)**	
Phenthoate	6.7	Chlorpropham	>10
		4,6-Dinitro-*o*-cresol	0.74
		Linuron	45
		Paraquat	>10
		Propachlor	>10
		Trifluralin	7.0

*Data taken from Vaccari and Saba, 1995; Vaccari et al, 1996. Adult rat striatal membranes.
**The following pesticides tested displayed low affinity for the [^{3}H]tyramine site, yelding K_i values >100 μM. INSECTICIDES: carbaryl, dichlorvos, dimethoate, methamidophos, nicotine, paraoxon, parathion, propoxur, TEPP. HERBICIDES: alachlor, alloxydim, amitrole, atrazine, barban, dalapon, 2,6-dichlorobenzonitrile, 2,4-D, diquat, glyphosate, 2,4,5-T. FUNGICIDES: chlorothalonil, diethyldithiocarbamate, methyl-isothiocyanate, metham, nabam, thiabendazole.

the estimated dietary intake of this hepatotoxic environmental pollutant in humans is set at 70–200 μg/day, it might be suspected to evoke neurotoxicity as well. Most of the presently tested inorganic and organic metal salts, the parkinsogenic manganese chloride included, act poorly on tyramine binding (Table 3), whereas organic tin derivatives, sodium dichromate and copper sulfate are potent binding inhibitors.

5. THE INFLUENCE OF AGEING ON NEUROTOXICITY

The ageing process, at least in rodents, involves the genetically-programmed progressive deterioration of most of the central neurotransmitter pathways, an effect coinciding with, but not necessarily correlated to, the recognized hypersensitivity to neurotoxins displayed by the aged central nervous system (Dawson et al., 1995). As a matter of fact, older rodents are dramatically more susceptible to the neurotoxic effects of several chemicals, such as the parkinsogenic agents MPTP and its metabolite MPP^+, than are younger animals (Jarvis & Wagner, 1985; Gupta et al., 1986; Ricaurte et al.,

Table 3. The [^{3}H]tyramine-labelled vesicular transporter for dopamine as a target of heterogeneous industrial products and metal ions*

Compounds	Affinity K_i (μM)
Solvents	
Arochlor 1254	2.7
1,2-Dibromo-3-chlorpropane	>100
1,2-Dibromoethane	>100
1,2-Dichloroethylene	>100
1,2-Dichloropropane	>100
1,3-Dichloropropene	>100
1,1,2,2-Tetrachloroethane	42
1,1,1-Trichloroethane	>100
Vinylidene chloride	>100
Metal ions	
$CuSO_4$	1.0
Trimethyltin	1.2
Dimethyltin	2.5
Sodium dichromate	2.7
$PbSO_4$	15
$AlCl_3$	>50
$NiCl_2$	>50
$ZnSO_4$	>100
$MnCl_2$	>100
$CdCl_2$	>100
$BaCl_2$	>100
LiCl	>100
$LaCl_3$	>100
RbCl	>100

*Adult rat striatal membranes.
Triplicate membrane aliquots (approx. 100 μg protein) were incubated for 10 min at 37°C with 4 nM [^{3}H]tyramine (spec.activity ≈ 45 Ci/mmol, ARC, St.Louis, MO) in the absence or presence of 10 concentrations of competing drugs. Dopamine 10 μM was used for measurement of non-specific binding (Vaccari, 1986; Vaccari et al., 1993). Values are means from 2 experiments.

1987a; Irwin et al., 1988; Finnegan et al., 1995). Thus, MPTP treatment in 21 month-old mice markedly decreased the presence and intensity of fluorescence in dopaminergic and noradrenergic neurons, extensively damaged the s.nigra, and also provoked physical disability in aged mice, much less evidently than in young MPTP-intoxicated animals (Gupta et al., 1986). These findings have generated a lot of interest in the underlying mechanisms, since the risk of developing Parkinson's disease increases with advancing age (Rajput, 1984). An additional ethiogenetic factor in the age-associated impairment of selected cognitive and motor functions might also involve oxidative molecular damage within different regions of the brain (Forster et al., 1996) caused by free radicals produced during toxin metabolism. Aged animals are supposed, indeed, to form free radicals more easily than their younger counterparts. Thus, it is interesting that in year-old mice, a single dose of MPTP caused early and lasting increases in reactive oxygen species production, and a subsequent fall in striatal dopamine concentrations. On the other hand, younger mice were protected from MPTP-neurotoxicity, without any significant modification in free radical concentrations or in dopamine homeostasis (Ali et al., 1994). The latter finding might be explained by the fact that older animals demonstrate increased monoamine oxidase (MAO-A and MAO-B) activities (Walsh & Wagner, 1989; Irwin et al., 1992; Finnegan et al., 1995). MAO's produce hydrogen peroxide during MPTP transformation to MPP^+, and the subsequent metal-catalyzed production of the highly toxic hydroxyl radicals in the presence of MPP^+ itself (Adams et al., 1993). The MPTP-associated depletion of striatal dopamine in older mice is an additional consequence of increased MAO-B activity (Finnegan et al., 1995). This is consistent with the finding that, when the dosage of MPTP is kept constant, the striatal concentrations of MPP^+ resulting from MAO-B-triggered endogenous production increase with ageing. The total exposure of the striatum in 8 month-old mice was approximately 3 times greater than that in their younger counterparts (Langston et al., 1987). On the other hand, conflicting results show that the brains of older mice either accumulate more MPP^+ than younger animals (Langston et al., 1987) or do not (Ricaurte et al., 1987b).

It is well known that age-associated alterations in mitochondrial DNA, following free radical-provoked oxydative stress, result in the continuously declining respiratory activity of the organelles. The inhibition of complex I in mitochondria (Ramsay et al., 1991) exerted by MPP^+ potentiates the physio-pathological decline of complexes I, II and IV with age, thus resulting in an even more severe, neurotoxic block of energy production in older animals (Desai et al., 1996). Aged rats are also more sensitive to the actions of the noradrenergic toxin DSP-4, which produces more severe biochemical and functional deficits than in younger animals (Sirviö et al., 1991; Riekkinen et al., 1987). This effect seemingly correlates with the impairment of compensatory responses following dopaminergic and noradrenergic degeneration (Ricaurte et al., 1987b; Sirviö et al., 1991). The extent of brain damage after prolonged seizures is strongly age-dependent. The accompanying cell-death bearing an excitotoxic origin due to excessive extracellular increases of glutamate may underlie the stronger neurotoxicity in selected brain regions of older rats, whereas immature animals are relatively resistant to glutamate-induced cell death (Liu et al., 1996). This age-dependent excitotoxicity is probably related to ageing-related loss in the number of transport sites for glutamate (Saransaari & Oja, 1995), and the consequent impairment of the neuroprotective mechanism against supraphysiological, extracellular concentrations of excitatory aminoacids (Rothstein et al., 1993). Of course, the nigro-striatal dopaminergic system and the related motor functions are affected by ageing (see Amenta et al., 1991; Emerich et al., 1993)), though apparently more mildly in rodents than in humans. There is, indeed, a continuous age-related decline in the levels of dopamine, plus its postmortem metabolite 3-methoxytyramine, in the human caudate nucleus and putamen (Carlsson & Winblad, 1976). In rats and mice there are more controversial and seemingly modest, though significant, age-related decreases (5% to 20% over the life-span) of actual striatal dopamine concentrations (Pradhan, 1980; Osterburg et al., 1981; Irwin et al., 1992; Huang et al., 1995; Kabuto et al., 1995), and in the density of nigro-striatal dopaminergic cells (Milgram et al., 1990). The accompanying loss in synaptic transport functions, however, appears more dramatic overall. In humans, there is indeed a marked loss in presynaptic, dopamine-containing nerve terminals as indicated by the decreasing density with ageing of caudate plasma membrane reuptake sites for dopamine (Zelnik et al., 1986), and the precipitous fall in the nigral content of dopamine transporter mRNA (Bannon et al., 1992). Similarly, ageing affects the synaptosomal uptake of dopamine and norepinephrine in the rodent striatum and hypothalamus (Jonec & Finch, 1975; Pradhan, 1980). In senescent mice the synaptosomal resting potential is also decreased, with concomitantly impaired activity of the electrogenic pump Na^+,K^+-ATPase, which means the depressed transport of pertinent neurotransmitters (Ando & Tanaka, 1990). This event may functionally reflect on the K^+-stimulated, exocytotic release of dopamine, which is decreased by more than 50% in striatal slices of 12 month-old rats, compared to their younger counterparts (Wustmann et al., 1983; Dobrev et al., 1995). Similarly, *in vivo* basal, K^+- and amphetamine-evoked dopamine releases from the basal ganglia of 22–24 month-old rats were greatly decreased, compared to younger animals (Huang et al., 1995; Nakano & Mizuno, 1996). Deterioration in synaptic functions does not appear to involve the physico-chemical characteristics of the neuronal and vesicular transporters. In fact, the affinity of monoamine uptake inhibitors for dopamine and serotonin transporters was similar in the striata of rat fetuses and mature (6 month-old) rats (Hyde & Bennett, 1994), and the binding affinities of the vesicular markers [^{3}H]tyramine and [^{3}H]dihydrotetra- benazine in striatal and cerebellar membranes, respectively, did not differ with age (see chapter 5.1). In more general terms, it appears that agent-induced depletion of brain dopamine and additional transmitters may be greater in aged than in young rats. Thus, *p*-chlorophenylalanine, a selective depletor for brain 5-HT in young rats, also markedly decreases the concentrations of dopamine and norepinephrine in aged animals (Riekkinen et al., 1996).

Overall, the programmed and free radical-induced neuronal loss in the aged brain, with the associated nervous hypofunction and fading of the compensatory responses (Ricaurte et al., 1987b; Sirviö et al., 1991; Riekkinen et al., 1992), may be amplified by the effects of specifically-targeted neurotoxins (Dawson et al., 1995).

5.1. Is the Interaction of MPP$^+$ with the Vesicular Transporter for Dopamine an Ageing-Sensitive Process?

The monoamine transporter in MPP$^+$-resistant, adrenal chromaffin granules is thought to play, at least at the very beginning of exposure, a neuroprotective role, since it would sequester toxins from the cytoplasm, therefore avoiding their noxious interaction with mitochondria (Reinhard et al., 1990; Edwards, 1993). This hypothesis has been strongly supported by the recent finding (Liu et al., 1992) that rat pheochromocytoma PC12 cells derived from the adrenal medulla express a VMAT which makes them relatively resistant to MPP$^+$ toxicity, unlike CHO fibroblast cells lacking the membrane monoamine transporter. When CHO cells were transfected with a cDNA expression library from PC12 cells, a reserpine-sensitive clone was isolated that accumulated MPP$^+$ and was extremely resistant to its toxicity. Thus, the gene transfer of a vesicular amine transporter protected against the toxin by sequestering it. The vesicular monoamine transporter ($VMAT_2$) in synaptic nerve endings, however, does not seem to entirely reflect the working mechanisms of its adrenal counterpart. In fact, purified synaptic vesicles obtained from the highly MPP$^+$-sensitive striatum take up [^{3}H]MPP$^+$ with an energy-dependent process. Conversely, vesicles prepared from the cerebellum, a very poorly dopaminergic, and an MPP$^+$-resistant brain region, do not (Del Zompo et al., 1993). Nevertheless, MPP$^+$ binds with high affinity at the [^{3}H]tyramine-labelled transporter for norepinephrine in cerebellar vesicles (Vaccari et al., 1993), an event that probably does not lead to significant vesicular sequestration/retention of the toxin in the organelles. According to the theory of neuroprotective sequestration, even modest reductions in vesicular uptake with ageing might divert cytoplasmic MPP$^+$-like toxins towards sensitive targets. On the other hand, if it is true that the storage compartment might first concentrate trace amounts, subsequently releasing significant amounts of toxins, in the long run this might result in neuronal damage (Johannessen, 1991). In humans, indeed, there is evidence for the progression of MPP$^+$-induced nigro-striatal lesions following only a transient exposure to the

Table 4. Binding characteristics of the vesicular transporter ligands [^{3}H]tyramine and [^{3}H]dihydrotetrabenazine in young-adult and aged rats[a]

	[^{3}H]Tyramine binding Striatum		[^{3}H]Dihydrotetrabenazine binding Cerebellum	
	Young[b]	Old[b]	Young[c]	Old[c]
K_{Dhigh}	7.1 ± 0.5	7.2 ± 0.8	7.7 ± 1.0	5.7 ± 0.8
K_{Dlow}	26.2 ± 3.6	29 ± 2.4	55.0 ± 13.1	49.6 ± 6.2
$B_{MAXhigh}$	3.0 ± 0.2	2.4 ± 0.3	0.095 ± 0.01	0.083± 0.01
B_{MAXlow}	9.1 ± 0.8	5.1 ± 0.6**	0.460 ± 0.06	0.460± 0.09

[a]Charles-River CD young (45-60 days) and old (18-22 months), male rats.
K_D (*nM*) and B_{MAX} (*pmol/mg protein*) values are means ± s.e.m. from [b]*N*=7 and [c]*N*=10 rats. Triplicate aliquots of approx. 100 µg proteins of freshly-prepared striatal membranes, or ≈ 600 µg proteins of frozen (-70°C) cerebellar membranes were incubated for 10 min at 37°C, and 30 min at 30°C with [^{3}H]tyramine (0.4-40 nM) or [^{3}H]dihydrotetrabenazine (0.2-25 nM) (sp.activity ≈ 145 Ci/mmol, Amersham Int., U.K.), respectively. Binding assays for [^{3}H]tyramine (Vaccari, 1986; Vaccari et al., 1993) and [^{3}H]dihydrotetrabenazine (Scherman et al., 1988) were run as detailed elsewhere. Binding parameters were calculated with the RADLIG v.4 (Biosoft) program.
**$P < 0.01$ vs the respective young counterpart (ANOVA followed by Newman-Keuls test).

toxin (Vingerhoets et al., 1994). More convincingly, it has been shown that even trace amounts of systemically-administered MPTP may affect brain dopamine neurons in mice (Bagchi, 1992).

We wanted to ascertain whether the interaction of MPP^+ with the [^{3}H]tyramine-labelled, striatal vesicular transporter for dopamine, and the [^{3}H]dihydrotetrabenazine-labelled monoamine transporter in cerebellar vesicles, differed between old (18–22 months) and young adult (45–60 days) Charles-River CD rats. Both ligands associate with striatal and cerebellar membranes, respectively, with double-profile Scatchard-kinetics, denoting attachment at two binding sites of their respective vesicular transporter. Consistently, the vesicular marker tetrabenazine inhibited both binding components of striatal [^{3}H]tyramine (unpublished), and reserpine displaced specifically-bound tyramine with two-site kinetics (Vaccari, 1986). There were no age-related changes in K_D and B_{MAX} values for either high- or low-affinity components of cerebellar [^{3}H]dihydrotetrabenazine binding and in the K_D's of striatal [^{3}H]tyramine binding (Table 4). However, the maximum number of binding sites for tyramine in the low affinity component fell by 43%, a likely further index for the ageing-related loss of dopaminergic nerve endings in the striatum described above. This finding reflects the previously shown (Scherman et al., 1989) decrease with age in human caudate tetrabenazine-labelling of the dopamine transporter. Unchanged dihydrotetrabenazine binding denoted, on the other hand, that the density of monoaminergic (noradrenergic) terminals in the cerebellum was not affected by ageing. To our knowledge, this is the first demonstration that there is a brain-region selectivity in age-related effects at the vesicular level, depending on the neurotransmitter specificity of innervation. In fact, as with the neuronal transporter (see chapter 5), the *striatal* vesicular carrier for dopamine is sensitive to ageing, whereas the *cerebellar* monoamine carrier is not. MPP^+displaced specifically bound [^{3}H]tyramine according to two-component kinetics, and [^{3}H]dihydrotetrabenazine with a one-site profile. The affinity values of MPP^+ for both ligands in the old striata and cerebella (Fig.1) did not differ from those in young counterparts, thus indicating that the molecular characteristics of the vesicular transporter(s) do not change with ageing.

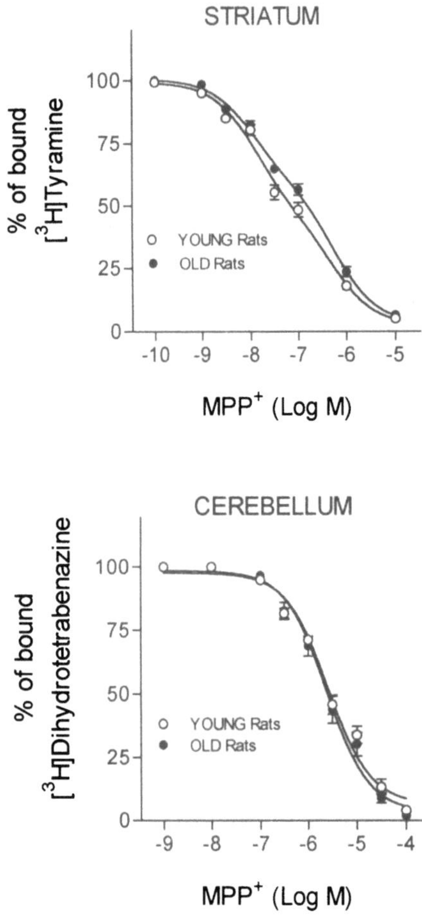

Figure 1. Similar inhibition by MPP^+ of specifically bound [^{3}H]tyramine (4 nM) in *striatal* membranes, and of [^{3}H]dihydrotetrabenazine (4 nM) in *cerebellar* membranes obtained from young-adult (45–60 days) and old (18–22 months) rats. Two-site analysis revealed the following affinity values in young and old striatal samples, respectively: K_{iHIGH}= 10.3 ± 1.9 and 9.1 ± 1.8 nM; K_{iLOW}= 428 ± 52 and 465 ± 47 nM (*N*=7). One-site analysis for MPP^+ competition on [^{3}H]dihydro-tetrabenazine binding gave K_i= 3.02 ± 0.37 and 2.97 ± 0.48 μM (*N*=11) in young and old cerebella, respectively. Displacement curves were analysed with the RADLIG v.4 (Biosoft) program.

6. CONCLUDING REMARKS

There is abundant evidence that the aged brain in rodents, non-human primates and humans is more sensitive than the younger brain to the effects of neurotoxins. The intrinsic neurotoxicity of brain-targeted pollutants may overlap in age-related, programmed and oxidative stress-provoked neuronal loss, changes in the rate of uptake/release functions for neurotransmitters and in their metabolism. The transport system for dopamine in synaptic vesicles is a high affinity target for a plethora of chemically heterogeneous agents, including both recognized and putative neurotoxins, pesticides, industrial pollutants and therapeutic drugs (Vaccari, 1993; Vaccari & Saba, 1995; Vaccari et al., 1996). Their interaction is, mostly, non-competitive in type, which renders their sequestration in vesicles unlikely. Taking for granted that, at least in the early phase of exposure, the vesicular sequestration of selected neurotoxins such as the parkinsogenic MPP^+ helps to lower their actual cytoplasmic levels and attenuates the toxic impact on sensitive mitochondria, there are two likely scenarios.

a. Transported toxins, which substitute for the legitimate transmitter, may initially disrupt the neuronal deposits which are physiologically necessary for nervous homeostasis. Subsequently, their endovesicular accumulation may become a source for further chronic leakage out into the cytoplasm, or for sudden release in toxically significant amounts.

b. Non-transported xenobiotics, because of their ability to influence the vesicular transporter for dopamine, may again impair the routine transmitter uptake/release functions and also hinder toxins from being internalized in vesicles. This putative neuroprotective mechanism, if any, would thus be compromised.

There are no evident ageing-related changes in the affinity of the prototype parkinsogenic toxin MPP^+ for the vesicular transporter for monoamines. The marked fall in the maximum number of binding sites for the vesicular marker tyramine, however, denotes the reduction in neuronal deposits in the aged striatum, a putative reflection of the demonstrated decrease in the density of striatal nerve endings. In conclusion, the continuous bombardment by events at the synaptic level throughout the life span, coinciding with the inescapable age-dependent decay in the neuronal framework, strongly suggest that old brains may be more easily affected by neurotoxins than are their younger counterparts.

ACKNOWLEDGMENTS

This work was supported by grants from the Italian Ministry for University and Scientific Research (MURST, 40% and 60% 1995 and 1996 to A.V.).

REFERENCES

Adams, J.D., Klaidman, L.K. & Leung, A.C. MPP^+ and $MPDP^+$ induced oxygen radical formation with mitochondrial enzymes. *Free Radical Biol.Med.* 15, 181–186 (1993).

Ali, S.F., David, S.N., Newport, G.D., Cadet, J.L. & Slikker, W. MPTP-induced oxidative stress and neurotoxicity are age-dependent: evidence from measures of reactive oxygen species and striatal dopamine levels. *Synapse* 18, 27–34 (1994).

Amenta, F., Zaccheo, D. & Collier, W.L. Neurotransmitters, neuroreceptors and aging. *Mech.Ageing.Devl.* 61, 249–273 (1991).

Ando, S. & Tanaka, Y. Synaptic membrane aging in the central nervous system. *Gerontology* 36 (S1), 10–14 (1990).

Bagchi, S.P. Trace dosages of the neurotoxins MPTP and MPP^+ may affect brain dopamine in vivo. *Life Sci.* 51, 389–396 (1992).

Bannon, M.J., Poosch, M.S., Xia, Y., Goebbel, D.J., Cassin, B. & Kapatos, G. Dopamine transporter mRNA content in human substantia nigra decreases precipituously with age. *Proc.Natl.Acad.Sci. USA* 89, 7095–7099 (1992).

Barbeau, A., Roy, M., Bernier, G. *et al.* Ecogenetics of Parkinson's disease: prevalence and environmental aspects in rural areas. *Can.J.Neurol.Sci.* 14, 36–48 (1987).

Borowsky, B. & Hoffman, B.J. Neurotransmitter transporters: molecular biology, function, and regulation. *Int.Rev.Neurobiol.* 38, 139–199 (1995).

Breese, G.R. & Taylor, T.D. Effects of 6-hydroxydopamine on brain norepinephrine and dopamine: evidence of selective degeneration of catecholamine neurons. *J.Pharmacol.Exp.Ther.* 174, 413–420 (1970).

Calne, D.B. Neurotoxins and degeneration in the central nervous system. *Neurotoxicology* 12, 335–340 (1991).

Carlsson, A. & Winblad, D.B. Influence of age and time interval between death and autopsy on dopamine and 3-methoxytyramine levels in human basal ganglia. *J.Neural Transm.* 38, 271–276 (1976).

Collins, M. A possible neurochemical mechanism for brain and nerve damage associated with chronic alcoholism. *Trends Pharmacol.Sci.* 3, 373–375 (1982).

Darchen, F., Scherman, D. & Henry, J.-P. Reserpine binding to chromaffin granules suggests the existence of two conformations of the monoamine transporter. *Biochemistry* 28, 1692–1697 (1989).

Dawson, R.Jr., Beal, M.F., Bondy, S.C., Di Monte, D.A. & Isom, G.E. Excitotoxins, aging, and environmental neurotoxins: implications for understanding human neurodegenerative diseases. *Toxicol.Appl.Pharmacol.* 134, 1–17 (1995).

Del Zompo, M., Piccardi, M.P., Ruiu, S., Quartu, M., Gessa, G.L. & Vaccari, A. Selective MPP^+ uptake into synaptic dopamine vesicles: possible involvement in MPTP toxicity. *Br.J.Pharmacol.* 109, 411–414 (1993).

Desai, V.G., Feuers, R.J., Hart, R.W. & Ali, S.F. MPP^+-induced neurotoxicity in mouse is age-dependent: evidence by the selective inhibition of complexes of electron transport. *Brain Res.* 715, 1–8 (1996).

Dobrev, D., Bergsträsser, E., Fischer, H.-D. & Andreas, K. Restriction and functional changes of dopamine release in rat striatum from young adult and old rats. *Mech.Ageing Devl.* 80, 107–119 (1995).

Edwards, R.H. Neural degeneration and the transport of neurotransmitters. *Ann.Neurol.* 34, 638–643 (1993).

Emerich, D.F., McDermott, P., Krueger, P. *et al.* Locomotion of aged rats: relationship to neurochemical but not morphologic changes in nigrostriatal dopaminergic neurons. *Brain Res.Bull.* 32, 477–486 (1993).

Erickson, J.D., Schäfer, M.K.-H., Bonner, T.I., Eiden, L.E. & Weihe, E. Distinct pharmacological properties and distribution in nervous and endocrine cells of two isoforms of the human vesicular monoamine transporter. *Proc.Natl.Acad.Sci.USA* 93, 5166–5171 (1996).

Finnegan, K.T., Irwin, I., Delanney, L.E. & Langston, J.W. Age-dependent effects of the 2'-methyl analog of 1-methyl-4-phenyl-1,2,3,6-tetrahydropyridine: prevention by inhibitors of monoamine oxidase-B. *J.Pharmacol.Exp.Ther.* 273, 716–720 (1995).

Forster, M.J., Dubey, A., Dawson, K.M., Stutts, W.A., Lal, H. & Sohal, R.S. Age-related losses of cognitive function and motor skills in mice are associated with oxidative protein damage in the brain. *Proc.Natl.Acad.Sci.USA* 93,4765–4769 (1996).

Golbe, L.I. The genetics of Parkinson's disease: a reconsideration. *Neurology* 40 (S3) 7–14 (1990).

Gupta, M., Gupta, B.K., Thomas, R., Bruemmer, V., Sladek, J.R. & Felten, D.L. Aged mice are more sensitive to 1-methyl-4-phenyl-1,2,3,6-tetrahydropyridine than young adults. *Neurosci.Lett.* 70, 326–331 (1986).

Henry, J.-P. & Scherman, D. Radioligands of the vesicular monoamine transporter and their use as markers of monoamine storage vesicles. *Biochem.Pharmacol.* 38, 2395–2404 (1989).

Huang, R.-L., Wang, C.-T., Tai, M.-Y., Tsai, Y.-F. & Peng, M.-Y. Effects of age on dopamine release in the nucleus accumbens and amphetamine-induced locomotor activity in rats. *Neurosci.Lett.* 200, 61–64 (1995).

Hyde, C.E. & Bennett, B.A. Similar properties of fetal and adult amine transporters in the rat brain. *Brain Res.* 646, 118–123 (1994).

Irwin, I., Ricaurte, G.A., Delanney, L.E. & Langston, J.W. The sensitivity of nigro-striatal dopamine neurons to MPP^+ does not increase with age. *Neurosci.Lett.* 87, 51–56 (1988).

Irwin, I., Finnegan, K.T., Delanney, L.E., Di Monte, D. & Langston, J.W. The relationships between aging, monoamine oxidase, striatal dopamine and the effects of MPTP in C57BL/6 mice: a critical reassessment. *Brain Res.* 572, 224–231 (1992).

Jarvis, M.F. & Wagner, G.C. Age-dependent effect of 1-methyl-4-phenyl-1,2,3,6-tetrahydropyridine (MPTP). *Neuropharmacology* 24, 581–583 (1985).

Johannessen, J.N. A model of chronic neurotoxicity: long-term retention of the neurotoxin 1-methyl-4-phenylpyridinium (MPP^+) within catecholaminergic neurons. *Neurotoxicology* 12, 285–302 (1991).

Johnson, R.G.Jr. Accumulation of biological amines into chromaffin granules: a model for hormone and neurotransmitter transport. *Physiol.Rev.* 68, 232–307 (1988).

Jonec, V. & Finch, C.E. Senescence and noradrenaline uptake by subcellular fractions of the C57BL/6J male mouse brain. *Brain Res.* 91, 197–203 (1975).

Kabuto, H., Yokoi, I., Mori, A., Murakami, M. & Sawada, S. Neurochemical changes related to ageing in the senescence-accelerated mouse brain and the effect of chronic administration of nimodipine. *Mech.Ageing Devl.* 80, 1–9 (1995).

Koller, W., Vetere-Overfield, B., Gray, C. *et al.* Environmental risk factors in Parkinson's disease. *Neurology* 40, 1218–1221 (1990).

Langston, J.W., Irwin, I. & Delanney, L.E. The biotransformation of MPTP and disposition of MPP^+: the effects of aging. *Life Sci.* 40, 749–754 (1987).

Lilienfeld, D., Chan, E., Ehland, J. *et al.* Two decades of increasing mortality from Parkinson's disease among the US elderly. *Arch.Neurol.* 47, 731–734 (1990).

Liu, Y., Peter, D., Roghani, A., Schuldiner, S., Privé, G.G., Eisenberg, D., Brecha, N. & Edwards, R.H. A cDNA that suppresses MPP^+ toxicity encodes a vesicular monoamine transporter. *Cell* 70, 539–551 (1992).

Liu, Z., Stafstrom, C.E., Sarkisian, M., Tandon, P., Yang, Y., Hori, A. & Holmes, G.L. Age-dependent effects of glutamate toxicity in the hippocampus. *Devl.Brain Res.* 97, 178–184 (1996).

Makino, Y., Ohta, S., Tachikawa, O. & Hirobe, M. Presence of tetrahydroisoquinoline and 1-methyl-tetrahydroisoquinoline in foods: compounds related to Parkinson's disease. *Life Sci.* 43, 373–378 (1988).

Milgram, N.W., Racine, R.J., Nellis, P., Mendonca, A. & Ivy, G.O. Maintenance on L-deprenyl prolongs life in aged male rats. *Life Sci.* 47, 415–420 (1990).

Myers, R.D. Isoquinolines, beta-carbolines and alcohol drinking: involvement of opioid and dopaminergic mechanisms. *Experientia* 45, 436–443 (1989).

Nakano, M. & Mizuno, T. Age-related changes in the metabolism of neurotransmitters in rat striatum: a microdialysis study. *Mech.Ageing Devl.* 86, 95–104 (1996).

Naudon, L., Raisman-Vozari, R., Edwards, R.H., Leroux-Nicolet, I., Peter, D., Liu, Y. & Costentin, J. Reserpine affects differentially the density of the vesicular monoamine transporter and dihydrotetrabenazine binding sites. *Eur.J.Neurosci.* 8, 842–846 (1996).

Osterburg, H.H., Donahue, H.G., Severson, J.A. & Finch, C.E. Catecholamine levels and turnover during aging in brain regions of male C57BL/6J mice. *Brain Res.* 224, 337- 352 (1981).

Osuntokun, B.O., Ogunniyi, A.O., Lekwauka, G.U. *et al.* Epidemiology of age-related dementias in the third-world and etiologic clues of Alzheimer's disease. *Tropical Geogr.Med.* 43, 345–351 (1991).

Pradhan, S.N. Central neurotransmitters and aging. *Life Sci.* 26, 1643–1656 (1980).

Rajput, A.H. Epidemiology of Parkinson's disease. *Can.J.Neurol.Sci.* 11S, 154–161 (1984).

Rajput, A.H., Uitti, R.G., Stern, W. *et al.* Geography, drinking water chemistry, pesticides and herbicides and the etiology of Parkinson's disease. *Can.J.Neurol.Sci.* 14, 414–418 (1987).

Rajput, A.H. Frequency and cause of Parkinson's disease. *Can.J.Neurol.Sci.* 19, 103–107 (1992).

Ramsay, R.R., Krueger, M.J., Youngster, S.K. & Singer, T.P. Evidence that the inhibition sites of the neurotoxic amine 1-methyl-4-phenylpyridinium (MPP^+) and the respiratory chain inhibitor piericidin A are the same. *Biochem.J.*, 273, 481–484 (1991).

Reinhard, J.F.Jr., Carmichael, S.W. & Daniels, A.J. Mechanisms of toxicity and cellular resistance to 1-methyl-4-phenyl-1,2,3,6-tetrahydropyridine and 1-methyl-4-phenyl- pyridinium in adrenomedullary chromaffin cell cultures.*J.Neurochem.* 55, 311–320 (1990).

Ricaurte, G.A., Irwin, I., Forno, L.S., Delanney, L.E., Langston, E.B. & Langston, J.W. Aging and 1-methyl-4-phenyl-1,2,3,6-tetrahydropyridine-induced degeneration of dopaminergic neurons in the substantia nigra. *Brain Res.* 403, 43–51 (1987a).

Ricaurte. G.A., Delanney, L.E., Irwin, I. & Langston, J.W. Older dopaminergic neurons do not recover from the effects of MPTP. *Neuropharmacology* 26, 97–99 (1987b).

Riekkinen, P.Jr., Riekkinen, M., Valjakka, A., Riekkinen, P. & Sirviö, J. DSP-4, a noradrenergic neurotoxin, produces more severe biochemical and functional deficits in aged than young rats. *Brain Res.* 570, 293–299 (1992).

Riekkinen, M., Aroviita, L., Kivipelto, M., Taskila, K. & Riekkinen, P.Jr. Depletion of serotonin, dopamine and noradrenaline in aged rats decreases the therapeutic effect of nicotine, but not of tetrahydroaminoacridine. *Eur.J.Pharmacol.* 308, 243–250 (1996).

Ross, S.B. Long-term effects of N-2-chloroethyl-N-ethyl-2-bromobenzylamine hydrochloride on noradrenergic neurones in the rat brain and heart. *Br.J.Pharmacol.* 58, 521–527 (1976).

Rothstein, J.D., Jin, L., Dykes-Hoberg, M. & Kuncl, R.W. Chronic inhibition of glutamate uptake provides a model of slow neurotoxicity. *Proc.Natl.Acad.Sci.USA* 90, 6591–6595 (1993).

Saransaari, P. & Oja, S.S. Age-related changes in the uptake and release of glutamate and aspartate in the mouse brain. *Mech.Ageing Devl.* 81, 61–71 (1995).

Scherman, D., Raisman, R., Ploska, A. & Agid, Y. [^{3}H]Dihydrotetrabenazine, a new in vitro monoaminergic probe for human brain. *J.Neurochem.* 50, 1131–1136 (1988).

Scherman, D., Desnos, C., Darchen, F., Pollak, P., Javoy-Agid, F. & Agid, Y. Striatal dopamine deficiency in Parkinson's disease: role of aging. *Ann.Neurol.* 26, 551- 557 (1989).

Schoenberg, B.S. Environmental risk factors for Parkinson's disease: the epidemiologic evidence. *Can.J.Neurol.Sci.* 14, 407–413 (1987a).

Schoenberg, B.S. Descriptive epidemiology of Parkinson's disease: disease distribution and hypothesis formulation. *Adv.Neurol.* 45, 277–283 (1987b).

Semchuk, K.M., Love, E.J. & Lee, R.G. Parkinson's disease and exposure to agricultural work and pesticide chemicals. *Neurology* 42, 1328–1335 (1992).

Simantov, R. Neurotransporters: regulation, involvement in neurotoxicity, and the usefulness of antisense nucleic acids. *Biochem.Pharmacol.* 50, 435–442 (1995).

Sirviö, J., Riekkinen, P.Jr., Valjakka,A., Jolkkonen, J. & Riekkinen, P.J. The effects of noradrenergic neurotoxin, DSP-4, on the performance of young and aged rats in spatial navigation task. *Brain Res.* 563, 297–302 (1991).

Slivka, A. & Cohen, G. Hydroxyl radical attack on dopamine. *J.Biol.Chem.* 260, 15466–15472 (1985).

Tanner, C.M. The role of environmental toxins in the etiology of Parkinson's disease. *Trends Neurosci.* 12, 49–54 (1989).

Tipton, K.F. & Singer, T.P. Advances in our understanding of the mechanisms of the neurotoxicity of MPTP and related compounds. *J.Neurochem.* 61, 1191–1206 (1993).

Vaccari, A. High-affinity binding of [^{3}H]-tyramine in the central nervous system. *Br.J.Pharmacol.* 89, 15–25 (1986).

Vaccari, A. The tyramine binding site in the central nervous system: An overview. *Neurochem.Res.* 18, 861–868 (1993).

Vaccari, A. & Gessa, G.L. [^{3}H]Tyramine binding: a comparison with neuronal [^{3}H]dopamine uptake and [^{3}H]mazindol binding processes. *Neurochem.Res.* 14, 949–955 (1989).

Vaccari, A., Del Zompo, M., Melis, F., Gessa, G.L. & Rossetti, Z.L. Interaction of 1-methyl-4-phenylpyridinium ion and tyramine with a site putatively involved in the striatal vesicular release of dopamine. *Br.J.Pharmacol.* 104, 573–574 (1991).

Vaccari, A., Saba, P.L., Gessa, G.L. & Del Zompo, M. Differential interaction of 1-methyl-4-phenylpyridinium ion with the putatively vesicular binding site of [^{3}H]tyramine in dopaminergic and nondopaminergic brain regions. *J.Neurochem.* 60, 758–760 (1993).

Vaccari, A. & Saba, P.L. The tyramine-labelled vesicular transporter for dopamine: a putative target of pesticides and neurotoxins. *Eur.J.Pharmacol.(ETPS)* 202, 309–314 (1995).

Vaccari, A., Saba, P.L., Ruiu, S., Collu, M. & Devoto, P. Disulfiram and diethyldithiocar- bamate intoxication affects the storage and release of striatal dopamine. *Toxicol.Appl.Pharmacol.* 139, 102–108 (1996).

Vieregge, P. Genetic factors in the etiology of idiopathic Parkinson's disease. *J.Neural Transm. (P-D Sect.)* 8, 1–37 (1994).

Vingerhoets, F.J., Snow, B.J., Tetrud, J.W., Langston, J.W., Schulzer, M. & Calne, D.B. Positron emission tomographic evidence for progression of human MPTP-induced dopaminergic lesions. *Ann.Neurol.* 36, 765–770 (1994).

Walsh, S.L. & Wagner, G.C. Age-dependent effects of 1-methyl-4-phenyl-1,2,3,6-tetrahydropyridine (MPTP): correlation with monoamine oxidase B. *Synapse* 3, 308–314 (1989).

Wustmann, C., Schmidt, J., Ihle, W., Gross, J. & Fischer, H.-D. Dopamine release from striatum slices of rats at different age: influence of hypoxia. *Biomed.Biochem.Acta* 42, 265–273 (1983).

Zelnik, N., Angel, I., Paul, S.M. & Kleinman, J.E. Decreased density of human striatal dopamine uptake sites with age. *Eur.J.Pharmacol.* 126, 175–176 (1986).

17

ALZHEIMER DISEASE

From Molecular Biology to Therapy

Ezio Giacobini

HUG, Belle-Idée, Department of Geriatrics
University Hospitals of Geneva
Route de Mon-Idée
CH-1226 Thonex, Geneva, Switzerland

1. PHARMACOTHERAPY OF ALZHEIMER DISEASE

The systematic development of Alzheimer Disease (A.D.) therapy is only ten year old. It was initiated on a large scale following the publication in New England Journal of Medicine of the first successful results obtained with the cholinesterase inhibitor (ChEI) tacrine, (THA, tetrahydroaminoacridine) by Summers et al. (1986). Numerous studies (cf. Becker et al. 1991) had been performed previously, particularly in USA, in small groups of patients, with physostigmine (physo) alone or in combination with lecithine. Physostigmine, like tacrine, showed definite but only shortlasting improvements of cognitive symptoms (attention, concentration, memory) which were accompanied with severe peripheral and central cholinergic side effects. These consisted mainly of gastro-intestinal symptoms and drowsiness but in the case of tacrine also of liver toxicity. Physostigmine and tacrine represent important milestones in AD therapy as they supported in the patient, the pharmacological hypothesis formulated in the experimental animal (Mattio et al. 1986; Giacobini, 1987; Hallak and Giacobini, 1986; Giacobini et al. 1986) that a treatment improving function of the central cholinergic system obtained through an increase of brain ACh would also improve cognition in AD patients. Targeting the cholinergic system for AD therapy does not necessarily limit itself to use a cholinesterase inhibitor (ChEI). Two other classes of cholinergic drugs might represent valid alternatives such as nicotinic and muscarinic agonists alone or in combination with ChEI (Fig 1). A second, non-cholinergic approach, is based on the classic pathological landmarks of the disease aiming to decrease beta-amyloid (beta-A4) deposition and amyloidogenic APP (Amyloid Precursor Protein) release in brain (Fig 1). A third way to AD therapy aims to correct events which are probably secondary to the disease process by means of estrogens, anti-oxidants, free-radical scavengers and anti-inflammatory agents (Fig.1). Based on the present knowledge of beta-A4 synthesis, processing, accumulation and deposition in cortex and related genetic factors (cf Che-

Brain Plasticity, edited by Filogamo *et al.*
Plenum Press, New York, 1997

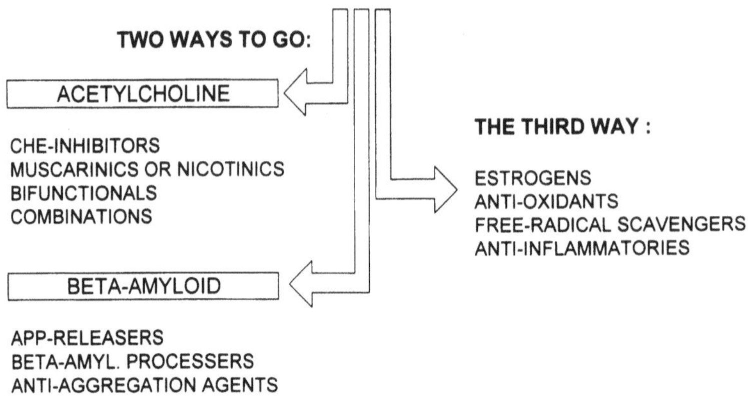

Figure 1. Treatment of Alzheimer disease. Which way to go?

cler, 1995; Van Broeckhoven, 1995; Sandbrink et al. 1996; Vito et al. 1996) (Fig 2) we may suggest at least five pharmacological interventions to selectively decrease extracellular concentration and deposition of beta-A4 in brain and to slow down the disease process (Fig.3). In theory, the same result could be achieved by blocking beta-A4 processing or slowing down secretion or deposition of amyloidogenic forms of APP such as the 42-aminoacid peptide beta-42. Processing and secretion of APP could be modified by inhibiting specifically one or several involved proteases (alpha, beta or gamma secretases) (Checler, 1995). One might also think of reducing aggregation by enhancing removal of deposited beta-A4 (Fig.4). There is evidence for considering apolipoproteins such as APOE 3 and APOE 2 as factors reducing beta-A4 aggregation and APOE4 as promoting it (Fig.4) (cf Roses, 1995 and Giacobini, 1996a). Presenilin 1 on the other hand, promotes both clearance and degradation of beta-A4 (Fig. 4).(cf. Van Broeckhoven, 1995; Yankner, 1996). The final goal is to block the progressive and irreversible degeneration of synapses and neurons which represents the main feature of the disease. A better understanding of the mechanism of action of presinilins and APOE could lead to new therapeutical strategies developing drugs interacting with these processes (Roses, 1995) or preventive and protective measures based on early identification of genetic risk factors comparable to a low-fat diet and control of hypertension for cardiovascular diseases. Recent data suggest an involvement of the cholinergic system in beta-A4 processing both experimentally and clinically (Soininen & Riekkinen, 1996) (Fig. 5). Chemical cholinergic lesions in the nucleus basalis of the rat forebrain decrease ACh cortical levels and release and increase accumulation of APP in CSF and cortex (Table I) (Haroutunian et al. 1995, 1996). The effect of the lesion is mimicked by muscarinic antagonists and reversed by treatment with ChEI. Amyloid beta peptide 1–42 supresses ACh synthesis in cholinergic neurons at nM concentrations (Hoshi et al. 1997) (Table I). The data of Soininen et al. (1995) and Poirier et al. (1995) show that the APOE-4 genotype exherts a deteriorating effect on the cholinergic system and might facilitate selective neuronal damage. This cholinergic hypofunction might condition the response to cholinergic drugs in AD patients (Poirier et al. 1995), (Fig. 5). Other data suggesting an interaction between cholinesterase inhibition and brain metabolism as clinical evidence of neuroprotection are summarized in Table II. These include the effect of a ChEI (tacrine) on cortical glucose metabolism and cerebral blood flow monitored in AD patients with imaging techniques (PET and SPECT) utilizing DFG (deoxy-fluoroglucose) and 11C-nicotine as biological markers of metabolic and choliner-

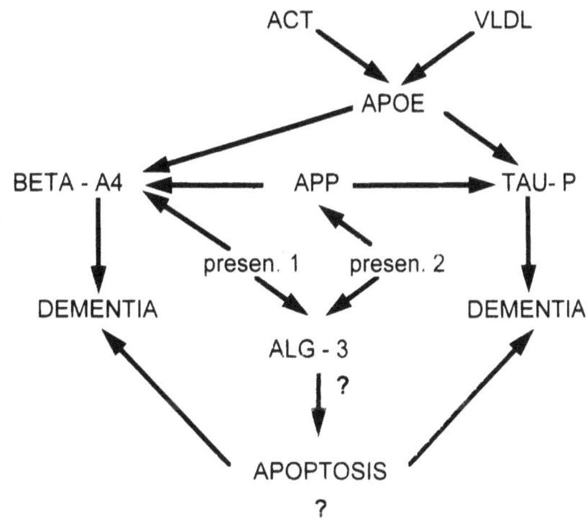

Figure 2. Processing of the beta-amyloid (beta-A4) precursor protein (APP), its regulation in Alzheimer disease and its relation to dementia and apoptosis. ACT = alpha-antichymotrypsin, VLDL = very low density lipoprotein receptor, APOE = apolipoprotein-E, ALG-3 = apoptosis-linked gene, TAU-P = phosphorylated tau protein. Presen = presenilin.

gic activity respectively (Nordberg 1993, 1995). A delay in time to nursing home placement is chosen as a clinical marker of deterioration of the patient (Gracon et al. 1996). Based on an extensive knowledge of basic and clinical pharmacology of ChEI we may think of a number of possible interactions with other drugs particularly significant for the second generation of ChEI now being tested in the patient (Table III). It is important to note that ChEI show marked individual differences with regard to side effects and tolerability.

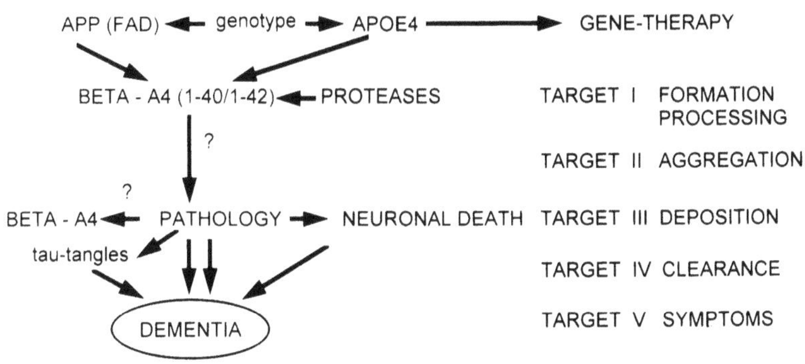

Figure 3. Five pharmacological interventions to selectively decrease extracellular concentration and deposition of beta-A4 in brain and to slow down dementia.

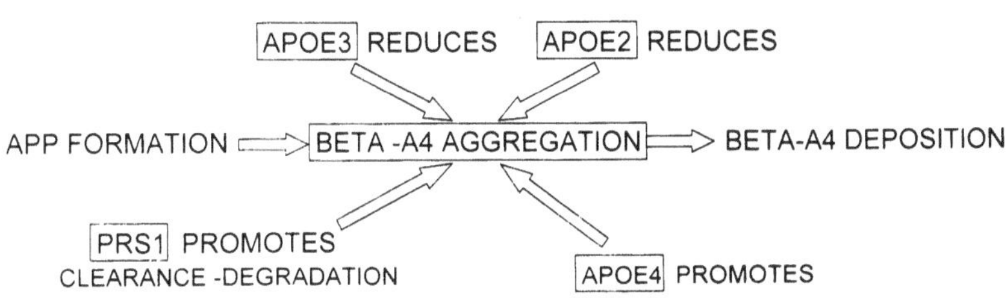

Figure 4. Effect of APOEs (apolipoproteins) and presenilin1 on beta-4 aggregation.

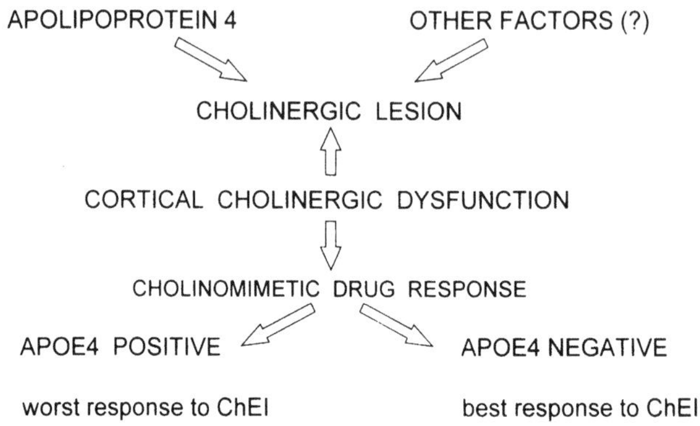

Figure 5. Relation between apolipoprotein 4, cholinergic lesions and cholinomimetic drug response to ChEI.

2. EFFECTS OF CHOLINESTERASE INHIBITORS ON CORTICAL NEUROTRANSMITTERS

Experimental as well clinical evidence indicate involvement and interactions between the cholinergic system and the biogenic amine system in the cognitive impairment observed in AD (cf. Giacobini and Cuadra, 1994 and Giacobini, 1996 b).

Table IV compares the effects on ACh, norepinephrine (NE) and dopamine (DA) levels as well as AChE inhibition after systemic administration of six clinically tested ChEI studied in our laboratory. Our results show a significant increase in cortex for all three neurotransmitters and for all six ChEIs investigated. In interpreting these effects we should keep in mind findings indicating the fact that cathecolaminergic afferents can differentially modulate forebrain cholinergic neurons as well as cortical cholinergic synapses (Zaborsky and Cullinan, 1996, Giacobini and Cuadra, 1994, Giacobini, 1996).

The results reported in Table IV suggest that extracellular ACh levels in cortex are related not only to ChE inhibition, as shown in previous microdialysis studies showing comparable elevations of ACh levels in spite of different magnitudes of ChE inhibition, but also to other effects (Messamore et al. 1993). This consideration may be of importance in predicting clinical effects as side effects, drug interactions (see dopaminergic effects and parkinsonism) and in setting dosages of various ChEI. Combinations of ChEI with

Table I. Cholinergic lesions and APP processing

A chemical lesion in the nucleus basalis of the rat[1] produces:
A) a decrease in ACh cortical level and release
B) a rapid and persistent accumulation in cortical and CSF APP
This response is:
A) age dependent
B) mimicked by muscarinic antagonists
C) reversed by treatment with ChEI[2]
The amyloid beta peptide (1-42) suppresses ACh synthesis in cholinergic neurons at nM co concentrations[3]

[1](Wallacc et al. 1991)
[2](Haroutunian et al. 1995, 1996)
[3](Hoshi et al. 1997)

Table II. Cholinesterase inhibitors: clinical evidence of neuroprotection

1.	Increased cortical glucose metabolism and rCBF (18 mo THA, PET DFG, Nordberg 1993, 1995)
2.	Increased cerebral nicotinic binding (12 moTHA, PET-11C nicotine, Nordberg 1995)
3.	Stable regional cerebral rCBF (14 mo THA, SPECT 133 Xe, Minthon et al. 1995)
4.	Delay in time to nursing home placement (500 days THA, Gracon et al. 1996)

monoamine-receptor agonists and antagonists are suggested by these studies (Giacobini and Cuadra, 1994 and Giacobini, 1996).

3. CHOLINESTERASE INHIBITION: A POSSIBLE MECHANISM TO SLOW DOWN AD DETERIORATION

The β-amyloid peptide (βA4), one of the major constituent proteins of neuritic plaques in the brain of AD patients, originates from a larger polypeptide denominated APP (Kang et al. 1987). APP is widely distributed throughout the mammalian brain including rat brain with a prevalent neuronal localization (Beeson et al. 1994). APP can be processed by several alternative pathways. A secretory pathway is believed to generate non-amyloidogenic soluble derivatives (APPs) following cleavage within the βA4 segment (Sisodia et al. 1990). Cholinergic agonists regulating processing and secretion of

Table III. Interactions of cholinesterase inhibitors with other drugs

References	Inhibitor	Interacting drug	Side effects	Mechanisms
Maany, 1996 McSwain and Forman, 1995	Tacrine	haloperidol	parkinsonism	Dopamine antagonism cholin. hyperactivity
Feldman et al. 1996	Tacrine other-ChEI	neuro-muscular blockers	prolonged action	neuro-musc. transm. muscle contraction
Paddle and Dowling, 1996	Tacrine Physostigmine	nicotinics	arrythmia	nicotinic block
Allain et al. 1996	Tacrine other ChEI	hypertensive drugs	acute hypertension	cholinergic/ adrenergic
Biyah et al. 1996	Tacrine	pentamidine	potentiation of bronchocostr.	muscarinic activity
Lebert et al. 1996	Tacrine other ChEI	epileptogenic agents	generalized seizures	cholinergic potentiation
	All ChEI	cholinomimetics	potentiation	cholinergic potentiation
	All ChEI	anti-muscarinic (phenotiazines, anti-histamin. tricyclic-anti-depressants)	reversal	inhibition of effect

Table IV. ChEI effects on ACh, NE, DA levels and ChE activity in rat brain cortex after s.c.administration

Compound (Ref.)	Dose mg/kg	ChE max.% inhib.	ACh	NE	DA
Physostigmine (1)	0.3	60	4000	75	120
Heptyl-physost.(2)	2	75	2500	25	75
E 2020 (3)	2	35	2100	100	80
MF-268 (4)	2	40	2500	100	60
Metrifonate (6)	80	70	1700	60	75
MDL 73,745 (5)	2	65	1020	120	370

E 2020 = (R,S)-1benzyl-4-(5,6 dimethoxy-1-idanon)-2-yl-methylpiperidine (1; 2; 3. Giacobini et al. 1996); MF-268 = 2, 6-dimethylmorfolin-octyl-carbamoyl eseroline (4. Zhu et al. 1996); Metrifonate =0,0-dimethyl-(1-hydroxy-2,2,2 trichloroethyl-phosphate) (6. Mori et al. 1995b); MDL 73,745=2,2,2-trifluoro-1-(3-trimethylsilylphenyl)ethanone (5. Zhu et al. 1995).

APPs by increasing, as demonstrated *in vitro*, protein kinase C (PKC) activity of target cells (Nitsch et al. 1992; Buxbaum et al. 1992; Buxbaum, 1995; Nitsch and Growdon, 1994) could decrease potentially amyloidogenic derivatives. We suggested that long-term inhibition of ChE increasing levels of synaptic ACh may result into activation of normal APP processing in AD brain (Giacobini, 1994). This effect could slow down the formation of amyloidogenic APP fragments. Lahiri et al. (1994) using nerve cell cultures found that the level of secretion of APP derivatives into conditioned media were inhibited by treating with 100 ug/ml tacrine (Table V). Chong and Suh (1996) found a dose-dependent effect of tacrine on APP processing (Table V). Tacrine at low concentrations (.02-.5mM) enhanced whereas concentrations above .5 mM blocked APP 770 processing in vitro (Table V). Haroutunian et al. (1996) reported that one week treatment with the ChEI phenserine decreased the levels of secreted beta-APP in the CSF of forebrain cholinergic lesioned rats suggesting that secretion of beta-APP into the CSF and neurons can be modulated by ChEI (Table I). To determine whether ChEI could alter the release of APP in brain we used superfused cortical slices of the rat (Mori et al. 1995) following the method described by Nitsch et al. (1993). Both short- and long-acting ChEI were tested for their ability to enhance the release of non-amyloidogenic soluble derivatives of APP (Mori et al. 1995a). These included: physo, heptastigmine and DDVP (dichlorvos, a metabolite of metrifonate)

Table V. Effect of ChEI on b-Amyloid processing *in vitro* and *in vivo*

Drug	Ref.	Conc. (mM)	APP Accumul.	APP Secret.	Preparation
Tacrine	(1)	500	–	+	APP 770
Tacrine	(2)	<500	+	–	APP 770
Tacrine	(3)	100mg/ml	+	–	Cell Lines
Phenserine	(4)	2.5mg/kg	+	–	Rat CSF
Tacrine	(5)	.5	+	–	Rat Cortical Slices
Tacrine	(6)	.1	–	+	Rat Cortical Slices
Physostigmine	(7)	.1	–	+	Rat Cortical Slices
Eptastigmine	(8)	.1	–	+	Rat Cortical Slices
DDVP (Metrifonate)	(9)	.02	–	+	Rat Cortical Slices

1; 2. Chong and Suh, 1996, 3. Lahiri et al. 1994, 4. Haroutunian et al. 1996, 5; 6; 7; 8; 9. Mori et al. (1995a)

Table VI. Drug-stimulated changes of basal APPs release and APP-KPI mRNA from rat brain (Mori et al. 1995a)

Drug	Conc. (μM)	Increase (% of basal)	ChE Act. (% Inhib.)	APP-KPI mRNA (% of basal)
Bethanechol	1	48	0	–
	100	53	0	–
Physostigmine	.1	48	25	–
Heptyl-physostygmine	.1	41	61	–35*
Dichlorvos	.02	33	95	–
Phorbol myristate	.1	–	–	+50

*From Giacobini et al. 1995 (5 mg/kg s.c, 48 hrs).

at concentrations producing ChE brain inhibitions ranging from 5% to 95%. All three ChEI elevated APPs release significantly above control levels (Table VI).

Using two different doses of tacrine (.5uM and .1uM) we found that only the lower dose elevated the release of APPs in the cortical slices which supports the data of Chong and Suh (1996) of a dose dependent modulation of APP secretion by AChE inhibition (Table VI). In our study, electrical field stimulation significantly increased the release of APPs within 50 min. Similar increase was observed after muscarinic receptor stimulation with bethanechol (BETHA) supporting results from in vitro experiments (Buxbaum, 1995) (Table VI). Tetrodotoxin (TTX) completely blocked the effect of electrical stimulation (Mori et al. 1995a).

The level of total APP RNAs in rat cortical slices did not change after incubation with BETHA, DDVP and physo, but activation of PKC with phorbol 12-myristate-13-acetate (100 nM) increased the level of total APP mRNA by 50% (Table VI) (Giacobini et al. 1995). Physo and DDVP administration (0.3 mg/kg and 80 mg/kg s.c., respectively) for 3–48 hrs did not significantly change the levels of APP 695 and APP-KPI (Kunitz-type protease inhibitor) mRNAs (Table VI). Heptastigmine administration (5 mg/kg s.c., 3–48 hrs) decreased by 35% the level of APP-KPI mRNA in rat cerebral cortex (Giacobini et al. 1995) (Table VI). AD pathology has been associated with an increase of the KPI-containing forms of APP and the propensity across species to develop neuritic plaques in cortical regions (Anderson et al. 1989). Our findings suggest that administration of ChEI to AD patients by increasing secretion of APP and inhibiting formation of specific APP mRNAs may exert a neuroprotective effect activating normal APP processing through a muscarinic mechanism and decreasing amyloid deposition in brain cells.

This effect of ChEI on APP processing could be reflected clinically by slowing down cognitive deterioration as suggested by clinical data of long term treatments with tacrine (Table VII).

4. CONCLUSIONS: THE FUTURE OF AD THERAPY

Beta-amyloid deposits, amyloidogenic APP release, tau protein hyperphosphorylation and formation of neurofibrillary tangles in brain constitute the primary targets of AD therapy attempts (Fig.2). Since beta-amyloid is derived from APP by proteolysis, selective inhibition of the proteases involved may prevent beta-amyloid formation. The identification and characterization of such enzymes has proven to be a difficult task. Also, APP

Table VII. Long-term effects of tacrine and eptastigmine on Alzheimer Disease progression

Reference	Daily dose (mg)	Test	Difference between treated and controls 11-15 mo.	17-24 mo.
Amberla et al. 1993	80-160	Buschke	no change	—
Wilcock et al. 1994	80	MMSE, CAMCOG, ADAS	no change	—
Eagger et al. 1994	50-150	AMTS*	no change	—
Solomon et al. 1996	160	ADAS-Cog	no change	2.8
Knopman et al. 1996	80-120	NHP**		delay
Imbimbo et al. 1997***	30	ADAS-Cog		6.4
		MMSE		3.9
		IADL		4.8

AMTS* = Abbreviated Mental Test Score
NHP** = Nursing Home Placement
*** = Eptastigmine, IADL = Instrumental Activity of Daily Leaving

fullfill a physiological function (Checler, 1995). Equally difficult has been to identify protein kinases specifically phosphorylating tau. Blocking enzymes such as kinases which regulate cell metabolism may produce severe side effects It is believed that even if none of these approaches could entirely prevent AD they could still slow down progression and delay onset of symptoms significantly. Cholinesterase inhibitors are presently the drugs of choice for AD. In less than ten years, starting from non-specific first generation drugs they have been developed into a second generation of selective molecules. The latest data from clinical trials suggests that optimization and maintainance of clinical effects for at least one year represents an achievable goal . Cholinesterase inhibitors, particularly second generation, (compounds post-physo and tacrine), affect cortical as well as sub-cortical neurotransmitters other than ACh. Their effects on NE and dopamine (DA) are of particular clinical interest. A newly demonstrated feature of ChEI is their ability to enhance the release of non-amyloidogenic soluble derivatives of APPs in vitro and in vivo and possibly to slow down formation of amyloidogenic compounds in brain. This process might also slow down cognitive deterioration of the patient as indicated by the analysis of clinical data of recent long-term trials. A critical question is: how long the effect of ChEI will be clinically significant? Five clinical studies extended beyond 12 mo show that patients treated with tacrine (high dose) do not deteriorate cognitively (ADAS-Cog) if compared to a placebo-treated group (Table VII) . At 17–21 mo, the difference between the two groups is still significant (Solomon et. al. 1996). This suggest a long term effect of ChEI. It remains to be demonstrated whether this effect is symptomatic or neuroprotective i.e. directed toward progression of disease pathology. Cholinomimetic alternatives other than ChE inhibition are being explored pharmacologically and clinically. Drugs most investigated are those showing direct stimulation of postsynaptic M1 and M3 muscarinic receptors. This line of therapy has not produced convincing results so far but may be more promising as a combination therapy. Multicenter studies with estrogens, anti-inflammatory drugs and anti-oxidants are in progress in USA.

ACKNOWLEDGMENT

The author thanks Christine Mesmer for typing and editing the manuscript.

REFERENCES

Allain, H. Maruelle, L., Beneton, C., Rouault,G., and Belliard, S. (1996). Acute episodes of high blood pressure with tacrine. Presse Medicale 25:30 pp 1388–1389.

Amberla, K., Nordberg, A., Viitanen, M., and Winblad, B. (1993). Long-term treatment with tacrine (THA) in Alzheimer's disease - evaluation of neuropsychological data. Acta Neurologica Scandinavica Supplement 149 pp 55–57.

Anderson, J.P., Refolo, L.M., Wallace, W., Mehta, P., Krishnamurthi, M., Gotlib, J., Bierer, L., Haroutunian, V., Perl, D. and Robakis, N.K. (1989). Differential brain expression of the Alzheimer's amyloid precursor protein. EMBO Journal 8: pp 3627–3632.

Becker, R., Moriearty, P., and Unni, L. (1991). The second generation of cholinesterase inhibitors: Clinical and pharmacological effects In Cholinergic basis for Alzheimer Therapy pp 263–296. Edited by Becker, R. and Giacobini, E.. Boston: Birkhäuser.

Beeson, J.G., Shelton, E.R., Chan, H.W. and Gage, F.H. (1994). Differential distribution of amyloid protein precursor immunoreactivity in the rat brain studied by using five different antibodies. Journal of Comparative Neurology 342: pp 78–96.

Biyah, K., Molimard, M, Naline, E., Bazelly, B., and Advenier, C. (1996). Indirect muscarinic receptor activation by pentamidine on airway smooth muscle British Journal of Pharmacology 119:6 pp 1131–1136.

Buxbaum, J.D., Oishi, M., Chen, H.I., Pinkas-Kramarski, R., Jaffe, E.A., Gandy, S.E. and Greengard, P. (1992). Cholinergic agonists and interleukin 1 regulate processing and secretion of the Alzheimer βA4 amyloid protein precursor. Proceedings of National Academy of Science, USA 89: pp 10075–10078.

Buxbaum, J.D. (1995). Post-translational control of the amyloid b-protein precursor processing. In Pathobiology of Alzheimer's Disease, Academic Press Limited, pp 98–114.

Checler, F. (1995). Processing of the b-amyloid precursor protein and its regulation in Alzheimer's disease. Journal of Neurochemistry 65:4 pp 1431–1444.

Chong,Y.H. and Suh, Y.H. (1996). Amyloidogenic processing of Alzheimer's amyloid precursor protein in vitro and its modulation by metal ions and tacrine. Life Science 59/7 pp 545–557.

Eagger, S., Richards, M., and Levy, R. (1994). Long-term effects of tacrine in Alzheimer's disease: an open study. International Journal of Geriatric Psychiatry 9 pp 643–647.

Feldman, S., and Karalliedde, L. (1996). Drug interactions with neuromuscular blockers. Drug-Safety 15:4 pp 261–273.

Giacobini, E. (1987). Models and strategies of cholinomimetic therapy of Alzheimer disease. In Cellular and molecular basis of cholinergic function, pp 882–901. Edited by Dowdall , M.J. and Hawthorne, J.N. England: Ellis Horwood.

Giacobini, E.(1996a). Closer to the truth about ApoE-4. Alzheimer Insights 2/2 pp 1–4.

Giacobini, E. (1996b). Cholinesterase inhibitors do more than inhibit cholinesterase. In Alzheimer Disease: From Molecular Biology to Therapy (pp 187–204). Edited by Becker, R. and Giacobini, E.. Boston: Birkhäuser.

Giacobini, E., Becker, R., Elble, T., Mattio, M., McIlhany, M. and Scarsella, G. (1987). Brain acetylcholine - A view from the cerebrospinal fluid. In Neurobiology of Acetylcholine (pp 85–101). Edited by Dun, N., New York: Plenum Press.

Giacobini, E. and Cuadra, G. (1994). Second and third generation cholinesterase inhibitors: From preclinical studies to clinical efficacy. In: Alzheimer Disease: Therapeutic Strategies (pp. 155–171). Edited by Giacobini, E. and Becker, R., Boston: Birkhäuser.

Giacobini, E., Mori, F., Buznikov, A. and Becker, R. (1995). Cholinesterase inhibitors alter APP secretion and APP mRNA in rat cerebral cortex. Society of Neuroscience Abstracts **21** pp 988.

Giacobini, E., Zhu, X.D., Williams, E., and Sherman, K.A. (1996). The effect of the selective reversible acethylcholinesterase inhibitor E2020 on extracellular acetylcholine and biogenic amines levels in rat cortex. Neuropharmacology **35** (2) pp 205–211.

Gracon, S., Smith, F. and Hoover, T. (1996). Long-term tacrine treatment: Effect on nursing home placement and mortality. *In Alzheimer Disease:* From Molecular Biology to Therapy (pp 205–209). Edited by Becker, R. and Giacobini, E.. Boston: Birkhäuser.

Hallak, M. and Giacobini, E. (1986). Relation of brain regional physostigmine concentrations to cholinesterase activity, and acetylcholine and choline levels in rat. Neurochemistry Research 11 pp 1037–1048.

Haroutunian, V., Greig, N., Pei, X.F., Utsuki, L., Acevedo, L.D., Gluck, R., Davis, K.L. and Wallace, D. (1996). Pharmacological modulation of Alzheimer's b-amyloid precursor protein levels in the CSF of rats with forebrain cholinergic system lesions. Society for Neuroscience 22 461.7 pp 1169.

Haroutunian, V., Greig , N.H., Gluck, R., Fiber, E., Davis, K.L. and Wallace, W.C. (1995). Selective attenuation of lesion-induced increases in secreted b-APP by acetylcholinesterase inhibitors. Society for Neuroscience 21 208.1 pp 5.

Hoshi, M, Takashima, A., Murayama, M., Yasutake, K., Yoshida, N., Ishiguro, K., Hoshino, T. and Imahori K. (1997). Nontoxic amyloid beta peptide (1–42) suppresses acetylcholine synthesis. Possible role in cholinergic dysfunction in Alzheimer's disease. Journal of Biological Chemistry 272/4 pp 2038–2041.

Imbimbo B.P., Perini,M.,Verdelli,G., and Troetel ,W.M. (1997). Two year treatment of Alzheimers's disease with eptastigmine. International Congress of Neurology (Buenos Aires)Abstracts.

Kang, J., Lemaire, H.G., Unterbeck, A., Salbaum, J.M., Master, C.L., Grzeschil, K.H., Multaup, G., Beyreuther, K. and Muller-Hill, B. (1987). The precursor of Alzheimer disease amyloid A4 protein resembles a cell-surface receptor. Nature 325 pp 733–736.

Knopman, D., Schneider, L., Davis, K., Talwalker, S., Smith, F., Hoover, T. and Gracon, S. (1996). Long-term tacrine treatment. Neurology 47 pp 166–177.

Lahiri, D.K., Lewis, S. and Farlow, M.R. (1994). Tacrine alters the secretion of the beta-amyloid precursor protein in cell lines. Journal of Neuroscience Research 37 pp 777–787.

Lebert, F., Hasenbroekx, C., Pasquier, F., and Petit , H. (1996). Convulsive effects of tacrine. Lancet 347 pp 1339–1340.

Maany, I. (1996). Adverse interaction of tacrine and haloperidol. Americann Journal of Psychiatry 153:11 pp1504.

Mattio, T., McIlhany, M., Giacobini, E. and Hallak, M. (1986). The effects of Physostigmine on acetylcholinesterase activity of CSF, plasma and brain. A comparison of intravenous and intraventricular administration in beagle dogs. Neuropharmacology *25* pp 1167–1177.

McSwain, ML., and Forman LM. (1995). Severe parkinsonian symptom development on combination treatment with tacrine and haloperidol (letter). Journal of Clinical Psychopharmacology 15 pp:284.

Minthon, L, Nilsson, K., Edvinsson, L., Wendt, P.E. and Gustafson, L. (1995). Long-term effects of tacrine on regional cerebral blood flow changes in Alzheimer's Disease. Dementia ***6*** pp 245–251.

Mori, F., Cuadra, G. and Giacobini, E. (1995b). Metrifonate effects on acetylcholine and biogenic amines in rat cortex. Neurochemistry Research 20 (9) pp 1081–1088.

Mori, F., Lai, C.C., Fusi, F. and Giacobini, E. (1995a). Cholinesterase inhibitors increase secretion of APPs in rat brain cortex. Neuro Report 6 (4) pp 633–636.

Nitsch, R.M., Farber, S.A., Growdon, J.H. and Wurtman, R.J. (1993). Release of amyloid beta-protein precursor derivatives by electrical depolarization of rat hippocampal slices. Proceeding of National Academy of Science, USA, 90 pp 191–193.

Nitsch, R.M. and Growdon, J.H. (1994). Role of neurotransmission in the regulation of amyloid beta-protein precursor processing. Biochemical Pharmacology 47 (8):1275.

Nordberg, A. (1993). Clinical studies in Alzheimer patients with positron emission tomography. Elsevier Science Publishers, Behavioural Brain Research 57 pp 215–224.

Nordberg, A. (1995). Long term treatment effects on progression of Alzheimer's disease as determined by functional brain studies. In Research Advances in Alzheimer's disease and Related Disorders (pp 293–298). Edited by Ikbal, K., Mortimer, J.A., Winblad, B., Wiesniewski, H.M.. Chichester: John & Sons.

Poirier, J., Delisle, M.C., Quirion, R., Aubert, I., Farlow, M., Lahiri, D., Hui, S., Bertrand, P., Nalbantoglu, J., Gilfix, B.M. and Gauthier, S. (1995). Apolipoprotein E4 allele as a predictor of cholinergic deficits and treatment outcome in Alzheimer disease. Proceeding of National Academy of Science 92 pp 12260–1226.

Paddle, B. and Dowling, M. (1996). Blockade of nicotinic responses by anticholinesterases. General Pharmacology 27:5 pp 861–872.

Roses, A. (1995). Perspective on the metabolism of apolipoprotein E and the Alzheimer disease. Experimental Neurology *132* pp 149–156.

Sandbrink, R., Hartmann, T., Masters, C.L. and Beyreuther, K. (1996). Genes contribution to Alzheimer's disease. Molecular Psychiatry 1 pp 27–40.

Soininen, H., Kosunen, O., Helisalmi, S., Mannermaa, A., Paljärvi, L., Talasniemi, S., Ryynänen, M. and Riekkinen, P., Sr. (1995). A severe loss of choline acetyltransferase in the frontal cortex of Alzheimer patients carrying apolipoprotein e4 allele. Neuroscience Letters 187 pp 79–82.

Soininen, H., and Riekkinen, J. Sr (1996). Apolipoprotein E, memory and Alzheimer's disease. Trends in Neursoscience *19* pp 224–228.

Solomon,P., Knapp, M., Gracon, S., Groccia , M. and Pendlebury, W. (1996). Long-term tacrine treatment in patients with Alzheimer's disease. Lancet 348 pp 275–276.

Van Broeckhoven, C. (1995). Presenilins and Alzheimer disease. Nature Genetics 11 pp 230–232.

Vito, P., Lacaná, E. and D'Adamio, L. (1996). Interfering with apoptosis: Ca^{2+} -binding protein ALG-2 and Alzheimer's disease gene *ALG*-3. *Science 271* pp 521–525.

Wilcock, G., Scott, M. and Pearsall, T. (1994). Long-term use of tacrine. Lancet 343 pp 294.

Yankner, B (1996). New clues to Alzheimer's disease: Unraveling the roles of amyloid and tau. Nature Medicine 2.8 pp 850–852.

Zaborsky L. and W.E. Cullinan, (1996). Direct cathecholaminergic-cholinergic interactions in the basal forebrain.I. Dopamine-beta-hydroxylase input to cholinergic neurons. Journal of Comparative Neurology 374 pp 5345–554.

Zhu, X.D., Giacobini, E. and Hornsperger, J.M. (1995). Effect of MDL 73,745 on acetylcholine and biogenic amine levels in rat cortex. European Journal of Pharmacology 276 pp 93–99.

Zhu, X.D., Cuadra, G., Brufani, M., Maggi, T., Pagella, P.G., Williams, E. and Giacobini, E. (1996). Effects of MF268, a new cholinesterase inhibitor, on acetylcholine and biogenic amines in rat cortex. Journal of Neuroscience Research 43 pp 120–126.

Part IV

Neurotrophins and Neuromodulators, Neurohormones, Cell Adhesion Molecules

18

STEROID AND PROTEIN REGULATORS OF GLIAL CELL PROLIFERATION

Luis Goya

Research Associate at the Instituto de Bioquímica (CSIC-UCM)
Facultad de Farmacia, Ciudad Universitaria
28040-Madrid, Spain

1. INTRODUCTION TO CELL GROWTH REGULATION

Normal development, differentiation and proliferation of tissues and cell types within multicellular organisms are stringently regulated by a complex combination of environmental factors that include systemic steroid and protein hormones, local and systemic acting growth factors, extracellular matrix components and cell-cell interactions. In most tissues, cell differentiation is accompanied by an arrest of proliferation and the maintenance of the growth arrested state is controled by specific hormonal signals which inhibit the expression or activity of growth stimulatory factors and/or induce growth suppressor gene products (Aaronson, 1991). However, for certain physiological processes, cell proliferation is neccessary to maintain populations of specific cell types. For instance, stem cells in the bone marrow provide a continuous source of hematopoietic cells in animals (Zipori, 1992). Quiescent cells will proliferate during tissue regeneration in a damaged organ, such in the liver (Fausto and Weber, 1993), whereas mammary epithelial cells are hormonally stimulated to proliferate during pregnancy and lactation (Ceriani, 1974). Finally, glial proliferation in response to brain neuronal damage has been well documented (Landis, 1994). The maintenance of these cells in the proliferative state, or the triggering of quiescent cells to proliferate, requires the selective stimulation of growth stimulatory gene products and/or suppression of growth inhibitory gene products.

Although steroid and protein growth regulators act through different classes of receptors, the final targets of their respective signal transduction pathways are components of the cell proliferation machinery which control the ability of a cell to divide.

1.1. Cell Cycle Regulation

In recent years, several classes of proteins have been characterized which drive the cell through selective steps within the cell cycle that culminates in cell division (Fridovich et al. 1990). These cell cycle regulators include the various cyclins (A, B, C, D1, D2, D3

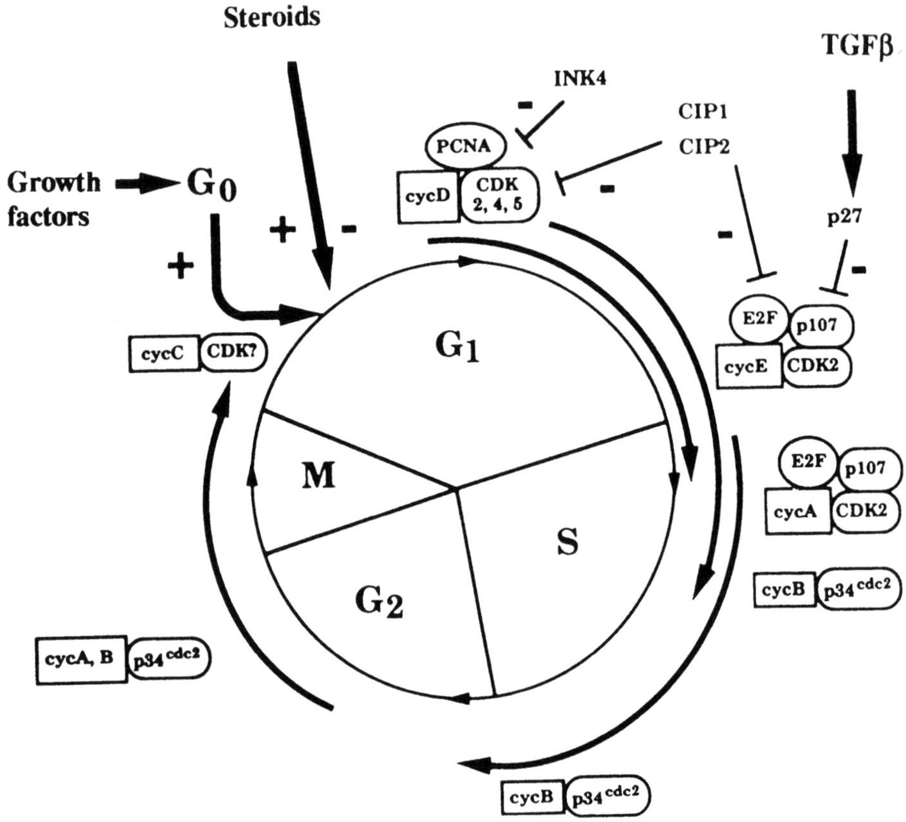

Figure 1. Cell cycle control by steroid and protein growth regulators. In mammalian cells most growth factors and steroids act early in G1, whereas TGF-beta inhibits cell cycle progression late in G1. The diagram shows the temporal appearance of cyclin complexes which form at specific steps within the cell cycle. cycA: cyclin A; cycB: cyclin B; cycC: cyclin C; cycD: cyclin D; cycE: cyclin E; CDK: cyclin dependent kinase; p34cdc2: cyclin dependent kinase-1; p107: retinoblastoma related protein; PCNA: proliferative cell nuclear antigen; E2F: E2F transcription factor; p27: TGF-beta regulated cell cycle inhibitor; CIP: CDK-interacting protein; and INK4: 16,000 Da inhibitor of cyclin dependent kinase. Reprinted with permission from Firestone et al. Steroid and protein regulators of normal and abnormal cell proliferation, in Hormones and Aging, Eds. P.S. Timiras, W.D. Quay and A. Vernadakis; 1995. Copyright CRC Press, Boca Raton, Florida.

and E) and several cyclin-dependent kinases (cdk2, cdk4, cdk5 and p34cdc2) which form specific cyclin-cdk complexes depending on the phase of the cell cycle (Sherr, 1993) (figure 1). Several other components of the cell cycle have been identified by their ability to interact with specific cyclin-cdk complexes (Pines, 1993), and recently, some of the cyclin-cdk complex-associated proteins have been shown to act as inhibitors of the kinase activity, causing an arrest of the cell cycle progression (Pines, 1993) (figure 1).

In normal and untransformed cells, cellular proliferation has been proposed to be mediated by the concerted action of cell cycle regulated genes which drive the cells through G1 and the rest of the cell cycle. In contrast, dysregulation of the G1 phase of the cell cycle has been involved in uncontrolled proliferation observed in many types of neoplastically transformed cells (Firestone et al., 1995).

Expression and activity of cell cycle regulatory proteins is timed and modulated by transcriptional, translational and post-translational mechanisms. For example, D cyclins have been shown to be transcriptionally regulated by trophic factors (Ross, 1996). While mRNAs for cyclins D2 and D3 are detected throughout the cell cycle, protein levels peak near the G1-S transition, indicating the importance of post-trancriptional regulation, which is likely to occur at the level of translation. Consistent with this notion, several cyclins are known to possess disproportionately large 3′-untranslated regions (UTRs), which might have destabilizing influences on the message, since truncation of cyclin D1 3′-UTR leads to increased mRNA levels in tumors (Ross, 1996). Therefore, the cell cycle is poised to respond rapidly to fine alterations in transcription or translation rates, as well as to mechanisms that enhance protein degradation. In addition, there is evidence that expression of at least the D cyclins is cell-type specific both in somatic cells and developing brain. Thus, there is ample opportunity for subtle modulation of the cell cycle in a stage-specific and regional-anatomic manner during CNS embryogenesis.

1.2. Control of Cell Proliferation in the CNS

The importance of mitotic control to development is particularly evident in the central nervous system (CNS), which is among the earliest organotypic specializations in the embryo. In mammals, regulation of mitotic rate coincides with major morphogenetic events in CNS development. Examples include the acceleration of the cell cycle (from 7 to 3 h), largely because of contraction of G1 and G2, in cells that traverse the region of the primitive streak (Ross, 1996). Moreover, there is progressive lengthening of cell cycle times within the neuroepithelium over the period of neural tube closure. Later in development, variation in mitotic rates within the ventricular germinal epithelium appears to contribute to the regional differences in neuronal number in the cerebral cortex (Ross, 1996).

However, mitotic control of cell proliferation in the CNS is not only important during development but also in response to specific neuronal damage during adulthood. After an injury that destroys neuronal tissue, changes occur in the residual brain occupied by astrocytes and the loss of the neuronal volume is compensated for by both hyperthrophy of preexisting astrocytes and the proliferation of astrocytes (Landis, 1996). The signal for astrocytic division after injury is still uncertain, and the nature of the dividing cell is not yet known. Astrocytes may arise in injured brain from the division of previously differentiated astrocytes, from the division of a precursor population, or from both. Astrocytes (or astrocytic precursors) probably proliferate at low rates in normal adult brain; this is likely to be accompanied by a balancing rate of astrocyte cell death. There is virtually no information about the cessation of division after injury, except that astrocytes in injured brain do not continue to divide to the point that their volume exceeds that of the original brain volume (Landis, 1996).

Current evidence indicates that cell division and cycle arrest are controlled by both intrinsic characteristics of the CNS progenitor and its response to signals provided by surrounding cells. For example, oligodendrocyte precursor cells in culture can be stimulated to divide by combinatorial influences of the protein growth factors, platelet-derived growth factors and insulin-like growth factor-I (Barres et al. 1994). However, oligodendrocytic presursors will stop dividing after approximately eight rounds, even in the presence of these mitogenic factors (Barres et al. 1994). This phenomenon appears to depend upon the concerted actions of growth factors and small hydrophobic molecules such as thyroid hormones, glucocorticoids and retinoid acid, because in the absence of these hydrophobic molecules, growth factors will sustain precursor-cell division up to 12 or more times. All three effector molecules (thyroid hormones, glucocorticoids and retinoid acid) are known to act through receptors that can, depending on the molecular context, activate or repress transcription of specific genes. Therefore, in the absence of external signals, intrinsic levels of transcription factors will decrease over time to halt division (Barres et al. 1994). In the presence of mitogens, the activity of transcription factors that have a positive regulatory effect on the cell cycle will be maintained above the critical threshold, while in the presence of hydrophobic effectors, this activity will drop to levels resulting in cell cycle arrest.

1.3. Cell Cycle Regulation by Steroids, Protein Hormones, and Growth Factors

In mammalian cells, control of the G1 phase by steroid hormones, protein growth factors and other biological regulators appears to control cell proliferation and differentiation (Pardee, 1989). The location within G1 where extracellular signals act is dependent on the cell type, state of differentiation and transformation, and the signal regulating the cell (figure 1). Steroid and steroid-like hormones which act through intracellular DNA binding receptors (see later section) have well documented effects on the proliferation of a variety of cells, including those derived from brain, liver, pancreas, ovary, prostate and mammary gland. Current evidence has shown that steroids exert their positive or negative growth effects relatively early in the G1 phase of the cell cycle (Goya et al. 1993b; Sanchez et al. 1993), or, in some cases, by regulating entry in the S phase (Berns et al. 1990). Growth factors generally act by inducing quiescent G0 cells to enter into and progress through G1 or by driving cells to cycle past a particular restriction point in G1. As part of the growth stimulatory mechanism, particular sets of cell cycle-regulated G1 marker genes (such as c-fos and c-myc) are temporally expressed in a specific pattern based on movement of cells through G1 and into S phase (Pardee, 1989). Depending on the cell type, growth factors either synergize with or override the effects of other growth regulatory hormones. For example, EGF or transforming growth factor-alpha (TGFα), which acts through the EGF receptor, can override the growth suppression effects of glucocorticoids in rat mammary tumor cells (Goya et al. 1993a).

Transforming growth factor-beta (TGFβ) is the most extensively characterized of the protein growth inhibitors (Massague, 1992). TGFβ inhibits proliferation of a variety of epithelial cells by blocking cell cycle progression late in the G1 phase of the cell cycle (figure 1). Much less is known about the function of true protein hormones in cell cycle control. Insulin stimulates cell cycle progression in certain breast cancer cells (Alexander et al. 1993), and override the growth suppressive effects of glucocorticoids or TGFβ in some hepatoma cells (Hung et al. 1993).

2. GROWTH FACTORS, RECEPTORS, AND SIGNAL TRANSDUCTION PATHWAYS

Cellular responses to polypeptide growth factors, differentiation factors, morphogens and hormones primarily stem from a cascade of signals that emanate from the cell surface, are transduced within the cytoplasm, and eventually diffuse into the nucleus resulting in the expression of critical genes that dictate specific functions. The majority of growth factors, via autocrine and paracrine mechanisms, interact with tyrosine kinase cell surface receptors, which induce a wave of secondary messengers, protein-protein interactions, and activation of downstream serine/threonine protein kinases (Ullrich and Schlessinger, 1990). One of the final targets is the transcriptional machinery in the nucleus with the direct phosphorylation of sequence specific DNA binding proteins culminating in the induction/repression of particular sets of target genes (Hunter and Karin, 1992). Alterations in the program and rates of gene expression affect cell proliferation and differentiation.

2.1. Growth Factor Receptors

Growth factors exert their pleiotropic actions by binding to cell surface receptors. These receptors can be categorized into three major functional groups: (1) seven-pass transmembrane receptors coupled to GTP-binding proteins (G proteins) in the cytoplasmic side of the plasma membrane; (2) transmembrane receptors that lack protein kinase activity, but associate with cytoplasmic tyrosine kinases; and (3) cell surface spanning receptors which contain a tyrosine kinase domain (figure 2). In addition, the TGFβ

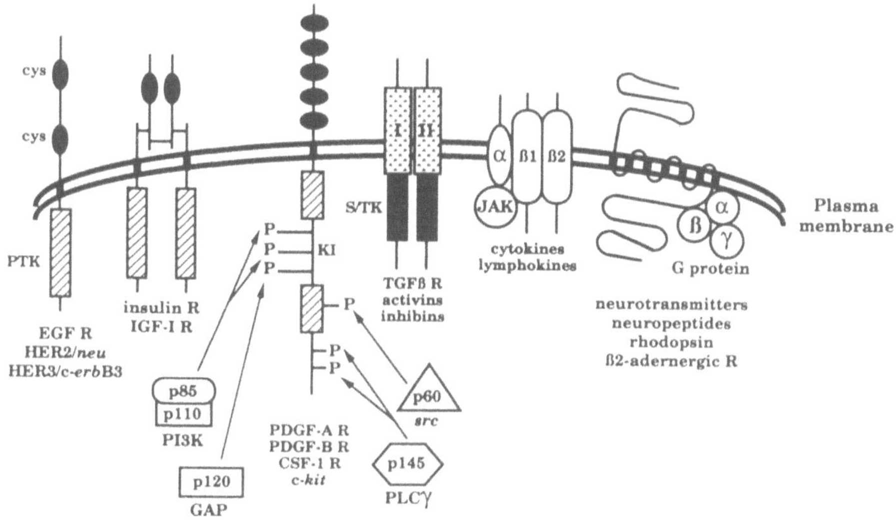

Figure 2. Structural organization and topology of growth factor receptor signalling molecules. Generalized structures are shown for transmembrane tyrosine protein kinase receptors (EGF, insulin and PDGF receptors), transmembrane serine/threonine kinase receptors (TGF-β receptor), transmembrane receptors lacking a kinase domain (cytokines), and seven pass transmembrane receptors (neurotransmitters). Extracellular domains of both monomeric (EGF) and dimeric (insulin and IGF) receptors contain two cysteine rich regions (oval with dots), whereas, other transmembrane tyrosine kinases (PDGF receptor) have five immunoglobulins-like repeats (filled oval) in the extracellular domain. The cytoplasmic protein tyrosine kinase (PTK) domain (hatched bars) can be either interrupted by a kinase insert (KI) region (e.g. PDGF receptor), or remain uninterrupted. Phosphatidylinositol-3-kinase (PI3K), rasGTPase-activating protein (GAP), Phospholipase C γ (PLCγ), and p60src are proteins containing SH2 domains that interact with autophosphorylated tyrosine kinase receptors. Transmembrane serine/threonine kinase (S/TK) receptors (dark bars) contain type I and type II receptor components (e.g. TGF-β receptor). Receptors for cytokines, lymphokines, growth hormone and prolactin are transmembrane heteromeric complexes (α/β) lacking any protein kinase domain and are associated with cytoplasmic tyrosine kinases (e.g. JAK). The seven-pass transmembrane receptors interact with G proteins comprised of α, β and γ subunits (e.g. neurotransmitters). Reprinted with permission from Firestone et al. Steroid and protein regulators of normal and abnormal cell proliferation, in Hormones and Aging, Eds. P.S. Timiras, W.D. Quay and A. Vernadakis; 1995. Copyright CRC Press, Boca Raton, Florida.

transmembrane receptor family contains a serine/threonine kinase domain on the cytoplasmic side (figure 2). Each class of receptors uses distinct signal transduction pathways which effectively transmit and transduce the extracellular signal to the nuclear apparatus resulting in the regulation of different sets of specific genes involved in cell growth, development and differentiation.

Members of the seven-pass transmembrane receptors subfamily are glycoproteins that bind ligands such as neurotransmitters, neuropeptides and polypeptide hormones and growth factors (Firestone et al. 1995). Upon receptor stimulation, the alpha subunit of the G protein, which has GTPase activity, dissociates from the membrane associated complex and by stimulating or inhibiting specific enzymes generate an array of second messengers including cAMP. Moreover, other targets of receptor activated G proteins are guanylate cyclase and phospholipase C-B (PLC-B) which produces second messengers such as diacylglycerol (DAG) and inositol triphosphates (IP3) that mobilize intracellular calcium and activate protein kinase C (PKC) (Simon et al. 1991).

A second class of transmembrane receptors bind ligands such as growth hormone, prolactin and a variety of lymphokines and cytokines (figure 2). These receptors do not posses any intrinsic tyrosine or serine/threonine protein kinase activity and are not associated with G proteins (Bazan, 1989).

2.2. Tyrosine Kinase Receptors and Their Substrates

Many different growth factors, including EGF, IGF-I and platelet-derived growth factor (PDGF) evoke cellular responses by binding to and activating cell surface receptors with intrinsic protein tyrosine kinase activity. Members within this family posses a cytoplasmic domain that contains tyrosine kinase catalytic activity (figure 2). The tyrosine kinase domain transmits the extracellular signal intracellularly and thereby mediates the biological responses (Ullrich and Schlessinger, 1990).

Both monomeric (EGF and TGFα) and dimeric (PDGF, CSF-I, NGF) growth factors activate their cognate receptors by inducing receptor oligomerization after binding to the extracellular domain (Williams, 1989). Ligand binding is accompanied by an alteration in conformation of the extracellular domain, facilitating interactions between cytoplasmic domains of adjacent receptors. This process leads to activation of kinase function which is essential for signal transduction and generation of both early and delayed cellular responses.

The tyrosine kinase receptor substrates are characterized by the presence of unique sequence motifs known as src homology or SH domains (Koch et al. 1991). The src protooncogene family of cytoplasmic receptor substrates contains three regions of distinct structural and functional modules termed SH1, SH2 and SH3. SH2 domains serve as docking sites for phosphotyrosine containing proteins and, therefore, recruit receptor substrates and other receptor binding proteins to specific autophosphorylated regions of growth factor receptor tyrosine kinases. Tyrosine phosphorylation creates high affinity sites for the binding of SH2 domain containing proteins such as rasGTPase Activating Protein (rasGAP), phosphatidyl inositide 3-kinase (PI3-kinase), phospholipase C gamma (PLCgamma) and pp60src (figure 2).

Tyrosine phosphorylation of PLCgamma has been observed in response to several different mitogenic factors such as EGF, PDGF, FGF and NGF (Carpenter, 1992). Phosphorylation of PLCgamma leads to rapid hydrolysis of phosphoinositides, generating second messengers such as DAG and IP3, which then activate PKC and mobilize intracellular calcium. PKC activation leads to phosphorylation of downstream targets that result in transcriptional activation and generation of biological responses. PI3-kinase, a dimeric protein, is also tyrosine phosphorylated by growth factor addition (figure 2). Another substrate for receptor tyrosine kinases is rasGTPase activating protein (GAP), a molecule that modulates the GTPase activity of ras (figure 2). Ras is a low molecular weight GTP binding protein that plays an important role in growth factor induced mitogenesis. Another important kinase identified in growth factor stimulated cells is the Mitogen Activated Protein Kinase (MAP kinase)(Thomas, 1992). In general, phosphorylation of transcription factors serves to propagate the signal transduced in the cytoplasm into the nucleus, enables the nuclear localization of latent cytoplasmic transcription factors, and provides positive and negative regulation of their DNA binding activity.

2.3. Role of Growth Factors on Glial Cell Growth and Differentiation

Hormones and growth factors mediate a variety of signals in the developing central nervous system (CNS) both in neurons and glial cells. They can signal cells to proliferate, quiesce, migrate or differentiate. Several growth factors which are mitogenic for various cell types also regulate cell division and proliferation of glial cells in culture (Goya et al. 1996; Morrison and De Vellis, 1981). Growth factors that evoke oligodendrocytic expression in glial progenitor cells are: fibroblast growth factor (FGF)(Pruss et al. 1982), epidermal growth factor (EGF)(Simpson et al. 1982), human platelet-derived growth factor (PDGF)(Noble et al. 1988) and some other neuronal factors (Sakerallidis et al. 1984). Factors and growth regulators shown to increase astrocytic expression in glial progenitor cells include: insulin, transferrin, platelet-activating factor (Kentroti et al. 1990), muscle-derived factors (Brodie and Vernadakis, 1991), nucleotides as well as glucocorticoids (Vernadakis et al. 1992).

While glial cells responsiveness to these signals has been reported in vivo, further study of the specific effect of each factor on glial growth and differentiation may benefit from in vitro models. Primary cell cultures and non-tumorigenic established cell lines remain subject to growth control and display some characteristic features including anchorage and serum dependence. In addition, in these cell lines, growth and proliferation are closely associated to differentiation, and promotion of differentiation usually evokes inhibition of growth. However, transformed tumorigenic cell lines are able to bypass growth control, and may proliferate in a highly differentiated state. Differentiation markers, other than cell growth, have been described in order to phenotypically characterize the cultures and to develop in vitro models for cell differentiation and aging.

2.3.1. Differentiation Markers for Glial Cell Models. Two enzyme activities which are characteristic of either oligodendrocytes or astrocytes are often used as markers to study glial cell differentiation and aging. The enzyme activity marker for oligodendrocytes is 2',3'-cyclic nucleotide 3'-phosphohydrolase (CNP), which breaks down the cyclic structure of the cyclic nucleotides so that the free phosphate groups can be released by a phosphatase (Poduslo, 1975). The activity of this enzyme is high in oligodendrocytes because the free phosphate groups are needed during the process of myelination, the main function of oligodendrocytes.

The enzyme activity marker for astrocytes is glutamine synthetase (GS), which converts glutamic acid into glutamine (Martinez-Hernandez et al. 1977). The high GS activity present in astrocytes provides a means of eliminating an excess of the excitatory amino acid glutamate (with high cell toxicity), thereby contributing to the control of neuronal death.

Because of its similarities with primary glial cells in culture, C6 rat glioma cell line has provided a useful model to study glial cell properties, glial growth factors and sensitivity of glial cells to various substances and conditions (Vernadakis et al. 1992). C6 rat glioma cells (2B clone) exhibit differential enzyme expression with cell passage (Parker et al. 1980), the activity of CNP is markedly high and that of GS very low in early passages (up to 30 passages). This relationship is reversed in late passage cells (beyond passage 70). This finding suggests that early passage C6 glioma are less differentiated, more glioblastic in nature, and that with succesive passages they differentiate into a primarily astrocytic population (Parker et al. 1980). Therefore, C6 glioma cell line has been frequently used as a model in vitro to study changes in glial cells with aging (Vernadakis et al. 1992).

Two important questions have been addressed recently in this model in our laboratory: (1) the effect of several growth factors on the growth properties of two different populations of C6 glioma cell line, early (young cultures) and late (aged cultures) passage, and (2) the dynamic of the process of differentiation from oligodendrocytic to astrocytic expression, as well as changes in the responsiveness of glial cells with aging in culture by measuring CNP and GS activities as markers for the phenotypically different cell populations.

2.3.2. Role of Insulin on Glial Cell Growth and Differentiation. Insulin, as well as insulin-like growth factors are present in the brain. Insulin increases astrocytic expression in cultures of glial cells (Lee et al. 1992). Rat C6 glioma cells express high levels of insulin-like growth factor-I (IGF-I) as well as IGF-I receptors (Kiess et al. 1989; Lowe et al. 1992). The effects of insulin in brain may be mediated through the IGF-I receptors. In C6 glioma cell line, IGF-I expression may be coupled to cellular proliferation and tumorigenicity. Our results support the physiological function of insulin and IGFs as differentiating factors for C6 glioma.

GS activity is a well characterized marker for astrocytes, therefore, any change in its activity may reflect a variation in the differentiation state of the cell. The remarkable increase in GS activity observed after treatment of early passage C6 with insulin strongly suggests that this factor induces differentiation of early passage C6 glioma (figure 3, left pannel). Furthermore, we had previously shown that insulin significantly reduced DNA synthesis on early passage C6 cells (Goya et al. 1996). All together, it can be concluded that insulin shows a clear effect on the phenotypic differentiation of glioma C6 cells in vitro.

2.3.3. Role of PDGF on Glial Cell Growth and Differentiation. Although the function of PDGF in the CNS is still unclear, a specific role in the regulation of glial cell development has been proposed (Richardson et al. 1990). In the developing rat CNS, glial progenitor cells (O-2A progenitors) are capable of giving rise to either oligodendrocytes or type 2 astrocytes in vitro (Raff et al. 1983). When O-2A progenitors are cultured in the absence of growth factors, they stop dividing and differentiate rapidly into oligodendrocytes (Raff et al. 1983). When cultured in medium supplemented with PDGF, O-2A progenitors are stimulated to divide (Noble et al. 1988; Richardson et al. 1988), and oligodendrocytes are produced over a period of several weeks in vitro just as in vivo (Raff et al. 1985; Raff et al. 1988). Thus, PDGF may be an important factor in the regulation and control of the final number of oligodendrocytes, and the time and rate of their production during development.

Data from our laboratory have shown that both early and late passage glioma C6 cells can be stimulated to grow by PDGF and indicate that the regulation of glial growth

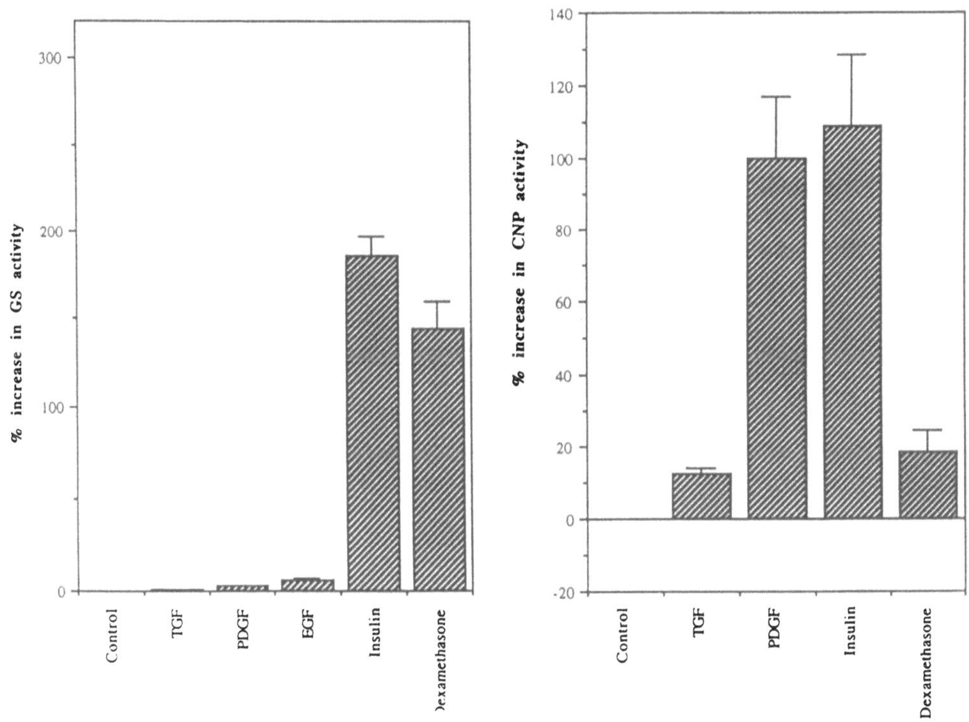

Figure 3. Effect of growth factors on the GS and CNP activities of C6 glioma cells. Early passage C6 glioma cells were plated at 0.7X106 cells per 100mm culture dish, incubated with TGFα (20ng/ml), PDGFβ (20ng/ml), EGF (20ng/ml), insulin (10ng/ml) or dexamethasone (10uM), for 7 days, and then harvested and prepared for assay of GS and CNP activities. For the GS assay (left pannel), three samples of each condition were used per assay, which was repeated three times and the results pooled. Statistical differences were found between untreated (control) and insulin and dexamethasone treated cells ($p<0.001$). For the CNP assay (right pannel), four samples of each condition were used per assay, and this was repeated with similar results. PDGFβ and insulin treatment significantly increased ($p<0.001$) CNP activity in early passage C6 glioma cells.

by this factor is age-independent and unrelated to the stage of differentiation of the cell (Goya et al. 1996). In fact, late passage cells respond faster than early passage to growth promotion, suggesting that with age, glial cells become more sensitive to factors affecting oligodendrocytes in order to compensate for the marked decrease in the number of this type of cells in the culture. However, the fact that CNP activity was increased by PDGF only in early passage glioma cells, seems to indicate that oligodendrocyte-like cultures (early passage glioma) are responsive to the differentiating effects of the growth factor (figure 3, right pannel), while the astrocytic-like culture (late passage) is only responsive to the growth-promoting effect of PDGF (Goya et al. 1996).

3. STEROID HORMONE REGULATION OF GROWTH

Steroid hormones, which include estrogens, progestins, androgens, mineralocorticoids, glucocorticoids and members of the vitamin D family, have a wide range of biological activities which include the control of reproduction, stress, salt and water balance, metabolism and development. After diffusion through the plasma membrane, steroids bind to and potentiate a functional change in their cognate receptor resulting in either stimulated or repressed transcription of steroid regulated genes (Truss and Beato, 1993). In the case of glucocorticoid receptors, ligand binding causes release of the 90-KDa heat shock protein from the inactive receptor complex in the cytoplasm which allows translocation to the nucleus. Other steroid receptors, such as estrogen, progestin and androgen receptors, are localized in the nucleus prior to ligand binding. The potent effects of steroids on gene transcription occur by specific binding of steroid receptor complexes to DNA transcriptional enhancer elements known as hormone response elements (HREs) within promoters of steroid regulated genes, and/or interfering or synergizing with other transcription factors (Truss and Beato, 1993).

Depending on the tissues, glucocorticoids can either stimulate or suppress growth of cells by altering gene expression of oncogenes such as c-jun (Goya et al. 1993b) or growth factors such as PDGF (Haraguchi et al. 1991), and/or by disrupting an autocrine loop of TGFα in a rat mammary tumor cell line (Alexander et al. 1993; Goya et al. 1993a).

3.1. Role of Glucocorticoids on Glial Cell Growth and Differentiation

Dexamethasone is a synthetic glucocorticoid with a potent cell growth-inhibitory effect, which is frequently used in the treatment of specific neoplastic growth such as brain and breast tumors. Its effects include multiple actions in the nervous system ranging from effects on appetite and behavior to effects on cell metabolism, cell growth, and differentiation (Goya et al. 1995). Glucocorticoid receptors have a widespread distribution in the CNS and have been found in glial cells (Goya et al. 1996; Reul and DeKloet, 1985; Vielkind et al. 1990). The inhibitory effect of dexamethasone previously observed in our laboratory in both early and late passage cells indicates that glucocorticoid receptors may be present in both oligodendrocytes and astrocytes (Goya et al. 1996). In summary, the increase in GS activity (figure 3, left pannel), together with the significant inhibition of DNA synthesis observed after treatment of early passage C6 with dexamethasone, strongly suggests that glucocorticoids induce phenotypic differentiation of early passage C6 glioma (Goya et al. 1996).

4. SUMMARY AND CONCLUSIONS

Any variation in CNP and GS activity in vitro under the effect of growth factors with different mechanisms of action, may indicate a general process of differentiation in the culture. An increase in the GS activity in early passage glioma may be interpreted as an induction of the differentiating phenotype involving cell transformation into a more astrocytic-like culture. Hence, under the above experimental conditions, dexamethasone and insulin are the two factors with the most potent differentiating effect on rat C6 glioma cell line.

Comparative studies with growth regulators using different receptor pathways and mechanisms of action may add valuable data to the current knowledge in the field. The data presented in this chapter show the plasticity of C6 glioma to the differentiating effect of various growth factors, especially insulin and dexamethasone, and confirms C6 glioma cell line as a useful model for studies on glial cell properties and proliferation in vitro.

ACKNOWLEDGMENT

I am grateful to Dr. Paola S. Timiras for the stimulating discussion and critical reading of the manuscript.

REFERENCES

Aaronson S.A. 1991. Growth factors and cancer, Science 254:1146.

Alexander D.B., Goya L., Webster M.K., Haraguchi T. and Firestone G.L. 1993. Glucocorticoids coordinately disrupt a transforming growth factor alpha autocrine loop and suppress the growth of 13762NF-derived Con8 rat mammary adenocarcinoma cells, Cancer Research 53:1816.

Barres B.A., Raff M.C., Gaese F., Bartke I., Dechant G. and Barde Y.-A. 1994. A crucial role for neurotrophin-3 in oligodendrocyte development. Nature 367:371–375.

Bazan J.F. 1989. A novel family of growth factor receptors: a common binding domain in the growth hormone, prolactin, erythropoietin and IL-6 receptors, and the p75 IL-2 receptor beta-chain, Biochem. Biophys. Res. Commun. 164:788.

Berns E.M., Schuurmans A.L., Bolt J., Lamb D.J., Foekens J.A. and Mulder E. 1990. Antiproliferative effects of suramin on androgen responsive tumour cells, Eur. J. Cancer 26.470.

Brodie C. and Vernadakis A. 1991. Muscle-derived factors induce proliferation and astrocytic phenotypic expression in C6 glial cells. Glia 4:269.

Carpenter G. 1992. Receptor tyrosine kinase substrates: src homology domains and signal transduction, FASEB J. 6:3283.

Ceriani R.C. 1974. Proceedings: hormones and other factors controlling growth in the mammary gland: a review, J. Invest. Derm. 63:93.

Fausto N. and Weber E.M. 1993. Mechanisms of growth regulation in liver regeneration and hepatic carcinogenesis, Progress in liver diseases 11:115.

Firestone G.L., Maiyar A.C. and Ramos R.A. 1995. Steroid and protein regulators of normal and abnormal cell proliferation, in Hormones and Aging, Eds. P.S. Timiras, W.D. Quay and A. Vernadakis, pp. 325–359.

Fridovich K.J., Hansen L.J., Keyomarsi K. and Pardee A.B. 1990. Progression through the cell cycle: an overview, Am. Rev. Resp. Dis. S3.

Goya L., Alexander D.B., Webster M.K., Kern F.G., Guzman R.C., Nandi S. and Firestone G.L. 1993a. Overexpression of transforming growth factor alpha overrides the glucocorticoid-mediated suppression of Con8 mammary tumor cell growth in vitro and in vivo, Cancer Res. 53:1816.

Goya L., Maiyar A.C., Ge Y. and Firestone G.L. 1993b. Glucocorticoids induce a G1/G0 cell cycle arrest of Con8 rat mammary tumor cells that is synchronously reversed by steroid withdrawal or addition of transforming growth factor-alpha, Mol. Endocrinol. 7:1121.

Goya L., Rivero F. and Pascual-Leone A.M. 1995. Stress, glucocorticoids and aging, in Hormones and Aging, Eds. P.S. Timiras, W.D. Quay and A. Vernadakis, pp. 249–264.

Goya L., Feng P-T., Aliabadi S. and Timiras P.S. 1996. Effect of growth factors on the in vitro growth and differentiation of early and late C6 glioma cells, Int. J. Devl. Neuroscience 14:409–417.

Haraguchi T., Alexander D.B., King D.S., Edwards C.P. and Firestone G.L. 1991. Identification of the glucocorticoid suppressible mitogen from rat hepatoma cells as an angiogenic platelet-derived growth factor A-chain homodimer, J. Biol. Chem. 266:18299.

Hung W.C., Chuang L.Y., Tsai J.H. and Chang C.C. 1993. Effects of insulin on TGF-beta 1-induced cell growth inhibition in the human hepatoma cell lines, Biochem. Mol. Biol. Int. 30:655.

Hunter T. and Karin M. 1992. The regulation of transcription by phosphorylation, Cell 70:375.

Kiess W., Lee L., Graham D.E., Greenstein L., Tseng L.Y., Richler M.M. and Nissely S.P. 1989. Rat C6 glial cells synthesize insulin-like growth factor (IGF-I) and express IGF-I receptors and IGF-II/mannose 6-phosphate receptors. Endocrinology 124: 1727.

Koch C.A., Anderson D., Moran M.F., Ellis C. and Pawson T. 1991. SH2 and SH3 domains: elements that control interactions of cytoplasmic signaling proteins, Science 252:668.

Landis D.M.D. 1994. The early reactions of non-neuronal cells to brain injury. Annu. Rev. Neurosci. 17:133–51.

Lee K., Kentroti S. and Vernadakis A. 1992. Comparative biochemical, morphological and immunocytochemical studies between C6 glial cells of early and late passages and advanced passages of glial cells derived from aged mouse cerebral hemispheres. Glia 6:245–257.

Lowe Jr.W.L., Meyer T., Karpen C.W. and Lorentzen L.R. 1992. Regulation of insulin-like growth factor I production in rat C6 glioma cells: possible role as an autocrine/paracrine growth factor. Endocrinology 130:2683.

Martinez-Hernandez A., Bell K.P. and Norenberg M.D. 1977. Glutamine synthetase-glial localization in the brain. Science 195:1356.

Massague J. 1992. Receptors for the TGF-beta family, Cell 69:1067.

Morrison R.S. and De Vellis J. 1981. Growth of purified astrocytes in chemically defined medium, Proc. Natl. Acad. Sci. USA 78:7205.

Noble M., Murray K., Stroobant P., Waterfield M.D. and Riddle P. 1988. Platelet-derived growth factor promoted division and motility and inhibits premature differentiation of the oligodendrocyte/type-2 astrocyte progenitor cell, Nature 333:556.

Pardee A.B. 1989. G1 events and regulation of cell proliferation, Science 246:603.

Parker K.K., Norenberg M.D. and Vernadakis A. 1980. Transdifferentiation of C6 glial cells in culture, Science 208:179.

Pines J. 1993. Arresting developments in cell-cycle control, Trends Biochem. Sci. 19:143.

Poduslo S.E. 1975. The isolation and characterization of a plasma membrane and myelin fraction derived from oligodendroglia of calf brain, J. Neurochem. 24:647.

Pruss R.M., Bartlett P.F., Gavridovic J., Lisak R.P. and Rattray S. 1982. Mitogens for glial cells: a comparison of the response of cultured astrocytes, oligodendrocytes and Schwann cells, Devel. Brain Res. 2:19.

Raff M.C., Abney E.R. and Fok-Seang J. 1985. Reconstitution of a developmental clock in vitro: a critical role for astrocytes in the timing of oligodendrocyte differentiation, Cell 42: 61.

Raff M.C., Lilien L.E., Richardson W.D., Burne J.F. and Noble M. 1988. Platelet-derived growth factor from astrocytes drives the clock that times oligodendrocyte development in culture, Nature 333:562.

Raff M.C., Miller R.H. and Noble M. 1983. A glial progenitor cell that develops in vitro into an astrocyte or an oligodendrocyte depending on the culture medium, Nature 274:813.

Reul J. and De Kloet E.R. 1985. Two receptor systems for corticosterone in rat brain: microdistribution and differential occupation, Endocrinology 117:2505.

Richardson W.D., Pringle N., Mosleuy M.J., Westermark R. and Dubois-Dalcq M. 1988. A role for platelet-derived growth factor in normal gliogenesis in the central nervous system, Cell 53:310.

Richardson W.D., Raff M. and Noble M. 1990. The oligodendrocyte-type-2 astrocyte lineage, Semin. Neurosci. 2:451.

Ross M.E. 1996. Cell division and the nervous system: regulating the cycle from neural differentiation to death. Trends Neurosci. 19:62–68.

Sakellaridis N., Mangoura D. and Vernadakis A. 1984. Glial cell growth in culture: influence of living cell substrata, Neurochem. Res. 9:1477.

Sanchez I., Goya L., Vallerga A.K. and Firestone G.L. 1993. Glucocorticoids reversibly arrest rat hepatoma cell growth by inducing an early G1 block in cell cycle progression, Cell Growth Differ. 4:215.

Sherr C.J. 1993. Mammalian G1 cyclins, Cell 73:1059.

Simon M.I., Strathmann M.P. and Gautam N. 1991. Diversity of G proteins in signal transduction, Science 252:802.

Simpson D.L., Morrison R., De Vellis J. and Herschman J.R. 1982. Epidermal growth factor binding and mitogen activity on purified populations of cells from the central nervous system, J. Neurosci. Res. 8:453.

Thomas G. 1992. MAP kinase by any other name smells just as sweet, Cell 68:3.

Truss M. and Beato M. 1993. Steroid hormone receptors: interaction with deoxyribonucleoic acid and transcription factors, Endocr. Rev. 14:459.

Ullrich A. and Schlessinger J. 1990. Signal transduction by receptors with tyrosine kinase activity, Cell 61:203.

Vernadakis A., Lee K., Kentroti S. and Brodie C. 1992. Role of astrocytes in aging: late passage primary mouse brain astrocytes and C6 glial cells as models, Prog. Brain Res. 94: 391.

Vielkind U., Walencewicz A., Levine J.M. and Churchill-Bohn M. 1990. Type II glucocorticoid receptors are expressed in olygodendrocytes and astrocytes, J. Neurosci. Res. 27:360.

Williams L.T. 1989. Signal transduction by the platelet-derived growth factor receptor, Science 243:1564.

Zipori D. 1992. The renewal and differentiation of hematopoietic stem cells, FASEB J. 6:2691.

19

ESTROGENS INFLUENCE GROWTH, MATURATION, AND AMYLOID β-PEPTIDE PRODUCTION IN NEUROBLASTOMA CELLS AND IN A β-APP TRANSFECTED KIDNEY 293 CELL LINE

David Chang, Judy Kwan, and Paola S. Timiras

Department of Molecular and Cell Biology
University of California
Berkeley, California 94720-3202

ABSTRACT

During development *in vivo* and *in vitro*, estrogens: a) increase brain excitability, particularly in limbic structures; b) are responsible for the maturation and cyclicity of limbic-hypothalamic interrelations; c) enhance myelinogenesis; and d) may act with NGF to stimulate neurite formation. In senescence, estrogen administration would improve memory in postmenopausal women. The absence or low levels of estrogens after menopause would increase prevalence of Alzheimer's dementia (AD) more in women than men, irrespective of age or ethnicity. In the present study, addition of 17-ß estradiol to cultured human neuroblastoma cells affected growth slightly, but stimulated cell maturation as shown by increased tyrosine hydroxylase activity. The extracellular deposition in brain tissue and around blood vessels of the amyloid ß-peptide (Aß), a 4.3 kD fragment of the larger integral membrane protein, ß-amyloid precursor protein (ß-APP), is considered an important characteristic of AD. We investigated whether 17-ß estradiol may influence the formation of the Aß (thus the abnormal accumulation of amyloid proteins) in neuroblastoma cells and in a ß-APP transfected human kidney 293 cell line. Two doses of 17 ß-estradiol were added to the cultures of both cell lines. Cells were grown until confluence, metabolically labeled with ^{35}S-methionine, immunoprecipitated with the rabbit antiserum R1282, gel electrophoresed and autoradiographed in order to compare levels of Aß under the different estradiol concentrations. While in neuroblastoma cells, levels of Aß were only slightly reduced after estradiol and a dose-effect relationship with the hormone could not be demonstrated, in the 293 cells, Aß band intensity decreased as concentration of estradiol increased. These data support the role of estrogen in normal and abnormal brain metabo-

Brain Plasticity, edited by Filogamo *et al.*
Plenum Press, New York, 1997

lism and suggest potential hormonal interventions which may reduce or prevent the formation of amyloid deposits that occur in AD.

1. INTRODUCTION

Estrogens have well-established actions on maturation and function of the brain during development and adulthood, but little is known of their effects on the aging brain. Recent studies suggest that even in old age, estrogens may significantly affect several aspects of brain function.

Among the well-established effects of estrogens described in early studies from a number of laboratories, our own studies show that, in rats, estrogens:

1. promote higher brain excitability in females than in males, with a peak during estrus (Heim and Timiras, 1963; Timiras, 1971, 1982; Vernadakis and Timiras, 1963; Woolley et al., 1961; Woolley and Timiras, 1962a,b);
2. regulate spontaneous/evoked excitability in limbic nuclei (amygdala, hippocampus) and development of their cyclicity (Curry and Timiras, 1972; McGowan-Sass and Timiras, 1975; Terasawa and Timiras, 1968a, b, 1969);
3. promote maturation of pathways from amygdala to hypothalamus and influence onset of estrous cyclicity at puberty (Pardey-Borrero et al., 1985; Sherwood and Timiras, 1974).

These actions are mediated through specific estrogen receptors abundant in the brain (Pfaff and Keiner, 1973; McEwen, 1979, 1995). Estrogen receptors are members of the superfamily of steroid/thyroid hormone/vitamin D_3/retinoic acid receptors capable of activating genes by directly binding DNA sites containing hormone-specific regulatory elements (Evans, 1988). Estrogen actions on neuronal electrical activity are mediated/associated with changes in brain metabolism, namely, water and electrolyte distribution (Valcana et al., 1967; Woolley and Timiras, 1964); myelinogenesis and synaptogenesis (Arai and Matsumoto, 1978; Casper et al., 1967; Frankfurth et al., 1990; Gould et al., 1990; Leedom et al., 1994); neurotransmitter enzyme activity and development (Vaccari et al., 1977); and amino acid transport and protein synthesis (Frankfurt et al., 1984; Hudson et al., 1970; Lagrange et al., 1996; Litteria, 1977; Litteria and Timiras, 1970). Estrogen receptor distribution and binding in the nervous system varies with development (Woolley et al., 1969) and may be modulated, more in females than males, by catecholaminergic stimulation (primarily dopaminergic) of synaptic remodeling and by membrane changes (Garcia-Segura et al., 1985; Olmos et al., 1987; Woolley et al., 1994).

Observations of the effects of estrogens on the brain of senescent animals and elderly people have stimulated interest on the potentially beneficial role of these hormones in preventing or delaying some of the age-associated neurodegenerative changes. For example, hormonal replacement therapy in post-menopausal women would enhance short-term memory and delay the onset or reduce the severity of cognitive impairment in Alzheimer's dementia (AD) (Barrett-Connor and Kritz-Silverstein, 1993; Fillit et al., 1986; Henderson et al., 1994; Honjo et al., 1995; Ohkura et al., 1994; Paganini-Hill and Henderson, 1994; Phillips and Sherwin, 1992; Smalheiser and Swanson, 1996; Tang et al., 1996). This beneficial action may be mediated by several mechanisms presumed to ameliorate dementia. They include: a) anti-depressant effect, b) improvement of cerebral blood flow, c) direct neuronal and/ or glial stimulation, d) anti-oxidant, anti-excitotoxic and anti-amyloid activity, and e) suppression of apolipoprotein E4.

Cessation of estrogen secretion by the ovary at menopause would be responsible for age-adjusted higher prevalence of degenerative diseases of old age, such as AD, in women than in men world wide and across ethnic groups (Brenner, 1994; Paganini-Hill and Henderson, 1994; Tang et al., 1996). Additionally, estrogen receptors are more numerous in the same locations where the pathologic correlates of AD (e.g., neurofibrillary tangles and neuritic plaques) are also localized. Thus, it has been suggested that estrogens can protect the brain from the appearance and progression of neurodegenerative diseases. The lower incidence of such diseases in male rats may be related to the high levels of aromatization enzymes of testosterone to estrogens in the brain (Roselli et al., 1996), hence, males, in whom testosterone persists throughout life, would have higher levels of estrogens in the brain compared to postmenopausal women. A number of studies with cultured neurons and neuronal cell lines have demonstrated a synergistic action of estrogens with the well-known action of nerve growth factor (NGF) for maintenance/ survival/arborization of basal forebrain/hypothalamic neurons in culture (Toran-Allerand et al., 1992; Sohrabji et al., 1994).

In the present study, 17-ß estradiol was added to neuroblastoma cells and kidney 293 cells: a) to assess its action on cell maturation as determined by the activity of the catecholaminergic enzyme, tyrosine hydroxylase; b) to explore whether estrogens may influence, in neuroblastoma cells, the formation of the amyloid ß-peptide leading to amyloidosis resembling that occurring in the brain of AD patients, and c) to compare the effects of estrogens on the formation of amyloid β-peptide in neuroblastoma cells and β-amyloid protein precursor (β-APP) transfected renal cells.

2. MATERIALS AND METHODS

2.1. Cell Culture

Cells are human neuroblastoma cells from the SH-SY5Y line (Sidell and Horn, 1985) which expresses tyrosine hydroxylase activity as well as the ß-APP_{695} form of the amyloid precursor protein (Weidmann et al., 1989). For comparison of tyrosine hydroxylase activity, rat pheochromocytoma cells from the PC 12 line were also cultured. Another cell line used was the human kidney 293 cell line, which was stably transfected with the cDNA encoding the ß-APP_{695} gene (Selkoe et al. 1988). This cDNA contained a double mutation (Lys to Asn at residue 595 plus Met to Leu at positon 596) and was derived from a Swedish familial Alzheimer's disease family (Citron et al., 1992) with an autosomal dominant form of the disease (Murrell et al., 1991).

Human neuroblastoma cells were routinely grown in Dulbecco's modified Eagle's medium (DMEM)/Ham's F-12 (50:50) supplemented with 10% fetal bovine serum and 1% antibiotic antimycotic solution. All cells were grown in Corning tissue culture plates at 37°C in a humidified atmosphere of air and CO_2 (95:5), and the medium was changed every 3 days. Cells were allowed to reach 90% confluency and were harvested using trypsinversene, replated in multiwells, and allowed to attach for at least 24 hours prior to any experiment. 17β-estradiol was added to the medium on day zero and the medium with the hormone was changed every 3 days. For pheochromocytoma cells, 10% bovine serum was replaced by 5% horse serum. The duration of the experiment was 8 days for the study of growth and tyrosine hydroxylase activity and 12–15 days for the study of amyloid-ß peptide production.

Renal 293 cells were cultured in the same DMEM medium containing fetal bovine serum and antibiotic antimycotic solution. Similarly, the medium was changed every 3 days for a duration of 15 days and in the treated cultures, 17β-estradiol was added on days 5, 11, and 15.

2.2. Differentiation

To induce differentiation of neuroblastoma cells, retinoic acid (10 μmol/L) from a stock solution (10 mmol/L) was added to the medium every other day for the duration of the experiment (Ino et al., 1986).

2.3. Hormone

Estrogen was 17β-estradiol from Sigma.

2.4. Cell Proliferation Assay

Neuroblastoma cells were routinely plated at the density of 2.5×10^5 cells per cm^2, which had been found to be best for cell proliferation. Cells were allowed to attach for 24 hours prior to adjustment of conditions. At the time of the assay the ethanol was aspirated, and the cells were allowed to dry under a slight vacuum for 1.5 hours, after which 0.5 ml DABA (0.4 mol/L) was added to each well and the multiwell plate incubated at 60°C for 1 hour. The reaction was terminated with 1 mol HCl/L. Total DNA content was then determined using a Turner fluorometer (Hinegardner, 1971).

2.5. Tyrosine Hydroxylase Activity

Tyrosine hydroxylase activity was measured according to Waymire et al. (1971) modified routinely for the frozen storage of cultured cells prior to assay.

2.6. Amyloid β-Peptide Determination

This determination involves a number of steps, which include: labelling of the cells with ^{35}S-methionine; immunoprecipitation by the addition of rabbit antisera (Haass et al., 1991, 1992; Weidemann et al., 1989), gel electrophoresis using a 10–20% Tris-Tricene gradient gel, gel fixation with 45% methanol, 45% ddH_20 and 10% glacial acetic acid, silver staining using the silver staining kit from Accurate Chemicals & Scientific Corporation; use of En^3Hance (New England Nuclear) treatment to maximize band visualization, and finally drying and autoradiography for one week.

3. RESULTS

As demonstrated in previous studies, addition of retinoic acid to the culture induced differentiation of the neuroblastoma cells (Reynolds et al., 1986; Ino et al., 1986). Morphologically, addition of retinoic acid to the medium transformed the neoplastic neuroblasts from immature, round, clustered cells with sparse and short processes to differentiated cells with more elongated cell bodies and more spread out and polar processes showing varicosities, as previously published (John et al, 1991). This morphological differentiation was associated with a significant decrease in proliferation rate suggestive of differentiation and further verified in the present experiments (Figures 1 and 2). Neurochemically, as

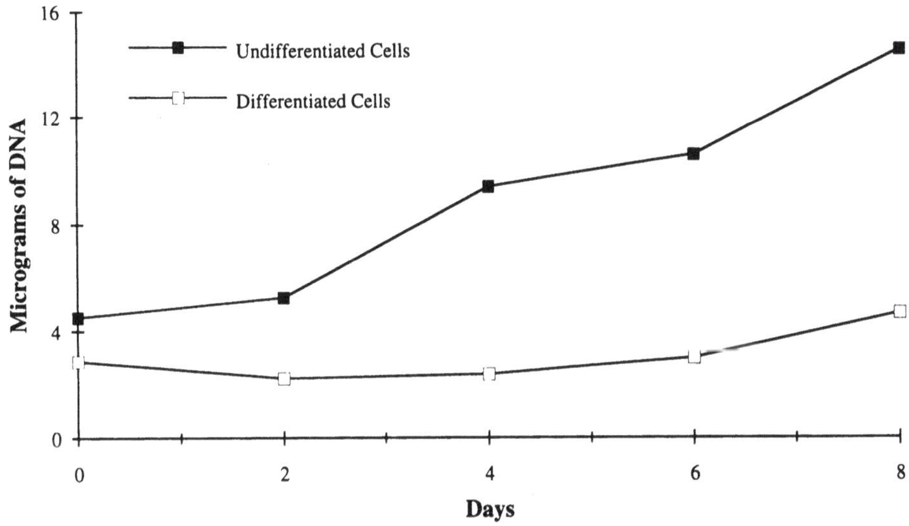

Figure 1. Retinoic acid (20 μM) added to the culture medium inhibits proliferation of human neuroblastoma cells. Each point represents the average of six experiments. Retinoic acid was added to the culture medium on day O and DNA content was measured until day 8. In the mediun without retinoic acid, cells showed with time a progressively high increase in DNA despite a slow initial period. In the medium with retinoic acid, cells had approximately the same DNA content as the controls without retinoic acid at the beginning of the experiments but did not show further growth thoughout the experimental period.

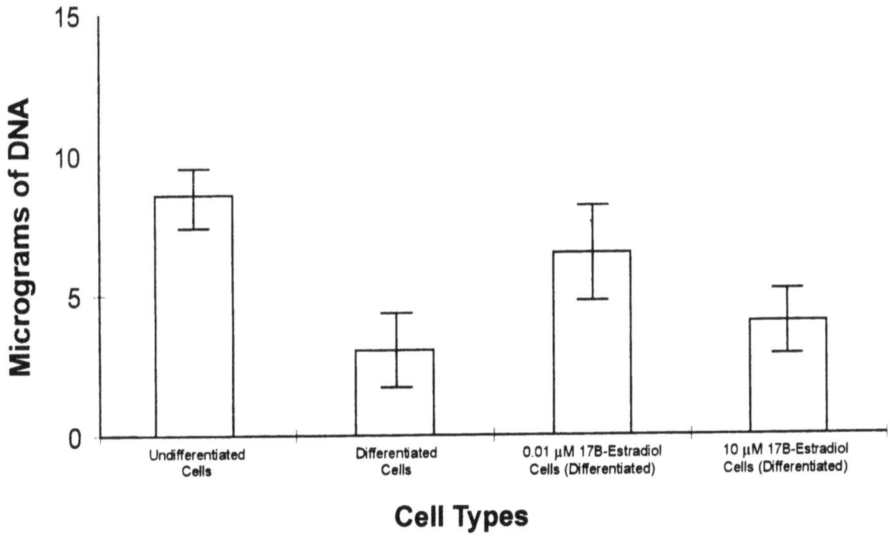

Figure 2. Effects of two doses of 17 ß-estradiol on inhibition of proliferation of human neuroblastoma cells induced by retinoic acid (20 μM). Bar graphs represent the mean DNA levels on the last (8th) day in six experiments, and bracketed lines the standard error of the mean. DNA content was significantly (Student's *t* -test) lower in the differentiated as compared to the undifferentiated cells and in the differentiated cells with the higher estradiol concentration. In the differentiated cells, those grown in the medium with the lower dose of estradiol had higher DNA content than in the other differentiated cells, but it was still lower when compared to the undifferentiated cells.

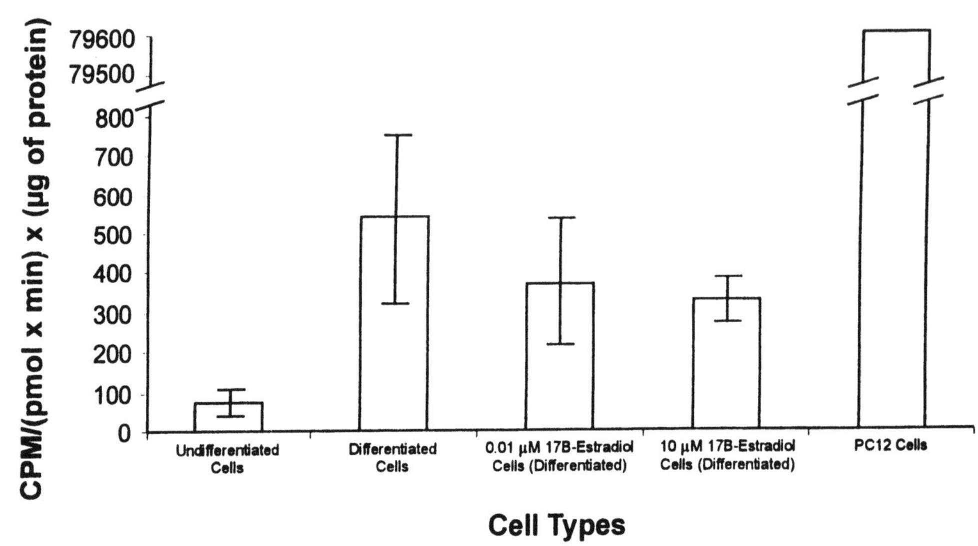

Figure 3. Tyrosine hydroxylase (TH) activity in undifferentiated cells and in differentiated (with retinoic acid) cells with two doses of estradiol and comparison with that in pheochromocytoma cells (PC12). Bar graphs represent the mean DNA levels on the last (8th) day from a total of six experiments, and bracketed lines the standard error of the mean. Differentiation induced a significant (Student's *t*--test) increase in TH activity in cells with or without estradiol. PC12 cells known to possess high adrenergic properties were used here as a verification of TH measurements and, as expected, had a very high TH activity.

in previous studies, differentiation was accompanied by a marked increase in the activity of the adrenergic enzyme, tyrosine hydroxylase (TH) significantly above that of undifferentiated controls; however, the increased activity did not reach the high TH levels found in pheochromocytoma cells used here as verification of the adequacy of our analytical technique and as comparison with cells extremely rich in adrenergic transmitters (Figure 3).

The addition of a higher level of estradiol (10 µM) to the culture medium decreased the proliferation rate of differentiated compared to undifferentiated cells; this action as well as that on TH activity is mediated through steroid hormone receptors (Evans ,1988) present in neurons (McEwen, 1979; Toran-Allerand et al, 1992) and in kidney cells (Sharma and Timiras, 1988). With the lower estradiol concentration (0.01 µM), DNA levels were higher than in differentiated (without estradiol) controls (Figure 2), in agreement with previous studies that demonstrated both an inhibitory (at higher doses) and stimulatory (at lower doses) action of the hormone on thymidine incorporation, particularly in the differentiated (as compared to undifferentiated) neuroblastoma cells (Timiras, 1995). TH activity was higher in the differentiated cells cultured with estradiol than in undifferentiated cells, with no significant differences among the differentiated cells with or without estradiol (Figure 3). It is possible that TH activity was already maximally stimulated by the differentiation process and could not be enhanced further by hormonal treatment.

In the neuroblastoma cells, Aß peptide content was low and could be detected only when the immunoprecipitates were not subjected to repeated washes, in which case the background staining remained quite high (Figure 4: lanes 3–4, 6–7, 9–10)). Thus, Aß peptide seems to be constitutively secreted at a very low rate by cells during their normal metabolism. Estradiol added to the medium did not significantly influence the Aß levels in

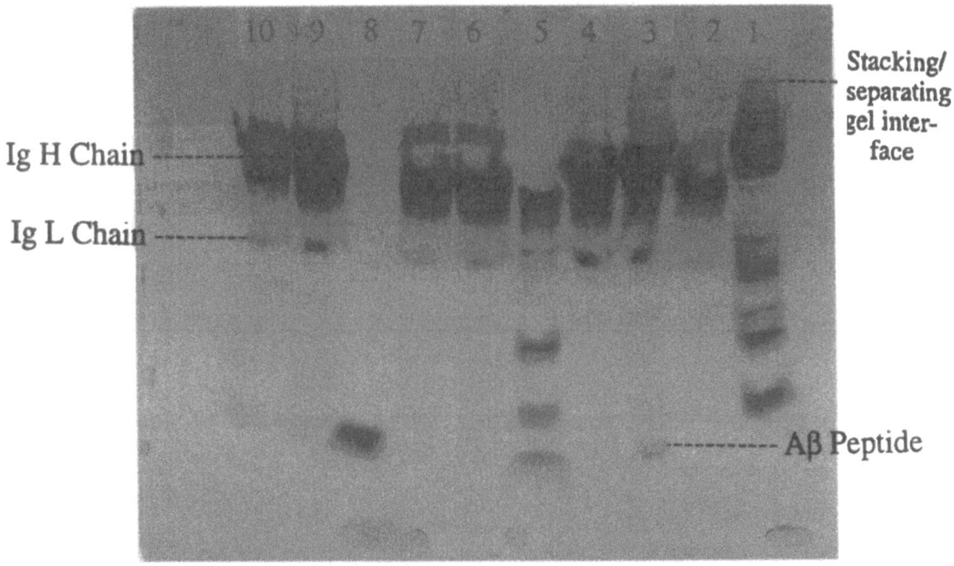

Figure 4. Aß peptide determination in differentiated human neuroblastoma cells. Lane 5 was loaded with standards of known molecular weight. There are 6 bands; the topmost one in this lane is an artifact. The last band, at 4.30 cm from the stacking-separating gel interface, is insulin with a mw of 3 kD. Lane 8 was loaded with synthetic Aß peptide (positive control) from Bachem California with a known mw of 4.3 kD. Except for lanes 1 and 2 (negative controls), which received no antibody and normal rabbit serum (NRS) (from unimmunized rabbits) respectively, all lanes that received α-Aß antisera (lanes 3, 4, 6, 7, 9, and 10) exhibited a light band at the same distance from the stacking-separating gel interface as lane 8 (loaded with control Aß).

the neuroblastoma cells. In the transfected kidney 293 cells, Aß immunoprecipitates were easily identifiable in cells not treated with any estrogen (Figure 5: lane 3), and Aß production was considerably decreased after 17 ß-estradiol was added to the medium, the higher estradiol concentration (2 x 10^{-9} M) (lanes 6, 7) having a greater effect than the lower concentration (2 x 10^{-11} M) (Figure 5: lanes 9, 10).

4. DISCUSSION

The effects of estradiol on neuronal proliferation and differentiation previously described in vivo and in vitro are further confirmed in the present experiments. Some of these effects may be mediated through or be synergistic with NGF (Toran-Allerand et al., 1992).

As expected, Aß levels were low in neuroblastoma cells (not transfected with the ß-APP gene) but high in the transfected 293 kidney cells. The low levels of neuronal Aß reflect a very modest production of this protein during normal metabolism. Nevertheless, the finding that soluble Aß is present in extracellular fluid in vitro suggests that, in cases of overproduction, Aß peptide may diffuse throughout the extracellular spaces affecting brain areas distant from the original site of Aß production and accumulation.

The observation that estradiol was capable of inhibiting the production of Aß *in vitro,* especially in those cells with high Aß levels, may be correlated with the apparently beneficial effects of estrogens in some conditions such as Alzheimer's disease where there

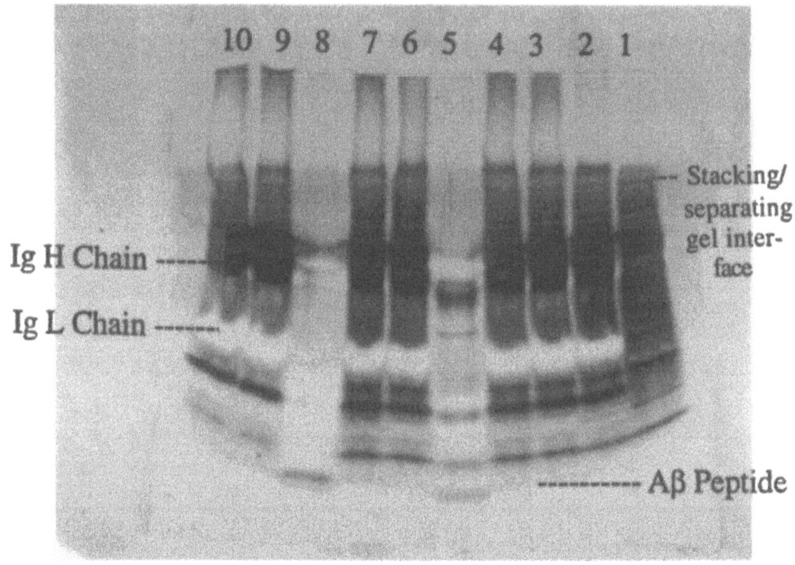

Figure 5. Aß peptide determination in ß-APP-transfected kidney 293 cells. Lane 5 was loaded with protein markers. There are 6 bands; however, the third one from the top (ß-lactoglobulin) is not visible here because this protein is known to silver stain at times with very low efficiency. The last band, at 5.78 cm from the top, is insulin ß chain with a mw of 2.3 kD. Unlike the gel from the neuroblastoma cells, both chains of insulin presented themselves as separate bands this time. Lane 8 was loaded with synthetic Aß peptide (positive control) from Bachem California with a known mw of 4.3 kD. Except for the two negative controls--lane 1 which received no antibody and lane 2 which received NRS--all lanes that received α-Aß antisera (lanes 3, 4, 6, 7, 9, and 10) exhibited a light band at the same distance from the top as lane 8 (loaded with control Aß).

is an overproduction and accumulation of this protein (amyloidosis) . Although this effect appears minimal in the neuronal cells where the Aß levels are very low, it may be assumed that, if these levels had been higher as in the ß-APP transfected cells, etstradiol might have had a clear inhibitory effect . This assumption is supported by the presence of a dose-effect response in the transfected kidney cells with the higher doses having a more effective action in reducing Aß levels. Studies have shown that Aß is the product of the amyloidogenic ß-secretase pathway of ß-APP (Shoji et al., 1992) and that estradiol promotes the non-amyloidogenic processing of ß-APP by increasing/inducing the activity and production of alpha-secretase (Gandy et al., 1994). Estrogens may also prevent or reduce the formation of mutations leading to the production of presenilin, found in some forms of familial AD (Selkoe, 1997).

In conclusion, the effects of the estrogen 17-ß estradiol *in vitro* on growth and differentiation of neuroblastoma cells and reduction of Aß production in ß-APP transfected kidney cells support the observations in humans of the beneficial effects of these hormones on normal and abnormal aging of the brain.

ADDENDUM

After sending the manuscript to the Editors, we became aware of two recent publications closely related and supportive of our research. In both publications, estradiol or its derivatives reduce the production of amyloid peptide in neuroblastoma (Green, P S.,

Gridley, K.E., Simpkins, J.W. Estradiol protects against beta-amyloid (25–35)-induced toxicity in SK-N-SH human neuroblastoma cells. *Neurosci. Lett.* 218:168, 1996.) and hippocampal cells (Behl, C., Skutella, T., Lezoualc'h, F., et al. Neuroprotection against oxidative stress by estrogens: Structure-activity relationship. *Mol. Pharmacol.* 51:535–541, 1997.) and this beneficial action would be mediated by the antioxidant activity of these hormones.

ACKNOWLEDGMENTS

The authors wish to thank Drs. D.J. Selkoe, C. Haass, and M. Citron of Harvard Medical School, Brigham and Women's Hospital, Center for Neurologic Diseases, for their generosity in providing the α-Aß antisera, some of the ß-APP transfected human kidney 293 cells, and for their invaluable advice on several techniques used in the experiment. Thanks also to Dr. John F. Reinhard of Wellcome Research Laboratories in Chapel Hill for samples of BH4 tetrahydrobiopterin enzyme and instructions for the TH assay.

REFERENCES

Arai, Y. and Matsumoto, A. Synapse formation of the hypothalamic arcuate nucleus during post-natal development in the female rat and its modification by neonatal estrogen treatment. *Psychoneuroendocrinology* 3:31–45, 1978.

Barrett-Connor, E. and Kritz-Silverstein, D. K. Estrogen replacement therapy and cognitive function in older women, *JAMA* 269:2637–2641, 1993.

Brenner, D. E., Kukull, W.A., Stergachis, A., et al. Postmenopausal estrogen replacement therapy and the risk of Alzheimer's disease: A population-based case-control study. *Am.J.Epidemiol.* 140: 262–267, 1994.

Casper, R., Vernadakis, A., and Timiras, P.S. Influence of estradiol and cortisol on lipids and cerebrosides in the developing brain and spinal cord of the rat. *Brain Res.* 5:524–526, 1967.

Citron, M., Oltersdorf, T., Haass, C. et al. Mutation of the Beta Amyloid precursor protein in familial Alzheimer's disease increases Beta protein production. *Nature* 360:672–674, 1992.

Curry, J.J. and Timiras, P.S. Development of evoked potentials in specific brain systems after neonatal administration of estradiol. *Exp. Neurol.* 34:129–139, 1972.

Evans, R. M. The steroid and thyroid hormone receptor superfamily. *Science* 240:889–895, 1988.

Fillit, H., Weinreb, H., Cholst, I., et al. Observation in a preliminary open trial of estradiol therapy for senile dementia-Alzheimer's type. *Psychoneuroendocrinology* 11:337–345, 1986.

Frankfurt, M., Fuchs, E., and Wuttke, W. Sex differences in gamma-aminobutyric acid and glutamate concentrations in discrete rat brain nuclei. *Neurosci. Lett.* 50:245–250, 1984.

Frankfurt, M., Gould, E., Woolley, C.S., and McEwen, B.S. Gonadal steroids modify dendritic spine density in ventromedial hypothalamic neurons: A Golgi study in the adult rat. *Neuroendocrinology* 51:530–535, 1990.

Gandy, S.E., Jaffe, A.B., Toran-Allerand, C.D., et al. Estrogen regulates metabolism of alzheimer amyloid beta precursor protein. *J. Biol. Chem.* 248:13065–13068, 1994.

Garcia-Segura, L.M., Baetens, D., and Naftolin, F. Sex differences and maturational changes in arcuate nucleus neuronal plasma membrane organization. *Dev. Brain Res.* 19:146–149, 1985.

Gould, E., Woolley, C.S., Frankfurt, m., McEwen, B.S. Gonadal steroids regulate dendritic spine density in hippocampal pyramidal cells in adulthood. *J. Neurosci.* 10(4):1286–1291, 1990.

Haass, C., Huang, A.Y., and Selkoe, D.J. Processing of beta-amyloid precursor protein in microglia and astrocytes favors an internal localization over constitutive secretion. *J. Neurosci.* 11(12):3783–3793, 1991.

Haass, C., Schlossmacher, M.G., Hung, A.Y., et al. Amyloid beta-peptide is produced by cultured cells during normal metabolism. *Nature* 359:322–325, 1992.

Heim, L.M. and Timiras, P.S. Gonad-brain relationship: Precocious brain maturation after estradiol in rats. *Endocrinology* 72:598–606, 1963.

Henderson, V.W., Paganini-Hill, A., Emanuel, C.K., et al. Estrogen replacement therapy in older women. Comparisons between Alzheimer's disease cases and nondemented control subjects. *Arch. Neurol.* 51:896–900, 1994.

Hinegardner, R.T. An improved fluorometric assay for DNA. *Anal. Biochem.*39:197–201, 1971.

Honjo, H., Tanka, K., Kashiwagi, T., et al. Senile dementia-Alzheimer's type and estrogen. *Hormone and Metabolic Research* 27:204–207, 1995.

Hudson, D.B., Vernadakis, A., and Timiras, P.S. Regional changes in amino acid concentration in the developing brain and the effects of neonatal administration of estradiol. *Brain Res.* 23:213–222, 1970.

Ino, M., Cole, G.M., and Timiras, P.S. Tyrosine hydroxylase and monamine oxidase-A activity increases in differentiating human neuroblastoma after elimination of dividing cells. *Dev. Brain Res.* 30:120–123, 1986.

John, N.J., Lew, G.M., Goya, L., et al. Effects of Seratonin on tyrosine hydroxylase and tau protein in a human neruoblastoma cell. In: Plasticity and Regeneration of the Nervous System (*Advances in Experimental Medicine and Biology*, Vol. 296. Timiras, P.S., Privat, A., Giacobini, E., et al. (Eds.), Plenum Press, New York, pp. 69–80, 1991.

Lagrange, A.H., Wagner, E.J., Ronnkleiv, O.K., and Kelly, M.J. Estrogen rapidly attenuates a GAGA-B response in hypothalamic neurons. *Neuroendocrinology* 64:114–123, 1996.

Leedom, L., Lewis, C., Garcia-Segura, L. M., and Naftolin, F. Regulation of arcuate nucleus synaptology by estrogen. *Ann. N.Y.Acad. Sci.* 743:61–71, 1994.

Litteria, M. Effects of neonatal estrogen on in vivo transport of α-aminoisobutyric acid into rat brain. *Exp. Neurol.* 57:817–827, 1977.

Litteria, M. and Timiras, P.S. *In vivo* inhibition of protein synthesis in specific hypothalamic nuclei by 17β-estradiol. *Proc. Soc. Exp. Biol. Med.* 134:256–261, 1970.

McEwen, B.S. Oestrogens and the structural and functional plasticity of neurons: Implications for memory, ageing and neurodegenerative processes. In: *CIBA Foundation Syposium, 191. Non-reproductive actions of sex steroids*, Bock, G. J. and Goode, J.A. (Eds.), John Wiley and Sons, Ltd., New York, pp. 52–73, 1995.

McEwen, B.S. Steroid hormone interactions with the brain: Cellular and molecular aspects. In: *Reviews of Neuroscience* Vol. 4, Schneider, D.M. (Ed.), Raven Press, New York, pp. 1–30, 1979.

McEwen, B.S. and Pfaff, D.W. Factors influencing sex hormone uptake by rat brain regions. I. Effects of neonatal treatment, hypophysectomy, and competing steroid on estradiol uptake. *Brain Res.* 21:1–16, 1970.

McGowan-Sass, B.K. and Timiras, P.S. The Hippocampus and hormonal cyclicity. In: *The Hippocampus*, Isaacson, R.L. and Pribram, K.H. (Eds.), Vol. 1, Plenum Press, New York, pp. 355–374, 1975.

Murrell, J., Farlow, M., Ghetti, B., and Bendon, M. A mutation in the amyloid precursor protein associated with hereditary Alzheimer's disease. *Science* 254:97–99, 1991.

Ohkura, T., Isse, K., Akazawa, K. et al. Evaluation of estrogen treatment in female patients with dementia of the Alzheimer type. *Endocr. J.* 41:361–371, 1994.

Olmos, G., Aguilera, P., Tranque, P., et al. Estrogen-induced synaptic remodelling in adult rat brain is accompanied by the reorganization of neuronal membranes. *Brain Res.* 425:57–64, 1987.

Paganini-Hill, A., and Henderson, V.W. Estrogen deficiency and risk of Alzheimer's disease in women. *Am. J. Epidemiol.* 140:256–261, 1994.

Pardey-Borrero, B.M., Tamasy, V., and Timiras, P.S. Circadian pattern of multiunit activity of the rat suprachiasmatic nucleus during the estrous cycle. *Neuroendocrinology* 40:450–456, 1985.

Pfaff, D.W. and Keiner, M. Atlas of estradiol-concentrating cells in the central nervous system of the female rat. *J. Comp. Neurol.* 151:121–158, 1973.

Phillips, S.M. and Sherwin, B.B. Effects of estrogen on memory function in surgically menopausal women. *Psychoneuroendocrinology* 17:485–495, 1992.

Reynolds, C.P., Biedler, J.L., Spengler, D.A., et al. Characterization of human neuroblastoma cell lines established before and after therapy. *J. Natl. Cancer Inst.* 76:375, 1986.

Roselli, C.E., Klosterman, S.A., and Fasasi, T.A. Sex differences in androgen responsiveness in the rat brain: Regional differences in the induction of aromatase activity. *Neurendocrinology* 64:139–145, 1996.

Selkoe, D.J. Alzheimer's disease: genotypes, phenotype, and treatments. *Science* 275:630–631, 1997.

Selkoe, D.J., Podlisny, M.D., Joachim, C.L., Vickers, E.A., et al. Beta amyloid precursor protein of Alzheimer's disease occurs as 110- to 135-kilodalton membrane associated proteins in neural and nonneural tissues. *Proc. Natl. Acad. Sci. U.S.A.* 85:7341–7345, 1988.

Sharma, R., and Timiras, P.S. Regulation of glucocorticoid receptors in the kidneys of immature and mature male rats. *Int. J. Biochem.* 20:141–145, 1988.

Sherwood, N.M. and Timiras, P.S. Comparison of direct-current and radio-frequency-current lesions in the rostral hypothalamus with respect to sexual maturation in the female rat. *Endocrinology* 94:1275–1285, 1974.

Shoji, M., Golde, T.E., Ghiso, J., et al., Production of the alzheimer amyloid beta protein by normal proteolytic processing. *Science 258*:126–129, 1992.

Sidell, N. and Horn, R. Properties of human neuroblastoma cells following induction by retinoic acid. In: *Advances in Neuroblastoma Research (Progress in Clinical and Biological Research*, Vol. 175). Evans, A.E., D'Angio, G.J., and Seeger, R.C. (Eds.), Alan Liss, New York, pp. 39–53, 1985.

Smalheiser, N.R. and Swanson, D.R. Linking estrogen to Alzheimer's disease: An informatics approach. *Neurology* 47:809–810, 1996.

Sohrabji, F., Miranda, R.C., and Toran-Allerand, C.D. Estrogen differentially regulates estrogen and nerve growth factor receptor mRNAs in adult sensory neurons. *J. Neurosci.* 14(2):459–471, 1994.

Tang, M.X., Jacobs, D., Stern, Y. et al. Effect of oestrogen during menopause on risk of age at onset of Alzheimer's disease. *Lancet* 348:429–32, 1996.

Teresawa, E. and Timiras, P.S. Cyclic changes in electrical activity of the rat midbrain reticular formation during the estrous cycle. *Brain Res.* 14:189–198, 1969.

Terasawa, E. and Timiras, P.S. Electrical activity during the estrous cycle of the rat: Cyclic changes in limbic structures. *Endocrinology* 83:207–216, 1968a.

Terasawa, E. and Timiras, P.S. Electophysiological study of the limbic system in the rat at onset of puberty. *Am. J. Physiol.* 215:1462–1467, 1968b.

Timiras, P.S. Education, homeostasis, and longevity. *Exp. Gerontol.* 30:189–198, 1995.

Timiras, P.S. Estrogens as 'organizers' of CNS function. In: *Influence of Hormones on the Nervous System*, Ford, D.H. (Ed.), S. Karger Basel, pp. 242–254, 1971.

Timiras, P.S. *Physiological Basis of Aging and Geriatrics*. CRC Press, Boca Raton, 1994.

Timiras, P.S. The timing of hormone signals in the orchestration of brain development. In: *The Development of Attachment and Affiliative Systems, Topics in Developmental Psychobiology*, Emde, R.N. and Harmon, R.J. (Eds.), Plenum Press, New York, pp. 47–63, 1982.

Toran-Allerand, C.D., Miranda, R.C., Bentham, W., et al. Estrogen receptors co-localize with low-affinity NGF receptors in cholinergic neurons of the basal forebrain. *Proc. Natl. Acad. Sci. U.S.A.* 89:4668–4672, 1992.

Vaccari, A., Brotman, S., Cimino, J., and Timiras, P.S. Sex differentiation of neurotransmitter enzymes in central and peripheral nervous systems. *Brain Res.* 132:176–185, 1977.

Valcana, T., Vernadakis, A., and Timiras, P.S. Influence of estradiol and cortisol on electrolytes in the central nervous system of developing rats. *Neuroendocrinology* 2:326–329, 1967.

Vernadakis, A. and Timiras, P.S. The effect of oestradiol on spinal cord convulsions in developing rats. *Nature* 197:906, 1963.

Waymire J.C., Bjur,R., and Weiner, N. Assay of tyrosine hydroxylase by coupled decarboxylation of DOPA formed from [1-^{14}C] tyrosine. *Anal Biochem.* 43:588–600, 1971.

Weidemann, A., Konig, G., Bunke, D., et al. Identification, biogenesis, and localization of precursors of Alzheimer's disease A4 amyloid protein. *Cell* 57:115–126, 1989.

Woolley, D.E. and Timiras, P.S. Estrous and circadian periodicity and electroshock convulsions in rats. *Am. J. Physiol.* 202:379–382, 1962b.

Woolley, D.E. and Timiras, P.S. The gonad-brain relationship: Effect of female sex hormones on electroshock convulsions in the rat. *Endocrinology* 70:196–209, 1962a.

Woolley, D.E. and Timiras, P.S. Water and electrolyte alterations in plasma, brain and liver of rats after castration and sex hormone administration. *Acta Endocrinol.* 46:12–24, 1964.

Woolley, D.E., Holinka, C.F., and Timiras, P.S. Changes in ^{3}H-estradiol distribution with development in the rat. *Endocrinology* 84:157–161, 1969.

Woolley, D.E., Hope, W.G., Thompson-Reece, M.A., Gietzen, D.W., and Conway, S.B. Dopaminergic stimulation of estrogen receptor binding in vivo: A re-examination. *Recent Progress in Hormone Research* 49:383–392, 1994.

Woolley, D.E., Timiras, P.S., Rozenweig, M.R., Krech, D., and Bennett, E.L. Sex and strain differences in electroshock convulsions of the rat. *Nature* 190:515–516, 1961.

20

ROLE OF TESTOSTERONE IN THE ACTIVATION OF SEXUAL BEHAVIOR AND NEURONAL CIRCUITRIES IN THE SENESCENT BRAIN

G. C. Panzica,[1*] E. García-Ojeda,[1†] C. Viglietti-Panzica,[1] N. Aste,[1‡] and M. A. Ottinger[2]

[1]Department Anatomy, Pharmacology, and Forensic Medicine
University of Torino
c.so M. D'Azeglio 52, I-10126 Torino, Italy
[2]Department Animal and Avian Science
University of Maryland
College Park, Maryland

1. THE BRAIN: A TARGET ORGAN FOR THE ACTION OF GONADAL STEROIDS

Several studies performed in mammals and, more recently, in other vertebrates, demonstrated that sex differences in reproductive behavior as well as in neuronal circuitries involved in its control largely depend on steroid hormones. The perinatal exposure to gonadal steroids and the presence of the appropriate gonadal hormones in the adulthood are necessary for the full expression of sexual behaviour (for a complete list of references see Panzica et al., 1995). More recently, a new group of studies indicated that gonadal hormones can cause changes in brain morphology and functions in the adult brain also in regions which are not directly related to sexual functions (i.e. the regulation of cholinergic neurons by estradiol in the rat forebrain according to a sexually dimorphic pattern). Furthermore, other steroid hormones other than gonadal hormones are also effective on neural structures which do not belong to the traditional neuroendocrine brain targets (Luine and

* Author for correspondence: Dr. G.C. **Panzica,** Department Anatomy, Pharmacology, and Forensic Medicine, c.so M. D'Azeglio 52, I-10126 Torino, Italy. Phone: +39-11-6707729; Fax: +39-11-6707732; e-mail: giancarlo.panzica@unito.it
† Present address: Departamento de Biologia Celular, Universidad de Salamanca, Salamanca (Spain).
‡ Present address: Institute of Molecular Neurobiology, Shiga University of Medical Science, Otsu, Shiga 520–21 Japan.

Harding, 1994). With aging, the gonads undergo anatomical, histological and vascular changes, and as a result of these changes all forms of circulating gonadal hormones decrease in both female and male. The magnitude of this decline is considerably dependent on individuals, age and health status (for a review see Timiras *et al.*, 1995). As a consequence, alterations that occur in neuroendocrine systems during aging can provide new insights into the general problem of how steroid modulate neuronal circuitries throughout the life.

The localization of gonadal steroid hormone sensitive regions in the brain has been accomplished in many vertebrate species. These studies established that receptor sites for estrogens, androgens, progestins, glucocorticoids, and mineralcorticoids exist in a phylogenetically stable manner in various regions of the brain (McEwen et al., 1979).

The immunocytochemical detection of steroid receptors at neuronal level of resolution gave strenght to the hypothesis that a specific action of these hormones in the differentiation and in the postnatal development of target cerebral circuits did occur. Indeed receptors for gonadal steroids and dimorphic structures were frequently observed in the same or in closely related regions, providing indirect evidence that the origin and the presence of these neuroanatomical dimorphisms might be determined by the action of gonadal steroids (for a list of reference see Kawata, 1995).

A more convincing demonstration of the significance of gonadal hormones in the regulation of sexual dimorphisms has been recently provided by the administration of antisense oligoprobes to estrogen receptor (ER). When these antisense oligoprobes were administered during a sensitive developmental period they prevent the establishment of the sexual dimorphism in the rat preoptic region (McCarthy et al., 1993).

2. EFFECTS OF GONADAL HORMONES ON BRAIN CIRCUITRIES

The physiological effects of steroid hormones (in particular those of gonadal hormones) are generally classified, on the basis of their latency and duration, as long-term or short-term effects (Fink et al., 1991). Long-term effects last for many days, weeks (reversible effects), or even lifetime (irreversible effects). The short-term effects (always reversible) take either only minutes (or seconds) to develop (rapid effects), or much longer time (intermediate effects) and may last for several hours. The very short-term effects of steroid hormones are probably depending on a direct membrane effect (Schumacher, 1990; McEwen, 1991; Ramirez et al., 1996). Although the fast non-genomic mechanism may trigger the influence of these hormones on brain circuitries, in the present review we will focus on the slower and more permanent genomic effects.

The genomic mechanism of gonadal hormone action within nerve tissue cells (both neurons and glial elements) includes the coupling to specific intracellular receptors and the subsequent binding of the activated hormone-receptor complex to steroid receptor binding sites known as hormone response elements in the regulatory region of genes [Ramirez et al. (1996) for review]. The mechanism of the steroid action on gene expression (mostly unknown) can cause either the activation or the repression of several genes and of the associated sequence of biochemical events. The effect of steroid on the neuronal genome may lead to several morphological and functional modifications of the brain structures. These changes include: modification of nuclear size, rearrangement of the ultrastructure of cellular compartments related to protein synthesis, changes in neuronal enzyme levels, modification of cell body size or shape, variations in the rate of synthesis of neurotransmitters or neuropeptides, alteration of the efferent circuits, sprouting of axonal projections, growth of dendritic arborization with increased number of spines, modification of afferent circuits, altered glial elements, increase in cell number, and volumetric changes of neuronal aggregations (brain nuclei).

The regulation of the number of cells in a selected region is one of the most significant mechanisms to determine long-term irreversible changes (currently called *organizational*) in the brain which are instrumental in the determination of the constitution of sexually dimorphic pathways. The variations in cell number are generally based on three different mechanisms : (1) the steroid hormone could stimulate neurogenesis in one sex, or (2) could prevent neuronal death, or (3) could regulate processes influencing the differentiation of neurons. These effects are not univocal and may differ according to the various brain regions (Arnold and Schlinger, 1993).

Long-term reversible effects of steroids on neural pathways (generally described as *activational* effects) occur in the adulthood regardless the sexual differentiation of the involved circuitries. [for reviews see Breedlove (1992); Panzica et al. (1995)].

Morphological changes in the brain due to the action of gonadal hormones do not require hormone involvement at every point. For example the stimulation of axonal growth (under the control of sex steroids) could stimulate the growth and differentiation of the target region which in itself is lacking of gonadal hormone receptors (domino theory: (Arnold and Schlinger, 1993). Therefore, the morphology and connections of a specific brain region is the result of an interaction of hormone-dependent and hormone-independent entities (Tobet and Fox, 1992).

Furthermore it is important to underline that steroid hormones can act on nervous tissue also through their metabolites. For example, testosterone (T) can be actively metabolized within the brain into a number of androgenic or estrogenic steroids. Two metabolic pathways of T are particularly relevant to the control of male sexual behavior: 5α-reduction and aromatization. The 5α-reductase catalyzes the reduction of T into 5α-dihydrotestosterone (5α-DHT), a steroid that binds with high affinity to the androgen receptor (Celotti et al., 1992). The enzymatic complex aromatase (ARO) catalyzes the aromatization of T producing 17β-estradiol (E_2). Estrogens bind with high affinity to ER and in this way activate physiological and behavioral responses that are different from those activated by androgens (Balthazart, 1993; Roselli and Resko, 1993).

3. SEXUALLY DIMORPHIC STRUCTURES IN THE VERTEBRATE BRAIN

Sexually dimorphic nuclei or regions have been described in several species of tetrapods and have been observed in different contexts including:

1. neuronal groups directly related to peripheral dimorphic organs (i.e. motoneurons of the spinal cord controlling the penis muscles or motoneurons controlling the avian syringeal musculature);
2. cerebral nuclei without any evident connection to sexually dimorphic structures or organs (i.e. the dimorphic nuclei of the preoptic and limbic regions in many vertebrate species including humans, the ventromedial hypothalamus, or the accessory olfactory pathway in the rat).

In many cases these sexually dimorphic structures are under the control of steroid hormones also during adulthood to maintain their sexually dimorphic characteristics. A complete revision of the literature concerning the sexually dimorphic structures that have been described in the vertebrate brain and their regulation by gonadal hormones is out of the aim of the present review. For further details and a complete list of references, the reader is addressed to some recent reviews (Tobet and Fox, 1992; Breedlove, 1992; Panzica et al., 1995).

It is interesting to note that, apart from motorneurons controlling dimorphic muscles, only in a few cases the sexually dimorphic structures have been directly related to sexual behavior. Since male sexual behavior is differentiated in many species (it is more easily activated by androgens in males than in females), it could be expected that sexually dimorphic structures would play a key role in the activation of this behavior. Moreover lesion and hormone implantation studies demonstrated that the preoptic region plays a key role in controlling copulatory and lordosis behaviors (Meisel and Sachs, 1994; Pfaff et al., 1994). However, it has been difficult to relate in a causal manner the sexually dimorphic nucleus observed in the rat preoptic area with the mechanism mediating male copulatory behavior (Arendash and Gorski, 1983; De Jonge et al., 1990). The same conclusion was reached in other mammalian species studied so far except in gerbils, in which the pars compacta of the medial preoptic nucleus was demonstrated to play a key role in the control of male sexual behavior (Yahr and Gregory, 1993). In humans there are studies that compared gender and sexual attitudes, demonstrating that some preoptic and limbic nuclei have volume values more similar to the psychological than to the genetic sex, but of course no experimental work has been done on this theme (Le Vay, 1991; Swaab et al., 1992; Zhou et al., 1995).

In birds two models have been investigated, one is that of the song controlling nuclei of oscine (singing birds, Arnold, 1990), and the second one is the medial preoptic nucleus (POM) of the Japanese quail (Panzica et al., 1996c). This last structure is a sexually dimorphic nucleus representing a key center in the action of steroids on male sexual (copulatory) behavior (Balthazart and Surlemont, 1990). Various aspects of the POM morphology change in relation to circulating gonadal hormones and therefore, the structure of this nucleus provides a model for the action of steroids in the brain. Several forms of neuronal plasticity are observed in the POM and this nucleus constitutes an ideal model for the study of sexual behavior in a context that permits a functional interpretation of the changes observed at the cellular level. In a following section we will therefore describe in details this model that has been also widely employed in researches concerning the neuroendocrinology of reproduction, as well as of aging [for a complete list of references see Panzica et al. (1996c)]

4. THE JAPANESE QUAIL BEHAVIORAL MODEL

Testosterone (T) is the molecule inducing the full expression of male copulatory behavior in quail, but, in contrast to what is observed in rodents, this behavior shows a quite extreme sexual dimorphism in this species: in laboratory test conditions, sexually mature males almost never fail to exhibit the complete copulatory sequence (grabbing, mount attempts, mounts, and cloacal contact movements). They disappear after castration and they appear when castration is followed by cronic T-treatment in male. Conversely, high doses of T on either intact or gonadectomized females are not sufficient to activate male behavioural sequence. On the contrary, the female-type receptive behavior can be activated in both the sexes by an appropriate treatment with estrogens. The sex difference in sensitivity to the activating effects of T on copulatory behavior results from the early exposure of female brain to higher levels of circulating estrogens. This evidence indicates that the neuronal circuits supporting male reproductive behavior are precociously sexually differentiated in this species [major details on these experiments are discussed elsewhere (Balthazart and Foidart, 1993; Panzica et al., 1996c)].

The activation of copulatory behavior in castrated male quail can be obtained by treatment with high doses of estrogens or by a combined treatment with physiological doses of E_2 and 5α-DHT. This suggests that, in physiological conditions, although T is the most effective steroid on copulatory frequencies both the androgenic and estrogenic product of T metabolism are responsible for the T-dependent behavioral activation. On the other hand, ARO inhibitors can completely suppress the effects of T on copulatory behavior, while 5α-reductase inhibitors have little or no effect. This observation demonstrates that the aromatization of T is a limiting step for the activation of sexual behavior by T in male quail to occur (Balthazart and Foidart, 1993). Therefore, studies on the location and response of the ARO-producing system are integral to understanding how the system works and what may become altered during aging.

5. NEURONAL CIRCUITS CONTROLLING MALE COPULATORY BEHAVIOR IN QUAIL

In the course of studies analyzing the dimorphic mechanisms involved in the activation of sexual behavior we discovered a specialized region (the POM) of the POA showing an evident dimorphism in the volume: the POM is significantly larger in adult male than in adult female quail (Viglietti-Panzica et al., 1986). Its volume is also steroid-sensitive in adulthood: it decreases when circulating levels of testosterone are low (castration, exposure to short-days) and it increases when testosterone levels are high (treatment with testosterone, exposure to long-days) (Panzica et al., 1987, 1991). The POM is a necessary and sufficient site of steroid action for the activation of male copulatory behavior (Balthazart and Surlemont, 1990). Cytoarchitectural and morphometrical studies provided evidence for the existence of two neuronal populations, located in the medial and in the dorso-lateral portions of the POM and distinguished by their size. These two populations are characterized by different sensitivities to T in adulthood. The medial population appears smaller and relatively insensitive to T in both sexes, whereas the dorso-lateral population (only in males) changes in size in correlation to the levels of T. The size of neurons in the dorsolateral part of POM appears to be irreversibly affected by embryonic exposure to steroids and this feature is therefore a good correlate of the behavioral sex difference (Aste et al., 1991; Panzica et al., 1991). Ultrastructural studies suggest that T influences specific cellular compartments involved in the synthesis and processing of proteins (Panzica et al., 1996a). The POM is characterized by the presence of a wide variety of neurotransmitters, neuropeptides and receptors [for a complete list of references see Panzica et al. (1992, 1996c)]. It can, in addition, be specifically distinguished from the surrounding POA by the presence of ARO-ir cells (Balthazart et al., 1990b), by high density of α_2-adrenergic receptors (Ball et al., 1989), and by a denser vasotocinergic innervation (Viglietti-Panzica et al., 1994). Some of these neurochemical markers of the dimorphic nucleus are themselves modulated by steroids. In particular, the ARO-ir cells of the lateral POM appear to be a key target for steroids in the activation of male copulatory behavior. The POM is bidirectionally connected to many brain areas (Balthazart et al., 1994). It receives inputs from a variety of sensory areas and fregulatory areas (i.e. catecholaminergic cell groups). This nucleus also sends outputs to "neurovegetative" centers and to brain regions directly connected to the motor pathways. These connections fully support the role of the POM as an integrative center for the control of male sexual behavior. The available data indicate that there is a high degree of steroid-induced neuronal plasticity in the POM including changes in neuronal function, in protein synthesis and in specific inputs. These phenomena are also directly related to a clear functional output, the activation of male sexual behavior (Panzica et al., 1996c).

5.1. Aromatase Activity and Aromatase-Immunoreactive Cells

Biochemical studies characterized and localized the enzyme aromatase in the quail brain, showing that a high level of ARO activity is present in the quail POA. Further studies employing microdissections by the "Palkovits" punch technique demonstrated that all the ARO activity of the POA was localized within the POM [for a review on these data see Balthazart et al. (1990a)].

Four main groups of ARO-ir cells are observed by immunocytochemistry, namely in the POA, the septal region, at the level of the nucleus of the stria terminalis (nST), and in the ventromedial hypothalamic nucleus. All preoptic ARO-ir cells are localized within the POM; the dense cluster of ARO-ir cells is preciselly associated to the nucleus throughout its rostral-caudal extent.

ARO-ir cells represent a large fraction of the total neurons present in the POM (about 40% of the total in the lateral POM; 20% in the medial part of the nucleus (Aste et al., 1994)). Only a small portion of this population (about 20%) show also immunopositivity for ER (Balthazart et al., 1991; Dellovade et al., 1995).

5.2. Vasotocin (VT) Immunoreactive System

The distribution of VT-ir cells and fibres has been studied in several avian species and the characteristics of the vasotocinergic system has been recently reviewed (Viglietti-Panzica and Panzica, 1991; Jurkevich et al., 1996a). In quail, the majority of large VT-ir or VT-gene expressing cells (Viglietti-Panzica, 1986; Aste et al., 1996a) is located laterally or periventricularly in the preoptic and anterior hypothalamic regions, however a specific parvocellular cell group (located in the nST) has also been observed (Aste et al., 1995, 1996c). VT-ir fibers are observed in several intra- and extrahypothalamic regions involved in the control of reproduction (Jurkevich et al., 1996a), as the POM (control of copulatory behavior), the lateral septum [site of the largest population of neurons producing gonadotropin-releasing hormone (GnRH) in birds], and nucleus intercollicularis (ICo, a mesencephalic center controlling vocalizations). In a recent study we revealed the existence of widespread anatomical interactions between estrogen-synthesizing neurons and vasotocinergic fibers in the quail brain. Close relationships between VT-ir fibers and ARO-ir cell bodies or processes were in fact observed (Balthazart et al., 1997). These data suggest that VT may be involved in the regulation of several aspects of reproduction in birds.

5.3. Effects of Gonadectomy and Testosterone Therapy on Sexually Differentiated Neuronal Circuitries

ARO activity, ARO-ir elements, VT-ir innervation of the POM and lateral septum, and VT-ir or VT gene expressing elements of the nST show a sexually dimorphic distribution pattern in the quail brain. These systems were therefore investigated in the male quail to understand whether gonadectomy and T replacement therapy could influence their distribution pattern, or the production and storage of the involved neurochemical markers. A detailed description of all the experiments done on this subject has been recently published [for the complete review of the literature see Panzica et al. (1996c)], we will try to summarize here the most important results.

ARO synthesis is strongly dimorphic and is stimulated by circulating T only in the male (Schumacher and Balthazart, 1986). In the POM, a loss of ARO-ir cells coincident with the loss of male sexual behavior was observed in young castrated males. The decrease in the number of ARO-ir elements was presumably linked to the drop in the levels of available T since subcutaneous administration of T by means of silastic implants restored aromatase staining in castrate males. These data demonstrate that sexual behavior and ARO-immunoreactivity show parallel changes in young male quail (Balthazart et al., 1992). As already mentioned, in male quail ARO-ir cells are more numerous in the dorsolateral than in the medial part of the POM. The number of lateral ARO-ir cells decreases in castrated males to 10 percent of the initial number. On the contrary, the positive neurons in the medial part of the POM are reduced only to 30 percent of their initial number (Aste et al., 1994). These results suggest hence the presence of two types of aromatase cells in the quail POA in young male bird. A small population of ARO-ir cells is present in the brain independently of the steroid environment and this represents at least one third of the cells that are found in the medial part of the POM. Other larger ARO-ir cells are sensitive to steroid stimulation and they almost completely disappear in castrated birds. Most of the observed effects of T on ARO-ir cells could be reproduced by treatment with either E_2 alone or E_2 in combination with 5α-DHT (Aste et al., 1994).

As already mentioned, the VT-ir system in the nST is sexually dimorphic in galliforms as was recently demonstrated by means of immunocytochemistry and *in situ* hybridization studies (Aste et al., 1996c; Jurkevich et al., 1996b). In mammals, several studies have elucidated the role of the nST and of other limbic structures (the medial amygdaloid nucleus) as source of the majority of extraneurohypophyseal sexually dimorphic T-dependent vasopressinergic projections [i.e. the projection to the lateral septum, for a review see (De Vries et al., 1994)]. Released directly into the brain interstitial fluid vasopressin (VP) derived from limbic structures may account for its own neuromodulator and/or neurotransmitter-like effects including regulation of sex specific behavior (De Vries, 1990). Sexually dimorphic VT-ir fibers were also observed in quail, specifically in the lateral septum (Viglietti-Panzica et al., 1992), in the POM (Aste et al., 1996b), and in the ICo (unpublished results). Specific experiments performed on intact, castrated, and castrated plus T young male quail have demonstrated that in POM and lateral septum the extent of vasotocinergic innervation is strictly dependent by circulating T levels and parallels the changes in sexual behavior (Viglietti-Panzica et al., 1992, 1994; Aste et al., 1996b). Very recent experiments have finally demonstrated that VT is exerting an inhibitory role in the expression of male quail sexual behavior when peripherally or intracerebroventricularly injected (Castagna et al., 1996).

6. AGING IN THE JAPANESE QUAIL

6.1. Behavioral Aspects

There is a large variability in the sexual behavior of old male quail. Some individuals become spontaneously sexually inactive (senescent) from the age of 18 months. Plasma levels of androgen decrease in all middle-aged birds (18–30 months of age) and do not differ significantly between active and inactive males. Once the animal has become completely senescent, the testes cease both spermatogenesis and steroid production. The senescence is typically observed very late in aging (after 30 months or older). At this time, plasma androgen levels are in concert with reproductive status (Ottinger, 1992; Ottinger et al., 1995). At this time few individuals still retain sexual activity and they have larger testis and increased levels of circulating androgens (see Fig. 1A-C).

These data suggest that the simplistic idea that the loss of gonadal steroids triggers the process of behavioral reproductive aging in the male Japanese quail is contradictory. Similar to data from mammals, plasma androgen levels do not significantly change during aging until the male has become reproductively senescent (Ottinger, 1992). In both rats and quail, aging males that remain sexually active retain higher plasma androgen levels and these levels are independent from the intensity of sexual activity. More intriguing is the fact that the sexual behavior can be restored in senescent male quail by administration of exogenous T (Ottinger and Balthazart, 1986), whereas mammals do not fully respond to this treatment (Taylor et al., 1996). If we assume that sexual behavior is primarly "written" in neuronal circuiteries, this means that compared to mammals, quail maintain a neuronal machinery that retains the capability to respond, even in aging. It has to be stressed however that the dose of T used to stimulate sexual behavior in senescent males was double the effective dose for young or in adult birds. This suggests that part of the machinery does deteriorate during aging and the recovery of function may be based on a certain degree of plasticity still remaining in the old quail brain, or to on an overstimulation of the persisting system. Hence, we have been investigating the neuroendocrine systems affected by the process of aging and which specific systems are involved in the recovery of behavior in these old senescent males.

6.2. Neuronal Circuitries

6.2.1. The ARO-ir System. Age-related changes in this system were studied in 3 groups of quail: 6 month-old adult sexually active males, 36 month-old senescent males, and 36 month-old sexually active males (Dellovade et al., 1995). In old senescent males the decrease in the number of ARO cells was significant compared with the other two groups. Old active males had an intermediate number of ARO-ir cells as compared with the other 2 groups (see Fig. 1D). A significant decrease of the immunoreactive population which is located in the nST was observed only in old senescent males. An interesting finding was that the percentage of neurons co-expressing ER increased with age, rising from the 19% of adult birds to 25% of old inactive male quail. This means that the reduction in the number of ARO-ir cells is primarily interesting in those elements that do not show ER. These data confirmed hence the importance of the ARO-ir system for the activation of sexual behavior: old active males show at the same time sexual behavior, larger testis, and a number of ARO-ir neurons higher than old inactive birds. A more detailed morphometrical study (Panzica et al., 1994) demonstrated that in contrast to young castrated males, the ARO-ir loss in old males (both active and inactive) was more marked in the medial vs the lateral POM. In old senescent males the decrease in number was paralleled by a similar decrease in the cell size, indicating a loss in synthetic structures similar to young castrated males. In old active males the ARO-ir cells were more numerous in the lateral than in the medial POM and had a cell size which was significantly larger than in the adult sexually active and old inactive males. Two conclusions can be drawn from these experiments: a) the ARO system is specifically influenced by the reduction of circulating T both in the young and in the old animal, and loss of part of this neuronal system resulted in a loss of

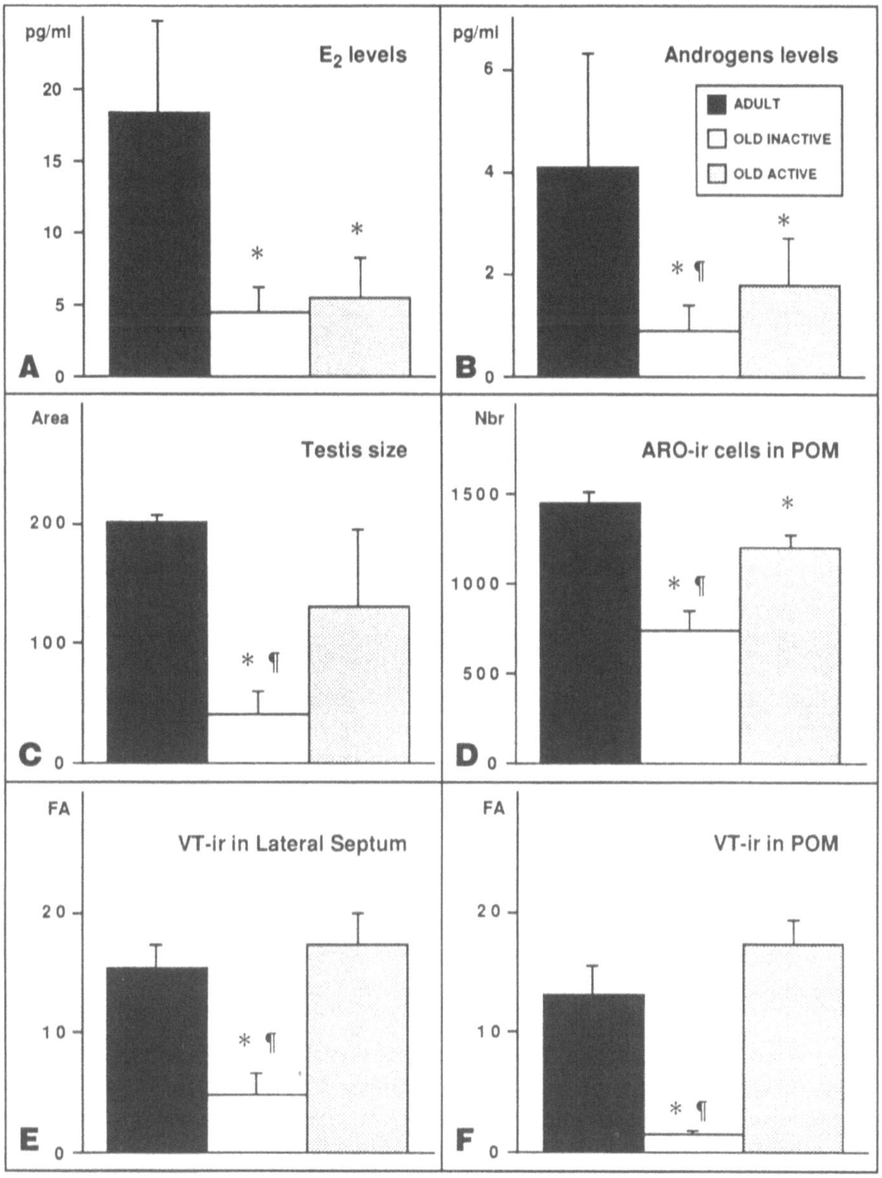

Figure 1. Histograms representing changes in endocrine, morphological, and neuronal characteristics during reproductive aging in male quail. **A-D** Old active males, are 36 month old male quail spontaneusly active; data are from Dellovade et al. (1995). **E-F** Old active males are 36 months senescent males treaed for 2 weeks with silastic implants of testosterone; data are from Panzica et al. (1996b). **A-B** Average levels of plasma estradiol (E2) and androgens in the 3 considered groups. Androgens in old active males are significantly lower than in adult, but significantly higher than in old inactive male quail.. **C**. Changes in testis size. Old inactive males have a significant lower area of testis in comparison with adult and old active males. **D**. Changes in the number of aromatase immunoreactive (ARO-ir) neurons within the POM during aging. Old active males have significantly less positive cells than adult, but significantly more positive cells than old inactive males. **E-F**. Vasotocin immunoreactive (VT-ir) fiber density in lateral septum and POM. Old inactive males have significantly less fractional area (FA) covered by immunopositive structures. * statistically significant difference ($p<0.05$) in comparison to adult group; ¶ statistically significant difference ($p<0.05$) in comparison to old active group.

the sexual performances; b) the specific mechanisms that the hormone decline provokes on the ARO system may vary in the young and in the old male quail. Moreover, the increase in cell size observed in old sexually active males has been already observed in other neurochemically identified systems during aging and it is generally considered as a signal of overstimulation of the surviving portion of the system which is trying to compensate for the partial loss of cells (Brody, 1992). Accordingly, it can be considered as a compensatory effect to minimize the deficient production of aromatase (due to the reduction in number of these elements) in order to maintain a production of E_2 sufficient to induce sexual activity.

6.2.2. The Vasotocin System. In recent studies we demonstrated an age-related decrease in VT-immunoreactivity in several brain areas (POM, lateral septum, and nST) important for the regulation of both endocrine and behavioral aspects of reproduction (Panzica et al., 1996b). Other portions of the VT-ir system (i.e. the magnocellular system) do not show the same decrease in positive structures (Viglietti-Panzica et al., 1996). Further, this age-related decrease in VT-immunoreactivity was restored to levels similar to those observed in adult male quail by T replacement therapy, a treatment that restores the sexual behavior in old senescent males (Fig. 1E-F). Similarly to what previously described for young males (Viglietti-Panzica et al., 1992, 1994) we have confirmed in the aged birds a direct correspondence between VT innervation of septo-preoptic region and sexual behavior. The decline of sexual behavior induced by castration or photoregression in the young male or by physiological events in old senescent males is paralleled by a decline in VT-immunoreactivity in specific cerebral areas. Moreover, these data indicate that the VT system retains a high degree of plasticity in old male quail. The source of the VT innervation of POM and lateral septum is still unknown in quail. However, the sexually dimorphic population of weakly immunostained cells located in the nST (showing itself a T-dependent decrease in the immunostaining during aging) represents a very good candidate as a source of this innervation.

A similar plasticity was observed in the mammalian VP system. A decrease in VP-ir fibers is present during aging and subcutaneous implants of T in 33 month-old Norway male rats restored VP innervations in sex steroid sensitive regions (Goudsmit et al., 1988), however, in this case, no behavioral recovery has been described.

6.2.3. The Gonadotropin-Releasing Hormone (GnRH) System. The process of reproductive senescence is ultimately leading to a complete loss of both endocrine and behavioral components of reproductive function. The study of age-related changes in the GnRH system is obviously central to the comprehension of variations in the endocrine regulation of reproduction. Moreover, several studies have suggested a direct involvement of GnRH in the control of sexual behavior (Dornan and Malsbury, 1989). In rats, hypothalamic GnRH is reduced in old sexually inactive males as compared to old sexually active males (Dorsa et al., 1984). Aged rats had a decreased number of neurons expressing GnRH mRNA compared to young males (Gruenewald and Matsumoto, 1991) and morphological alterations of the synaptology of GnRH neurons (Witkin, 1989).

No morphological data are at moment available on this system in aged birds, however, several endocrinological studies have been performed on both young and aged male quail [for a review see Ottinger et al., (1997)]. GnRH concentrations decrease in the preoptic-septal area and in the median eminence to very low levels in the senescent male. Moreover, the loss of GnRH occurred first in the preoptic-septal region and then later in the median eminence, perhaps indicating the loss of synthetic capability and then later depletion of stored hormone. The GnRH release by longitudinal hypothalamic slices (containing a large portion of the GnRH system) from old males was consistently lower than from young males. There was a qualitative difference in GnRH release in slices taken from young reproductive and old inactive males with reduced amplitude of release. Lastly, slices from young reproductive males are strongly responsive to catecholaminergic stimulation of the GnRH secretion, whereas slices from old senescent males were significantly less responsive. These data provide evidence that the GnRH system is capable of response even in senescent males, but the qualitative response becomes altered during aging.

Even if direct experiments on the senescent male quail have not been performed it has to be considered the fact that E_2 stimulates not only the sexual behavior, but also, directly, the GnRH production (Ottinger et al., 1997). Therefore, changes in the ARO-producing system are probably integral to alterations of GnRH system during aging. Moreover, immunocytochemical studies on serial sections demonstrated that VT-ir fibres have a spatial distribution very close to that of the septal group of GnRH-ir neurons (Panzica et al., 1992). These observations suggest that the VT-ir system might have a primary role in regulating the activity of GnRH neurons in quail.

7. CONCLUSION

Data reviewed in this paper demonstrate that the aging brain still maintains a certain degree of plasticity when stimulated by exogenous supply of testosterone. Moreover, in the quail model, the sensitivity of neuronal circuitries to T is paralleled by a recovery of copulatory behavior to levels comparable to those observed in adult sexually mature male quail.

Male copulatory behavior declines during aging and ceases in old senescent males. Old senescent male quail (36 months) have in fact greatly reduced plasma levels of T, Moreover, decreased levels of T, loss in testis weight and loss of behavioural capacity have been observed in photoregressed and castrated young males. However, it is not clear whether the mechanisms underlying loss of male quail sexual behavior are the same for these three conditions.

Several morphological and biochemical features of the quail brain are modified by the reduction in circulating levels of T resulting from aging, exposure to short day or surgical gonadectomy. In most of cases these modifications are complementary to the loss of sexual behavioural.

The VT-ir system shows similar changes in the young and in the old male quail. The decrease in VT-ir structures in the lateral septum, POM, and nST depends on the loss of T due to gonadectomy, exposure to short-day photoperiod (young birds) or aging process (old senescent individuals). When the birds (old and young) are supplied with exogenous T a full recovery of the immunoreactive structures is observed in all the mentioned regions. Data obtained in quail are very similar to those reported for the VP system in rat (Goudsmit et al., 1988). It should be noted that both in rodents (Fliers et al., 1985) and in quail (Viglietti-Panzica et al., 1996) the magnocellular system is not depressed by aging. On the contrary the cell size is slightly increased suggesting an activation of the system. Only the structures that are present in sex steroid dependent areas show a significant decrease in VT/VP immunoreactivity during aging. It appears hence that the VT (or VP) system in itself should not be considered the direct responsible of the loss of sexual behavior which starts when circulating levels of T are still high (and hence presumably still stimulating the system).

A loss of ARO-ir cells in the POM presumably linked to decreased levels of available T was observed in both old and young (castrated or photoregressed) males. However, this loss was more marked in young castrated than in old senescent males, and the two different populations of ARO-ir elements which are present in the POM (lateral vs medial) were affected in opposite directions in the young and in the old quail. We have not data concerning the ARO system in old senescent birds treated with exogenous T, however we investigated the distribution of this system in spontaneusly active old birds which showed rather high levels of circulating T. The ARO-ir system of these birds is different (in terms of number of cells) from that of both the young and the old senescent males, moreover, the reported increase of the cell size suggests that the surviving portion of the system was trying to recover the functions (presumably the production of E_2) of the whole ARO system. Only a few studies have considered changes in brain aromatase activity during aging in mammals, and no immunocytochemical study has been performed. ARO activity decreases with age, and can be restimulated by supply of exogenous T (Chambers et al., 1991). This treatment is not fully active in recovering the sexual function in old rat (only the 25% of experimental old rats demonstrated ejaculation).

The peculiar situation of the quail, in which exogenous T can re-stimulate the presence of sexual behavior in very old males could be related to the peculiarity of the brain circuitery which is involved in the control of male copulatory behavior in this species. VT-ir fibers are innervating the POM and the ARO-ir cells (that play a key role in the control of sexual behavior), and the septal region where a large cluster of GnRH neurons is present. The specific way of interaction between ARO and VT and its functional implications remain unclear at present. It is, however, evident that some of the functional outputs of the brain are affected both by estrogens and VT. For example, appetitive and consummatory aspects of male sexual behavior in quail are activated by estrogens, but inhibited by the central action of VT that probably plays a tonic inhibition on these behaviors (Castagna et al., 1996).

In conclusion, it may be speculated that the sensitivity of the VT system to T (lasting also in the aged animal) could indirectly determine the restoration and regulation of the functions of the ARO system in the POM and of the GnRH system. If this is the case, the putative center of origin of these VT fibers (the nucleus of the stria terminalis) should be regarded as the most important regulatory center of the sexual behavior in quail.

ABBREVIATIONS

5a-DHT: 5a-dihydrotestosterone, AR: androgen receptor, ARO: aromatase, E2: estradiol, ER: estrogen receptor, GnRH: gonadotropin-releasing hormone, ICo: nucleus intercollicularis, ir: immunoreactive, nST: nucleus of the stria terminalis, POA: preoptic anterior region, POM: nucleus preopticus medialis, T: testosterone, VP: vasopressin, VT: vasotocin.

ACKNOWLEDGMENTS

This study was supported by grants from MURST (40% to A. Fasolo, 60% to CVP and GCP), CNR (93.00372, 94.02462 to GCP), NATO (CRG 92.1267 to GCP and MAO), EU (94–0472 to GCP), Maryland Agriculture Experimental Station, USDA and NRI to MAO. EGO was a fellow of the EU at the Dept. of Anatomy, Pharmacology, and Forensics Medicine of Torino.

REFERENCES

Arendash, G.W. and Gorski, R.A. (1983). Effects of discrete lesions of the sexually dimorphic nucleus of the preoptic area or other medial preoptic regions on the sexual behavior of male rats. Brain Res.Bull. **10**, 147–154.

Arnold, A.P. (1990). The passerine bird song system as a model in neuroendocrine research. J.Exp.Zool. **4 (suppl)**, 22–30.

Arnold, A.P. and Schlinger, B.A. (1993). The puzzle of sexual differentiation of the brain and behavior in Zebra finches. Poultry Science Rev. **5**, 3–13.

Aste, N., Panzica, G.C., Viglietti-Panzica, C., and Balthazart, J. (1991). Effects of *in ovo* estradiol benzoate treatments on sexual behavior and size of neurons in the sexually dimorphic medial preoptic nucleus of Japanese quail. Brain Res.Bull. **27**, 713–720.

Aste, N., Panzica, G.C., Aimar, P., Viglietti-Panzica, C., Harada, N., Foidart, A., and Balthazart, J. (1994). Morphometric studies demonstrate that aromatase-immunoreactive cells are the main target of androgens and estrogens in the quail medial preoptic nucleus. Exp.Brain Res. **101**, 241–252.

Aste, N., Panzica, G.C., Viglietti-Panzica, C., Absil, P., Balthazart, J., Mühlbauer, E., and Grossmann, R. (1995). The vasotocin system of the nucleus of the stria terminalis in the Japanese quail. Soc.Neurosci.Abstr. **21**, 357.

Aste, N., Mühlbauer, E., and Grossmann, R. (1996a). Distribution of AVT gene expressing neurons in the prosencephalon of japanese quail and chicken. Cell Tissue Res. **286**, 365–373.

Aste, N., Viglietti-Panzica, C., Balthazart, J., and Panzica, G.C. (1996b). Influence of gonadal hormones on peptidergic pathways of the quail brain. Poultry and Avian Biology Reviews **6**, 275.

Aste, N., Viglietti-Panzica, C., Mühlbauer, E., Grossmann, R., and Panzica, G.C. (1996c). Sexual dimorphism of vasotocinergic structures in quail nucleus of stria terminalis. Italian J.Anatomy Embriol. **101, Suppl.1**, 136–137.

Ball, G.F., Nock, B., McEwen, B.S., and Balthazart, J. (1989). Distribution of α_2-adrenergic receptors in the brain of the Japanese quail as determined by quantitative autoradiography: implication for the control of sexually dimorphic reproductive processes. Brain Res. **491**, 68–79.

Balthazart, J., Foidart, A., Surlemont, C. , and Harada, N. (1990a). Preoptic Aromatase in Quail: Behavioral, Biochemical and Immunocytochemical Studies. In *Hormones, Brain and Behaviour in Vertebrates. Vol.2. Behavioural activation in males and females - Social interactions and reproductive endocrinology. Comp. Physiol. Vol. 9* (ed Balthazart, J.) pp. 45–62. Karger, Basel, New York

Balthazart, J., Foidart, A., Surlemont, C., Vockel, A., and Harada, N. (1990b). Distribution of aromatase in the brain of the Japanese quail, ring dove, and zebra finch: an immunocytochemical study. J.Comp.Neurol. **301**, 276–288.

Balthazart, J. and Surlemont, C. (1990). Copulatory behavior is controlled by the sexually dimorphic nucleus of the quail preoptic area. Brain Res.Bull. **25**, 7–14.

Balthazart, J., Foidart, A., Surlemont, C., and Harada, N. (1991). Neuroanatomical specificity in the co-localization of aromatase and estrogen receptors. J.Neurobiol. **22**, 143–157.

Balthazart, J., Surlemont, C., and Harada, N. (1992). Aromatase as a cellular marker of testosterone action in the preoptic area. Physiol.Behav. **51**, 395–409.

Balthazart, J. (1993). Brain aromatase and reproductive function. In *Local systems in reproduction* (eds Magness, R.R. and Naftolin, F.) pp. 13–31. Raven Press, New York

Balthazart, J. , and Foidart, A. (1993). Neural bases of behavioral sex differences in quail. In *The development of sex differences and similarities in behavior* (ed Haug, M.) pp. 51–75. Kluwer Acad.Publ., Amsterdam

Balthazart, J., Dupiereux, V., Aste, N., Viglietti-Panzica, C., Barrese, M., and Panzica, G.C. (1994). Afferent and efferent connections of the sexually dimorphic medial preoptic nucleus of the male quail revealed by *in vitro* transport of DiI. Cell Tissue Res. **276**, 455–475.

Balthazart, J., Absil, P., Viglietti-Panzica, C., and Panzica, G.C. (1997). Vasotocinergic innervation of areas containing aromatase-immunoreactive cells in the quail forebrain. J.Neurobiol. (in press)

Breedlove, S.M. (1992). Sexual dimorphism in the vertebrate nervous system. J.Neurosci. **12**, 4133–4142.

Brody, H. (1992). The aging brain. Acta Neurol.Scand. **85**, 40–44.

Castagna, C., Absil, P., and Balthazart, J. (1996). Central effects of vasotocin on appetitive and consummatory sexual behavior in male quail. Soc.Neurosci.Abstr. **22**, 2068.

Celotti, F., Melcangi, R.C., and Martini, L. (1992). The 5α-reductase in the brain: molecular aspects and relation to brain function. Front.Neuroendocrinol. **13**, 163–215.

Chambers, K.C., Thornton, J.E., and Roselli, C.E. (1991). Age-related deficits in brain androgen binding and metabolism, testosterone, and sexual behavior of male rats. Neurobiol.Aging **12**, 123–130.

De Jonge, F.H., Louwerse, A.L., Ooms, M.P., Evers, P., Endert, E., and Van De Poll, N.E. (1990). Lesions of the SDN-POA inhibit sexual behavior of male Wistar rats. Brain Res.Bull. **23**, 483–492.

De Vries, G.J. (1990). Sex differences in neurotransmitter systems. J.Neuroendocrinol. **2**, 1–13.

De Vries, G.J., Al Shamma, H.A. , and Zhou, L. (1994). The sexually dimorphic vasopressin innervation of the brain as a model for steroid modulation of neuropeptide transmission. In *Hormonal restructuring of the adult brain. Basic and clinical perspectives. Ann. New York Acad.Sciences. Vol. 743* (eds Luine, V.N. and Harding, C.F.) pp. 95–120. New York Acad.Sciences, New York

Dellovade, T.L., Rissman, E.F., Thompson, N., Harada, N., and Ottinger, M.A. (1995). Co-localization of aromatase enzyme and estrogen receptor immunoreactivity in the preoptic area during reproductive aging. Brain Res. **674**, 181–187.

Dornan, W.A. and Malsbury, C.W. (1989). Neuropeptides and male sexual behavior. Neurosci.Biobehav.Rev. **13**, 1–15.

Dorsa, D.M., Smith, E.R., and Davidson, J.M. (1984). Immunoreactive β-endorphin and LHRH levels in the brains of aged male rats with impaired sex behavior. Neurobiol.Aging **5**, 115–120.

Fink, G., Rosie, R., Sheward, W.J., Thomson, E., and Wilson, H. (1991). Steroid Control of Central Neuronal Interactions and Function. J.Steroid Biochem.Mol.Biol. **40**, 123–132.

Fliers, E., De Vries, G.J., and Swaab, D.F. (1985). Changes with aging in the vasopressin and oxytocin innervation of the rat brain. Brain Res. **348**, 1–8.

Goudsmit, E., Fliers, E., and Swaab, D.F. (1988). Testosterone supplementation restores vasopressin innervation in the senescent rat brain. Brain Res. **473**, 306–313.

Gruenewald, D.A. and Matsumoto, A.M. (1991). Age-related decreases in serum gonadotropin levels and gonadotropin-releasing hormone expression in the medial preoptic area of the male rat are dependent upon testicular feddback. Endocrinol. **129**, 2442–2450.

Jurkevich, A., Barth, S.W., Aste, N., Panzica, G.C., and Grossmann, R. (1996a). Intracerebral sex differences in the vasotocin system in birds: possible implication on behavioral and autonomic functions. Horm.Behav. (in press)

Jurkevich, A., Barth, S.W., and Grossmann, R. (1996b). Sexual dimorphism of arg-vasotocin gene expressing neurons in the telencephalon and dorsal diencephalon of the domestic fowl. An immunocytochemical and in situ hybridization study. Cell Tissue Res. **287**, 69–77.

Kawata, M. (1995). Roles of steroid hormones and their receptors in structural organization in the nervous system. Neurosci.Res. **24**, 1–46.

Le Vay, S. (1991). A Difference in Hypothalamic Structure Between Heterosexual and Homosexual Men. Science **253**, 1034–1037.

Luine, V.N. , and Harding, C.F. (1994). *Hormonal restructuring of the adult brain: Basic and clinical perspectives. Annals New York Academy of Sciences. Vol. 743*, New York Acad.Sci., New York.

McCarthy, M.M., Schlenker, E.H., and Pfaff, D.W. (1993). Enduring consequences of neonatal treatment with antisense oligodeoxynucleotides to estrogen receptor messenger ribonucleic acid on sexual differentiation of rat brain. Endocrinol. **133**, 433–439.

McEwen, B.S., Davis, P., Parson, B., and Pfaff, D.W. (1979). The brain: target for steroid hormone action. Ann.Rev.Neurosci. **2**, 65–112.

McEwen, B.S. (1991). Steroid affect neural activity by acting on the membrane and the genome. Trends Pharmacol.Sci. **12**, 141–147.

Meisel, R.L. , and Sachs, B.D. (1994). The physiology of male sexual behavior. In *The Physiology of reproduction* (eds Knobil, E. and Neill, J.D.) pp. 3–105. Raven Press, New York.

Ottinger, M.A. and Balthazart, J. (1986). Altered endocrine and behavioral responses with reproductive aging in the male Japanese quail. Horm.Behav. **20**, 83–94.

Ottinger, M.A. (1992). Altered neuroendocrine mechanisms during reproductive aging. Poultry Science Rev. **4**, 235–248.

Ottinger, M.A., Nisbet, I.C.T., and Finch, C.E. (1995). Aging and reproduction: Comparative endocrinology of the Common Tern and Japanese quail. Am.Zool. **35**, 299–306.

Ottinger, M.A., Thompson, N., Viglietti-Panzica, C., and Panzica, G.C. (1997). Neuroendocrine regulation of GnRH and behavior during aging in birds. Brain Res.Bull. (in press)

Panzica, G.C., Viglietti-Panzica, C., Calcagni, M., Anselmetti, G.C., Schumacher, M., and Balthazart, J. (1987). Sexual differentiation and hormonal control of the sexually dimorphic medial preoptic nucleus in quail. Brain Res. **416**, 59–68.

Panzica, G.C., Viglietti-Panzica, C., Sánchez, F., Sante, P., and Balthazart, J. (1991). Effects of testosterone on a selected neuronal population within the preoptic sexually dimorphic nucleus of the Japanese quail. J.Comp.Neurol. **303**, 443–456.

Panzica, G.C., Aste, N., Viglietti-Panzica, C., and Fasolo, A. (1992). Neuronal circuits controlling quail sexual behavior. Chemical neuroanatomy of the septo-preoptic region. Poultry Science Rev. **4**, 249–259.

Panzica, G.C., Aste, N., Dellovade, T.L., Rissman, E.F., Foidart, A., Balthazart, J., and Ottinger, M.A. (1994). Aromatase-containing cells of the medial preoptic nucleus respond differentially to testosterone in young and aged male quail. Ann.Endocrinol. **55**, 37.

Panzica, G.C., Aste, N., Viglietti-Panzica, C., and Ottinger, M.A. (1995). Structural sex differences in the brain: Influence of gonadal steroids and behavioral correlates. J.Endocrinol.Invest. **18**, 232–252.

Panzica, G.C., Castagna, C., Aste, N., Viglietti Panzica, C., and Balthazart, J. (1996a). Testosterone effects on the neuronal ultrastructure in the medial preoptic nucleus of male Japanese quail. Brain Res.Bull. **39**, 281–292.

Panzica, G.C., García-Ojeda, E., Viglietti Panzica, C., Thompson, N.E., and Ottinger, M.A. (1996b). Testosterone effects on vasotocinergic innervation of sexually dimorphic medial preoptic nucleus and lateral septum during aging in male quail. Brain Res. **712**, 190–198.

Panzica, G.C., Viglietti-Panzica, C., and Balthazart, J. (1996c). The sexually dimorphic medial preoptic nucleus of quail: a key brain area mediating steroid action on male sexual behavior. Front.Neuroendocrinol. **17**, 1–75.

Pfaff, D.W., Schwartz-Giblin, S., McCarthy, M.M. , and Kow, L.M. (1994). Cellular and molecular mechanisms of female reproductive behaviors. In *The physiology of reproduction* (eds Knowbil, E. and Neill, J.D.) pp. 107–220. Raven Press, New York

Ramirez, V.D., Zheng, J., and Siddique, K.M. (1996). Membrane receptors for estrogen, progesterone, and testosterone in teh rat brain: fantasy or reality? Cell.Mol.Neurobiol. **16**, 175–198.

Roselli, C.E. and Resko, J.A. (1993). Aromatase activity in the rat brain: hormonal regulation and sex differences. J.Steroid Biochem.Mol.Biol. **44**, 499–508.

Schumacher, M. and Balthazart, J. (1986). Testosterone-induced brain aromatase is sexually dimorphic. Brain Res. **370**, 285–293.

Schumacher, M. (1990). Rapid membrane effects of steroid hormones: an emerging concept in neuroendocrinology. Trends in Neuroscience **13**, 359–362.

Swaab, D.F., Gooren, L.J.G. , and Hofman, M.A. (1992). The human hypothalamus in relation to gender and sexual orientation. In *The human hypothalamus in health and disease. Progress in brain research, vol. 93* (eds Swaab, D.F., Hofman, M.A., Mirmiran, M., Ravid, R., and Van Leeuwen, F.W.) pp. 205–215. Elsevier, Amsterdam

Taylor, G., Bardgett, M., Farr, S., Humphrey, W., Womack, S., and Weiss, J. (1996). Aging of the brain-testicular axis: reproductive systems of healthy old male rats with or without endocrine stimulation. Proc.Soc.Exp.Biol.Med. **211**, 69–75.

Timiras, P.S. (1996). *Hormones and aging*, (UnPub).

Tobet, S.A. , and Fox, T.O. (1992). Sex differences in neuronal morphology influenced hormonally throughout life. In *Handbook of Behavioral Neurobiology, Vol. 11: Sexual differentiation* (eds Gerall, A.A., Moltz, H., and Ward, I.L.) pp. 41–83. Plenum Press, New York.

Viglietti-Panzica, C. (1986). Immunohistochemical study of the distribution of vasotocin reacting neurons in avian diencephalon. J.Hirnforsch. **27**, 559–566.

Viglietti-Panzica, C., Panzica, G.C., Fiori, M.G., Calcagni, M., Anselmetti, G.C., and Balthazart, J. (1986). A sexually dimorphic nucleus in the quail preoptic area. Neurosci.Lett. **64**, 129–134.

Viglietti-Panzica, C. and Panzica, G.C. (1991). Peptidergic neurons in the avian brain. Ann.Sci.Nat.Zool.,Paris **12**, 137–155.

Viglietti-Panzica, C., Anselmetti, G.C., Balthazart, J., Aste, N., and Panzica, G.C. (1992). Vasotocinergic innervation of the septal region in the Japanese quail: sexual differences and the influence of testosterone. Cell Tissue Res. **267**, 261–265.

Viglietti-Panzica, C., Aste, N., Balthazart, J., and Panzica, G.C. (1994). Vasotocinergic innervation of sexually dimorphic medial preoptic nucleus of the male Japanese quail: influence of testosterone. Brain Res. **657**, 171–184.

Viglietti-Panzica, C., García-Ojeda, E., Aste, N., Panzica, G.C., Thompson, N., and Ottinger, M.A. (1996). The vasotocin system in the quail brain: changes with age. Soc.Neurosci.Abstr. **22**, 1890.

Witkin, J.W. (1989). Aging changes in synaptology of luteinizing hormone-releasing hormone neurons in male rat preoptic area. Neuroendocrinology **49**, 344–348.

Yahr, P. and Gregory, J.E. (1993). The medial and lateral cell groups of the sexually dimorphic area of the gerbil hypothalamus are essential for male sex behavior and act via separate pathways. Brain Res. **631**, 287–296.

Zhou, J.N., Hofman, M.A., Gooren, L.J.G., and Swaab, D.F. (1995). A sex difference in the human brain and its relation to transsexuality. Nature **378**, 68–70.

ENDOGENOUS OPIOIDS AND PRENATAL DETERMINANTS OF NEUROPLASTICITY

Ian S. Zagon,[1] Steven W. Tobias,[2] and Patricia J. McLaughlin[1]

[1]Department of Neuroscience and Anatomy
[2]Department of Comparative Medicine
The Pennsylvania State University
The M.S. Hershey Medical Center
Hershey, Pennsylvania 17033

1. INTRODUCTION

An endogenous opioid system involved in the growth of developing, neoplastic, renewing, and healing cells and tissues was first postulated in the early 1980's. This concept arose from the demonstration that blockade of native opioids (i.e., enkephalins, endorphins) from opioid receptors in developing animals (Zagon and McLaughlin, 1983a,c) or tumors transplanted into mice (1983b) accelerated growth when the opioid antagonist utilized was continuously available. Such observations gave rise to the hypothesis that endogenous opioids serve as growth inhibitory molecules and function in an active, tonic fashion. At this time it also was learned that if opioids were disrupted from opioid receptors for only a short time each day, growth could be delayed. The explanation for this latter observation resided in knowledge that during the time of opioid receptor blockade there is a compensatory production of opioid peptides and receptors. During the interval when the opioid antagonist is not available, an increased concentration of peptide can interact with cells containing more receptors, and the functional sensitivity is heightened. Since opioids are growth inhibitory, the finding of suppression of development or oncogenesis was in keeping with this thesis. These defining principles set forth the hypothesis that native opioid peptides are associated with growth, in addition to the function of opioids/opioid receptors in neuromodulatory events (Akil et al., 1984).

In the past decade considerable progress has been made in understanding the role of opioids in neural development (see Zagon and McLaughlin, 1993). Particularly important is how one assesses whether an opioid peptide does indeed function in growth. A number of criteria have been established that serve as guidelines for identification of an opioid growth factor, and include: (a) stereospecificity, (b) sensitivity to opioid antagonists, (c) activity at physiologically relevant concentrations, (d) effects demonstrated within a relatively short time if cell replication is implicated, (e) lack of cross-reactivity to other receptor-selective

Brain Plasticity, edited by Filogamo *et al.*
Plenum Press, New York, 1997

compounds, (f) a naturally occurring opioid peptide, (g) *in vivo* activity, (h) spatial and temporal distribution consistent with specific growth-related effects, (i) distinction from other opioid growth factors, (j) effectiveness by acute, as well as chronic, application, (k) elimination of confounding variables such as physical dependence, tolerance, and/or withdrawal, (l) reversibility in action, (m) an opioid receptor responsible for mediating the effect of the opioid, and (n) association of both the peptide and the receptor with appropriate cells/tissues. If the opioid is a growth factor in humans, then it is incumbent for demonstration of the peptide (and receptor) to be shown in human cells/tissues *in vivo* and, at least in a tissue culture milieu and within the constraints of function, and physiological activity.

Although there has been considerable activity in the field of opioids and growth (Barg et al., 1993; Bartolome et al., 1991; McLaughlin, 1996; Meriney et al., 1991; Murgo, 1985; Seatriz and Hammer, 1993; Shahabi and Sharp, 1995; Villiger and Lotz, 1992; Zadina et al., 1985; Zagon et al. 1996a,b; Zagon and McLaughlin, 1991a,b,1993), and reports of the effects of opioids on growth either *in vivo* or *in vitro* are in the literature, the only candidate meeting all of the requirements set forth above for an opioid growth factor has been the pentapeptide, [Met^5]-enkephalin. Indeed, a screen of natural and synthetic opioids, including those selective for different opioid receptors (e.g., μ, δ, κ), showed that only [Met^5]-enkephalin influenced the growth of the brain (Zagon and McLaughlin, 1993), heart (McLaughlin, 1996), neuroblastoma (Zagon and McLaughlin, 1991a), or colon cancer (Zagon et al., 1996a) *in vivo*. Moreover, studies *in vitro* have revealed that [Met^5]-enkephalin was the only opioid peptide from a screening study that altered the growth of neuroblastoma (Zagon and McLaughlin, 1991a) and colon cancer cells (Zagon et al., 1996b). To signify the physiological properties of [Met^5]-enkephalin, this peptide has been termed the opioid growth factor (OGF). In the future, if other opioid peptides are proven to fulfill the requirements of an opioid related to growth, this nomenclature will have to be modified (e.g., OGF1, OGF2). At that point, rigorous, systematic, and extensive experimentation will be needed to determine if, indeed, other opioids participate in growth processes, and whether artifacts (e.g., tissue culture) are involved in observations of other opioids modulating growth.

OGF is a negative growth regulator in developing, renewing, healing, and neoplastic tissues, and is direct and rapidly acting, noncytotoxic, reversible in action, obedient to the intrinsic rhythms of the cell (e.g., circadian rhythm), and is neither species nor cell-tissue specific. This growth factor is especially targeted to events related to cell proliferation, although it appears to influence cell migration, differentiation, and tissue organization. OGF and its receptor, zeta (ζ), are in a state of continuous and unremitting balance with respect to growth modulation. Both OGF and the ζ receptor are associated with the cellular elements being governed.

Chronic blockade of endogenous opioids and their receptors during postnatal life, imposed by administration of the opioid antagonist naltrexone, has been shown to result in rat pups that are larger in size, and to have enhanced brain and organ weights, increased body weights, early acquisition of some physical characteristics, spontaneous motor and reflexive behaviors, and an acceleration in the proliferation and development of brain cells (including differentiation of spines and dendrites in the cerebellum and hippocampus) (Hauser et al., 1989). The duration of opioid receptor blockade governs response. Thus, a dose of 20 to 50 mg/kg naltrexone (up to 2.5 % of the LD_{50}; Braude and Morrison, 1976) or multiple doses of either naltrexone (a lower dosages) or naloxone that institute a state of continuous opioid receptor blockade will produce similar action.

Although we know a great deal about the function of opioids relative to the postnatally growing animal, very little is known about the role of opioids in prenatal life. Both opioids and opioid receptors have been reported in the fetus (e.g., Clendeninn et al., 1976; Coyle and Pert, 1976; Knodel and Richelson, 1980; Neale et al., 1978), suggesting that these elements are present early in life. A few studies have begun to examine the effects of administering opioid antagonists during pregnancy (Cohen et al., 1996; D'Amato et al., 1988, Harry and Rosecrans, 1979; Hauser et al., 1989; Hetta and Terenius, 1980; Keshet and Weinstock, 1995; Leng et al., 1985; Mayer et al., 1985; Monder et al., 1979; Pfeiffer et al., 1984; Seatriz and Hammer, 1993; Shepanek et al., 1989, 1995; Vorhees, 1981; Ward et al., 1986). Unfortunately none of these investigations disrupted opioid-receptor interaction continuously from fertilization to parturition and, with some exceptions (Cohen et al., 1996; Keshet and Weinstock, 1995), did not examine whether the opioid antagonist treatment utilized was sufficient to block opioid receptors. Therefore, it is difficult to discern from these reports whether the response(s) observed is(are) due to blockade of opioid receptors and/or reflects events following opioid-receptor interaction. Thus, a series of studies (McLaughlin et al., 1997a,b) were conducted to address the question if opioids function in the regulation of developmental processes during prenatal life. The paradigm chosen was to continuously interrupt opioid-receptor interfacing from fertilization to parturition with the use of the opioid antagonist, naltrexone (NTX); nociceptive tests were utilized to verify that opioid receptor blockade was maintained. To eliminate the influence of residual drug accumulation and problems of maternal care because of sequelae related to drug termination, pups were culled to 10 /litter and fostered to control mothers at birth. This chapter presents and discusses the evidence concerning opioid function in prenatal life as a determinant of neonatal outcome as well as postnatal well-being, and is focused on the shaping of the nervous system and its plasticity. For methodological details of the studies presented, along with tangential information obtained from these investigations, please refer to the reports of McLaughlin and coworkers (McLaughlin et al., 1997a,b).

2. CHRONIC OPIOID RECEPTOR BLOCKADE DURING PREGNANCY: MATERNAL AND OFFSPRING EFFECTS

2.1. Maternal Observations

Daily injections of 50 mg/kg NTX throughout gestation did not influence pregnancy or maternal behavior. The body weights of control and NTX-injected females were similar throughout gestation (Fig. 1), and the mothers of both the NTX and control groups appeared to clean and nurse their young in a comparable fashion.

On day 1 of pregnancy, nociceptive tests using a hot-plate showed no differences between the randomly assigned females, with the controls and NTX-treated rats having latencies of 19.8 ± 1.1 and 19.3 ± 1.0 seconds, respectively. To determine the duration of receptor blockade, as examined earlier for 21-day old rats (Zagon and McLaughlin, 1984), female rats were challenged with morphine 22 hr following NTX or saline injections on the day of parturition. Control and NTX-injected rats exhibited a baseline latency that was equivalent (Fig. 2). Following injection of 10 mg/kg morphine, the control animals had twice the latency response time than baseline levels, but NTX-injected rats had a mean latency time that was comparable to baseline values. Furthermore, the latency for the controls given morphine differed markedly ($p<0.03$) from the post-morphine level of rats in the NTX group.

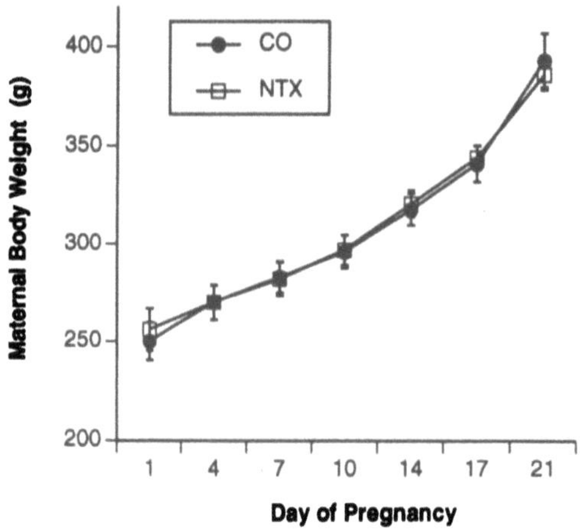

Figure 1. The body weight of pregnant rats receiving daily injections of 50 mg/kg naltrexone (NTX) (n = 11) or saline (CO) (n = 12) throughout gestation. No differences in body weights were detected between these groups during pregnancy. Data represent means ± SEM. (Reproduced from McLaughlin et al., 1997a, with permission of the publisher.)

2.2. Neonatal Observations

The length of time for gestation, size of litters, and mortality at birth was comparable for animals born to NTX- or saline-injected females (Table 1). The total pup mass of all animals (live or stillborn) from control mothers was 87.3 g, whereas the total weight of all pups born to NTX-treated females was 80.2 g; total pup mass between the two groups did not differ significantly.

The mean birth weight for NTX-treated rats was 8% greater than for control rats, with this difference being statistically significant ($p<0.01$) (Fig. 3). Pups exposed prenatally to NTX weighed significantly more than control rats on postnatal days 3, 6, 9, 12, 15, and 18 and, on postnatal day 21, rats subjected to NTX during prenatal life weighed over one third (36%) more than control subjects (32.9 ± 1.0 g). At weaning, the weights of male and females rats within either the saline- or NTX-treated groups were similar. Data were subjected to repeated measures analysis, and a substantial and significant difference in the growth patterns of the 2 groups over time was noted. For instance, at 21 days the predicted value of the NTX group was 55.9 g compared to 39.4 g of the control group.

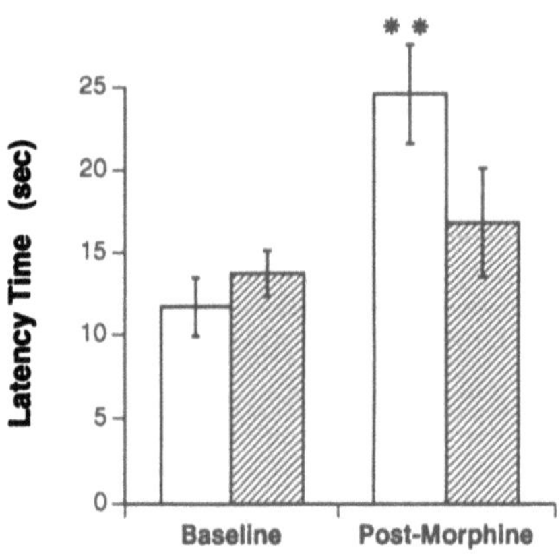

Figure 2. Latencies (sec) on the hot-plate (55°C) of rats treated daily during pregnancy with 50 mg/kg naltrexone (NTX) (n = 11) or saline (CO) (n = 12); animals were evaluated 22 hr following an injection of drug on the day of parturition. At least 10 rats in each group were tested. Data represent means ± SEM. The baseline values for NTX and CO groups did not differ. Significantly different from the baseline CO latencies, as well as baseline NTX levels, at $p<0.01$ (**). (Reproduced from McLaughlin et al., 1997a, with permission of the publisher.)

Table 1. Effects of daily injections of naltrexone or saline throughout pregnancy on the length of gestation, litter size, and number of stillborns

	Controls	Naltrexone
Gestation length (days)	22.5 ± 0.2	23.1 ± 0.1
Litter size	14.7 ± 0.8	12.1 ± 0.9
Stillborns/deaths	0.6 ± 0.2	1.1 ± 0.5

Values represent means ± S.E.M. for the litters of pregnant rats injected daily with 50 mg/kg NTX (n = 11) or saline (control) (n = 12) throughout pregnancy. No significant differences were noted between groups in any parameter assessed. (Reproduced from McLaughlin et al., 1997a, with permission of the publisher.)

Measurements of brain weight revealed that prenatal exposure to NTX resulted in an increase of over 40% in neonatal brain weight as compared to controls. This increase in weight of the brain was also noted at 10- and 21-days, with animals born to mothers receiving NTX during pregnancy have weights that reached over 30% and 18%, respectively, from control levels. At all three timepoints (i.e., 0, 10, and 21 days) the changes in brain weight for the NTX groups was statistically significant in comparison to control levels.

2.3. Discussion

Employing a strategy of disrupting the interaction of opioid peptides and opioid receptors in a sustained manner by application of the potent opioid antagonist NTX, we found that perturbation of endogenous opioid systems did not alter the length of gestation, course of pregnancy, litter size, embryo/fetal/neonatal viability, number of stillborns, or maternal weight gain. Maternal food and water intake also appeared to be unaffected by opioid receptor blockade during gestation. These data infer that opioids have little influence on maternal factors associated with pregnancy and parturition. Interference of opioid-receptor interfacing during prenatal life did have a significant impact on the offspring, with neonates weighing more than their control counterparts. The increased weight of newborns exposed *in utero* to NTX appeared to be inherent to intrinsic processes of

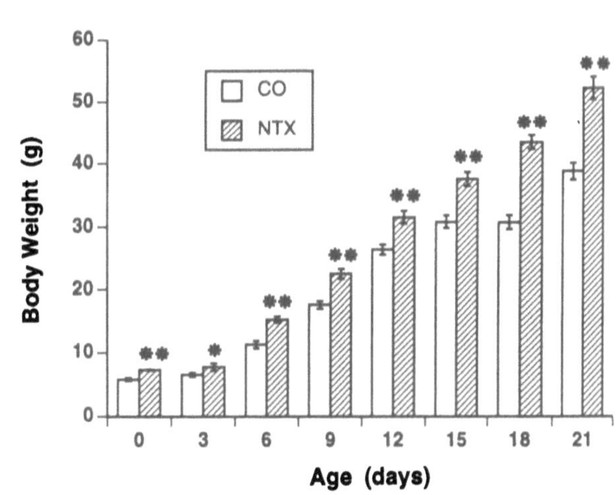

Figure 3. Body weights of offspring born to mothers receiving daily injections of either 50 mg/kg naltrexone (NTX) or saline (CO) throughout gestation. Pups were culled to litters of 10 each and cross-fostered at birth to untreated females. Values are means ± SEM for at least 40 pups per group at each age; at 21 days, no differences between males and females were noted and sexes were combined. Significantly different from controls at $p<0.05$ (*) and $p<0.01$ (**). (Reproduced from McLaughlin et al., 1997a, with permission of the publisher.)

growth, and not related to the size of the litter (e.g., fewer pups/mother) or length of gestation (e.g., longer gestation times). We also observed that the entire course of preweaning development with respect to the gain in body weight was markedly greater in animals subjected to opioid receptor blockade during pregnancy. These results document the importance of opioid-receptor interaction in the prenatal period as a determinant of offspring characteristics both in the neonate and throughout the preweaning period. The effects of NTX-induced changes in ontogeny as to the physical, mental, and emotional well-being of the neonate and infant, as well as the long-term impact on juvenile and adult animals, requires elucidation.

The effects of prenatal exposure to NTX on maternal and offspring characteristics have been examined in earlier reports (Cohen et al., 1996; D'Amato et al., 1988; Harry and Rosecrans, 1979; Hetta and Terenius, 1980; Keshet and Weinstock, 1995; Leng et al., 1985; Mayer et al., 1985; Monder et al., 1979; Nieder and Corder, 1982; Pfeiffer et al., 1984; Seatriz and Hammer, 1993; Shepanek et al., 1989, 1995; Vorhees, 1981; Ward et al., 1986). Despite the use of 2 different opioid antagonists – NTX and naloxone, and differences in species (rats, mice), method of exposure (e.g., oral, intraperitoneal), timing of exposure during gestation (e.g., prenatal days 2 or 3, last trimester), and the extent of exposure (days, weeks), a number of observations have emerged that allow us to understand the repercussions of opioid-receptor manipulation during prenatal life. Thus, with some exceptions (e.g., Hetta and Terenius, 1980), the reproductive capacity of adult females used for breeding (Pfeiffer et al., 1984), incidence of pregnancy (Pfeiffer et al., 1984), maintenance of pregnancy and/or maternal weight (D'Amato et al., 1988; Keshet and Weinstock, 1995; Monder et al., 1979; Pfeiffer et al., 1984; Seatriz and Hammer, 1993; Vorhees, 1981), maternal behavior (D'Amato et al., 1988), implantation (Pfeiffer et al., 1984), litter size (D'Amato et al., 1988; Keshet and Weinstock, 1995; Leng et al., 1985; Mayer et al., 1985; Monder et al., 1979; Pfeiffer et al., 1984; Seatriz and Hammer, 1993; Shepanek et al., 1995, Vorhees, 1981), gestation time (Keshet and Weinstock, 1995; Leng et al., 1985; Vorhees, 1981), ratio of males to females (Seatriz and Hammer, 1993; Shepanek et al., 1995; Vorhees, 1981), mortality of offspring (Monder et al., 1979; Vorhees, 1981), and body weight at birth and/or in the preweaning period (Harry and Rosecrans, 1979; Keshet and Weinstock, 1995; Leng et al., 1985; Monder et al., 1979; Seatriz and Hammer, 1993; Shepanek et al., 1995; Ward et al., 1986) are not influenced by NTX. This included results obtained from experimental paradigms in which nociceptive tests were administered to determine the activity of the opioid antagonist so as to insure opioid receptor blockade (Cohen et al., 1996; Keshet and Weinstock, 1995). The present study persistently blocked opioid-receptor interaction throughout gestation, eliminating the repercussions of a restricted timetable of blockade during pregnancy, and the confounding influences accompanying drug exposure and termination. Moreover, this investigation avoided continuing drug exposure and/or maternal dysfunction after parturition by using a cross-fostering procedure to normal mothers The results confirm and extend reports about the lack of effects of opioid-receptor interruption on pregnancy and parturition. However, we observed that opioid receptor blockade instituted in the pregnant female does have extensive ramifications with respect to body development in the neonate and preweaning animal. These results would suggest that opioids function in the embryonic and fetal rat to precisely regulate somatic properties.

It is interesting to compare the present study concerned with the effects of perturbing opioid-receptor relationship during prenatal life on postnatal development, with early data in which the influence of opioid antagonist exposure only during the preweaning period was examined (Zagon and McLaughlin, 1983a,c, 1984, 1985b, 1989). It appears that the effects of continuous opioid peptide-receptor blockade with an opioid antagonist during the prenatal period introduced changes in body development of greater magnitude than those occurring when drug exposure was restricted to the postnatal period. Thus, postnatal day 21 rats exposed to NTX *in utero* weighed 36% more than controls. On the other hand, animals exposed to NTX from birth to day 21 weighed 20% more than controls. These data would suggest that somatic size is determined both prenatally and postnatally, and that prenatal events play an important role in conferring a pattern of growth that extends (at least) into the early postnatal period.

3. OPIOID RECEPTOR BLOCKADE DURING PRENATAL LIFE: BEHAVIORAL DEVELOPMENT

3.1. Physical Characteristics

Rat pups exposed to NTX prenatally often exhibited a significantly earlier appearance than controls with respect to the timing of appearance of certain physical characteristics (Table 2). Eye opening and the appearance of hair covering began 1 day earlier in the pups exposed to NTX prenatally than in the control animals. At the time when 100% of the animals had hair covering in the NTX-treated group, and 70% of these rats had their eyes open, no control animal had exhibited either of these traits. Rats born of NTX-subjected mothers demonstrated a significantly earlier appearance than control offspring in the initial appearance of incisor eruption, hair covering, ear opening, and eye opening as revealed by rank order comparisons. Additionally, computation of the age (median age) when all the animals in both groups exhibited incisor eruption, ear opening, and eye opening showed that animals exposed to NTX during prenatal life displayed these attributes markedly earlier than the control animals.

Table 2. Effects of maternal exposure to naltrexone throughout gestation on the development of physical characteristics of the offspring

	Group	Range (days)	Incidence of initial appearance	Median age (day)	Incidence at median age
Incisor eruption	Control	7-12	21%	8	37%
	Naltrexone	7-13†	60%		90%*
Hair covering	Control	10-10	100%	10	100%
	Naltrexone	9-9†	100%		100%
Ear opening	Control	13-15	5%	14	58%
	Naltrexone	13-13†	100%		100%*
Eye opening	Control	13-15	16%	14	53%
	Naltrexone	12-13†	70%		100%*

Range indicates the age (days) at which a characteristic was initially observed for any rat in a given group and the age at which 100% of the animals in each group displayed a characteristic. Analysis of the day of initial appearance using Mann Whitney U rank order tests; $p<0.05$ (†) relative to control. Incidence of initial appearance is the percentage of animals in each group that displayed the characteristic on the initial day of appearance for the treatment group. Median age (days) is the time when 50% or more of rats from both groups displayed a physical characteristic. Incidence at the median age represents the proportion of rats with a characteristic in each treatment group at the median age (day) of all animals (i.e., includes rat pups from both the naltrexone-treated and saline-injected groups). Significantly different from controls at $p<0.05$ (*) by chi-square. Each group contained 40 animals representative of 12 litters. (Reproduced from McLaughlin et al., 1997b, with permission of the publisher.)

Table 3. Summary of spontaneous motor behavioral development in rats prenatally exposed to naltrexone

	Group	Range (days)	Incidence of initial appearance	Median age (day)	Incidence of median age
Unilateral head turn, no return	Control	1-4	63%	1	63%
	Naltrexone	1-1†	100%		100%*
Unilateral head turn with return	Control	1-5	21%	1	21%
	Naltrexone	1-3†	80%		80%*
Simultaneous movement of head, forelimbs	Control	1-3	63%	1	63%
	Naltrexone	1-4	90%		90%
Pivoting<360°	Control	2-8	10%	4	32%
	Naltrexone	3-3†	100%		100%*
Crawling	Control	4-11	21%	9	89%
	Naltrexone	4-9	39%		100%
Walking	Control	11-15	5%	13	89%
	Naltrexone	9-13†	5%		100%

Range indicates the age (days) at which a characteristic was initially observed for any rat in a given group and the age at which 100% of the animals in each group displayed a characteristic. Analysis of the day of initial appearance using Mann Whitney U rank order tests; $p<0.05$ (†) relative to control. Incidence of initial appearance is the percentage of animals in each group that displayed the characteristic on the initial day of appearance for the treatment group. Median age (days) is the time when 50% or more of rats from both groups displayed a physical characteristic. Incidence at the median age represents the proportion of rats with a characteristic in each treatment group at the median age (day) of all animals (i.e., includes rat pups from both the naltrexone-treated and saline-injected groups). Significantly different from controls at $p<0.05$ (*) by chi-square. Each group contained 40 animals representative of 12 litters. (Reproduced from McLaughlin et al., 1997b, with permission of the publisher.)

3.2. Spontaneous Motor Behaviors

Spontaneous motor development in rat pups prenatally exposed to NTX and saline is presented in Table 3. In general, the spontaneous motor abilities in rats born of females given NTX often were accelerated in comparison to controls. Although the day of initial appearance of a motor behavior for a single member in each group was comparable between the NTX-treated and control animals, computation of the rank order of appearance for unilateral head turn and no return, unilateral head turn with return, pivoting, and walking revealed a significantly earlier appearance in offspring exposed prenatally to NTX than for control subjects. Evaluation of the number of rats with each spontaneous behavior on the median day of appearance revealed that more rats exposed to NTX *in utero* exhibited unilateral head turn and no return, unilateral head turn with return, and pivoting than control animals.

3.3. Reflexive Tests

Rats subjected maternally to NTX throughout gestation displayed many sensory and motor reflexive behaviors prior to the initial appearance in the control group (Table 4). In fact, some animals exposed to NTX prenatally had positive responses to righting, falling, visual orientation, auditory orientation, and edge aversion 1 to 4 days earlier than offspring in the control group. From 40% to 50% of the rats in the NTX group exhibited these behaviors at a point when no offspring in the control group displayed any of these reflexive responses. Olfactory orientation and tail hanging, however, were observed to initiate in rats of NTX-treated mothers 1 day later than control animals. The initial appearance of reflexes related to falling, visual orientation, auditory orientation, olfactory

Table 4. Summary of sensorimotor development in rats prenatally exposed to naltrexone

	Group	Range (days)	Incidence of initial appearance	Median age (day)	Incidence at median age
Pain - withdrawal	Control	1-1	100%	1	100%
	Naltrexone	1-1	100%		100%
Olfactory orientation	Control	2-9	10%	6	84%
	Naltrexone	3-9†	10%		40%
Righting	Control	3-9	32%	4	53%
	Naltrexone	2-4	50%		100%*
Cross extensor	Control	3-11	10%	8	58%
	Naltrexone	3-3†	100%		100%*
Edge aversion	Control	4-10	16%	7	63%
	Naltrexone	3-9†	30%		90%
Tail hanging	Control	4-9	16%	6	47%
	Naltrexone	5-7	10%		80%
Bar grasping	Control	6-11	5%	8	53%
	Naltrexone	6-9	20%		90%*
Negative geotaxis	Control	7-13	32%	10	68%
	Naltrexone	8-11	10%		50%
Falling	Control	14-16	10%	15	84%
	Naltrexone	11-14†	50%		100%
Visual orientation	Control	17-18	47%	17	47%
	Naltrexone	14-17†	40%		100%*
Auditory orientation	Control	17-17	100%	17	100%
	Naltrexone	13-16†	50%		100%

Range indicates the age (days) at which a characteristic was initially observed for any rat in a given group and the age at which 100% of the animals in each group displayed a characteristic. Analysis of the day of initial appearance using Mann Whitney U rank order tests; $p<0.05$ (†) relative to control. Incidence of initial appearance is the percentage of animals in each group that displayed the characteristic on the initial day of appearance for the treatment group. Median age (days) is the time when 50% or more of rats from both groups displayed a physical characteristic. Incidence at the median age represents the proportion of rats with a characteristic in each treatment group at the median age (day) of all animals (i.e., includes rat pups from both the naltrexone-treated and saline-injected groups). Significantly different from controls at $p<0.05$ (*) by chi-square. Each group contained 40 animals representative of 12 litters. (Reproduced from McLaughlin et al., 1997b, with permission of the publisher.)

orientation, edge aversion, and cross extension was earlier in rats born to mothers receiving NTX during pregnancy than control offspring using a rank order comparison. Calculation of the age (median age) when all the animals in both groups exhibited reflexes associated with righting, visual orientation, cross extension, and bar grasping showed that rats exposed to NTX *in utero* exhibited these reflexes significantly earlier than controls.

3.4. Motor Activity and Related Behaviors

Rat pups exposed to NTX during prenatal life entered 25% fewer squares in the open field than control animals (Table 5) when examined on postnatal day 21. The frequency and incidence of face washing, rearing, grooming, wet-dog shakes, and boli in rats subjected to NTX or saline during gestation are presented in Table 5. Exposure to NTX prenatally decreased the frequency of rearing by 45%, and the number of boli by 71%, compared to control subjects. The frequency of face-washing, grooming, and wet-dog shakes did not differ between either to NTX or saline groups. Evaluation of the incidence of these activities showed subnormal levels in rearing, grooming, wet-dog shakes, and boli for animals in the NTX group.

Table 5. Activity levels and related behaviors of 21-day old rats exposed prenatally to naltrexone

Activity	Group	Frequency (range/animal)	Incidence of activity
Open field	Control	67.7 ± 5.6 (5-131)	100%
	Naltrexone	50.7 ± 6.4† (3-141)	100%
Face washing	Control	2.4 ± 0.3 (1-6)	97%
	Naltrexone	1.9 ± 0.2 (1-7)	90%
Rearing	Control	4.7 ± 0.7 (1-21)	93%
	Naltrexone	2.6 ± 0.4†† (1-9)	69%*
Grooming	Control	1.4 ± 0.0 (3-5)	60%
	Naltrexone	0.0 ± 0.0 (0-0)	0%*
Wet-dog shakes	Control	0.2 ± 0.0 (1-1)	10%
	Naltrexone	0.0 ± 0.0 (0-0)	0%*
Boli	Control	0.7 ± 0.2 (1-2)	37%
	Naltrexone	0.2 ± 0.1†† (1-2)	7%*

Frequency is the average number of squares entered in the open field, or the average number of occurrences for an activity. Data are expressed as means ± SEM and were analyzed using ANOVA and the Newman-Keuls tests; $p<0.05$ (†) or $p<0.01$ (††). The incidence of activity is the percentage of animals in each group that displayed the activity; significantly different from controls at $p<0.05$ (*) by chi-square. Each group contained 40 animals representative of 12 litters. (Reproduced from McLaughlin et al., 1997b, with permission of the publisher.)

3.5. Discussion

Sustained opioid blockade throughout gestation, with cross-fostering to untreated mothers at birth, yields offspring who have an acceleration in the timetable of physical and behavioral maturation. Such data would lead us to suggest that elimination of the interplay between opioids and opioid receptors during ontogeny has remarkable implications on physical and behavioral development in the postnatal period. Further reasoning would indicate that if blockade of opioid-opioid receptor interaction accelerates the development of physical and behavioral characteristics, then opioids must serve actively to regulate the ontogeny of these modalities by repressive mechanisms.

Prevention of opioid-opioid receptor interfacing during gestation also had considerable impact on measures of behavior that included locomotion and emotionality. The number of squares traversed in the open field, used to assess locomotor activity, showed that offspring delivered by mothers receiving NTX entered fewer squares than control rats. These data would suggest that the NTX-treated rats were less active than control animals. Because motor activity in the open field decreases with age in rats (Zagon et al., 1979), our results could indicate that the NTX-exposed pups were more advanced in this behavioral measure. The frequency of such activities as rearing, face washing, and defecation provided an index of "emotionality" (Hall, 1941; Werboff et al., 1962). Although face washing was comparable in animals subjected prenatally to NTX or saline, rats exposed to NTX *in utero* did exhibit a lower frequency and/or incidence of activities related to rearing, grooming, wet-dog shakes, and defecation. The reason(s) for this reduced activity in the NTX-exposed offspring is(are) unclear, but once again may be related to the advanced maturational state of these rats.

The effects of administering opioid antagonists during gestation on postnatal outcome with respect to behavior have been reported earlier. Vorhees (1981) dispensed the short-acting, low potency opioid antagonist naloxone to rats at 20 mg/kg once or twice daily on days 7–20 of gestation. Naloxone-treated offspring were accelerated in postweaning growth, upper incisor eruption, righting and startle development, home scent discrimination, and directional swimming, but as adults showed impaired Biel water maze learning. Shepanek and colleagues (1989, 1995) injected naloxone into pregnant rats from gestation day 7 to 20 and found that offspring exposed to 1 mg/kg habituated more rapidly in the open field and showed less activity than control pups. Bar pressing rates in a visual discrimination task were subnormal in male rats of the 10 mg/kg group, while 10 mg/kg males and females showed reduced bar pressing rates on differential reinforcement of low rates of responding. Maternal NTX administration (10 mg/kg/day) using minipumps implanted in rats on day 17 of gestation prevented some behavioral alterations induced in the offspring by prenatal stress (Keshet and Weinstock, 1995). Cohen et al. (1996) found that NTX given to mothers during the last 9 days of gestation produced offspring with no changes in swimming ability during preweaning life, but facilitated masculine behavior and suppressed feminine receptivity. Comparison of these studies with the present is difficult because we have blocked opioid receptors throughout gestation, and the antagonist was presented in such a manner (i.e., a dosage of 50 mg/kg) as to maintain a block of opioid-receptor interaction in the mother as tested by morphine challenge and measures. Thus, our model eliminated confounding variables introduced by opioid antagonist exposure and response that makes interpretation of other data difficult. It may be noted that fetal exposure to β-endorphin resulted in mild developmental delays of rats, supporting our finding that blockade of the endogenous opioids enhances neurobiological growth (Sandman and Yessaian, 1986).

Comparison of the behavioral repercussions from prenatal exposure to NTX with the effects of postnatal treatment (Zagon and McLaughlin, 1984, 1985a) to NTX is valuable in order to decipher the role of opioids in somatic and neurobiological development. In general, sustained disruption of opioid-opioid receptor interaction during prenatal life appears to have more ramifications in terms of physical and behavioral measures than exposure after birth. In regard to physical characteristics, for example, only eye opening was accelerated in rats given NTX postnatally, but incisor eruption, ear opening, and eye opening were all markedly advanced in initial appearance in animals exposed to NTX throughout gestation (using changes in the incidence of median age as a measure). Prenatal or postnatal exposure to NTX often had the same effects on the appearance of many spontaneous motor and sensorimotor responses (e.g., acceleration of pivoting, no effect on crawling), but differences in behaviors altered due to the timing of opioid receptor blockade were apparent. Thus, increases in the median age of righting, bar grasping, and cross-extensor reflex only were noted in the group exposed prenatally to NTX. However, auditory and olfactory orientation and negative geotaxis only were influenced in the group receiving injections of NTX postnatally. Evaluation of open field behavior revealed that offspring exposed to NTX either prenatally or postnatally were less active at weaning, but rearing, wet-dog shakes, and defecation (but not face washing) were markedly decreased only in animals of the prenatal NTX group compared to controls. Thus, the expression of physical traits and behavioral measures in developing rats is related to opioid-receptor interaction, and both prenatal and postnatal periods of neurogenesis contribute individually to postnatal development of behavior.

The influence of exogenously applied native opioid peptides during gestation remains relatively unexplored, and all of the reports (Kashon et al., 1992; Sandman and Yessaian, 1986; Zadina et al., 1985) have been concerned with β-endorphin and exposure during the last semester of pregnancy. β-endorphin given three times daily (33 μg/injection) alters the differentiation of some sexually dimorphic traits (Kashon et al., 1992), whereas daily injections of 100 μg of β-endorphin was not found to disturb the timetable of eye opening (Zadina et al., 1985; Sandman and Yessaian, 1986), open field rearing behavior in adults (Zadina et al., 1985), or "activity", and startle reflex (Sandman and Yessaian, 1986). Although these studies indicate that prenatal exposure to β-endorphin is only of some consequence to the sexual traits of offspring, the limited exposure (i.e., third semester of pregnancy) and dosages of this native peptide do not permit a full understanding of β-endorphin activity. Moreover, placed into the perspective of the present investigation in which receptor blockade was employed throughout pregnancy, it is premature to draw any conclusions about the role of NTX on β-endorphin's (or any other endogenous opioid peptide's) potential modulatory capabilities on developmental events.

4. PERSPECTIVES ABOUT PRENATAL OPIOIDS AND NEUROPLASTICITY

The prenatal organism appears to rely upon opioid peptides for "normal" growth, particularly in reference to somatic and neurobiological development. Deprivation of the signaling from opioids by the introduction of opioid receptor blockade permits an enhancement of these developmental events. Therefore, opioids must control embryogenesis by exerting a repressive effect on growth, and this influence must be of a tonic nature. At least one opioid peptide, [Met^5]-enkephalin - termed opioid growth factor (OGF), has been reported to directly govern cell proliferation (e.g., Zagon et al., 1996b, Zagon and McLaughlin, 1991a,b, 1993). OGF is an autocrine growth factor produced by both neural and non-neural cells in biological systems that are undergoing development, cellular renewal, or repair. In addition, the ζ opioid receptor has been identified as the mediator of OGF action with respect to growth. We have learned that blockade of opioids from interacting with opioid receptors has considerable meaning to the growth of the fetus and infant rat. Therefore, it could be postulated that daily administration of the opioid antagonist NTX interferes with the delicate equilibrium between OGF and the ζ receptor - both presumably associated with developing cells of the body and organs such as the brain. Indeed, OGF and the ζ receptor have been localized to a variety of developing and renewing cells and tissues, including those constituting the developing nervous system. Blockade of OGF-ζ receptor interaction would thereby permit cells to escape the regulatory nature of OGF and accelerate replication. This would have the net effect of increasing somatic development. Such a hypothesis merits further consideration in order to comprehend the importance of opioids in dictating the organization of the body and its contents.

The specific mechanism(s) as to how NTX modulates growth needs to be elucidated. A dosage of 50 mg/kg is 2.5% of the LD_{50} in adult rats (Braude and Morrison, 1976), and was not deemed adverse to any aspect of the present experiments. Furthermore, administration of a 50 mg/kg dosage of NTX in this and other studies is known to produce an opioid receptor blockade for 24 hours; after this point NTX's efficacy to block opioid receptors is dissipated, indicating that drug is no longer present. We do know that opioids are tonically active negative regulators of DNA synthesis, and that the opioid responsible for these activities, [Met^5]-enkephalin, is an autocrine produced growth factor in both neural and non-neural tissues. Given the marked effects on somatic and neurobehavioral development in offspring exposed to NTX only during embryogenesis, and the well-documented direct effects of NTX in influencing growth (e.g., organ and tissue culture studies), it may be postulated that daily interruption of opioid-opioid receptor interaction by NTX prevents opioids from their normal trophic activities in the developing nervous system of the prenatal rat. Behavioral alterations in evidence during preweaning ontogeny and at weaning in animals exposed to NTX *in utero* indicate that marked changes have occurred in the nervous system during early life which are manifested at later stages (and when drug is no longer present). Whether these changes are reflective and consequential to the alterations in somatic growth that occur, or if the timetable of physical and behavioral ontogeny is dictated by opioid-receptor interaction needs to be elucidated.

ACKNOWLEDGMENTS

This work was supported by NIH grants NS-20500 and HL-53557.

REFERENCES

Akil, H.; Watson, S.J.; Young, E.; Lewis, M.E.; Katchaturian, H.; Walter, J.M. 1984. Endogenous opioids: Biology and function. Annu. Rev. Neurosci. 7:223–255.

Barg, J.; Belcheva, M.; McHale, R.; Levey, R; Vogel, Z.; Coscia, C.J. 1993. β-endorphin is a potent inhibitor of thymidine incorporation into DNA via μ- and κ-opioid receptors in fetal rat brain cell aggregates in culture. J. Neurochem. 60:765–767.

Bartolome, J.V.; Bartolome, M.B.; Lorber, B.A.; Dileo, S.J.; Schanberg, S.M. 1991. Effects of central administration of beta-endorphin on brain and liver DNA-synthesis in preweaning rats. Neuroscience 40:289–294.

Braude, M.C.; Morrison, J.M. 1976. Preclinical toxicity studies of naltrexone. NIDA Res. Monog. 9:16–26.

Clendeninn, N.J.; Petraitis, M.; Simon, E.J. 1976. Ontogological development of opiate receptors in rodent brain. Brain Res. 118:157–160.

Cohen, E.; Keshet, G.; Shavit, Y.; Weinstock, M. 1996. Prenatal naltrexone facilitates male sexual behavior in the rat. Pharmacol. Biochem. Behav. 54:183–189.

Coyle, J.T.; Pert, C.B. 1976. Ontogenetic development of [^{3}H]naloxone binding in rat brain. Neuropharmacology 15:555–560.

D'Amato, F.R.; Castellano, C.; Ammassari-Teule, M.; Oliverio, A. 1988. Prenatal antagonism of stress by naltrexone administration: Early and long-lasting effects on emotional behaviors in mice. Develop. Psychobiol. 21:283–292.

De Cabo, C.; Colado, M.I.; Pujol, A.; Martin, M.I.; Viveros, M.P. 1994. Naltrexone administration effects on regional brain monamines in developing rats. Brain Res. Bull. 34:395–406.

De Cabo de la Vega, C.; Pujol, A.; Viveros, M.P. 1995. Neonatally administered naltrexone affects several behavioral responses in adult rats of both genders. Pharmacol. Biochem. Behav. 50:277–286.

Diggle, P.J.; Liang, K.Y.; Zeger, S.L. 1994. Analysis of longitudinal data. Oxford: Clarendon Press.

Hall, C.S. Temperament: a survey of animal studies. Pyschol. Bull. 38:909–943;1941.

Harry, G.J.; Rosecrans, J.A. 1979. Behavioral effects of perinatal naltrexone exposure: A preliminary investigation. Pharmacol. Biochem. Behav. 11 (Suppl.):19–22.

Hauser, K.F.; McLaughlin, P.J.; Zagon, I.S. 1989. Endogenous opioid systems and the regulation of dendritic growth and spine formation. J. Comp. Neurol. 281:13–22.

Hetta, J.; Terenius, L. 1980. Prenatal naloxone affects survival and morphine sensitivity of rat offspring. Neurosci. Letts. 16:323–327.

Kashon, M.L.; Ward, O.B.; Grisham, W.; Ward, I.L. 1992. Prenatal β-endorphin can modulate some aspects of sexual differentiation in rats. Behav. Neurosci. 106:555–562.

Keshet, G.I.; Weinstock, M. 1995. Maternal naltrexone prevents morphological and behavioral alterations induced in rats by prenatal stress. Pharmacol. Biochem. Behav. 50:413–419.

Knodel, E.L.; Richelson, E. 1980. Methionine-enkephalin immuno-reactivity in fetal rat brain cells in aggregating culture and in mouse neuroblastoma cells. Brain Res. 197:565–570.

Leng, G.; Mansfield, S.; Bicknell, R.J.; Dean, A.D.P.; Ingram, C.D.; Marsh, M.I.C.; Yates, J.O.; Dyer, R.G. 1985. Central opioids: A possible role in parturition? J. Endocrinol. 106:219–224.

Mayer, A.D.; Faris, P.L.; Komisaruk, B.R.; Rosenblatt, J.S. 1985. Opiate antagonism reduces placentophagia and pup cleaning by parturient rats. Pharmacol. Biochem. Behav. 22:1035–1044.

McLaughlin, P.J. 1994. Opioid antagonist modulation of rat heart development. Life Sci. 54:1423–1431.

McLaughlin, P.J. 1996. Regulation of DNA synthesis of myocardial and epicardial cells in developing rat heart by [Met^5]-enkephalin. Am. J. Physiol. 271:R122-R129.

McLaughlin, P.J.; Tobias, S.W.; Lang, C.M.; Zagon, I.S. 1997a. Chronic exposure to the opioid antagonist naltrexone during pregnancy: Maternal and offspring effects. Physiol. Behav., in press.

McLaughlin, P.J.; Tobias, S.W.; Lang, C.M.; Zagon, I.S. 1997b. Opioid receptor blockade during prenatal life modifies postnatal behavioral development. Pharmacol. Biochem. Behav., in press.

Meriney, S.D.; Ford, M.J.; Oliva, D.; Pilar, G. 1991. Endogenous opioids modulate neuronal survival in the developing avian ciliary ganglion. J. Neurosci. 11:3705–3717.

Monder, H.; Yasukawa, N.; Christian, J.J. 1979. Perinatal naloxone: When does naloxone affect hyperalgesia? Pharmacol. Biochem. Behav. 11:235–237.

Murgo, A.J. 1985. Inhibition of B16-BL6 melanoma growth in mice by methionine-enkephalin. J. Natl. Cancer Inst. 75341–344.

Najam, N.; Panksepp, J.. 1989. Effect of chronic neonatal morphine and naloxone on sensorimotor and social development of young rats. Pharmacol. Biochem. Behav. 33:539–544.

Neale, J.H.; Barker, J.L.; Uhl, G.R.; Snyder, S.H. 1979. Enkephalin-containing neurons visualized in spinal cord cultures. Science 201:467–469.

Nieder, G.L; Corder, C.N. 1982. Effects of opiate antagonists on early pregnancy and pseudopregnancy in mice. J. Reprod. Fert. 65:341–346.

Pfeiffer, D.G.; Nikolarakis, K.E.; Pfeiffer A. 1984. Chronic blockade of opiate receptors: Influence on reproduction and body weight in female rats. Neuropeptides 5:279–282.

Sandman, C.A.; Yessaian, N. 1986. Persisting subsensitivity of the striatal dopamine system after fetal exposure to beta-endorphin. Life Sci. 39: 1755–1763.

Seatriz, J.V.; Hammer, R.P. 1993. Effects of opiates on neuronal development in the rat cerebral cortex. Brain Res. Bull. 30:523–527.

Shahabi, N.A.; Sharp, B.M. 1995. Antiproliferative effects of δ opioids on highly purified CD4$^+$ and CD8$^+$ murine T cells. J. Pharmacol. Exp. Ther. 273:1105–1113.

Shepanek, N.A.; Smith, R.F.; Anderson, L.A.; Medici, C.N. 1995. Behavioral and developmental changes associated with prenatal opiate receptor blockade. Pharmacol. Biochem. Behav. 50:313–319.

Shepanek, N.A.; Smith, R.F.; Tyer, Z.; Royall, D.; Allen, K. 1989. Developmental, behavioral, and structural effects of prenatal opiate receptor blockade. New York Acad. Sci. 562:377–379.

Villiger, P.M.; Lotz, M. 1992. Expression of prepro-enkephalin in human articular chondrocytes is linked to cell proliferation. EMBO J. 11:135–143.

Vorhees, C.V. 1981. Effects of prenatal naloxone exposure on postnatal behavioral development of rats. Neurobehav. Toxicol. Teratol. 3:295–301.

Ward, O.B.; Monaghan, E.P.; Ward, I.L. 1986. Naltrexone blocks the effects of prenatal stress on sexual behavior differentiation in male rats. Pharmacol. Biochem. Behav. 25:573–576.

Werboff, J.; Havlena, H.; Sikov, M.R. 1962. Effects of prenatal X-irradiation on activity, emotionality, and maze-learning ability in the rat. Radiat. Res. 16:441–452.

Zadina, J.E.; Kastin, A.J.; Coy, D.H.; Adinoff, B.A. 1985. Developmental, behavioral, and opiate receptor changes after prenatal or postnatal β-endorphin, CRF, or Tyr-MIF-1. Psychoneuroendocrinology 10:367–383.

Zagon, I.S.; Hytrek, S.D.; Lang, C.M.; Smith, J.P.; McGarrity, T.J.; Wu, Y.; McLaughlin, P.J. 1996a. Opioid growth factor ([Met^5]-enkephalin) prevents the incidence and retards the growth of human colon cancer. Am. J. Physiol. 271:R780-R786.

Zagon, I.S.; Hytrek, S.D.; McLaughlin, P.J. 1996b. Opioid growth factor tonically inhibits human colon cancer cell proliferation in tissue culture. Am. J. Physiol. 271:R511-R518.

Zagon, I.S.; McLaughlin, P.J. 1983a. Increased brain size and cellular content in infant rats treated with opiate antagonist. Science 221:1179–1180.

Zagon, I.S., McLaughlin, P.J. 1983b. Naltrexone modulates tumor response in mice with neuroblastoma. Science 221:671–673.

Zagon, I.S.; McLaughlin, P.J. 1983c. Naltrexone modulates growth in infant rats. Life Sci. 33:2449–2454.

Zagon, I.S.; McLaughlin, P.J. 1984. Naltrexone modulates body and brain development in rats: A role for endogenous opioid systems in growth. Life Sci. 35:2057–2064.

Zagon, I.S.; McLaughlin, P.J. 1985a. Naltrexone's influence on neurobehavioral development. Pharmacol. Biochem. Behav. 22:441–448.

Zagon, I.S.; McLaughlin, P.J. 1985b. Opioid antagonist-induced regulation of organ development. Physiol. Behav. 34:507–511.

Zagon, I.S.; McLaughlin. P.J. 1989. Naloxone modulates body and organ growth of rats: Dependency on the duration of opioid receptor blockade and stereospecificity. Pharmacol. Biochem. Behav. 33:325–328.

Zagon, I.S.; McLaughlin, P.J. 1991a. The role of endogenous opioids and opioid receptors in human and animal cancers. In: Plotnikoff, N.P.; Murgo, A.J.; Faith, R.E.; Wybran, J., eds. Stress and Immunity, Caldwell, NJ:CRC, pp. 343–356.

Zagon, I.S.; McLaughlin, P.J. 1991b. Identification of opioid peptides regulating proliferation of neurons and glia in the developing nervous system. Brain Res. 542:318–323.

Zagon, I.S.; McLaughlin, P.J. 1993. Opioid growth factor receptor in the developing nervous system. In: Zagon, I.S.; McLaughlin, P.J., eds. Receptors in the Developing Nervous System. Growth Factors and Hormones. Volume 1. London:Chapman and Hall, pp. 39–62.

Zagon, I.S.; McLaughlin, P.J.; Thompson, C.I. 1979. Development of motor activity in young rats following perinatal methadone exposure. Pharmacol. Biochem. Behav. 10:743–749.

Zagon, I.S.; Sassani, J.W.; McLaughlin, P.J. 1995. Opioid growth factor modulates corneal epithelial outgrowth in tissue culture. Am. J. Physiol. 268:R942-R950.

22

BRAIN PLASTICITY AND THE NEURAL CELL ADHESION MOLECULE (NCAM)

Lars Christian B. Rønn,[1*] Nina Pedersen,[1] Henrik Jahnsen,[2]
Vladimir Berezin,[1] and Elisabeth Bock[1]

[1]The Protein Laboratory, Institute for Molecular Pathology
Panum Institute bld. 6.2.
Blegdamsvej 3, 2200 Copenhagen N
[2]Department for Neurophysiology
Panum Institute bld. 16.5. Blegdamsvej 3
2200 Copenhagen N., Denmark

1. INTRODUCTION

NCAM is a member of the Ig superfamily that is highly expressed in the nervous system. NCAM mediates cell-cell adhesion and adhesion of cells to the extracellular matrix. In addition to influencing the adhesive behaviour of cells, NCAM binding triggers the activation of intracellular signalling pathways. It is well established that NCAM is important for the formation of proper neuronal connections in the developing nervous system. Recently, plasticity of neuronal connections in the mature brain has also been shown to be dependent on NCAM function.

2. NCAM STRUCTURE

The neural cell adhesion molecule (NCAM) belongs to the cell adhesion molecules (CAMs) mediating adhesion between cells. NCAM is a member of the Ig superfamily, a large family currently found to comprise > 160 members (Brummendorf and Rathjen, 1995). The characteristic feature of the Ig superfamily is the presence of one or more immunoglobulin homology (Ig) domains (Williams and Barclay, 1988). The extracellular part of NCAM contains five Ig domains. In general, the Ig domains in cell adhesion molecules are believed to be of the C2 type which is an intermediate form between the constant

* Corresponding author: Lars Christian B. Rønn, The Protein Laboratory, Institute for Molecular Pathology, Panum Institute bld. 6.2., Blegdamsvej 3, 2200 Copenhagen N. Tel: (+45) 31 21 81 09; Fax: (+45) 35 36 01 16; E-mail: lcr@biobase.dk.

(C) type and the variable (V) type known from antibodies, but the N-terminal Ig domain in NCAM (Ig1) has recently been assigned to the so-called intermediate set (I-set) of Ig domains based on NMR spectroscopy (Thomsen et al., 1996). In addition to five Ig domains, NCAM contain two so-called fibronectin type III homology domains (FnIII domains). In brain, NCAM is found in three main forms. NCAM-180 (Mr 190) and NCAM-140 (Mr 135) are transmembrane forms while NCAM-120 (Mr 115) is attached to the membrane via a phosphatidylinositol anchor (see fig. 1). In addition, soluble forms of NCAM exist corresponding to approximately 1–2% of the total NCAM present in the CNS. Soluble NCAM is believed to be a mixture of various NCAM forms including extracellular parts of NCAM-180 or NCAM-140 released by proteolysis, NCAM-120 released from the lipid anchor and intact NCAM-180 and NCAM-140 forms released with preserved transmembrane domains (Bock et al., 1987; Olsen et al., 1993).

3. THE EXPRESSION OF NCAM

All NCAM forms are expressed from a single large gene. By alternative splicing (Murray et al., 1986; Walsh and Dickson, 1989) and selection between polyadenylation sites, four main classes of mRNA for NCAM are generated encoding NCAM-180 (exon 1–14, 16–19), NCAM-140 (exon 1–14, 16, 17, 19) and NCAM-120 (exon 1–15). These main mRNA species may be varied further. Notably, the insertion of the 30 bp variable-domain alternatively spliced exon (VASE) (Small et al., 1988) between exon 7 and 8 results in the transformation of the fourth Ig domain from a C2 type to a V type like Ig domain with major implications for the function of the NCAM protein. In the region encoding the FnIII domains (exon 11 to 14), the exons a, b and c and the triplet AAG may be inserted. The different mRNA classes are selectively regulated during the development of the nervous system. The fraction of NCAM mRNA containing VASE increases from approximately 3% in the embryonic brain to 50% in the adult brain (Small and Akeson, 1990). The diversity of the NCAM polypeptides is further increased by posttranslational modifications. All major NCAM forms are modified by N-linked glycosylation (Krog and Bock, 1992). Two of these sites on the fifth Ig domain (Nelson et al., 1995) are capable of carrying the highly unusual polysialic acid (PSA) moieties (Zuber et al., 1992; Rutishauser, 1996). PSA consists of homopolymers (up to >50) of N-acetyl neuraminic acid (NeuNAC) in an α2–8 binding (Rougon, 1993; Rutishauser and Landmesser, 1996). These polyanions make up metastable helices with a large hydrated volume. NCAM is the dominant carrier of PSA in vertebrates (Rougon et al., 1986), thus in homozygote NCAM knock-out mice a decrease of 85% in PSA expression is observed (Cremer et al., 1994). Two polysialylyl transferases capaple of synthesizing PSA-NCAM have recently been cloned (Nakayama et al., 1995; Nakayama and Fukuda, 1996; Scheidegger et al., 1995; Eckhardt et al., 1995; Yoshida et al., 1995). Polysialylation of NCAM is highly developmentally regulated in a manner reciprocal to the change in the insertion of VASE. Thus PSA positive NCAM (PSA-NCAM, also called embryonic NCAM), is prominent in the embryonic brain but decreases markedly in the adult animal (Edelman and Chuong, 1982; Rougon et al., 1982) apart from certain regions (Miragall et al., 1988; Seki and Arai, 1991) including the olfactory system and the dentate gyrus where the presence of PSA on NCAM is thought to reflect the preserved potential for structural remodulation in these regions. In the CNS, NCAM-180 is expressed by neurons while NCAM-140 and NCAM-120 are additionally expressed by glial cells at approximately 5–10% of the levels expressed by neurons (Noble et al., 1985). NCAM-120 is primarily expressed by glial

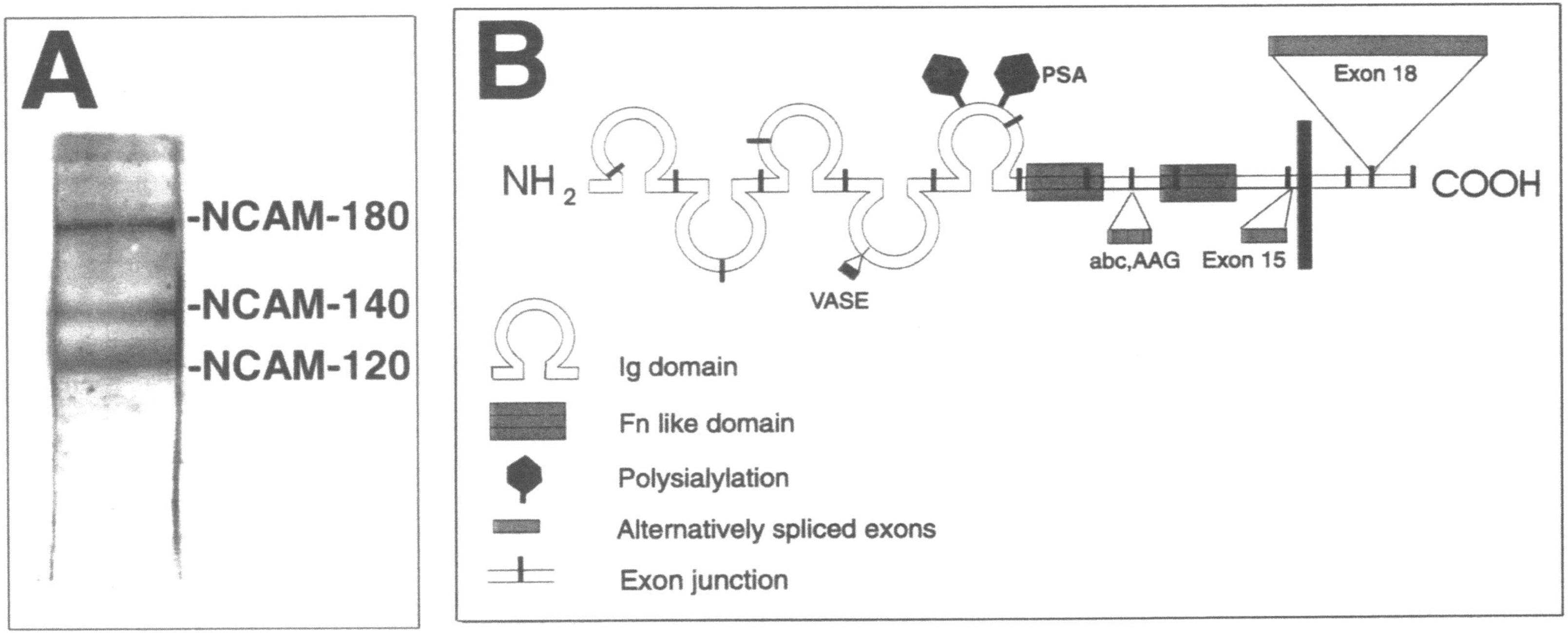

Figure 1. The neural cell adhesion molecule (NCAM). **A)** In the mature rat brain, NCAM is seen in three major forms: NCAM-180 (190 kDa), NCAM-140 (135 kDa) and NCAM-120 (115 kDa). Western blotting using the Ig fraction of a polyclonal rabbit-anti-rat NCAM antibody (Ab, 1:500) followed by an alkaline phosphatase coupled swine-anti-rabbit Ab (1:1000). **B)** The extracellular part of NCAM consists of 5 immunoglobulin (Ig) domains and 2 fibronectin like (FnIII) domains common to all forms. The exon 15 contains a stopcodon as well as two polyadenylation sites. The use of this exon results in the generation of NCAM-C, a glycosylphosphatidyl inositol (GPI) anchored form of NCAM. NCAM-B contains a transmembrane and a cytoplasmic domain coded by exons 16,17 and 19, while NCAM-A in addition contains the cytoplasmic part coded by exon 18.

cells. The NCAM-forms may apparently be selectively distributed in the cell membrane (Powell et al., 1991) and also have different lateral mobilities (Pollerberg et al., 1986). NCAM-180 has the lowest lateral mobility in the membrane and is believed to be coupled to spectrin in the cytoskeleton (Crossin and Hoffman, 1991; Woo and Murray, 1994). NCAM-140 has been shown to associate with the tyrosin kinase p59^{fyn}. This complex in turn can recruit p125^{FAK} (Beggs et al., 1997). In general however, little is known about intracellular binding partners of NCAM. Both NCAM-180 and NCAM-140 seems to be concentrated at sites of cell-cell contact (Crossin and Hoffman, 1991; Bloch, 1992). Furthermore, it has been reported that NCAM-180 is specifically concentrated at postsynaptic densities (Persohn et al., 1989).

4. NCAM BINDING

NCAM was originally described as a molecule mediating a homophilic (NCAM binding to NCAM) cell-cell adhesion independent of calcium (Thiery et al., 1977; Moran and Bock, 1988). NCAM binds NCAM with low affinity, thus the overall strength of the NCAM mediated cell adhesion is strongly dependent on the NCAM density on the interacting cell membranes. The presence of VASE appears to increase the strength of the homophilic NCAM binding (Chen et al., 1994) while the binding is markedly weakened by the presence of PSA (Sadoul et al., 1983) probably due to either steric hindrance from the large hydrated volume of PSA or due to repulsion between the negatively charged PSA chains on the opposing membranes. Alternatively, PSA may work as a receptor (Joliot et al., 1991) for homeoproteins due to a certain similarity with nucleic acid. The homophilic NCAM binding has been proposed to be mediated by either the third Ig domain (Rao et al., 1992; Rao et al., 1993; Rao et al., 1994) or by the simultaneous reciprocal interaction between all five Ig domains (Ranheim et al., 1996). However, recent evidence suggests that also reciprocal interactions between the first and the second Ig domains are involved in the homophilic NCAM binding (Kiselyov et al., 1997). This presumed interaction may be localized to a cluster of acidic residues in the first Ig domain and a cluster of basic residues in the second Ig domain (Thomsen et al., 1996).

In addition to the homophilic binding, NCAM also takes part in heterophilic interactions binding other molecules. The fourth Ig domain of NCAM is believed to bind to a carbohydrate group on a related cell adhesion molecule L1 (Kadmon et al., 1990; Horstkorte et al., 1993). This NCAM-L1 binding presumably strengthens the homophilic L1 binding (Kadmon et al., 1990), hence the name "assisted homophilic interaction". This binding may be inhibited by the presence of PSA on NCAM (Acheson et al., 1991; Tang et al., 1994). Recently, NCAM has been shown to interact with yet another member of the immunoglobulin superfamily of CAMs, the TAG-1/axonin-1 with high affinity (Milev et al., 1996). In addition, it has been proposed that NCAM may interact with the FGF receptor (FGF-R) through a "CAM homology domain" (CDH) located between the first and the second Ig domains of the FGF-R (Williams et al., 1994). NCAM not only functions as a CAM but also bind various extracellular matrix components including chondroitin sulphate proteoglycans (Grumet et al., 1993), collagens (Probstmeier et al., 1989), and heparin/heparan sulphate (Cole et al., 1986) including agrin (Storms et al., 1996), the latter being mediated primarily by the second Ig domain but possibly also by the first Ig domain (Kiselyov et al., 1997). Thus for any particular cell, the overall NCAM mediated adhesive behavior is a result of the balance between NCAM mediated cell-cell adhesion and NCAM mediated adhesion of cells to the extracellular matrix.

5. NCAM AND TRANSMEMBRANE SIGNAL TRANSDUCTION

There is increasing evidence that NCAM not only functions as a "cell glue" but also is involved in signal transduction (Doherty et al., 1995). When the extracellular part of NCAM binds a ligand, the event is reported across the membrane and influences the levels of second messenger within the cytosol. In PC12 cells as well as in certain primary neurons, the binding of NCAM antibodies or purified NCAM results in an increase in intracellular calcium, as well as in a decrease in the intracellular concentrations of inositol bisphosphate (IP_2) and inositol trisphosphate (IP_3) and a lower pH (Schuch et al., 1989; von Bohlen und Halbach et al., 1992). It has subsequently been shown that this signaling event is necessary for the stimulatory effect of NCAM on neurite extension *in vitro* (Doherty et al., 1991). It has been suggested that NCAM binding, possibly through cis-binding and activation of the FGF-R (Williams et al., 1994; Hall et al., 1996; Saffell et al., 1997), leads to stimulation of phospholipase C, which in turn mediates the production of diacyl glycerol (DAG). Through the action of DAG-lipase, arachidonic acid (AA) is released from DAG. Ultimately the AA (Williams et al., 1994), or a metabolite from AA, results in a calcium influx through N- and L-type calcium channels. This signaling cascade may involve the calcium-calmodulin dependent protein kinase II (CAMKII) (Williams et al., 1995). Other reports suggest a constitutive association between NCAM and the tyrosine kinase $p59^{fyn}$. After NCAM binding, this complex recruits and activates the focal adhesion kinase $p125^{fak}$ (Beggs et al., 1994; Beggs et al., 1997). $p125^{fak}$ activity may be a mediator for neuronal responses secondary to NCAM signalling including altered growth cone behaviour (Atashi et al., 1992; Klinz et al., 1995, Beggs et al., 1997). In addition to the presumed signalling effects secondary to the binding of membrane attached NCAM molecules on apposing cells, soluble NCAM forms may potentially act as intercellular messengers. Thus, released NCAM fragments may bind to NCAM on neighboring cells activating second messengers in these cells via the above mentioned signaling pathway. Such signalling may be regulated by extracellular proteolysis changing the balance between soluble and membrane bound NCAM. Similarly, phospholipase activity may release the extracellular part of NCAM-120 from the lipid anchor (Sadoul et al., 1986). By the action of the intracellular protease calpain, cytoskeletal attachment of NCAM-180 and NCAM-140 may be broken (Sheppard et al., 1991; Covault et al., 1991). In certain cell types, the expression of NCAM influences the secretion of matrix metalloproteinases (MMPs) (Edvardsen et al., 1993) which may affect cell adhesion, cell migration and neurite extension. In addition, it has been reported that NCAM is associated with an ecto ATPase activity (Dzhandzhugazyan and Bock, 1993).

6. THE ROLE OF NCAM IN NEURITE EXTENSION

The role of NCAM in neurite extension *in vitro* has been studied extensively. In addition to the importance of diffusible neurotrophic factors (Keynes and Cook, 1995) and various inhibitory signals (Luo and Raper, 1994), neurite extension is believed to depend on balanced interactions (Bixby and Harris, 1991) between adhesion molecules present on the neuron, notably integrins, NCAM, L1 and N-cadherin, and adhesive molecules presented by the extracellular matrix and by other cells. The neurite stimulatory effect of extracellular matrix components and fibroblasts primarily depends on integrins but probably also on heterophilic binding by NCAM expressed by the neuron, while neurite extension on astrocytes or muscle cells additionally is stimulated by homophilic interactions mediated by NCAM, N-cadherin and L1.

NCAM mediated binding has been shown to stimulate neurite extension on neurons from various tissues by maintaining neurons *in vitro* on simple substrata consisting of purified NCAM, recombinant NCAM domains or NCAM fusion proteins. The first NCAM Ig domain has been shown to increase neurite extension in rat moto- and cerebellar neurons. NCAM FnIII domains also stimulate neurite extension from cerebellar cells (Frei et al., 1992), but not motoneurons (Stahlhut et al., 1997). Moreover, NCAM FnIII domains increase the number and the branching of neurites formed and alters growth cone morphology in motoneurons (Stahlhut et al., 1997). In PC12 cells, NCAM results in a conversion from an adrenal to a neural phenotype (Doherty et al., 1991). In addition, NCAM upregulates the expression of choline acetyl transferase in sympathetic neurons (Rutishauser et al., 1988). NCAM in the substratum also influence fasciculation (Frei et al., 1992; Pollerberg and Beck-Sickinger, 1993).

To study NCAM interactions in neurite extension on a cellular substratum, neurons have been maintained on fibroblasts transfected with NCAM or other cell adhesion molecules. In this model, NCAM presented by fibroblasts increases neurite extension from neurons from various brain areas and neural cell lines differentially. NCAM-180 generally is a weak stimulator while NCAM-140 and NCAM-120 stimulate strongly (Doherty et al., 1995). Moreover, the stimulatory effect is developmentally regulated. NCAM-140 stimulates neurite extension from postnatal day 6 (P6) but not P11 retinal ganglion cells (Doherty et al., 1991), from P6 but not P8 cerebellar neurons (Doherty et al., 1992) and from embryonic day 18 (E18) but not P4 hippocampal neurons (Doherty et al., 1992). This change may reflect the upregulation of NCAM containing VASE on the neurons (Walsh et al., 1992). In addition, the presence of PSA-NCAM on the neuron is necessary for the NCAM response (Doherty et al., 1990). The stimulatory effect of NCAM has, as previously mentioned, been proposed to depend on activation of the FGF-R secondary to homophilic binding between NCAM presented by the monolayer fibroblasts and NCAM expressed by the neurons (Williams et al., 1994; Hall et al., 1996).

7. THE ROLE OF NCAM IN THE FORMATION OF NEURONAL CONNECTIONS DURING DEVELOPMENT

During embryogenesis, the expression of NCAM is regulated in a spatial and temporal pattern that is correlated with several major morphogenetic events implying a role for NCAM in morphogenesis (Edelmann and Crossin, 1991). NCAM is expressed already in the oocyte (Møller et al., 1991; Kimber et al., 1994). At early stages, NCAM presumably functions as a cell-cell adhesion molecule keeping cells together in the appropriate positions. During the induction of the mesoderm, the expression of NCAM is transiently downregulated, possibly to decrease cell-cell adhesion to allow the necessary cell migration. NCAM has also been proposed to be involved in the induction of the neural tube, whereafter NCAM expression mainly is restricted to the newly generated neural tissues. During migration of neural crest cells (Bronner-Fraser et al., 1992), NCAM expression is transiently decreased in analogy with the change observed during mesoderm induction. Expression of PSA-NCAM, which weakens NCAM mediated cell adhesion, has been shown necessary for the migration of various cell types including olfactory neurons (Tomasiewicz et al., 1993; Ono et al., 1994), oligodendocyte precursors (Wang et al., 1994) and LHRH neurons (Yoshida et al., 1995). In myogenesis, activity dependent PSA-NCAM expression appears necessary for the separation of myotubes preceeding the generation of individually activated muscle fibers (Fredette et al., 1993). Thus, in early development it may be a general principle that NCAM mediated cell adhesion is transiently down-regulated to "loosen up structures" allowing migration of cells during reorganizations while subsequent expression of mature NCAM forms stabilizes the newly formed structures.

Later in development, NCAM mediated adhesion between neural cells appear to promote fasciculation during axon growth whereas NCAM binding to the extracellular matrix promotes sprouting of the same neurites once the target is reached. During neurite growth, NCAM is generally strongly polysialylated which is believed to promote neurite outgrowth presumably by increasing the distance between the interacting neural cell membranes. PSA expression on neurons may either increase or decrease branching depending on PSA expression in the environment (Rutishauser, 1996). In contrast, NCAM expressed by mature neurons generally is not polysialylated. This results in strong NCAM mediated cell-cell adhesion, stabilizing the existing structures.

During the generation of the retino-tectal projection, NCAM is expressed on neuroepithelial cells contacting the growth cones from the extending optical axons. The NCAM expressing cells may constitute "stepping stones" for the growing axon-tip generating a pathway to be followed towards the tectal target cells. The expression of PSA-NCAM by the retinal axons and their substrate appear to stimulate fasciculation (Yin et al., 1995). Moreover, PSA-NCAM seems necessary for the activity dependent organization of retino-tectal connections in *Xenopus* and chick (Fraser et al., 1984; Williams et al., 1996). A basically similar role of PSA-NCAM is proposed in the innervation of the rat spinal cord by the corticospinal tract. Preceeding innervation of muscles, a transient polysialylation of motoneurons is reported to be necessary in defasciculation (Tang et al., 1994) and branching (Landmesser et al., 1990). In the generation of the neuromuscular endplate, NCAM initially is expressed in the entire cell membrane of myoblasts (Covault and Sanes, 1986; Bixby and Reichardt, 1987). However, after the formation of the synapse, NCAM expression is restricted to the neuromuscular endplate. Recently, NCAM has been shown to bind agrin (Storms et al., 1996) which is believed to be involved in the formation of the neuromuscular junction (Fallon et al., 1985; Nastuk and Fallon, 1993). The changes in NCAM expression may be dependent on contact between the nerve terminal and the target muscle cell (Bruses et al., 1995) and on activity (Dahm and Landmesser, 1991; Rafuse and Landmesser, 1996; Kiss et al., 1994).

The evidence suggests that NCAM is involved in multiple morphogenic events during development. It is thought that NCAM generally changes during development from being a plasticity promoter to being a stability promoter (Rutishauser, 1996). This functional change is thought to rely on a decrease in polysialylation and a concomitant increase in the fraction of NCAM copies containing the VASE insert. Despite the presumed importance of NCAM during development, "knockout mice" carrying a null mutation in the NCAM gene (Cremer et al., 1994) survive relatively well without major disturbances in the development of the nervous system apart from changes in the olfactory system and to a smaller extent in the hippocampus (Tomasiewicz et al., 1993). In contrast, a targeted mutation to produce only secreted NCAM forms (Rabinowitz et al., 1996) resulted in a lethal phenotype presumably mediated by heterophilic NCAM interactions indicating a vital role of NCAM.

8. SYNAPTIC PLASTICITY IN THE MATURE BRAIN

In the adult brain, connections between neurons are not fixed. In contrast, constant re-modelling takes place, especially during regeneration and probably, in a more restricted manner, during the establishment of memory. This persisting potential for structural re-modulations may be designated "synaptic plasticity". According to a prevailing hypothesis, plastic processes in the adult brain generally rely on the same mechanisms as those important during the development of the nervous system (Kandel and O'Dell, 1992; Aubert et al., 1995). Thus, plastic events in the adult brain may be seen as a controlled "replay" of embryonic programs. To study synaptic plasticity, it is therefore an obvious strategy to study mechanisms believed to be important during development including NCAM function (Doherty et al., 1995; Rose, 1995; Fields and Itoh, 1996).

9. NCAM IN REGENERATION

Extensive structural re-modulations in the adult brain are seen in regeneration following lesions and several reports suggest an involvement of NCAM in regenerative processes (Daniloff et al., 1986; Bonfati et al., 1996; Becker et al., 1993). Following lesions in cortex or hippocampus NCAM is up-regulated during regeneration (Muller et al., 1994; Miller et al., 1994; Styren et al., 1994; Poltorak et al., 1993). Similarly, NCAM expression increases in denervated muscle cells accompanying regeneration of motoneurons (Rieger et al., 1988). Moreover, denervated muscle cells re-express the presumed plasticity promoting PSA-NCAM (Figarella-Branger et al., 1990; Covault and Sanes, 1985; Illa et al., 1992). This response is weakened in aging rats probably reflecting a reduced regenerative potential (Olsen et al., 1994). Re-expression of glial PSA-NCAM (Le Gal La Salle et al., 1992) accompanied by increased total NCAM on neurons (Niquet et al., 1993) is observed in the hippocampus following kainate induced status epilepticus.

10. NCAM IN LEARNING

After learning, more subtle structural changes in neural connections are believed to be necessary for the establishment of long-term memory. A role for NCAM in learning and memory has been indicated by a number of experiments in which normal NCAM function has been inhibited. Hence, NCAM knockout mice show impaired ability to establish long-term memory for a spatial learning task (Cremer et al., 1994). In addition, interference with NCAM function by injected antibodies inhibit the establishment of certain forms of memory but only when the antibody injection is performed in a discrete time window after the learning session. Both in the chick and in the rat, intraventricular injection of NCAM antibodies six hours after a learning session has been shown to inhibit the subsequent establishment of memory for a passive avoidance response (Doyle et al., 1992; Scholey et al., 1993). Interestingly, a transient learning-associated increase in the synthesis of a number of glycoproteins, including NCAM, is observed in chicken six hours post-training (Mileusnic et al., 1995). NCAM antibodies infused six hours post-training also block glucocorticoid stimulation of learning (Sandi et al., 1995).

A role for NCAM in learning and memory is further indicated by observations that establishment of long-term memory is accompanied by changes in the biosynthesis and in the metabolism of NCAM. In rats, an increase in the fraction of cells expressing PSA-NCAM in the dentate gyrus has been shown following passive avoidance (Fox et al., 1995) or Morris water maze learning (Murphy et al., 1996) with a peak approximately 12 hours after the learning session. A temporally similar increase in polysialylation is seen in a discrete band in the entorhinal cortex (O'Connell et al., 1996). The increased sialylation in the dentate gyrus can be re-activated by subsequent learning tasks. The increase in polysialylation can primarily be localized to NCAM-180 (Doyle et al., 1992). In addition, enzymatic removal of PSA by endo-N from the hippocampi and the overlaying cortex inhibits spatial learning in the Morris water maze (Becker et al., 1996).

Additional evidence has been gained from studies on learning and memory in the marine snail *Aplysia*. In neurons from this organism, learning related structural remodulations *in vitro* depend on a down-regulation of apCAM (an *Aplysia* NCAM homolog) by decreased synthesis as well as increased endocytosis (Mayford et al., 1992; Bailey et al., 1992). This down-regulation of apCAM has been suggested to increase synaptic plasticity primarily by weakening the adhesive contact between cells. Interestingly, new synaptic contacts preferentially form at apCAM enriched sites (Zhu et al., 1994; Zhu et al., 1995). Similarly, an NCAM-like adhesion molecule, fasciclin II, appears to be crucial for the regulation of synapse formation in the neuromuscular junction in Drosophila (Schuster et al., 1996; Dawis et al., 1996; Schuster et al., 1996).

11. NCAM IN LONG-TERM POTENTIATION (LTP)

Long-term potentiation (LTP) is a long lasting increase in synaptic efficacy, which may represent a synaptic learning mechanism (Bliss and Collingridge, 1993; Larkman and Jack, 1995; Nicoll and Malenka, 1995). Following the induction of LTP in the dentate gyrus, an increased extracellular content of soluble NCAM (115 kD) has been reported (Fazeli et al., 1994) implying a role of NCAM in this form of synaptic plasticity. The soluble NCAM may be the result of increased proteolytic activity 2–3 h after LTP induction (Fazeli et al., 1990; Nedivi et al., 1993; Qian et al., 1993; Baudry and Lynch, 1980) serving to loosen up persisting structures thereby facilitating synaptic remodelling. Alternatively, soluble NCAM may function as an intercellular messenger in LTP as NCAM binding has been shown to increase intracellular calcium in certain cultured neurons. A postsynaptic increase in intracellular calcium is thought to be crucial in LTP induction. The binding of soluble NCAM to membrane attached NCAM may therefore modulate postsynaptic signalling following LTP induction. In addition, presynaptic binding of soluble NCAM may influence transmitter release. However, an effect of NCAM ligands on basal synaptic transmission would then be expected but is not observed (Lüthi et al., 1994). As yet another alternative, signalling across the synaptic cleft may be acheived through changes in transsynaptic NCAM binding.

In the hippocampal CA1 region, two independent studies (Lüthi et al., 1994; Rønn et al., 1995) have demonstrated a reducing effect of NCAM antibodies on LTP induction. When NCAM antibodies were injected in this region, subsequent induction of LTP was inhibited (see Fig 2). A more modest inhibition has been observed following application of recombinant proteins containing the NCAM FnIII domains while no effect was seen after injection of recombinant proteins containg the NCAM Ig1 domain. Moreover, the binding of NCAM to another neural cell adhesion molecule, L1, appears to be involved in LTP. Thus, LTP induction is inhibited by the presence of oligomannosides or synthetic peptides corresponding to the putative L1 binding site of NCAM (Luthi et al., 1994; Horstkorte et al., 1993).

The effect of NCAM antibodies in LTP may be due to interference with NCAM signalling (Fig 3). NCAM binding NCAM presumably stimulate a signalling cascade (Schuch et al., 1989) including arachidonic acid (AA (Williams et al., 1994)), a putative intercellular messenger in LTP (Williams et al., 1989), and CAMKII (Williams et al.,

1995), believed to be necessary for LTP-maintenance (Fukunaga et al., 1996). Thus, NCAM antibodies may block NCAM dependent signalling necessary for LTP. Alternatively, NCAM antibodies may stimulate NCAM dependent signalling simulating NCAM binding. If so, an "untimely" activation of this signalling pathway may inhibit subsequent LTP induction. It has previously been shown that NMDA-R activation prior to tetanization actually decreases potentiation (Coan et al., 1989). Accordingly, a postsynaptic stimulation of CAMKII triggered by NCAM antibodies, may inhibit subsequent LTP-induction in analogy with the reported occlusion of LTP in mice over-expressing calcium independent CAMKII (Pettit et al., 1994).

Another possible action of NCAM antibodies in LTP is to interfere with cell adhesion mediated by NCAM thereby inhibiting structural remodelling associated with LTP. This is supported by the recent observation that the presence of PSA-NCAM, thought to be a plasticity promoting NCAM form, is necessary for the induction of LTP. Thus, prior removal of PSA with endo-N blocks subsequent LTP induction in slices (Becker et al., 1996) or slice cultures (Muller et al., 1996) from the hippocampal CA1 region. Moreover, LTP is reduced in slices or slice cultures from NCAM knockout mice. As in learning, PSA-NCAM may be presumed to facilitate remodelling of synaptic connections necessary for the expression of LTP by weakening NCAM mediated cell adhesion.

12. THE ROLE OF NCAM IN THE FORMATION AND PLASTICITY OF NEURONAL CONNECTIONS

As a cell adhesion molecule highly expressed in neurons including in synaptic contacts, NCAM is well suited to take part in the formation of proper contacts between neurons in the developing and mature brain. Moreover, NCAM binding activates intracellular messengers believed to be important in synaptic plasticity including arachidonic acid, calcium and the CAMKII. *In vitro*, the activation of NCAM mediated signalling results in increased neurite outgrowth, a response predicted to be crucial in plasticity of neuronal connections. Indeed, interference with NCAM function by antibodies, enzymatic treatment or gene knockout inhibits LTP, learning or learning associated structural changes in

Figure 2. Induction of LTP after local pressure injection of NCAM antibodies or fusion proteins. **A)** Electrode setup in the hippocampal slice preparation. Schaffer collaterals (sch) were stimulated in the CA3 region. Evoked responses were recorded by two electrodes in the CA1 stratum radiatum. One electrode, the test electrode (antibody), was simultaneously used for pressure injection of antibody (Ab). The other electrode (reference) served as an internal control. **B)** Immunohistochemical detection of Ab in the CA1 stratum radiatum (rad) after pressure injection. Pyr: stratum pyramidale. **C)** Effect of NCAM Ab on the induction of LTP in CA1. Time course of LTP induced by high frequency stimulation (HFS) 20 min after injection of 0.5 pmol of NCAM antibody (Ig fraction) compared to LTP induced at a reference recording site. Datapoints represent means of recorded field EPSP slopes (n=13). **D)** LTP induced after injection of non immune serum (Ig fraction, n=9). **E)** Effect of NCAM recombinant proteins on the induction of LTP in CA1. Fusionproteins were injected through one recording electrode while control recombinant protein (MBP) was injected simultaneously through the other recording electrode serving as an internal control. Time course of LTP induced subsequent to injection of MBP coupled NCAM FnIII (MBP-fnIII) or MBP alone (n=6). **F)** Induction of LTP after injection of MBP coupled NCAM IgI domain (MBP-loop1) or MBP alone (n=8). Statistics: Mann Whitney U test. $^*p<0.05$, $^{**}p<0.01$. A and B and data in C and D are reprinted from Rønn et al (1995), Brain Research 677 p. 145–151 with kind permission from Elsevier Science - NL, Sara Burgerhartstraat 25, 1055 KV Amsterdam, The Netherlands.

various model systems. In addition, learning affects NCAM biosynthesis in a temporally controlled manner. Thus, by mediating changes in adhesion and transmembrane signalling, NCAM appears to be important in the conversion of transient activity patterns into stable structural changes in neuronal connectivity in the developing and mature nervous system.

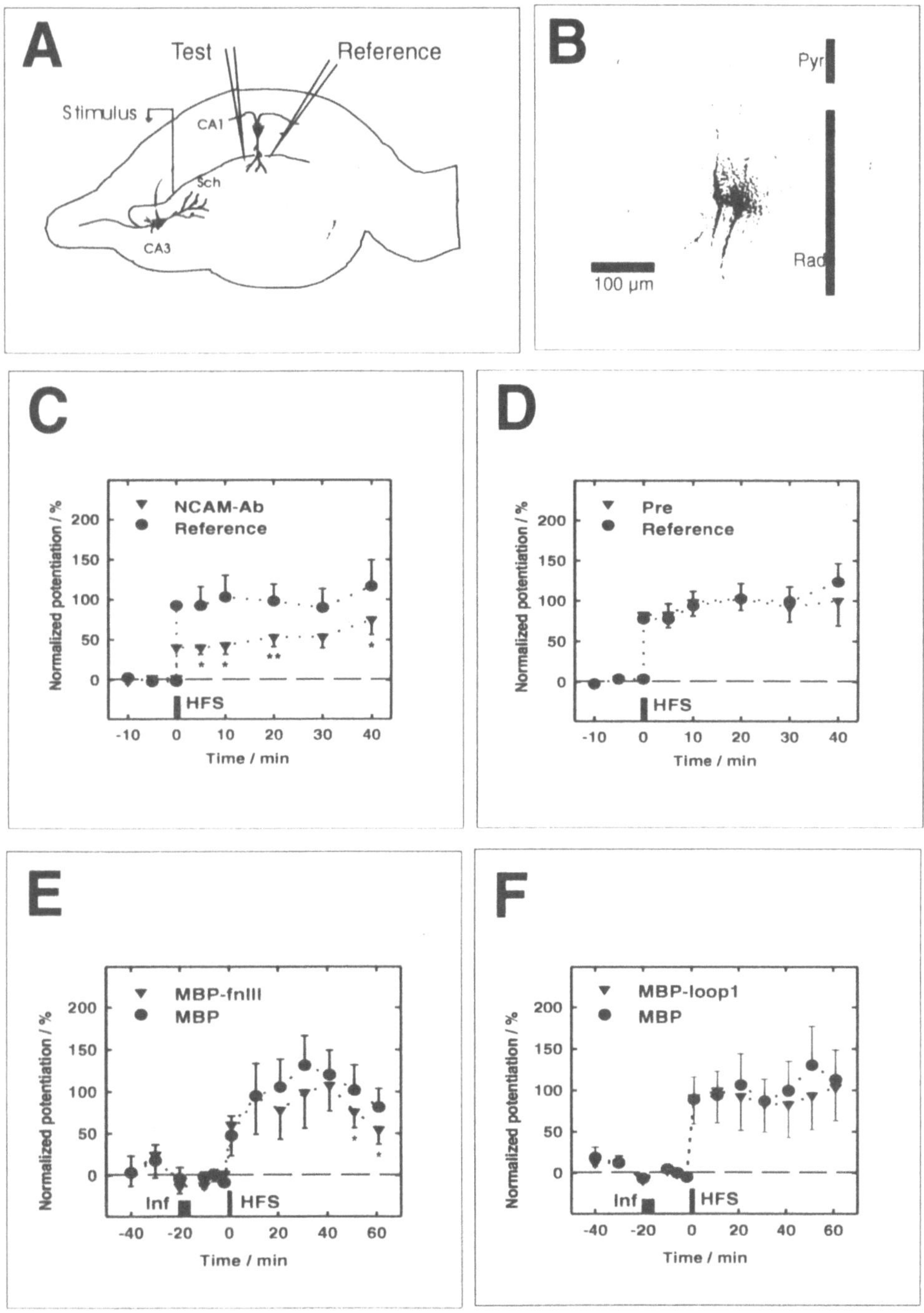

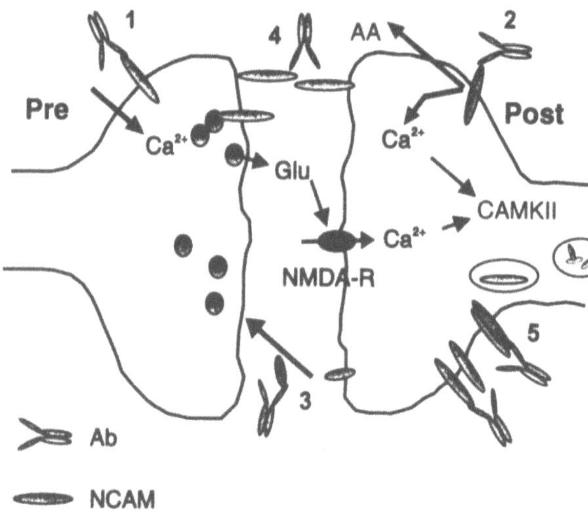

Figure 3. Possible effect of NCAM antibodies in LTP. Binding of antibodies may stimulate or inhibit NCAM mediated signal transduction in the presynaptic terminal **(1)** changing intracellular calcium levels and subsequently transmitter release. Alternatively, a similar postsynaptic **(2)** influence on NCAM mediated signalling may affect activation of the calcium calmodulin dependent protein kinase type II (CAMKII) or release of arachidonic acid (AA), a putative retrograde messenger in LTP. A retrograde signal may also consist of released soluble NCAM **(3)** or changes in transsynaptic NCAM mediated adhesion **(4)**. In addition, NCAM antibody may interfere with structural remodulations in LTP by interfering with endocytosis of NCAM **(5)**. Alternatively, NCAM antibody may influence clustering of other proteins necessary for LTP-induction and/or maintenance.

REFERENCES

Acheson, A., Sunshine, J.L., and Rutishauser, U. (1991). NCAM polysialic acid can regulate both cell-cell and cell-substrate interactions. Journal of Cell Biology *114*, 143–153.

Atashi, J.R., Klinz, S.G., Ingraham, C.A., Matten, W.T., Schachner, M., and Maness, P.F. (1992). Neural cell adhesion molecules modulate tyrosine phosphorylation of tubulin in nerve growth cone membranes. Neuron *8*, 831–842.

Aubert, I., Ridet, J.L., and Gage, F.H. (1995). Regeneration in the adult mammalian CNS: guided by development. Current Opinion in Neurobiology *5*, 625–635.

Bailey, C.H., Chen, M., Keller, F., and Kandel, E.R. (1992). Serotonin-mediated endocytosis of apCAM: an early step of learning-related synaptic growth in Aplysia. Science *256*, 645–649.

Baudry, M. and Lynch, G. (1980). Regulation of hippocampal glutamate receptors: evidence for the involvement of a calcium-activated protease. Proceedings of the National Academy of Sciences of the United States of America *77*, 2298–2302.

Becker, C.G., Artola, A., Gerardy-Schahn, R., Becker, T., Welzl, H., and Schachner, M. (1996). The polysialic acid modification of the neural cell adhesion molecule is involved in spatial learning and hippocampal lon-term potentiation. Journal of Neuroscience Research *45*, 143–152.

Becker, T., Becker, C.G., Niemann, U., Naujoks-Manteuffel, C., Gerardy-Schahn, R., and Roth, G. (1993). Amphibian-specific regulation of polysialic acid and the neural cell adhesion molecule in development and regeneration of the retinotectal system of the salamander Pleurodeles waltl. Journal of Comparative Neurology *336*, 532–544.

Beggs, H.E., Baragona, S.C., Hemperly, J.J., and Maness, P.F. (1997). NCAM140 interacts with the focal adhesion kinase p125fak and the SRC-related tyrosine kinase p59fyn. Journal of Biological Chemistry *272*, 8310–8319.

Beggs, H.E., Soriano, P., and Maness, P.F. (1994). NCAM-dependent neurite outgrowth is inhibited in neurons from Fyn-minus mice. Journal of Cell Biology *127*, 825–833.

Bixby, J.L. and Harris, W.A. (1991). Molecular mechanisms of axon growth and guidance. Annual Review of Cell Biology *7*, 117–159.

Bixby, J.L. and Reichardt, L.F. (1987). Effects of antibodies to neural cell adhesion molecule (N-CAM) on the differentiation of neuromuscular contacts between ciliary ganglion neurons and myotubes in vitro. Developmental Biology *119*, 363–372.

Bliss, T.V. and Collingridge, G.L. (1993). A synaptic model of memory: long-term potentiation in the hippocampus. Nature *361*, 31–39.

Bloch, R.J. (1992). Clusters of neural cell adhesion molecule at sites of cell-cell contact. Journal of Cell Biology *116*, 449–463.

Bock, E., Edvardsen, K., Gibson, A., Linnemann, D., Lyles, J.M., and Nybroe, O. (1987). Characterization of soluble forms of NCAM. FEBS Letters *225*, 33–36.

Bonfati, L., Merighi, A., and Theodosis, T. (1996). Dorsal rhizotomy induces transient expression of the highly sialylated isoform of the neural cell adhesion molecule in neurons and astrocytes of the adult rat spinal cord. Neuroscience *74*, 619–623.

Bronner-Fraser, M., Wolf, J.J., and Murray, B.A. (1992). Effects of antibodies against N-cadherin and N-CAM on the cranial neural crest and neural tube. Developmental Biology *153*, 291–301.

Brummendorf, T. and Rathjen, F.G. (1995). Cell adhesion molecules 1: immunoglobulin superfamily. Protein Profile *2*, 963–108.

Bruses, J.L., Oka, S., and Rutishauser, U. (1995). NCAM-associated polysialic acid on ciliary ganglion neurons is regulated by polysialytransferase levels and interaction with muscle. Journal of Neuroscience *15*, 8310–8319.

Chen, A., Haines, S., Maxson, K., and Akeson, R.A. (1994). VASE exon expression alters NCAM-mediated cell-cell interactions. Journal of Neuroscience Research *38*, 483–492.

Coan, E.J., Irving, A.J., and Collingridge, G.L. (1989). Low-frequency activation of the NMDA receptor system can prevent the induction of LTP. Neuroscience Letters *105*, 205–210.

Cole, G.J., Loewy, A., and Glaser, L. (1986). Neuronal cell-cell adhesion depends on interactions of N-CAM with heparin-like molecules. Nature *320*, 445–447.

Covault, J., Liu, Q.Y., and el-Deeb, S. (1991). Calcium-activated proteolysis of intracellular domains in the cell adhesion molecules NCAM and N-cadherin. Brain Research *11*, 11–16.

Covault, J. and Sanes, J.R. (1985). Neural cell adhesion molecule (N-CAM) accumulates in denervated and paralyzed skeletal muscles. Proc. Natl. Acad. Sci. USA *82*, 4544–4548.

Covault, J. and Sanes, J.R. (1986). Distribution of N-CAM in synaptic and extrasynaptic portions of developing and adult skeletal muscle. Journal of Cell Biology *102*, 716–730.

Cremer, H., Lange, R., Christoph, A., Plomann, M., Vopper, G., Roes, J., BrownR., Baldwin, S., Kraemer, P., Scheff, S., and et al. (1994). Inactivation of the N-CAM gene in mice results in size reduction of the olfactory bulb and deficits in spatial learning. Nature *367*, 455–459.

Crossin, K.L. and Hoffman, S. (1991). Expression of adhesion molecules during the formation and differentiation of the avian endocardial cushion tissue. Developmental Biology *145*, 277–286.

Dahm, L.M. and Landmesser, L.T. (1991). The regulation of synaptogenesis during normal development and following activity blockade. Journal of Neuroscience *11*, 238–255.

Daniloff, J.K., Levi, G., Grumet, M., Rieger, F., and Edelman, G.M. (1986). Altered expression of neuronal cell adhesion molecules induced by nerve injury and repair. Journal of Cell Biology *103*, 929–945.

Dawis, G.W., Schuster, C.M., and Goodman, C.S. (1996). Genetic dissection of structural and functional components of synaptic plasticity. III. CREB is necessary for presynaptic functional plasticity. Neuron *17*, 669–679.

Doherty, P., Ashton, S.V., Moore, S.E., and Walsh, F.S. (1991). Morphoregulatory activities of NCAM and N-cadherin can be accounted for by G protein-dependent activation of L- and N-type neuronal calcium-channels. Cell *67*, 21–33.

Doherty, P., Cohen, J., and Walsh, F.S. (1990). Neurite outgrowth in response to transfected N-CAM changes during development and is modulated by polysialic acid. Neuron *5*, 209–219.

Doherty, P., Fazeli, M.S., and Walsh, F.S. (1995). The neural cell adhesion molecule and synaptic plasticity. Journal of Neurobiology *26*, 437–446.

Doherty, P., Moolenaar, C.E., Ashton, S.V., Michalides, R.J., and Walsh, F.S. (1992). The VASE exon downregulates the neurite growth-promoting activity of NCAM 140. Nature *356*, 791–793.

Doherty, P., Rowett, L.H., Moore, S.E., Mann, D.A., and Walsh, F.S. (1991). Neurite outgrowth in response to transfected N-CAM and N-cadherin reveals fundamental differences in neuronal responsiveness to CAMs. Neuron *6*, 247–258.

Doherty, P., Skaper, S.D., Moore, S.E., Leon, A., and Walsh, F.S. (1992). A developmental regulated switch in neuronal responsiveness to NCAM and E-cadherin in the rat hippocampus. Development *115*, 885–892.

Doyle, E., Nolan, P.M., Bell, R., and Regan, C.M. (1992). Intraventricular infusions of anti-neural cell adhesion molecules in a discrete posttraining period impair consolidation of a passive avoidance response in the rat. Journal of Neurochemistry 59, 1570–1573.

Doyle, E., Nolan, P.M., Bell, R., and Regan, C.M. (1992). Hippocampal NCAM180 transiently increases sialylation during the aquisition and consolidation of a passive avoidance response in the adult rat. Journal of Neuroscience Research *31*, 513–523.

Dzhandzhugazyan, K. and Bock, E. (1993). Demonstration of (Ca(2+)-Mg2+)-ATPase activity of the neural cell adhesion molecule. FEBS Letters *336*, 279–283.

Eckhardt, M., Muhlenhoff, M., Bethe, A., Koopman, J., and Frosch, M. (1995). Molecular characterization of eukaryotic polysialyltransferase-1. Nature *373*, 715–718.

Edelman, G.M. (1987). CAMs and Igs: cell adhesion and the evolutionary origins of immunity. Immunological Reviews *100*, 11–45.

Edelman, G.M. and Chuong, C.M. (1982). Embryonic to adult conversion of neural cell adhesion molecules in normal and staggerer mice. Proceedings of the National Academy of Sciences of the United States of America *79*, 7036–7040.

Edelmann, G.M. and Crossin, K.L. (1991). Cell adhesion molecules: implications for a molecular histology. Ann. Rev. Biochem. *60*, 155–190.

Edvardsen, K., Chen, W., Rucklidge, G., Walsh, F.S., Obrink, B., and Bock, E. (1993). Transmembrane neural cell-adhesion molecule (NCAM), but not glycosyl-phosphatidylinositol-anchored NCAM, down-regulates secretion of matrix metalloproteinases. Proceedings of the National Academy of Sciences of the United States of America *90*, 11463–11467.

Fallon, J.R., Nitkin, R.M., Reist, N.E., Wallace, B.G., and McMahan, U.J. (1985). Acetylcholine receptor-aggregating factor is similar to molecules concentrated at neuromuscular junctions. Nature *315*, 571–574.

Fazeli, M.S., Breen, K., Errington, M.L., and Bliss, T.V.P. (1994). Increase in extracellular NCAM and amyloid precursor protein following induction of long-term potentiation in the dentate gyrus of anaesthetized rats. Neuroscience Letters *169*, 77–80.

Fazeli, M.S., Errington, M.L., Dolphin, A.C., and Bliss, T.V. (1990). Increased efflux of a haemoglobin-like protein and an 80 kDa protease into push-pull perfusates following the induction of long-term potentiation in the dentate gyrus. Brain Research *521*, 247–253.

Fields, R.D. and Itoh, K. (1996). Neural cell adhesion molecule in activity-dependent development and synaptic plasticity. Trends in Neurosciences *19*, 473–480.

Figarella-Branger, D., Nedelec, J., Pellissier, J.F., Boucraut, J., and Bianco, N. (1990). Expression of various isoforms of neural cell adhesive molecules and their highly polysialylated counterparts in diseased human muscles. Journal of the Neurological Sciences *98*, 21–36.

Fox, G.B., O'Connell, A.W., Murphy, K.J., and Regan, C.M. (1995). Memory consolidation induces a transient and time-dependent increase in the frequency of neural cell adhesion molecule polysialylated cells in the adult rat hippocampus. Journal of Neurochemistry *65*, 2796–2799.

Fraser, S.E., Murray, B.A., Chuong, C.M., and Edelman, G.M. (1984). Alteration of the retinotectal map in Xenopus by antibodies to neural cell adhesion molecules. Proceedings of the National Academy of Sciences of the United States of America *81*, 4222–4226.

Fredette, B., Rutishauser, U., and Landmesser, L. (1993). Regulation and activity-dependence of N-cadherin, NCAM isoforms, and polysialic acid on chick myotubes during development. Journal of Cell Biology *123*, 1867–1888.

Frei, T., von Bohlen und Halbach, F., Wille, W., and Schachner, M. (1992). Different extracellular domains of the neural cell adhesion molecule (N-CAM) are involved in different functions. Journal of Cell Biology *118*, 177–194.

Fukunaga, K., Muller, D., and Miyamoto, E. (1996). CaM kinase II in long-term potentiation. Neurochemistry International *4*, 343–358.

Grumet, M., Flaccus, A., and Margolis, R.U. (1993). Functional characterization of chondroitin sulfate proteoglycans of brain: interactions with neurons and neural cell adhesion molecules. Journal of Cell Biology *120*, 815–824.

Hall, H., Walsh, F.S., and Doherty, P. (1996). A role for the FGF receptor in the axonal growth response stimulated by cell adhesion molecules. Cell Adhesion & Communication *3*, 441–450.

Horstkorte, R., Schachner, M., Magyar, J.P., Vorherr, T., and Schmitz, B. (1993). The fourth immunoglobulin-like domain of NCAM contains a carbohydrate recognition domain for oligomannosidic glycans implicated in association with L1 and neurite outgrowth. Journal of Cell Biology *121*, 1409–1421.

Illa, I., Leon-Monzon, M., and Dalakas, M.C. (1992). Regenerating and denervated human muscle fibers and satellite cells express neural cell adhesion molecule recognized by monoclonal antibodies to natural killer cells. Annals of Neurology *31*, 46–52.

Joliot, A.H., Triller, A., Volovitch, M., Pernelle, C., and Prochiantz, A. (1991). alpha-2,8-Polysialic acid is the neuronal surface receptor of antennapedia homeobox peptide. New Biologist *3*, 1121–1134.

Kadmon, G., Kowitz, A., Altevogt, P., and Schachner, M. (1990). The neural cell adhesion molecule N-CAM enhances L1-dependent cell-cell interactions. Journal of Cell Biology *110*, 193–208.

Kadmon, G., Kowitz, A., Altevogt, P., and Schachner, M. (1990). Functional cooperation between the neural adhesion molecules L1 and N-CAM is carbohydrate dependent. Journal of Cell Biology *110*, 209–218.

Kandel, E.R. and O'Dell, T.J. (1992). Are adult mechanisms also used for development?. Science *258*, 243–244.

Keynes, R. and Cook, G.M. (1995). Axon guidance molecules. Cell *83*, 161–169.

Kimber, S.J., Bentley, J., Ciemerych, M., Møller, C.J., and Bock, E. (1994). Expression of N-CAM in fertilized pre- and periimplantation and parthenogenetically activated mouse embryos. European Journal of Cell Biology *63*, 102–113.

Kiselyov, V.V., Berezin, V., Maar, T.E., Soroka, V., Edvardsen, K., Schousboe, A., and Bock, E. (1997). The first Ig-like NCAM domain is involved in double reciprocal interaction with the second Ig-like NCAM domain and in heparin binding. Journal of Biological Chemistry. Journal of Biological Chemistry *272*, 10125–10134.

Kiss, J.Z., Wang, C., Olive, S., Rougon, G., Lang, J., Baetens, D., and Harry, D. (1994). Activity-dependent mobilization of the adhesion molecule polysialic NCAM to the cell surface of neurons and endocrine cells. EMBO Journal *13*, 5284–5292.

Klinz, S.G., Schachner, M., and Maness, P.F. (1995). L1 and N-CAM antibodies trigger protein phosphatase activity in growth cone-enriched membranes. Journal of Neurochemistry *65*, 84–95.

Krog, L. and Bock, E. (1992). Glycosylation of neural cell adhesion molecules of the immunoglobulin superfamily. APMIS *Supplementum*. *27*, 53–70.

Landmesser, L., Dahm, L., Tang, J.C., and Rutishauser, U. (1990). Polysialic acid as a regulator of intramuscular nerve branching during embryonic development. Neuron *4*, 655–667.

Larkman, A.U. and Jack, J.J. (1995). Synaptic plasticity: hippocampal LTP. Current Opinion in Neurobiology *5*, 324–334.

Le Gal La Salle, G., Rougon, G., and Valin, A. (1992). The embryonic from of neural cell adhesion molecule (E-NCAM) in the rat hippocampus and its reexpression on glial cells following kainic acid-induced status epilepticus. The Journal of Neuroscience *12*, 872–882.

Luo, Y. and Raper, J.A. (1994). Inhibitory factors controlling growth cone motility and guidance. Current Opinion in Neurobiology *4*, 648–654.

Lüthi, A., Laurent, J.P., Figurov, A., Muller, D., and Schachner, M. (1994). Hippocampal long-term potentiation and neural cell adhesion molecules L1 and NCAM. Nature *372*, 777–779.

Mayford, M., Barzilai, A., Keller, F., Schacher, S., and Kandel, E.R. (1992). Modulation of an NCAM-related adhesion molecule with long-term synaptic plasticity in Aplysia. Science *256*, 638–644.

Mileusnic, R., Rose, S.P., Lancashire, C., and Bullock, S. (1995). Characterisation of antibodies specific for chick brain neural cell adhesion molecules which cause amnesia for a passive avoidance task. Journal of Neurochemistry *64*, 2598–2606.

Milev, P., Maurel, P., Haring, M., Margolis, R.K., and Margolis, R.U. (1996). TAG-1/axonin-1 is a high-affinity ligand of neurocan, phosphacan/protein-tyrosine phosphatase-zeta/beta, and N-CAM. Journal of Biological Chemistry *271*, 15716–15723.

Miller, P.D., Styren, S.D., Lagenaur, C.F., and DeKosky, S.T. (1994). Embryonic neural cell adhesion molecule (N-CAM) is elevated in the denervated rat dentate gyrus. Journal of Neuroscience *14*, 4217–4225.

Miragall, F., Kadmon, G., Husmann, M., and Schachner, M. (1988). Expression of cell adhesion molecules in the olfactory system of the adult mouse: presence of the embryonic form of N-CAM. Developmental Biology *129*, 516–531.

Moran, N. and Bock, E. (1988). Characterization of the kinetics of neural cell adhesion molecule homophilic binding. FEBS Letters *242*, 121–124.

Muller, D., Stoppini, L., Wang, C., and Kiss, J.Z. (1994). A role for polysialylated neural cell adhesion molecule in lesion-induced sprouting in hippocampal organotypic cultures. Neuroscience *61*, 441–445.

Muller, D., Wang, C., Skibo, G., Toni, N., Cremer, H., Calaora, V., Rougon, G., and Kiss, J.Z. (1996). PSA-NCAM is required for activity-induced synaptic plasticity. Neuron *17*, 413–422.

Murphy, K.J., O'Connell, A.W., and Regan, C.M. (1996). Repetitive and transient increases in hippocampal neural cell adhesion molecule polysialylation state following multitrial spatial training. Journal of Neurochemistry *67*, 1268–1274.

Murray, B.A., Hemperly, J.J., Prediger, E.A., Edelman, G.M., and Cunningham, B.A. (1986). Alternatively spliced mRNAs code for different polypeptide chains of the chicken neural cell adhesion molecule (N-CAM). Journal of Cell Biology *102*, 189–193.

Møller, C.J., Byskov, A.G., Roth, J., Celis, J.E., and Bock, E. (1991). NCAM in developing mouse gonads and ducts. Anatomy & Embryology *184*, 541–548.

Nakayama, J. and Fukuda, M. (1996). A human polysialyltransferase directs in vitro synthesis of polysialic acid. Journal of Biological Chemistry *271*, 1829–1832.

Nakayama, J., Fukuda, M.N., Fredette, B., Ranscht, B., and Fukuda, M. (1995). Expression cloning of a human polysialyltransferase that forms the polysialylated neural cell adhesion molecule present in emryonic brain. Proc. Natl. Acad. Sci. USA *922*, 7031–7035.

Nastuk, M.A. and Fallon, J.R. (1993). Agrin and the molecular choreography of synapse formation. Trends in Neurosciences *16*, 72–76.

Nedivi, E., Hevroni, D., Naot, D., Israeli, D., and Citri, Y. (1993). Numerous candidate plasticity-related genes revealed by differential cDNA cloning. Nature *363*, 718–722.

Nelson, R.W., Bates, P.A., and Rutishauser, U. (1995). Protein determinants for specific polysialylation of the neural cell adhesion molecule. Journal of Biological Chemistry *270*, 17171–17179.

Nicoll, R.A. and Malenka, R.C. (1995). Contrasting properties of two forms of long-term potentiation in the hippocampus. Nature *377*, 115–118.

Niquet, J., Jorquera, I., Ben-Ari, Y., and Represa, A. (1993). NCAM immunoreactivity on mossy fibers and reactive astrocytes in the hippocampus of epileptic rats. Brain Research *626*, 106–116.

Noble, M., Albrechtsen, M., Moller, C., Lyles, J., Bock, E., Goridis, C., and Rutishauser, U. (1985). Glial cells express N-CAM/D2-CAM-like polypeptides in vitro. Nature *316*, 725–728.

O'Connell, A.W., Fox, G.B., Barry, T., Foley, A.G., Murphy, K.J., Fichera, G., Kelly, J., and Regan, C.M. (1997). Spatial learning activates neural cell adhesion molecule polysialylation in a cortico-hippocampal pathway within the medial temporal lobe. Journal of Neurochemistry, *In Press*.

Olsen, M., Krog, L., Edvardsen, K., Skovgaard, L.T., and Bock, E. (1993). Intact transmembrane isoforms of the neural cell adhesion molecule are released from the plasma membrane. Biochemical Journal *295*, 833–840.

Olsen, M., Zuber, C., Roth, J., Linnemann, D., and Bock, E. (1995). The ability to reexpress polysialylated NCAM in soleus muscle after denervation is induced in aged rats compared to young adult rats. International Journal of Developmental Neuroscience. *13*. 97–104.

Ono, K., Tomasiewicz, H., Magnuson, T., and Rutishauser, U. (1994). N-CAM mutation inhibits tangential neuronal migration and is phenocopied by enzymatic removal of polysialic acid. Neuron *13*, 595–609.

Persohn, E., Pollerberg, G.E., and Schachner, M. (1989). Immunoelectron-microscopic localization of the 180 kD component of the neural cell adhesion molecule N-CAM in postsynaptic membranes. Journal of Comparative Neurology *288*, 92–100.

Pettit, D.L., Perlman, S., and Malinow, R. (1994). Potentiated transmission and prevention of further LTP by increased CaMKII activity in postsynaptic hippocampal slice neurons. Science *266*, 1881–1885.

Pollerberg, G.E. and Beck-Sickinger, A. (1993). A functional role for the middle extracellular region of the neural cell adhesion molecule (NCAM) in axonal fasciculation and orientation. Developmental Biology *156*, 324–340.

Pollerberg, G.E., Schachner, M., and Davoust, J. (1986). Differentiation state-dependent surface mobilities of two forms of the neural cell adhesion molecule. Nature *324*, 462–465.

Poltorak, M., Herranz, A.S., Williams, J., Lauretti, L., and Freed, W.J. (1993). Effects of frontal cortical lesions on mouse striatum: reorganization of cell recognition molecule, glial fiber, and synaptic protein expression in the dorsomedial striatum. Journal of Neuroscience *13*, 2217–2229.

Powell, S.K., Cunningham, B.A., Edelman, G.M., and Rodriguez-Boulan, E. (1991). Targeting of transmembrane and GPI-anchored forms of N-CAM to opposite domains of a polarized epithelial cell. Nature *353*, 76–77.

Probstmeier, R., Kuhn, K., and Schachner, M. (1989). Binding properties of the neural cell adhesion molecule to different components of the extracellular matrix. Journal of Neurochemistry *53*, 1794–1801.

Qian, Z., Gilbert, M.E., Colicos, M.A., Kandel, E.R., and Kuhl, D. (1993). Tissue-plasminogen activator is induced as an immediate early gene during seizure, kindling and long-term potentiation. Nature *361*, 453–457.

Rabinowitz, J.E., Rutishauser, U., and Magnuson, T. (1996). Targeted mutation of Ncam to produce a secreted molecule results in a dominant embryonic lethality. Proceedings of the National Academy of Sciences of the United States of America *93*, 6421–6424.

Rafuse, V.F. and Landmesser, L. (1996). Contractile activity regulates isoform expression and polysialylation of NCAM in cultured myotubes: involvement of Ca2+ and protein kinase C. Journal of Cell Biology *132*, 969–983.

Ranheim, T.S., Edelman, G.M., and Cunningham, B.A. (1996). Homophilic adhesion mediated by the neural cell adhesion molecule involves multiple immunoglobulin domains. Proceedings of the National Academy of Sciences of the United States of America *93*, 4071–4075.

Rao, Y., Wu, X.F., Gariepy, J., Rutishauser, U., and Siu, C.H. (1992). Identification of a peptide sequence involved in homophilic binding in the neural cell adhesion molecule NCAM. Journal of Cell Biology *118*, 937–949.

Rao, Y., Wu, X.F., Yip, P., Gariepy, J., and Siu, C.H. (1993). Structural characterization of a homophilic binding site in the neural cell adhesion molecule. Journal of Biological Chemistry *268*, 20630–20638.

Rao, Y., Zhao, X., and Siu, C.H. (1994). Mechanism of homophilic binding mediated by the neural cell adhesion molecule NCAM. Evidence for isologous interaction. Journal of Biological Chemistry *269*, 27540–27548.

Rieger, F., Nicolet, M., Pincon-Raymond, M., Murawsky, M., Levi, G., and EdelmanGM. (1988). Distribution and role in regeneration of N-CAM in the basal laminae of muscle and Schwann cells. Journal of Cell Biology *107*, 707–719.

Rose, S.P. (1995). Cell-adhesion molecules, glucocorticoids and long-term-memory formation. Trends in Neurosciences *18*, 502–506.

Rougon, G. (1993). Structure, metabolism and cell biology of polysialic acids. Eur J Cell Biol *61*, 197–207.

Rougon, G., Deagostini-Bazin, H., Hirn, M., and Goridis, C. (1982). Tissue- and developmental stage-specific forms of a neural cell surface antigen linked to differences in glycosylation of a common polypeptide. EMBO Journal *1*, 1239–1244.

Rougon, G., Dubois, C., Buckley, N., Magnani, J.L., and Zollinger, W. (1986). A monoclonal antibody against meningococcus group B polysaccharides distinguishes embryonic from adult N-CAM. Journal of Cell Biology *103*, 2429–2437.

Rutishauser, U. (1996). Polysialic acid and the regulation of cell interactions. Current Opinion in Cell Biology *8*, 679–684.

Rutishauser, U., Acheson, A., Hall, A.K., Mann, D.M., and Sunshine, J. (1988). The neural cell adhesion molecule (NCAM) as a regulator of cell-cell interactions. Science *240*, 53–57.

Rutishauser, U. and Landmesser, L. (1996). Polysialic acid in the vertebrate nervous system: a promoter of plasticity in cell-cell interactions. Trends in Neurosciences *19*, 422–427.

Rønn, L.C., Bock, E., Linnemann, D., and Jahnsen, H. (1995). NCAM-antibodies modulate induction of long-term potentiation in rat hippocampal CA1. Brain Research *677*, 145–151.

Sadoul, K., Meyer, A., Low, M.G., and Schachner, M. (1986). Release of the 120 kDa component of the mouse neural cell adhesion molecule N-CAM from cell surfaces by phosphatidylinositol-specific phospholipase C. Neuroscience Letters *72*, 341–346.

Sadoul, R., Hirn, M., Deagostini-Bazin, H., Rougon, G., and Goridis, C. (1983). Adult and embryonic mouse neural cell adhesion molecules have different binding properties. Nature *304*, 347–349.

Saffell, J.L., Williams, E.J., Mason, I.J., Walsh, F.S., and Doherty, P. (1997). Expression of a dominant negative FGF receptor inhibits axonal growth and FGF receptor phosphorylation stimulated by CAMs. Neuron *18*, 231–242.

Sandi, C., Rose, S.P., Mileusnic, R., and Lancashire, C. (1995). Corticosterone facilitates long-term memory formation via enhanced glycoprotein synthesis. Neuroscience *69*, 1087–1093.

Scheidegger, E.P., Sternberg, L.R., Roth, J., and Lowe, J.B. (1995). A human STX cDNA confers polysialic acid expression in mammalian cells. Journal of Biological Chemistry *270*, 22685–22688.

Scholey, A.B., Rose, S.P.R., Zamani, M.R., Bock, E., and Schachner, M. (1993). A role for the neural cell adhesion molecule (NCAM) in a late, consolidating phase of glycoprotein synthesis 6 hours following passive avoidance training of the young chick. Neuroscience *55*, 499–509.

Schuch, U., Lohse, M.J., and Schachner, M. (1989). Neural cell adhesion molecules influence second messenger systems. Neuron *3*, 13–20.

Schuster, C.M., Davis, G.W., Fetter, R.D., and Goodman, C.S. (1996). Genetic dissection of structural and functional components of synaptic plasticity. I. Fasciclin II controls synaptic stabilization and growth. Neuron *17*, 641–654.

Schuster, C.M., Dawis, G.W., Fetter, R.D., and Goodman, C.S. (1996). Genetic dissection of structural and functional components of synaptic plasticity. II. Fasciclin II controls presynaptic structural plasticity. Neuron *17*, 655–667.

Seki, T. and Arai, Y. (1991). The persistent expression of a highly sialylated NCAM in the dentate gyrus of the adult rat. Neuroscience research *12*, 503–513.

Sheppard, A., Wu, J., Rutishauser, U., and Lynch, G. (1991). Proteolytic modification of neural cell adhesion molecule (NCAM) by the intracellular proteinase calpain. Biochimica et Biophysica Acta *1076*, 156–160.

Small, S.J. and Akeson, R. (1990). Expression of the unique NCAM VASE exon is independently regulated in distinct tissues during development. Journal of Cell Biology *111*, 2089–2096.

Small, S.J., Haines, S.L., and Akeson, R.A. (1988). Polypeptide variation in an N-CAM extracellular immunoglobulin-like fold is developmentally regulated through alternative splicing. Neuron *1*, 1007–1017.

Stahlhut, M., Berezin, V., Bock, E., and Ternaux, J. (1997). NCAM-fibronectin-type-III-domain substrata with and without a six amino acid long proline rich insert increase the dendritic arborization of spinal motoneurons. Journal of Neuroscience Research *48*, 112–121.

Storms, S.D., Kim, A.C., Tran, B.T., Cole, G.J., and Murray, B.A. (1996). NCAM-mediated adhesion of transfected cells to agrin. Cell Adhesion & Communication *3*, 497–509.

Styren, S.D., Lagenaur, C.F., Miller, P.D., and DeKosky, S.T. (1994). Rapid expression and transport of embryonic N-CAM in dentate gyrus following entorhinal cortex lesion: ultrastructural analysis. Journal of Comparative Neurology *349*, 486–492.

Tang, J., Rutishauser, U., and Landmesser, L. (1994). Polysialic acid regulates growth cone behavior during sorting of motor axons in the plexus region. Neuron *13*, 405–414.

Thiery, J.P., Brackenbury, R., Rutishauser, U., and Edelman, G.M. (1977). Adhesion among neural cells of the chick embryo. Progress in Clinical & Biological Research *15*, 199–206.

Thomsen, N.K., Soroka, V., Jensen, P.H., Berezin, V., Kiselyov, V.V., and Bock, E. (1996). The three-dimensional structure of the first domain of neural cell adhesion molecule. Nature Structural Biology *3*, 581–585.

Tomasiewicz, H., Ono, K., Yee, D., Thompson, C., Goridis, C., Rutishauser, U., and Magnuson, T. (1993). Genetic deletion of a neural cell adhesion molecule variant (N-CAM-180) produces distinct defects in the central nervous system. Neuron *11*, 1163–1174.

von Bohlen und Halbach, F., Taylor, J., and Schachner, M. (1992). Cell type specific effects of the neural cell adhesion molecules L1 and N-CAM on diverse second messenger systems. European Journal of Neuroscience *4*, 896–909.

Walsh, F.S. and Dickson, G. (1989). Generation of multiple N-CAM polypeptides from a single gene. Bioessays *11*, 83–88.

Walsh, F.S., Furness, J., Moore, S.E., Ashton, S., and Doherty, P. (1992). Use of the neural cell adhesion molecule VASE exon by neurons is associated with a specific down-regulation of neural cell adhesion molecule-dependent neurite outgrowth in the developing cerebellum and hippocampus. Journal of Neurochemistry *59*, 1959–1962.

Wang, C., Rougon, G., and Kiss, J.Z. (1994). Requirement of polysialic acid for the migration of the O-2A glial progenitor cell from neurohypophyseal explants. Journal of Neuroscience *14*, 4446–4457.

Williams, A.F. and Barclay, A.N. (1988). The immunoglobulin superfamily - domains for cell surface recognition. Annual Review of Immunology *6*, 381–405.

Williams, D.K., Gannon-Murakami, L., Rougon, G., and Udin, S.B. (1996). Polysialylated neural cell adhesion molecule and plasticity of ipsilateral connections in Xenopus tectum. Neuroscience *70*, 277–285.

Williams, E.J., Furness, J., Walsh, F.S., and Doherty, P. (1994). Activation of the FGF receptor underlies neurite outgrowth stimulated by L1, NCAM, and N-cadherin. Neuron *13*, 583–594.

Williams, E.J., Mittal, B., Walsh, F.S., and Doherty, P. (1995). A Ca2+/calmodulin kinase inhibitor, KN-62, inhibits neurite outgrowth stimulated by CAMs and FGF. Molecular & Cellular Neurosciences *6*, 69–79.

Williams, E.J., Walsh, F.S., and Doherty, P. (1994). The production of arachidonic acid can account for calcium channel activation in the second messenger pathway underlying neurite outgrowth stimulated by NCAM, N-cadherin, and L1. Journal of Neurochemistry *62*, 1231–1234.

Williams, J.H., Errington, M.L., Lynch, M.A., and Bliss, T.V.P. (1989). Arachidonic acid induces a long-term activity-dependent enhancement of synaptic transmission in the hippocampus. Nature *341*, 739–742.

Woo, M.K. and Murray, B.A. (1994). Solid-phase binding analysis of N-CAM interactions with brain fodrin. Biochimica et Biophysica Acta *1191*, 173–180.

Yin, X., Watanabe, M., and Rutishauser, U. (1995). Effect of polysialic acid on the behavior of retinal ganglion cell axons during growth into the optic tract and tectum. Development *121*, 3439–3446.

Yoshida, K., Tobet, S.A., Crandall, J.E., Jimenez, T.P., and Schwarting, G.A. (1995). The migration of luteinizing hormone-releasing hormone neurons in the developing rat is associated with a transient, caudal projection of the vomeronasal nerve. Journal of Neuroscience *15*, 7769–7777.

Yoshida, Y., Kojima, N., Kurosawa, N., Hamamoto, T., and Tsuji, S. (1995). Molecular cloning of Sia alpha 2,3Gal beta 1,4GlcNAc alpha 2,8-sialyltransferase from mouse brain. Journal of Biological Chemistry *270*, 14628–14633.

Zhu, H., Wu, F., and Schacher, S. (1994). Aplysia cell adhesion molecules and serotonin regulate sensory cell-motor cell interactions during early stages of synapse formation in vitro. Journal of Neuroscience *14*, 6886–6900.

Zhu, H., Wu, F., and Schacher, S. (1995). Changes in expression and distribution of Aplysia cell adhesion molecules can influence synapse formation and elimination in vitro. Journal of Neuroscience *15*, 4173–4183.

Zuber, C., Lackie, P.M., Catterall, W.A., and Roth, J. (1992). Polysialic acid is associated with sodium channels and the neural cell adhesion molecule N-CAM in adult rat brain. Journal of Biological Chemistry *267*, 9965–9971.

INDEX

ERRATUM

The orientation of Figure 3 on page 43 is incorrect. Please note the correct orientation and figure part labeling below:

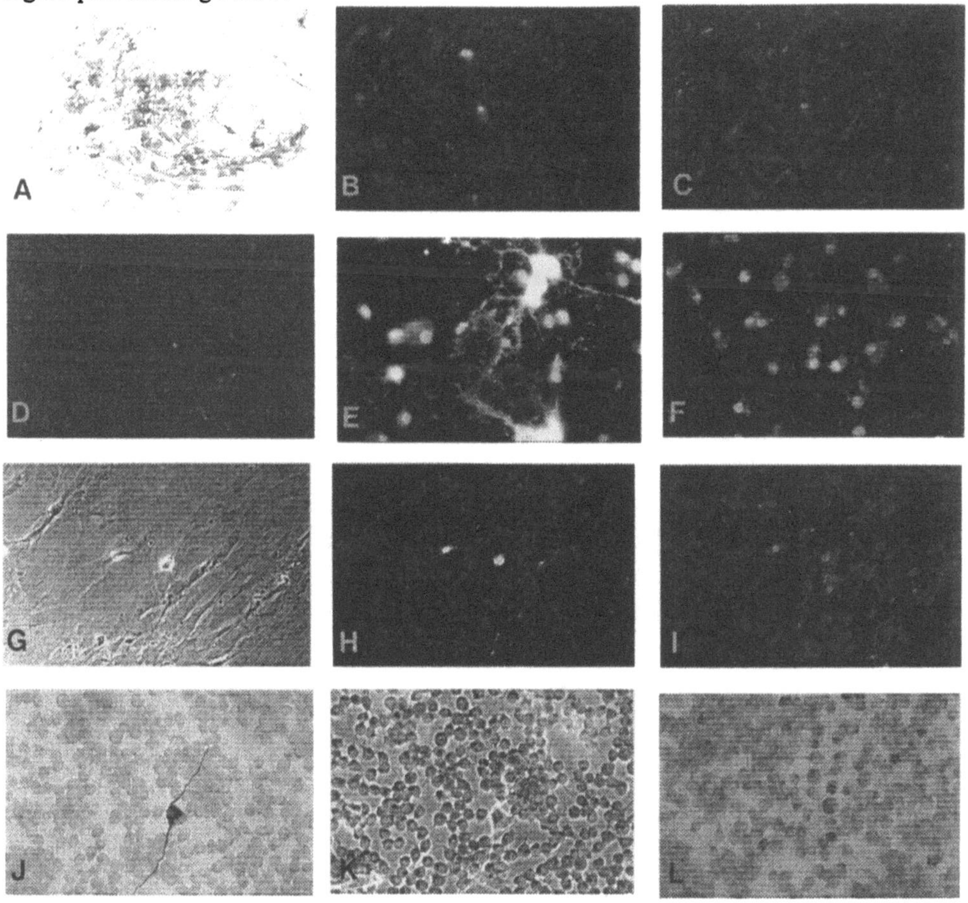

Figure 3. E15CH astrocytic culture system. Culture day 8 to 12. (A) E15CH stained for GFAP (antibodies from Dr. D. Dahl, detection with peroxidase reaction). (B) E15CH stained for myelin basic protein (antibodies from Dr. T. Campagnoni, detection with fluorescein (FITC) immunofluorescence), 0–2 cells per four optic fields are positive. (C) Same field as in B, viewed under rhodamine filter to reveal counter stain with propidium iodide. (D) E15CH stained for GalCer (antibodies Dr. C. Dyer, detection with FITC), 0–1 cells per optic field are positive. (E) Positive control for B: rat oligodendrocytes stained for myelin basic protein (FITC) and counter stained with propidium iodide (brighter staining on nuclei). (F) Positive control for D: rat oligodendrocytes stained for GalCer (FITC) and counter stained with propidium iodide (brighter staining on nuclei). (G) Phase photomicrograph of E15CH cultures depicting two phase bright, small round cells with processes (microglia). (H) The same field viewed under fluorescein filter reveals that the two small cells are positive for α_1 trypsin (antibodies from Dr. F. Gremo), a microglia marker. (I) Same field as in H, viewed under rhodamine filter to reveal counter stain with propidium iodide. (J) Human microglia cultures (from Dr. F. Gremo) stained for GFAP, reveals positive staining of an astrocyte, while all microglia are negative. (K) The same field as in J under phase for better viewing of the microglia cell processes. (L) Positive control for H: human microglia cells are positive for α_1 trypsin (detection with peroxidase). Notes: (1) Similar results were obtained with E15CB (cerebellum) astrocytic cultures. (2) A–D, G–I were photographed under a 25× Neofluar lens; J–L were under a 40× Neofluar. (3) G–L: similar results were obtained with staining for lysozyme or RCA, two other microglia markers.

Brain Plasticity: Development and Aging
Filogamo *et al.*, eds.

0-306-45765-2
Plenum Press, New York, 1997

MIX
Papier aus verantwortungsvollen Quellen
Paper from responsible sources
FSC® C105338

If you have any concerns about our products,
you can contact us on
ProductSafety@springernature.com

In case Publisher is established outside the EU,
the EU authorized representative is:
Springer Nature Customer Service Center GmbH
Europaplatz 3, 69115 Heidelberg, Germany

Printed by Libri Plureos GmbH
in Hamburg, Germany